Goat Production

Goat Production

Edited by

C. GALL

*Institute for Animal Breeding and Genetics,
Veterinary School, Hannover,
Federal Republic of Germany*

1981

ACADEMIC PRESS

A Subsidiary of Harcourt Brace Jovanovich, Publishers

London New York Toronto San Francisco Sydney

ACADEMIC PRESS INC. (LONDON) LTD
24–28 Oval Road,
London NW1

U.S. Edition published by
ACADEMIC PRESS INC.
111 Fifth Avenue,
New York, New York 10003

British Library Cataloguing in Publication Data

Goat production.
1. Goats
I. Gall, C.
636.3′908 SF383

ISBN 0-12-273980-9

LCCCN 81-66393

Filmset by Northumberland Press Ltd, Gateshead, Tyne and Wear
Printed in Great Britain by Fletcher and Sons Ltd, Norwich

Contributors

P. Charlet *Institut National Agronomique Paris-Grignon, Chaire de Zootechnie (Centre de Paris), 16, Rue Claude-Bernard, 75231 Paris Cedex 05, France*

J. M. Corteel *Ing. Principal de Recherches, Laboratoire de Neuroendocrinologie Sexuelle, I.N.R.A., B.P. 1, 37 Nouzilly, France*

C. Devendra *Head, Animal Science Programme Branch, MARDI, Bag Berkunci No. 202, Pejabat Pos Universiti Pertanian, Serdang, Selangor, Malaysia*

C. Gall *Veterinary School, Hannover, Institute for Animal Breeding and Genetics, Bünteweg 17p, 3000 Hannover 71, Federal Republic of Germany*

Omar García B. *FONAIAP, Estación Experimental "El Cuji", Apartado 592, Barquisimeto, Venezuela*

H. S. Hofmeyr *Animal and Dairy Science Research Institute, Private Bag X2, Irene 1675, Republic of South Africa*

J. C. Le Jaouen *Institut Technique de L'Elevage Ovin et Caprin, I.T.O.V.I.C., Maison Nationale des Eleveurs, 149, Rue de Bercy, 75595 Paris Cedex 12, France*

I. L. Mason *Animal Breeding Consultant, 8 Ramsay Garden, Edinburgh EH1 2NA, U.K.*

L. B. Merrill *Texas A&M University, Agriculture Research Station at Sonora, Box 918, Sonora, Texas 76950, U.S.A.*

P. Morand-Fehr *Laboratory of Department of Animal Sciences (I.N.R.A.), Institut National Agronomique, 16, Rue Claude-Bernard, 75231 Paris Cedex 05, France*

R. T. Naudé *Animal and Dairy Science Research Institute, Private Bag X2, Irene 1675, Republic of South Africa*

G. Ricordeau *Centre de Recherches de Toulouse, Laboratoire de Génétique des Petits Ruminants, B.P. 12, 31320 Castanet-Tolosan, France*

M. Shelton *Texas A&M University, Agricultural Research and Extension Center, Rt. 2, Box 950, San Angelo, Texas 76901, U.S.A.*

T. Skjevdal *Agricultural University of Norway, Department of Animal Nutrition, P.O. Box 25, 1432 Ås-NLH, Norway*

C. A. Taylor Jr. *Texas A&M University, Agriculture Research Station at Sonora, Box 918, Sonora, Texas 76950, U.S.A.*

T. A. Steine *Department of Animal Genetics and Breeding, Agricultural University of Norway, 1432 Ås-NLH, Norway*

S. Trodahl *Statskonsulenttjenesten, Landbruksdepartementets opplysningstjeneste, Kontoret for informasjon og rettleiing, Moerv 12, 1430 Ås, Norway*

Christine S. F. Williams *Director, Laboratory Animal Care Service, C Building—Clinical Center, Michigan State University, East Lansing, Michigan 48824, U.S.A.*

Contributors to Chapter 14

H. E. Adler *School of Veterinary Medicine, Department of Avian Medicine, University of California, Davis, California 95616, U.S.A.*

M. K. Ashfaq *Department of Microbiology, N.Y.S. College of Veterinary Medicine, Cornell University, Ithaca, New York 14853, U.S.A.*

M. de Camp *U.S. Department of Agriculture, Animal and Plant Health Inspection Service, Veterinary Services, Scrapie Field Trial, Mission, Texas 78572, U.S.A.*

S. G. Campbell *Department of Microbiology, N.Y.S. College of Veterinary Medicine, Cornell University, Ithaca, New York 14853, U.S.A.*

G. R. Carter *Microbiology and Public Health, Michigan State University, East Lansing, Michigan 48824, U.S.A.*

W. W. Clark *U.S. Department of Agriculture, Animal and Plant Health Inspection Service, Veterinary Services, Scrapie Field Trial Mission, Texas 78572, U.S.A.*

T. B. Crawford *Goldmount Veterinary Center, Sherburn, Minnesota 56171, U.S.A.*

F. M. Hamdy *Diagnostic Research, U.S. Department of Agriculture, Agricultural Research, Northeastern Region, Plum Island Animal Disease Center, P.O. Box 848, Greenport, New York 11944, U.S.A.*

J. L. Hourrigan *5101 S. 8th Rd., Apt. 308, Arlington, Virginia 22204, U.S.A.*

L. M. Jones *Department of Veterinary Science, University of Wisconsin-Madison, 1655 Linden Drive, Madison, Wisconsin 53706, U.S.A.*

A. L. Klingsportn *1206 Tweed Lane, Bowie, Maryland 20715, U.S.A.*

A. B. Larson *National Animal Disease Labs, Ames, Iowa 50010, U.S.A.*

J. W. McVicar *U.S. Department of Agriculture, Federal Research, Northeastern Region, Plum Island Animal Disease Center, P.O. Box 848, Greenport, New York 11944, U.S.A.*

H. E. Metcalf *U.S. Department of Agriculture, Animal and Plant Health Inspection Service, c/o Arthropod-borne Animal Disease Research Lab, P.O. Box 25327, Denver Federal Center, Denver, Colorado 80225, U.S.A.*

Annie Rodolakis *Ministère de l'Agriculture, Institut National de la Recherche Agronomique, Centre de Recherches de Tours, Station de Pathologie de la Reproduction, Nouzilly—B.P. 1, 37380 Monnaie, France*

M. Roguinsky *Ministère de l'Agriculture, Institut National de la Recherche Agronomique, Centre de Recherches de Tours, Station de Pathologie de la Reproduction, Nouzilly—B.P. 1, 37380 Monnaie, France*

T. W. Schillhorn van Veen *Large Animal Surgery and Medicine, Michigan State University, A154 Veterinary Clinic, East Lansing, Michigan 48824, U.S.A.*

G. Scott *Department of Tropical Animal Health, Royal (Dick) School of Vet. Studies, Summerhall, Edinburgh EH 9 1 QH, U.K.*

Mary C. Smith *Department of Clinical Sciences, New York State College of Veterinary Medicine, Ithaca, New York 14850, U.S.A.*

P. Yvore *Laboratoire de Parisitologie, Centre de Recherches de Tours, I.N.R.A., Nouzilly, 37380 Monnaie, France*

Preface

In recent years, interest in goats has increased for various diverse reasons: pleasure in breeding goats like other pets; production of goat milk in response to a market which pays premium prices for goat cheese; scientific research on lactating ruminants for which the goat is a model; development of goat production as a source of food and income in developing countries. This interest created a demand for more information on goats, which was satisfied by Mackenzie's "Goat Husbandry", Palazon's "Ganado Cabrio", Devendra and Burns' "Goat Production in the Tropics", French's "Observations on The Goat", Quittet's "La Chèvre" and Guss' "Management and Diseases of Dairy Goats", to name some of the more recent books. There are also many booklets, guidebooks and extension bulletins which give practical advice to goat keepers on local breeds and conditions. Furthermore, in recent years proceedings of international conferences have been published which, in addition to the publications in scientific journals, are useful sources of reference. However, there is a great deal more information which is not readily available.

The publishers felt that there was a need to compile a comprehensive analysis of the literature and make the information available to a wider audience. It was decided that this should not take the form of another text-book on goat breeding, because the subject has been adequately covered in the books mentioned, but rather a summary of current information on particular aspects of goat production where there has been recent research. It was felt that this would be done best by several authors who are actively engaged in research with goats and have contributed to knowledge in their respective fields. As a result, we present a selection which attempts to cover all basic and applied aspects of the subject using both experimental and practical experience.

In the chapters on temperate and tropical climate we wanted to discuss the particular aspects of goat production which differ greatly from one zone to the other. This could not be done without some overlap between these and other chapters. If every possible redundancy had been eliminated, many of the valuable, and sometimes contradictory, views of the authors would have been lost.

Many of the results of research with goats are scattered in journals which are only locally distributed. Information which was not considered worth publishing in journals has been hidden away in experimental station, progress and project reports. Although it has not been possible to include all these results, some of them have been analysed in the book. Information on the physiological reactions of goats has accumulated as a by-product of biological research which used goats as experimental animals, but much of this is still to be analysed for its relevance to goat production.

The book is written mainly for those engaged in research, teaching, extension, planning and development. Practical breeders, veterinarians and students may find it useful, but they will have to supplement their reading with one of the textbooks mentioned above.

June 1981 C. Gall

Contents

1. Goats in Agriculture: Distribution, Importance and Development

C. Gall

A. Introduction

Goats were among the first domestic animals kept by man for the production of meat, milk, skins and fibre. Originating in Asia goats have spread over all the continents and inhabit almost all climatic zones from the arctic circle to the equator. They are managed under all types of animal husbandry, from the intensive and sophisticated to the most extensive forms of nomadic grazing and are even exploited where living as feral animals in Australia and New Zealand. Although in some areas they produce in intensive systems with high inputs for special markets, the vast majority of goats supply the poorer part of the agricultural population with meat and milk for auto-consumption. This is particularly important in tropical countries but in times of need goats have helped people to survive in industrialized countries as well. Hence in most countries goat-keeping is associated with poverty and consequently goats have been neglected in development plans. Furthermore the goat has the reputation of being a danger to the vegetation and it is often said that goats left free to graze will destroy all the greenery. This view is discussed in Chapters 7 and 16. In order to avoid damage some countries have severely restricted goat-keeping (Cyprus, Greece, Israel, Morocco, Pakistan, Portugal, Syria, Tunisia, Turkey, Yugoslavia) or even tried to eradicate goats completely (Venezuela: see Garcia *et al.*, 1972a).

However, since the late sixties there has been a gradual change in this attitude. Planners are becoming aware of the virtues of the goat, its capacity to survive and even produce under environmental conditions where to the superficial observer it may be difficult to see how goats feed at all: in high mountains goats use vegetation which does not seem to be accessible to any larger animal and in some Mediterranean countries goats are almost starved through the winter with little feed except the branches shepherds lop off oaks and yet kid and start lactating with the onset of vegetation in spring. Encouraged mainly by observations in extreme situations of scarce grazing, as in the Sahel Zone of Africa, planners are less concerned to eliminate goats and more interested in how to make the best use of the goats' potential while avoiding disadvantages by adequate management. On the other hand, there is widespread renewed interest in intensive goat-keeping, making use of the high efficiency of goats and taking advantage of a special market for goat products.

This development is reflected in the scientific literature and in scientific meetings. Mackenzie's book with editions in 1956 and 1965 was about the first comprehensive documentation of knowledge about goats, apart from a number of booklets and management guides of more regional interest. In 1962 a review article appeared in *Herbage Abstracts* (Campbell *et al.*, 1962) giving research results which threw a new light on the goats' role in range deterioration. This discussion was continued with a review on recent research on goats in *Nutrition Abstracts* (Harvey and Rigg, 1964) which gave a fairly encouraging outlook for the future of goats. In 1966 in *Animal Breeding Abstracts* Maule discussed the role of dairy goats in the tropics. In 1970 Devendra and Burns' "Goat production in the tropics" and French's "Observation on the goat" appeared. A series of conferences have been held since 1960, which are listed on p. 34.

A number of international meetings were held between 1970 and 1980 in Venezuela, Mexico and in Europe during the annual meetings of the European Association of Animal Productions (EAAP), proceedings of which have not been published. The American Dairy Science association held two symposia on dairy goats (ADSA, 1978, 1980), the Latin American Association for Animal Production (ALPA) held a symposium on goat breeding in difficult environments and in France since 1975 annual meetings have been held on research in sheep and goats (INRA/ITOVIC 1975–1979).

B. Goat Distribution

Statistics on goats are not very reliable since they are generally not based on actual census but on estimates. Often goat-keepers are interested in underestimating goat numbers because of the general tendency to under-value property, fear of taxation or cultural barriers, as in Muslim societies where counting of stock is regarded as an offence (Dahl and Hjort, 1979). In addition, goat numbers tend to oscillate because of changing climatic situations, the status of vegetation and disease, more than large stock.

Figure 1 shows the density of goats in the world. The concentration in the tropical and dry zones is evident. The underlying country statistics do not

Fig. 1. Goat density in the world (Fao, 1978). Each dot represents 500 000 goats.

Table I. Goat numbers in relation to human population (FAO, 1978).

Country	Population 1000	Goats 1000	Goats per 1000 inhabitants
Africa			
Somalia	3446	16400	4759
Namibia	965	2100	2176
Mauritania	1527	3200	2095
Botswana	750	1150	1533
Niger	5006	6700	1338
Mali	6146	5629	915
Libya	2625	2100	800
Lesotho	1226	925	754
Sudan	16693	11974	717
Ethiopia	30350	17120	564
Chad	4157	2254	542
Swaziland	511	260	508
Nigeria	68724	24188	351
Guinea Bissau	553	182	329
Cent. Afr. Emp.	1912	588	307
Kenya	14658	4415	301
Morocco	19168	5700	297
Rhodesia	6960	2050	294
Tanzania	16886	4700	278
Togo	2449	650	265
Comoros	330	84	254
Benin	3341	850	254
Cameroon	6855	1636	238
Djibouti	114	593	201
South Africa	26904	5270	195
Senegal	5364	1000	186
Ghana	10775	1900	176
Uganda	12436	2144	172
Madagascar	8776	1500	170
Gambia	541	90	166
Gabon	536	89	166
Ivory Coast	7248	1150	158
Malawi	5297	818	154
Tunisia	6213	950	152
Rwanda	4631	682	147
Algeria	17530	2519	143
Burundi	4068	578	142
Angola	6908	925	133
Liberia	1677	185	110
Zaire	26445	2731	103
N/C America			
Haiti	4784	1200	250
St Kitts, etc	66	14	212
Jamaica	2115	350	165
Mexico	65416	8193	125
Barbados	249	27	108
Antigua	74	8	108

Table I. (cont.)

Country	Population 1000	Goats 1000	Goats per 1000 inhabitants
South America			
Bolivia	4889	2946	602
Peru	16719	2070	123
Argentina	26394	3200	121
Asia			
Mongolia	1577	4411	2797
Yemen AR	5800	7800	1344
Yemen DR	1815	1286	708
Cyprus	698	450	644
Qatar	101	49	485
Turkey	43065	18276	424
Iran	33945	13500	397
UA Emirates	690	240	347
Pakistan	77732	25597	329
Iraq	12269	3620	295
Oman	843	201	238
Jordan	2156	490	227
Saudi Arabia	7860	1700	216
Nepal	13528	2450	181
Afghanistan	20882	3000	143
Syria	7990	1039	130
India	660976	70704	106
Lebanon	3152	330	104
Bangladesh	79900	8122	101
Europe			
Greece	9370	4489	479
Albania	2687	665	247
Oceania			
Tonga	95	11	115

reveal the importance of goats in some areas because they do not consider regional differences within countries. For example, in Venezuela goats do not seem of much importance if related to the whole country; in actual fact they are the main resource in some states of the country. In Mexico the average goat density is 125 goats per 1000 inhabitants but in the three states with the highest goat densities, Coahuila, Nuevo Leon and Zacatecas, it is 1640, 1370 and 1060 goats per 1000 inhabitants respectively. The importance of goats in a country may be better assessed by relating their numbers and their products to the human population, although this still disregards regional differences. Tables I and II show that goat numbers and production per capita vary considerably even between countries with significant goat populations. There is a linear increase with geographical latitude of the ratio of goat numbers to human population and to cattle or sheep

Table II. Milk and meat production from goats in relation to human population (FAO, 1978).[a]

Country	Production (kg) per 1000 inhabitants		
	Milk	Meat	Milk and meat
Africa			
Somalia	83·575	15·467	99·042
Mauritania	44·532	3·202	47·734
Niger	24·571	3·907	28·478
Sudan	22·884	1·397	24·281
Djibouti	—	22·693	22·693
Botswana	4·000	4·800	8·800
Mali	5·695	2·894	8·589
Algeria	7·815	0·496	8·311
Libya	5·333	0·763	6·096
Namibia	—	5·223	5·223
Ethiopia	3·097	1·708	4·805
Swaziland	—	4·755	4·755
Tunisia	3·863	0·792	4·655
Upper Volta	3·007	1·223	4·230
Tanzania	2·961	1·110	4·071
Kenya	2·865	1·133	3·998
Guinea Bissau	1·808	1·009	2·817
Senegal	1·678	0·671	2·349
Benin	1·497	0·778	2·275
Morocco	1·356	0·792	2·148
Burundi	1·475	0·509	1·984
Rwanda	1·512	0·380	1·892
Chad	—	1·775	1·775
Lesotho	—	1·566	1·566
Nigeria	—	1·163	1·163
Cent. Afr. Emp.	—	1·145	1·145
Reunion	—	1·060	1·060
South Africa	—	1·059	1·059
N/C America			
Bahamas	18·265	0·329	18·594
Haiti	5·435	0·903	6·338
Mexico	4·265	0·308	4·573
St. Kitts etc.	—	1·061	1·061
South America			
Bolivia	2·864	1·282	4·146
Peru	1·196	0·596	1·792
Chile	0·923	0·432	1·355
Asia			
Qatar	99·010	2·772	101·782
Cyprus	53·009	5·857	58·866
Mongolia	24·096	12·682	36·778
Yemen AR	22·069	6·466	28·535

Table II. (cont.)

Country	Production (kg) per 1000 inhabitants		
	Milk	Meat	Milk and meat
Asia			
Turkey	14·629	2·437	17·066
Yemen DR	13·233	2·768	15·901
Oman	10·676	2·327	13·048
UA Emirates	7·246	3·965	11·211
Syria	9·136	0·974	10·110
Jordan	6·494	2·695	9·189
Saudi Arabia	7·125	1·489	8·614
Israel	8·082	0·389	8·471
Iran	6·540	1·317	7·857
Lebanon	5·711	1·015	6·726
Pakistan	4·425	1·956	6·381
Iraq	5·053	1·027	6·080
Kuwait	4·167	0·336	4·503
Bangladesh	3·905	0·356	4·261
Afghanistan	2·230	1·221	3·451
India	1·085	0·412	1·497
Europe			
Greece	44·077	4·771	48·848
Albania	10·048	2·456	12·504
Spain	8·096	0·355	8·451
France	7·504	0·220	7·724
Norway	6·158	0·065	6·223
Malta	6·006	0·114	6·120
Bulgaria	5·675	0·330	6·005
Portugal	4·055	0·338	4·393
Switzerland	3·805	0·089	3·894
Austria	2·531	0·059	2·590
Czechoslovakia	2·377	0·048	2·425
Italy	2·099	0·057	2·156
German DR	1·552	0·013	1·565
USSR	1·531	0·134	1·665
Developed market economies	1·791	0·134	1·925

[a] Selected countries with more than 1 kg of combined milk and meat production per 1000 inhabitants.

numbers (Shelton, 1978). This indicates that goats play a more important role in tropical than in temperate countries, which of course largely coincide with developing and industrialized countries. In some countries, mainly in Africa and Asia, goat products contribute considerably to milk and meat supply. In some countries this would probably be even more evident if it could be taken into account that much of the goat produce is not marketed but consumed by the producers; it may be more vital for their nutrition than is reflected by the contribution to the national product. Furthermore these statistical averages do not reflect the importance of goats for some groups of the population since others may not consume any goat milk or meat at all. At any rate, in the few countries with more than one goat per head of the human population and more than 10 kg of combined milk and meat production per head, goats are doubtless an important asset. But in other countries with lower averages, regional differences may conceal the importance of goats for some groups of the population. Thus, statistical analysis cannot replace individual judgement in assessing the importance of goat production in a country.

C. Conditions for Goat-keeping

1. Climate

Although goats may thrive under almost all climatic conditions, there is a clear tendency for higher goat concentrations in dry zones. These are indicated in Fig. 1 of Chapter 17 and it is conspicuous that the higher goat densities in Fig. 1 largely coincide with the dry zones. While goats are particularly well adapted to hot and dry conditions (see Chapter 17), it would be wrong to conclude that climatic conditions are favourable to goats only in these zones. The concentration is more due to the fact that in the dry zones there is less opportunity for alternative land use.

There seems to be one exception to the universality of the goat. In climatic zones where high temperatures are combined with high humidity, goats are rare if not kept individually and stall fed. This is often attributed to the susceptibility of goats to pulmonary diseases and parasites, but experimental evidence is lacking. It seems, however, very likely that diseases are involved because abundant feed supply and high population densities with a low standard of living would favour goat-keeping in these zones.

Goats in general dislike rain. Unlike cattle they avoid being exposed to rain and seem to be almost horrified by being hit by the first drops of rain. It is not easy to understand this behaviour in an animal originating in alpine zones, where sudden rains are frequent even in summer. It may, however, be a form of adaptive behaviour preventing goats from continuing their nimble movements between and on rocks when these become slippery from the wet. In the mountains the goat may easily find shelter under precipices and rocks. However, if feeding is reduced during extended spells of rain and cool weather, goats may be affected because rumen fermentation decreases and the animals may have difficulties in maintaining their body temperature. Breeders of Angora goats are afraid of cold spells after shearing when goats may become unable to move and may die from cold. This phenomenon is discussed in Chapter 12.

2. Vegetation

As could be expected from the prevalence of goats in dry areas, the type of vegetation most profitably used by goats is that of the bush and grassland vegetation of dry zones. In other climatic areas goats mainly use coarse or inaccessible vegetation like that of alpine pastures or waste vegetation. Goats can of course make use of better vegetation (see E3) but it is normally reserved for other species while the goats' capacity to use sparse vegetation efficiently is exploited.

Goats carefully select the more nutritious parts of fodder plants and thus are able to obtain an adequate diet even under very scarce grazing conditions (see Chapter 16). This habit is a disadvantage under intensive conditions because of high wastage. By contrast, goats are much less selective with respect to the kind of plants they feed on. Reports from various countries document the wide range of plants eaten by goats, e.g. in Mexico (Carrera and Cano, 1968), Venezuela (Martinez Gonzales *et al.*, 1972) and Kenya (Glover *et al.*, 1966). In the latter study it was found that in addition to the plants consumed by all ruminants the numbers of species selected by cattle, sheep and goats were 17, 20 and as much as 90, respectively.

Where the vegetation does not produce sufficient forage to feed stock the whole year migratory systems of animal husbandry have developed. In these, goats are typically used for grazing the sparsest vegetation.

Due to the climatic conditions very little use is made by goats of the abundant vegetation in humid and hot areas unless goats are stall fed with collected forage and possibly with various additional agricultural by-products.

3. Orography

The wild goat originated from mountainous areas on both sides of the timberline. Even today, domestic goats are extremely skilful at climbing rocks in alpine terrain; they are able to climb along almost vertical precipices, holding on to small projecting rocks, and can even zigzag down a chimney by leaping from wall to wall. Thus, they utilize the most exposed patches of fodder and therefore in broken terrain are preferred to other stock. They may be used for grazing pastures where important differences in altitude have to be regularly surmounted (E2b). In areas where goats are kept jointly with sheep, goats will be used to graze the alpine pastures or when mountain trails have to be used to reach the grazing grounds.

Although goats excel over other stock in alpine terrain, they are used in almost all other types of terrain. As well as climbing mountains they are able to travel and can thrive where large distances have to be covered to collect the maintenance requirement. Hence they are kept in all forms of desert and even on moorland.

4. Social Conditions

There are very few exceptions to the rule that goats are considered to be secondary to other stock and people will try whenever possible to exchange some or all of their goats for cattle. This limits goat-keeping predominantly to the rural poor who have to keep goats for subsistence production, and use their produce, chiefly milk, in the household. In addition to milk, goat meat may supplement their diet and animals may be slaughtered for festive occasions as well as for hair and skins. Alternatively, some income may be earned from these products. Thus goats play an important part in improving the nutrition and income of this population stratum. Goat-keeping is related to the absence of job opportunities offered by industry and commerce. People will tend to keep goats while there is excess labour available whereas labour shortage will lead to a reduction of goat numbers. However, the labour used for herding goats is often children because the specific behaviour of goats, fast moving, dispersing and climbing hillsides, requires great mobility in the herdsman. Wealthier farmers may keep some goats supplementary to cattle for easy procurement of milk when cows are away from the homestead or when they are dry. Carcasses of kids and goats provide a meal of meat without the need to organize consumption with neighbours, sales or preservation.

5. Disease

There is a popular saying that goats do not fall sick except in order to die. This may reflect that fact that goats are not seen suffering from minor disturbances and become visibly ill only if affected seriously. On the other hand, due to their smaller body size they may succumb to critical disease more easily than, say, cattle. Goats appear to be particularly susceptible to bronchopneumonia and this is held to be the cause of their inability to resist the combination of hot and humid climate.

Goats are susceptible to internal parasites. It has been argued that as predominantly browsing animals, goats are not adapted to withstand the infestation of grass with worm larvae. When goats cannot avoid grazing close to the ground they are exposed to a host of parasite larvae which they cannot combat. On the other hand, goats are less susceptible than other livestock to some diseases. They are said to be seldom affected by trypanosomiasis (Finelle, 1973) and have been recommended for use in tse-tse infested areas. Recent investigations have shown, however, that goats are susceptible to trypanosomiasis but that breed differences exist, local breeds from tse-tse areas suffering much less from the disease than imported breeds (Griffin and Allonby, 1979a). Hence, if goats are to be used in tse-tse areas care must be taken that their trypanosomiasis tolerance is ensured, otherwise economic losses will be incurred (Griffin and Allonby, 1979b).

There is a common belief that goats should be resistant to tuberculosis. While it is well proven that this is not true, the idea seems to be favoured by many. The author has not been able to trace the origin of this contention.

6. Agricultural Development

In areas of intensive agriculture there is generally not much room for grazing goats. While feed may be available at field perimeters, along roads, irrigation canals etc., agriculturists are afraid of the damage goats may inflict on their fields. So, unless goats are tethered or closely herded, these forage resources can only be used by zero-grazing. Use of agricultural by-products may be the reason for keeping goats in agricultural areas. It may enable small producers who could not afford cattle to convert some waste into animal products.

Where goats are exploited under intensive management for a specialized production such as the goat dairy operations in France, cultivated improved pastures and fodder crops are used (see Chapters 11 and 15). Under less favourable conditions care must be taken to evaluate the utilization of the resources by goats. Often other stock, cattle in particular, make better use of it. The goats' preference for browse necessitates their complete elimination from afforestation areas. Until trees have grown high enough so that goats will not reach their branches, there seems to be no possibility of making use of goats for cropping weeds between the trees. Even in orchards goats may not graze unless the lowest branches are out of reach, which in the taller breeds may be well above 2 m. But goats (like sheep) may be used for feeding on undergrowth in established rubber, oil palm or coconut plantations provided that any damage to the trees can be avoided. There are limited experiences in Sri Lanka, Indonesia and the Philippines (Devendra, 1976). Goats have been successfully used in rubber plantations in Malaysia (Lee *et al.*, 1978) where 4·4 yearling goats per hectare were grazed under three- and five-year-old trees but were in addition stall fed on cut grass and legumes cultivated under immature rubber trees. Goats (and sheep) effectively controlled weeds and, instead of expenditure on chemical control, provided some income which could be useful to small farmers in particular. Details of management based on experiments with this system are still lacking.

D. Problems in Goat-keeping

Many people, including planners, government officials, foresters and agriculturists, do not like goats or at least are sceptical about their virtues but quite conscious of their potential danger. Part of this attitude seems to originate in irrational prejudice, and part in the specific behaviour of goats, and in particular the possible impact of their feeding behaviour on vegetation.

1. Prejudice

Although some of the prejudice against goats may be understood from superficial observation of their looks and behaviour, especially of the male during the reproductive season, much of it probably is of mythical origin. In the Old Testament (e.g. Leviticus 16, 20–22) the Israelites were told to load all their sins on a buck and drive it out into the desert. The term "scapegoat" for someone who is blamed for others' mistakes has survived in general language and there seems to be an undertone in "buck" which prompts writers to avoid the word and speak of he-goats. Also, goats are often connected with evil in some form or other. In Europe it has been quite common to keep a goat in cattle and horse barns. It was believed to attract disease and evil and keep it away from the other stock. Similarly, in Lesotho a black goat is run with Angora goats (Joubert, 1973).

2. Ecological Balance

Goats' impact on vegetation is of serious concern to ecologists. The most striking example of their potential danger is provided by oceanic islands where goats have been introduced and developed into a feral population. There, plant societies have developed in isolation from the continental flora without herbivore interaction and thus probably have not evolved any resistance against defoliation by grazing and browsing animals. If, in the absence of natural enemies, goats also grew to considerable numbers, overstocking and overgrazing have led to disastrous results. It would, however, be wrong to deduce from these observations that the goat as livestock will of necessity destroy vegetation. Coblentz (1977), from his study of feral goats on Santa Catalina Island, concluded that the diet of goats is determined more by the availability of feed than by preference. "Goats are opportunistic generalists and tend to consume the most palatable and nutritious forage that is available." Similar conclusions have been drawn from investigations on the feeding behaviour of domestic goats, discussed in Chapters 7 and 16. If equilibrium is destroyed by excessive land use, agricultural use of marginal land in low-rainfall areas, elimination of forests by felling or by fire, the presence of goats may prevent any recuperation. However, it is different to argue that the vegetation of a certain area should be conserved from an ecologist's point of view than to decide whether the (maybe scarce) resources of an area should be used by goats for the benefit of people who may have no other means of survival. On an uninhabited island goats may well be eliminated in order to preserve the vegetation, but eliminating goats from pastoral areas where people derive their livelihood from them has not been successful.

With the same intention of environment conservation, some governments (Cyprus, Greece, Israel, Morocco, Pakistan, Portugal, Syria, Tunisia, Turkey, Yugoslavia) have severely restricted free grazing of goats or at least tried to discourage goat-keeping. This seems to be a reasonable measure if it is to support range rehabilitation or afforestation programmes.

In Cyprus before independence goats had been almost eliminated from forest areas. In agricultural areas a system was introduced by which communities could vote for or against goat admission on pastures. Conflicts between agriculturists and stock breeders easily assured a majority against goats and as early as 1913, 45% of all villages had abandoned goat-keeping. However, frequent wood fires bore witness to the continuous hostility of goat-keepers. Since independence, anti-goat legislation has been continually questioned. Syria has issued a decree banning goats from forest areas and limiting goat numbers per hectare in agricultural areas. Village councils are, following ancient traditions, entrusted with the regulation of grazing and stock limitation (Blanchemain, 1965).

Other governments have tried to ban goat-keeping entirely. In Venezuela an eradication campaign launched in 1952 had little success and was abandoned four years later (Garcia *et al.*, 1972). These experiences show that eliminating goats from deteriorated pastures seems to be an oversimplified approach to the protection and restoration of vegetation. Unless alternatives are offered to the people living from the goats, such programmes are liable to fail.

3. Land Tenure

One of the main problems in goat-keeping is the uncontrolled use of public or common land, resulting from the paradoxical combination of private ownership of stock with communal land use (Clawson *et al.*, 1971). The organization of traditional pastoral systems, which in the past did include legal rights of land use, have been upset by the increasing cultivation of marginal lands, a consequence of population growth. The result is defective production systems with overgrazing or waste and where, instead of the theoretical common grazing right, there is in reality inequality through appropriation by collectives or by powerful individuals (Bourbouze and Guessous, 1977). In Morocco the forest authorities may, according to the legislation, prohibit or permit but they must limit goat grazing in woods. But this concerns grazing only; any browsing or lopping is strictly prohibited. However, in actual fact during the winter months loppings regularly supply more than half of the goats' forage and stock keepers are continually challenging the claim of the government on the forest by insisting on communal rights.

In Dalmatia an attempt has been made to revive the "Jegar" system of utilizing public and private forest (Ziani, 1964). A village committee decides upon stock numbers and plot rotation; oak shoots are lopped every four or five years, and ash trees every three years. Some bushes under timber trees may be cut every year in summer. The whole canopy may be browsed during the winter months. Loppings are preserved by drying or ensiling. Such a complex system which has been functioning for decades in the past may well be an example for similar situations.

In Syria cooperatives are trying to revive an ancient system (Hema), based on Islamic tradition, by which grazing is regulated (Draz, 1969). Some approaches towards solving these problems have been summarized by Byerly *et al.* (1978). Aspects of this problem as they apply to the dry tropics are discussed in Chapter 16.

E. Types of Goat-keeping

1. Size of Herds

The number of goats kept by an owner is one of the main indicators of different forms of goat-keeping. It is linked to the utilization of different feed resources. Four main systems of goat-keeping may be distinguished (Le Jaouen, 1978):

(1) large herds of between 100 and 1000 head grazing on ranges, owned by pastoralists who may to some extent migrate with their herds;
(2) sedentary medium-to-large herds of 50 to 500 head utilizing pastures and additional cultivated fodder;
(3) intensive operations of medium size herds of 50 to 250 head or more;
(4) small family herds.

System (1) is orientated towards meat production, while milk may produce 50% of gross income in (2). (3) is orientated towards commercial milk production while herds of type (4) produce milk and meat for home consumption of small farmers or landless rural families.

In Greece, which has the most significant goat herd of Europe, the majority of the 4·6 million goats are in herds of less than 20 head (Boyazoglu and Zervas, 1977). But between 1966 and 1976 the number of intensively managed herds increased by 50% (from 0·6 to 0·9 million) while the number of extensive herds increased by 12% only (3·3 to 3·7 million). Similarly, in France there has been a tendency for the number of operations of type (3) to increase (Le Jaouen, 1978).

Goats kept individually or in small numbers by small farmers or landless families are fed on agricultural and household by-products, taken for grazing on waysides, field perimeters, along irrigation canals and the like. They may stray around the village, be taken to grazing, herded or tethered. The term "subsistence production" is not well defined. It is quite natural for people living in the country to make use of all possible resources to produce food or goods. Goats take their place with other small animals, rabbits, poultry, pigs, sheep or the like. All of them have their particular niche in the family environment and all contribute in some way to its nutrition. The goats' specific merit is that of being dairy animals, shared sometimes by ewes, while cows under these circumstances will often be too demanding.

Although these individual goats are kept as a sideline to other agricultural and non-agricultural activities, larger herds may be an important part of the family activities, producing a major contribution to its sustenance. In other cases large herds of goats may be the only source of income. The size of these herds is quite variable depending on the pressure on land by other families or land tenure, extent and quality of pastures, but most of them will vary between 100 and 1000 head. Even the large herds are normally grazed on public land. In semi-arid zones individually owned land will be used for goats only where grazing is not accessible, sufficient or suitable for cattle. The use of extensive grazing may be combined with the use of by-products, residues etc. Using improved pastures or cultivated land is the exception where intensive goat production mainly for cheese is established, as in France. Large herds of goats are often combined with sheep and sometimes with cattle or both. Migratory herds are often mixed with several species. Similarly, small herds kept as sidelines are often mixed. This is more or less accidental, cattle contributing most to wealth and pride while small stock serves for short-term supplies. In Mexico and in Argentina (Maubecín, 1979) ranchers may let their cowboys run goats with the cattle for their own meat supply.

2. Migratory Systems of Goat-keeping

Migratory pastoral systems are sophisticated and efficient ways of utilizing scarce vegetational resources. In an area where crops cannot be cultivated because of insufficient rainfall man has to derive his livelihood entirely from animal husbandry. In sedentary systems the space that can be used by a herd is determined by the distance which animals can travel daily between night shelter, pasture and water points. The number of animals that can be kept is determined by the carrying capacity of the pasture within this range. If the number of animals that can be kept in a sedentary system is not sufficient to sustain a family the area will either have to be abandoned or man has to turn to migration.

Another reason for migratory systems may be the possibility of making use of grazing grounds which do not yield forage except during a limited part of the year. Herds may be brought to these grounds during the summer season, thus taking advantage of crop residues or seasonal grass growth. This is typically the case in high mountain areas which may be snow-covered during a large part of the year. Similarly riverine areas may be flooded for part of the year but yield good grazing during another season. This may coincide with parasite and fly infestation which has to be avoided for part of the year. Animals used in these systems must be able to travel large distances, use scarce vegetation and sometimes climb rough alpine pastures. Goats fulfil these requirements quite well. A full discussion of the details of migratory husbandry systems will not be given here because these are not limited to goats. However, since goats are an important part of most of these systems, some of the main features will be mentioned.

(a) *Nomadism*

Nomadism is frequent in countries to the east and south of the Mediterranean basin, the Sudano-Sahelian zone of Africa, and the Near and Middle East. Nomads do not have permanent homesteads but change their camps continually following established routes. Normally they do not hold any legal land titles but may claim the area used as tribal land. The routes of animal movement, land and water use are agreed upon between different groups of nomads. However, the balance of this system is endangered by pressure on land, changes of social order and increasing demands for quality of life which all have one common result, namely overstocking of the range with all its consequence.

Nomads dedicate themselves almost exclusively to animal husbandry and stay on a site as long as grazing and water are available. Among the stock kept, goats have an important place. They are often milked. Milk is consumed fresh or processed into sour products and ghee; preparation of cheese or dried casein is not common. Sometimes, where there is a high density of herds not too far away from a highway or settlement, tradesmen will follow the nomads and buy milk for processing near the camps. Prices are settled between the tradesmen and the tribal leaders.

However, under these conditions prices are usually not very favourable for the producers since they are not in a position to bargain. Attempts have been made to establish mobile cheese-making units operated in cooperative form (Westergaard, 1972). However, as with many similar efforts, it is difficult for such installations to compete with the efficient system of tradesmen operating at low cost and using family labour.

(b) *Transhumance*

The classical form of transhumance is the movement of herds between the permanent dwellings in the plains and grazing grounds on mountain pastures during the summer months. It is practised in southern and south-east European countries, the Near and Middle East, central and south-east Asian countries and in many parts of Africa. To a minor extent we find it in Latin America and in almost all countries with high mountains which are snow covered part of the year or at least cold enough for the vegetation to become dormant in winter. People have permanent dwellings but in summer part of the family moves with the stock to the summer grazing grounds. Individual families may move with their herds or a group of families may join their stock to form one large herd. Family members may herd the animals or herdsmen may be hired. There may be one or more summer camps, sometimes with solid permanent buildings. At summer pasture, herds may be split according to species and goats taken to the highest elevations with the more broken pastures.

If milk is produced on the summer pasture it will mainly be processed to cheese or ghee. It was probably under this system that interest began in utilizing the whey by specific forms of cheese where whey protein is retrieved (see Chapter 11). As with the nomads, tradesmen may follow the herds to the summer pastures and purchase milk. Where a simple form of transhumance is practised in the European Alpine mountains with cattle being moved to the high summer pasture, the "Alps", goats are used to supply fresh household milk. While the cows are too high up in the mountains for fresh milk to be brought daily down to the village, milch goats may fit in. Small numbers of goats (2–5) from each household are collected together in the morning and taken by a young shepherd to grazing grounds which may be about 1000 m above the village (Fig. 2). They graze during the day and return at night to their individual homes to be milked. They spend the night in the barn and after morning milking are again collected and grazed. Goats are suited to this kind of management because they are able to cover the required distance and altitude and use grazing grounds which are not necessarily usable by cattle and sheep. However, this type of goat-keeping is disappearing because farmers do not tend to rely on auto-

Fig. 2. Goat herd in Switzerland taken from grazing in the morning ('Heim-Herde").

supply as in the past and readily buy the small quantities of milk used for home consumption together with other food, but it is mainly becoming limited by labour shortage. Children are not available any more and young herdsmen will not find this job sufficiently remunerative.

3. Intensive Goat Management

Goat management may be classified as intensive because of high expenses for (i) investments, (ii) labour or (iii) feed. We consider as intensive any system where goats are not left to fend for themselves with only a minimum of care. This may include:

—grazing improved pastures, feeding cultivated fodder and conserved forage in periods when pasture availability is limited;
—supplementary feeding in addition to what the goats consume on pasture;
—feeding agricultural by-products or concentrates;
—supervising grazing goats or tethering in order to permit feeding on limited areas where goats would damage adjacent fields if left grazing on their own;
—housing, i.e. protection against adverse environment.

Traditionally, herding is not normally considered as intensive management because children usually supervise the goats. However, socially this is highly undesirable since children are withheld from school and for realistic future planning labour in this form should not be taken for granted. If hired labour is to be used, goat numbers and levels of production have to be high enough to pay for it.

Simple forms of housing for protecting goats overnight are almost universal. But elaborate goat houses for intensive feeding and management are limited to specialized goat dairy production. In densely populated areas of the tropics with irrigated crop production where there are no opportunities for grazing, goats may be confined the whole year and fed with collected fodder, grass, weeds, tree leaves or loppings and straw. In other countries, like France, USA and parts of Mexico, goats may be confined in order to manage them intensively with minimum labour.

4. Feral Goats

Feral goats are discussed in detail in Chapter 2. They may be a nuisance because they reduce vegetation in some countries or islands; in some countries they are regarded as game animals and in Australia and New Zealand some use of feral goats is made for meat (Moorhouse and Batten, 1978). A plan is being considered to increase this production by applying a minimum of herd management to the feral goat herds (Batten, 1977).

F. Utility of Goats

1. Meat

While in Western countries goats are mainly known as dairy animals, their world-wide importance is chiefly as producers of meat. In most Asian countries and in Africa, where dwarf breeds prevail, goats are considered almost exclusively as meat animals. One advantage of goats as suppliers of meat is that they produce a small carcass which can be consumed by a family or a group of families without the need for retailing or conserving. Butchers may rapidly sell the meat even of larger goats. This is considered an advantage in the tropics in particular. Carcasses of young kids are highly valued in many countries while preference for meat of older goats is less frequent. In some countries it is accepted and even preferred to mutton but in others it is difficult to market. Religion may be a strong reason for a preference of goat meat over beef or pork.

In some Asian countries, like China, Korea, Taiwan and the Philippines, goat carcasses are not skinned but cooked with the skin, the hair of which has been removed by scraping after scalding in boiling water (Rumich, 1968). In some African countries women are not allowed to eat goat meat. Since they are not allowed to consume goats' milk either, it has been speculated that this may be a protective measure against brucellosis infection (Devendra and Burns, 1970).

The different forms of goat meat production are discussed in Chapter 9.

2. Milk

One of the main virtues of goats is their dairy potential. Milk yield varies between breeds but most of them are able to produce some milk in addition to the requirements of their kids. The size of the surplus milk yield depends largely on the available feed. The typical environment for keeping larger herds of goats does not provide sufficient nutrients for high milk yields and there dairy breeds are not appropriate. But dairy potential is desirable with goats kept individually which can be provided with sufficient feed for higher milk yields. Of course specialized dairy operations need productive dairy breeds. Central European breeds mainly derived from the Swiss stock are used all over the world.

Producing milk has the advantage of a continuous daily protein supply, while meat from goats and other small stock is produced at irregular intervals. Moreover meat tends to be prepared for men while milk is more likely to be fed to children, diseased and aged people who are the ones most in need of animal protein. It is difficult to differentiate clearly between goat-keeping for meat and for milk because in many cases where goats are kept primarily for meat, individual does are occasionally milked. This seems to be the case in most Asian countries where statistics do not record any milk production at all and yet goats' milk contributes considerably to family nutrition (Devendra, 1979). In northern Mexico large herds of several hundreds of goats may be kept by an individual. In years of poor pasture growth the does may just raise one kid for slaughter while in better years they may to some extent be milked as well.

Specialized goat dairy operations are established in many countries. They produce for a specialized market of cheese and milk in Europe and North America and of meat in Latin America. In northern Mexico there is increasing interest in intensive dairy goat operations. In 1979 there were about 18000 goats managed in herds of 100 to 2000 head on irrigated pasture or in stabulation (Juárez, 1979). In Israel, where keeping goats on the range is discouraged (1978: about 100000 range goats), a small but thriving dairy goat population of 18000 does is producing an average of about 650 kg milk which is processed by one dairy to cheese ("Fromes") (Laor, 1979).

In California there is a group of dairy goat operations of between 300 and 1000 animals feeding a dairy plant which processes dehydrated milk. Goat product prices in these countries are considerably higher than those for bovine products so that a high degree of intensification and use of resources which alternatively could be used for cattle or sheep production is justified. The goat industries in France and Norway are discussed in Chapters 11 and 15.

Milk is consumed liquid and used in foods for infants, aged and diseased people; it is sometimes considered particularly suitable as coffee whitener. It is also processed to butter, cheese, sweets and caseine. Processing and marketing of dairy products are discussed in Chapter 11.

3. Other Products and Uses of Goats

(a) Goat skins yield some of the finest leathers such as velours, suede, and chamois for clothes, chevreau for fine shoes, Saffian, Morocco, ecrasé and glazed kid (glacé) for gloves, book-binding and fine leatherware. FAO estimates the annual world production of fresh goat skins at 345000 tons. Much of it is exported either fresh or processed and for countries with large goat populations this is an important asset. In Niger, for example, goat skins in 1965 accounted for 80% of all exports (Robinet, 1967). However, since the cost of curing is high, prices paid to producers are low; as a rule, the value of the skin pays for the cost of slaughtering and dressing of the carcass but in the Red Sokoto of Niger the skin may be one sixth to one fifth the value of a six-month-old kid (Robinet, 1967); in Brazil the skin together with the offal represents about 30% of the value of the animal (Nogueira, 1979). Poor techniques of flaying and conserving (Jeannin *et al.*, 1971), as well as injuries or external parasites, may seriously affect the value of skins. As in other species, the value of the skin is connected with feeding, fat deposition and hair cover, short- and fine-haired animals yielding better skins.

Some breeds of goats, as in Niger, Nigeria, Ethiopia, Yemen, Pakistan and India, are particularly valued for their skins, which may fetch prices 10–15% above the average. Devendra and Burns (1970) have discussed reports on breed differences and the possibility of genetical improvement of skin quality.

Locally, goat-skins are used for producing utility goods such as clothing, bags, water sacks.

(b) Goat hair is still used by some pastoralists for producing threads and ropes. Cashmere and mohair, the hair of Angora goats, are highly valued special products which are discussed in Chapter 12. Cashmere, the fine undercoat of several Asian goat breeds, is produced mainly in high altitudes of above 3000 m in northern India, China, Mongolia, USSR, Iraq, Iran and Turkey. The total annual production is estimated at about 4000 tons (Burns *et al.*, 1962).

Although cashmere fetches premium prices, the main income from keeping cashmere goats is derived from meat and milk. Recently in Australia appreciable cashmere yields have been observed in feral goats (Smith *et al.*, 1973). In connection with a renewed interest in using goats in dense shrub vegetation, this valuable fibre may gain importance in the future. Raw cashmere is usually collected from the farms and marketed in central places. The proportion of cashmere fibre in raw cashmere varies between 50 and 75%. Hair has to be sorted out by hand, which is done mainly by the collectors or processors.

(c) Manure may be regarded as valuable asset where goats are kept in permanent confinement. In dry tropical countries manure accumulating in corrals and night krals is often wasted but within reasonable distance from urban centres it may be sold, e.g. in Venezuela in 1976 Bs120 to 160 (US $24–32) were paid for a truck load of 3–4 m^3.

(d) Medicinal or tonic properties are in some countries attributed to goat products. In Switzerland goat butter is used as component in ointments which are used against arthrosis, arthritis, rheumatism and neuritis (Koller, 1979).

In Korea goat meat is mainly used as a therapeutic agent or tonic, for pregnant women and aged people in particular. The best effects are expected from meat of 100-day-old kids of black colour consumed during the winter months and reproductive management is adjusted to these market requirements. Likewise blood may be regarded as a tonic. In the Philippines black goats are said to have the most effective blood in this respect (Rumich, 1968).

(e) Transport: it is quite unusual for goats to be used as pack animals but for the sake of completeness it should be mentioned that in Nepal the Northern Hill goat and in India the White Himalayan goat are used for transport of goods across the mountains. There are no exact reports available on details of pack loads or distances carried. No harness is used but half-filled bags are put on the withers so as to hang down on both sides (Chapter 3, Fig. 36). Goats have in the past been used as draft animals for light pulling in connection with the goat dairy business but mainly for pleasure. In 1960 in Canada and in the USA there were estimated to be 35 harness goats left (Holst, 1960). In many countries an occasional team of goats may still be seen but just as a curiosity. The alleged difficulty in obtaining suitable harness goats was their size. A wether should stand 9 hands (about 90 cm) weighing 115 to 135 kg or more.

(f) Bush control: use is made of the particular grazing habits of goats for controlling bush growth. Because of their preference for browse, they are able to open thick bush normally not accessible to cattle and allowing little grass growth. Continuous mixed grazing with cattle may be a method suitable for keeping a reasonable balance of pasture plants. These aspects are discussed in Chapter 7.

Use of goats for this purpose is known in Africa and details of goat management have been studied in several experiments (Joubert, 1973). Similarly goats are used in South Africa, North America (Texas), Australia and New Zealand (Batten, 1979).

In Europe use of goats for grazing marginal land which has fallen fallow because of economic and labour constraints is being considered. Goats may have some advantages over sheep on land where there is a problem of bush encroachment (Fehse, 1974). In a small experiment in southern Germany 30 dairy goats (German coloured improved) were managed for five years on waste land without being milked. Bush encroachment was prevented and the vegetational cover influenced favourably. The cost per ha of keeping goats for this purpose was DM330 per ha per year (Müller, 1977), slightly less than mechanical or chemical treatments.

(g) Experimental animals: goats are used as the model of lactating ruminants (Gall, 1964; Kessler, 1979; Larson, 1978). The advantage of its small size may be even greater with pygmy (West African dwarf) goats (Hoversland *et al.*, 1971). Advantage is taken of goats' docile temperament, and investigators often mention the cooperation of the animals, which quickly adapt to experimental conditions. Goats have been used for studying ruminant physiology, hormone action, lactation, environment physiology etc.

4. Marketing

As a rule goat products do not pass through organized markets. Most of them are consumed by the producers themselves. Surplus meat or milk is sold directly on local markets or to middle men. Butchers buy goats, slaughter and sell them by the road side (Fig. 3) or on the market. Larger numbers of kids may be processed in slaughterhouses. Shipping to the centres of consumption, capitals in particular, may be as live animals or carcasses. Both forms are practised in Mexico, Brazil and Argentina. Price fluctuations may be considerable, largely because of seasonal reproduction with reduced

Fig. 3. Butcher with carcass of goat wether at the road side, Venezuela.

yield following unfavourable seasons, e.g. in northern Mexico prices for cabritos at 5 to 6 weeks of age may in September, October and November increase two to three-fold (from US \$3·60 to US \$9·60 in 1971) as a consequence of the fact that few does are on heat between March and May. Some kid carcasses are exported from Argentina to Europe but generally the whole production is consumed locally and in the capitals. Some carcasses of feral goats from New Zealand are exported to the South Pacific, Far East and Mediterranean countries (Moorhouse and Batten, 1978) and from Australia to Mediterranean countries, the Far and Near East (Restall, 1976). While most of the goat meat is sold fresh, some is preserved by drying as, for example, the "landi" in Pakistan.

Milk which exceeds the requirements of the household is typically processed into sour milk products, ghee or cheese. Like meat it is mainly sold on local markets. In some African and Asian countries goats may be milked on the street in the presence of the customer. In countries with a developed goat dairy industry milk is sold to dairies and processed. This is discussed in Chapter 11.

As with meat, prices for milk tend to vary during the course of the year. For instance, in France 90% of the kiddings occur in January, February and March. In order to counterbalance the seasonal fluctuation price incentives are given for increased production towards the end of the year. In 1974 prices per litre in May, the month of highest production, and in October, the month of the lowest production, were F.fr. 1·10 and 1·55 against a weighted annual mean price of F.fr. 1·22 (Le Jaouen, 1976).

Skins obtained from individual slaughterings are mainly air dried and sold occasionally. Quality is as a rule poor due to poor flaying and drying techniques. Where goats and kids are slaughtered in larger numbers, skins are mainly salted or sold directly to tanners. There is one tannery in northern Mexico (Monterrey, N.L.) which in 1970, in addition to about one million kid and 350000 goat skins produced locally, processed about 200000 skins imported from India and the USA. In countries with a favourable infrastructure goat skins pass through the same channels as other skins and hides. France, Italy, Spain, West Germany, Poland, Mexico and the USA are among the larger importers of goat skins.

In order to ensure supply to the local processing industry the export of raw skins has been banned in Brazil (Nogueira, 1979) and in Peru (Nolte, 1979).

G. Development of Goat Production

If it is accepted that goats have their place in agriculture, then ways have to be found to improve their level and efficiency of production, the quality of their products and the social conditions of people keeping goats.

From discussions in this and other chapters in this book it emerges that the main limiting factors to goat development are: feed resources, land tenure, diseases, genetic disposition, education and management practices, supporting infrastructure, capital and input—to product–price relationships. The order of importance of these constraints may vary. Development efforts, both individual and institutional, will have to consider the possible constraints and decide from local analysis which factors should have priority. There is no doubt that goats are an important means of utilizing grazing resources, sparse vegetation in semi-arid zones in particular. They may be the sole or one of the few means of providing a living for people in those zones. Free grazing goats are certainly a potential danger to vegetation but with adequate management this danger can be avoided. If people living on goats are not to be deprived of their livelihood they should be assisted in improving goat management rather than forced to eliminate goats. Improved management will lead to more productive animals better able to resist disease and temporary underfeeding; hence the need to keep excessive numbers will be obviated. Because goats have a fairly well-defined but limited place in agriculture, opportunities for goat production and its development will have to be compared with alternative resource uses in any given area. In this comparison the specific merits of goats will have to be kept in mind, which are:

—ability to make use of specific types of vegetation more efficiently than other species;
—adaptation to adverse climatic conditions;
—the small animal size, which facilitates adaptation to given resources, obviates the need for heavy equipment and expensive facilities and eases animal handling;
—spread of risk over a larger number of animals as compared with cattle for a given resource use;
—early maturity and high reproductive rate which enable building up of a herd with low cash investment; fast capital turnover, adaptation to changing environmental conditions and recovery after animal losses;
—flexibility of herd management due to mobility, reproductive rate, reproductive season, possibility of alternating between milking and nursing kids;
—early slaughter age and small carcasses;
—lack of religious taboos on goat meat.

Disadvantages of goats, especially in comparison with large ruminants, are:

—small size makes them susceptible to predation and theft;
—small size is probably involved in disease risk in that the small animal has fewer reserves when seriously ill. In addition, animal health services have in the past devoted less attention to goats and veterinary skill is lacking in many countries;
—many breeds have a restricted breeding season which may impede the adaptation of reproductive management to the natural conditions and to market requirements;
—small size increases labour costs in commercial units.

Advantages and disadvantages together with the possible levels of production and economic returns will have to be compared with those of other species, particularly sheep, cattle and buffaloes and under intensive conditions with alternative land use for crop production. Furthermore, consideration will have to be given to: labour availability, infrastructure, capital availability, market outlets, consumer preference, social, psychological and ethnic aspects.

When goat development is attempted production targets will have to be related to the scarce production inputs, mainly land area or quantity of feed used. General productivity will have to be broken down into individual characteristics:

—quantity of lean meat, fertility, growth rate (milk yield), feed conversion;
—lactation milk yield, milk composition, lactation length, milkability;
—fleece weight and fibre quality;
—reproductive efficiency, adaptability and resistance to adverse conditions, feed consumption and feed efficiency especially with coarse feeds (particularly important when introducing exotic breeds).

One of the main problems in goat-keeping in dry areas is the use of common land by individually owned stock. Indiscriminate grazing with unlimited numbers of animals is the cause of overgrazing and prevents any measures of rational resource use such as range and herd management, herd improvement and disease control. Traditionally, some control over stock number and pasture utilization has been the rule on all common or public land. Any improvement in range and herd management is hard to envisage unless some of the former regulations are reinstated, either by imposing restrictions of stock numbers and pasture use on individually owned herds or by changing to collective herd ownership or at least to collective herd management.

1. Avenues for Development

(a) Feed improvement: under extensive conditions the most important area of improvement will be feeding. Since the feed resource chiefly exploited is the range, this would mean above all improved range management. The most simple and at the same time most difficult measure is adaptation of animal numbers to the capacity of the range. Under the prevailing grazing systems with the combination of private herd ownership and communal grazing rights, the necessary destocking has proved almost impossible to achieve. Similarly, adequate grazing management, establishing pasture reserves and deferred grazing, are hardly possible. The only way out of this dilemma is a communal approach to animal and range management. Group ranches or grazing and range management associations, possibly reviving traditional systems, may be one way (see also Chapter 16). With government support such associations could establish reserves for the dry season. Access to these reserves could be subject to a limitation of animal numbers.

In the reserves during the dry season feed resources could be improved by planting fodder trees, cacti etc. Where some degree of intensification is achieved, supplementary feeding may be envisaged, chiefly with minerals and agricultural by-products. It has been repeatedly shown, however, that such feeding may yield disappointing results unless the specific demand is well established and feeding is restricted to the groups of animals which need it most. The long-term effect has to be considered because short-term supplementation may just be sufficient to replenish depleted body reserves and not lead to increased production. Similarly, some fodder conservation may be appropriate.

(b) Animal health: though goats have a reputation for their general resistance, under particular circumstances heavy losses may occur from disease (Dahl and Hjort, 1979). Availability of veterinary services, therefore, will be an important precondition for development. Epidemiological services with diagnostic facilities will be the main tools. In view of the small value of the individual goat, preventive measures will have priority while individual treatment will be confined to particular circumstances. Disease control will have to be seen important for product quality and hygiene as well.

(c) Animal breeding: while production is more likely to be affected by all the aspects of animal management, the genetic potential may eventually become the limiting factor. In every advanced development programme, improvement of both management and genetic quality will, therefore, have to go hand in hand. It may often be quite difficult to decide which of the two should receive first priority, because improved management may not pay because of the limited genetic potential or otherwise genetic improvement may not lead to improved production, due to the environmental limitations. Although the first need will be for improved management, it may be necessary to start a breed improvement programme simultaneously because of the time delay involved in its realization. The possible improvements achieved by crossbreeding and breeding plans are discussed in Chapters 3 and 4. Artificial insemination as a field technique is not likely to be applied in such programmes but it may be useful within a breeding nucleus. Similarly, embryo transfer may in the future be useful for implanting exotic breeds (Moore and Eppleston, 1979).

(d) Supporting services: insufficient knowledge about animal management and limited education is a general problem and not one confined to goat development. Therefore specific measures cannot be envisaged. However, within extension services, educational programmes, investment projects and the like, due consideration should be given to goat development. Goats are likely to be overlooked because their contribution to the national product is small and sometimes they do not even produce visibly for the market. To the disadvantage of producers, pegging prices often does not include goat products. Establishing classification systems for goat carcasses may be a simple measure for ensuring fair producer prices. A major difficulty is credit because private and public lenders are concerned about the lack of collateral and the high administrative costs of small loans. In most circumstances it is very difficult to achieve the economic returns of goat production necessary for repaying credit and interest. Therefore, credit for goat development will mainly be extended within a general development programme where goats are only one component.

Dairy goats may be used as a tool for teaching dairying to both youths and inexperienced farmers. This is done in Israel where small goat dairy units are attached to agricultural schools (Laor, 1971).

Ethnic and cultural aspects of development may appear quite different in different societies. They will have to be considered particularly in pastoral societies or changes necessary for development cannot be expected. A merely technological and economic approach to development is not sufficient to overcome the constraints imposed by tradition and to induce rational decisions and modern management practices (Byerly *et al.*, 1978).

(e) Intensive production: the possibilities and justification for intensive goat production depend on various factors the most important of which are: market and prices, availability of capital, availability of production inputs and technical knowhow.

There are mainly two situations offering prospects for intensive goat production:

(i) a market paying premium prices for goat products allowing them to compete with cattle, buffalo and sheep products; as a rule, prices will have to be twice as high in order to justify inputs which could be used for, say, dairy cattle;
(ii) stratification, i.e. intensive management within an extensive system of the productive sector of the herd during the period of highest productivity, mainly lactation. Possibly this could also be done for finishing although with goats there is little scope for intensive fattening.

In the latter case the intention is not to compete with other branches of animal production but to increase the overall efficiency of goat operations in areas where goats have definite advantages over other ruminants. Additional inputs have to yield net returns and show better resource utilization. The resulting partial destocking will assist in range improvement. Intensification will be most justified in milk production, particularly where this is based on range grazing and only lactating goats are managed intensely, while during all other phases of the productive life when nutrient requirements are lower the animals will be managed on the range. Management systems for this kind of stratification have not as yet been developed, probably because in milk production the cattle model is followed unquestioningly. But conditions with goats are different because they have a shorter lactation period and will be dry almost half the year and are pregnant only while not lactating. This could be taken advantage of by circulating goats between the dry herd on the range and the milking herd under intensive conditions (Gall, 1975). Another way of intensification would be an accelerated breeding programme with an 8-month kidding interval.

Intensifying goat production through feeding may be done directly by supplying concentrates at moderate stocking rates, or by increasing stocking rates (which in France are 6·4 goats/ha) through higher fodder yields (Le Jaouen, 1977). However, intensification has clear-cut limits, because:

—the efficiency of meat production is as a rule inferior to that of cattle or sheep;
—with the exception of some special local markets (cabrito), goat meat is less appreciated by the quality market than mutton and beef;
—milk production is more expensive than with cows, due to high labour requirements.

Some economic calculations are discussed in Chapters 10 and 18. Calculations for Swiss conditions (Ammann, 1978) have shown that, at about average pasture production of 6·5 tonnes dry matter with 3000 Starch Equivalents per hectare, stocking rates per hectare of 1·4 dairy cattle and 8·1 goats, the contribution by both species to gross income is about the same but in relation to labour hours with goats it is about 14% less.

When simulated production systems of cattle and goats in different environments are compared, goats tend to be more efficient converters of energy in poor environment but less efficient in good environment (Fitzhugh, 1978).

(f) Product care and marketing: as a result of the prevailing home consumption, goat products are quite often not given the necessary care for successful marketing. Harvesting, conservation, processing, packaging, labelling—all the steps from the raw product to the consumer—may be defective. It seems to be very important to emphasize the particular qualities of goat products in contrast to other animal products. This is successfully done in all countries with a thriving goat industry, such as with goat cheese in France and some Mediterranean countries, milk products in North America, cabritos in Latin America, skins and fibres in many other countries. If goat products have to compete with cattle and even with sheep products on a mass scale they are at a disadvantage.

2. Development Planning

The objective of goat development is to improve the utilization of natural resources which cannot be used by other livestock and at the same time to conserve these resources for continuing use and protect the environment. In specific circumstances the primary objective may be to make the best use of market opportunities for specific goat products. Traditional goat-keeping has been established without any planning effort and mostly without any assistance. If goat production is to be developed purposefully, adequate planning is necessary. This will involve allocation of both natural resources and production inputs. Under extensive conditions in dry areas economic returns will be difficult to obtain if major investments are required. However, under the marginal conditions of these areas such investments often have to be seen not only with a view to the possible immediate returns but also to the long-term effects on the environment, infrastructure and social conditions of the people. For instance, building fences on dry ranges may be barely paid for by increased goat production. It may, however, be the only way to improve management and increase production and thereby arrest overgrazing and eventual desertification, and enable the people to live in acceptable social conditions. The desire to protect the environment from the adverse effects of uncontrolled goat grazing has in the past prompted some governments, as in Venezuela and in Mediterranean countries, to prohibit goat-keeping. This will not be a solution unless ways and means are offered simultaneously to the people for recouping the lost production from the goats, small though it may be. A long-term strategy is required which must consider the effect of adequately managed goats on the vegetation, possible interactions with other livestock, the contributions of goats to animal production, and the necessary input. Since regulation of range management will under most circumstances have the highest priority, cooperative approaches will be inevitable. Success is not possible unless all goat owners using the range in a given area unite in their efforts. While cooperatives will primarily try to organize grazing, they will in the long run assist the individual goat owner with procuring credit and input and marketing of products. There are successful examples in Africa, the Near East, India and, as far as marketing is concerned, in France and the USA.

While adequate range management will be the main measure for improving feeding, planned utilization of agricultural by-products and, in limited cases, some concentrates, may be a tool for increasing the overall efficiency of goat operations. Breed improvement, because of the time lag involved, requires careful advance planning. This will apply to breed evaluation and selection within autocthonous breeds as well as cross-breeding with improver breeds.

References

Ammann, P. (1978). Schriften Schweiz. *Verein. Tierzucht, Bern* **54**, 3–9.

ADSA (1978). Symposium: dairy goats. *J. Dairy Sci.* **61**, 988–1029.

ADSA (1980). Int. Symp.: Dairy Goats. 22nd Ann. Meet. Amer. Dairy Sci. Ass. Logan, Utah, USA, June 1979. *J. Dairy Sci.* **63**, 1591–1599.

Batten, G. J. (1977). *N.Z. J. Agr.* **135** (1), 5–9.

Batten, G. J. (1979). *Proc. 32nd Weed and Pest Control Conf.*, Dunedin, New Zealand, 292–296.

Blanchemain, A. (1965). *Bull. Techn. Inf. Ing. Serv. Agric.* No. 203, 829–845.

Bourbouze, A. and Guessous, F. (1977). *In* "Symposium on Goat Breeding in Mediterranean Countries", Malaga-Grenade-Murcia, Spain, Oct. 1977, EAAP and Spanish National Committee Animal Production, Madrid, pp. 207–227. *Rev. Elv. Méd. vét. Pays Trop.* **32**, 191–198 (1979).

Boyazoglu, J. G. and Zervas, N. (1977). *Elev. Ed. Bovin-Ovin-Caprin* **67**, 66–75.

Burns, R. H., Bergen, W. and Young, S. S. (1962). *J. Text. Inst.* (*Manchr.*) **53** (Trans.) T 45–T 68.

Byerly, T. C., Fitzhugh, H. A., Hodgson, B. J., Nguyen, T. D. and Scoville, O. J. (1978). *In* "The Role of Ruminants in Support of Man", pp. 85–95. Winrock International Livestock Research and Training Center, Morrilton, Arkansas, USA.

Campbell, Q. P., Ebersöhn, I. P. and von Broembsen, H. H. (1962). *Herb. Abstr.* **32**, 273–275.

Carrera, C. and Cano, J. (1968). *Rev. Mexicana Prod. Anim.* **1**(1), 17–18.

Clawson, M., Landsberg, H. H. and Alexander, L. T. (1971). "The Agricultural Potential in the Middle-East", 74, 124. American Elsevier, New York.

Coblentz, B. E. (1977). *J. Range Manage.* **30**, 415–419.

Coblentz, B. E. (1978). *Biol. Conserv.* **13**, 279–286.

Dahl, G. and Hjort, A. (1979). Having Herds. Pastoral herd growth and household economy, 335 S. Department of Social Anthropology, University of Stockholm.

Devendra, C. (1976). *In* "The Role of Sheep and Goats in Agricultural Development", pp. 24–26. Winrock International Livestock Research and Training Center, Morrilton, Arkansas, USA.

Devendra, C. (1979). *Wld Anim. Rev.* (FAO) No. 32, 33–41.

Devendra, C. and Burns, M. (1970). "Goat Production in the Tropics". Tech. Comm. No. 19, CBABG, Commonwealth Agricultural Bureaux, Farnham Royal, Bucks, England, 184 pp.

Draz, O. (1969). The "Hema" system of range reserves in the Arabian Peninsula. Its possibilities in range improvement and conservation projects in the Near East. FAO Document PL:PFC/13, 10 June 1969.

EAAP/CNEZ (1977). "Symposium on Goat Breeding in Mediterranean Countries", Malaga-Grenade-Murcia, Spain, Oct. 1977, EAAP and Spanish National Committee Animal Production, Madrid, 393 pp.

FAO (1960). Expert meeting on production and utilization of sheep and goat milk in the Mediterranean Region, Athens, Greece, Oct. 1960. FAO Meeting Report No. AN–1961.

FAO (1962). Expert meeting on production and utilization of sheep and goat milk in the Mediterranean Region, Roquefort, France, 1962. FAO Meeting Report No. AN–1962.

FAO (1964). Seminar on goat-raising policies in the Mediterranean and Near East Regions, Yugoslavia, Turkey, Greece, Sept.–Oct. 1964. FAO/For. Sem.

FAO (1969). Report of the first session of the expert panel on the utilization of sheep and goat milk. FAO, AN:1969/2, 25 pp.

Fehse, R. (1974). *Schweiz. Landw. Mh.* **52**, 337–349.

Finelle, P. (1973). *Wld Anim. Rev.* (FAO) No. 7, 1–6.

Fitzhugh, H. A. (1978). *J. Anim. Sci.* **46**, 797–806.

French, M. H. (1970). Observation on the goat. FAO Agric. Studies, No. 80, 204 pp. (English, French and Spanish editions).

Gall, C. (1964). *Tierärztl. Umsch.* **19**, 516–520.

Gall, C. (1975). *Wld Anim. Rev.* (FAO) No. 13, 1–8.

García, O., Castillo, J. and Gado, C. (1972). *Agr. Trop.* **22**, 239–250.

Glover, P. E., Stewart, J. and Gwynne, M. D. (1966) *E. Afr. Agric. Forest. J.* **32**, 184–191.

Griffin, L. and Allonby, E. W. (1979a) *Trop. Anim. Hlth Prod.* **11**, 133–142.

Griffin, L. and Allonby, E. W. (1979b). *Trop. Anim. Hlth Prod.* **11**, 127–132.

Harvey, D. and Rigg, J. C. (1964). *Nutr. Abstr.* **34**, 641–645.

Holst, A. C. (1960). "The Harness Goat", The AMGRA Handbook, 138–139. Star Press, Shelby N.C.

Hoversland, A. S., Dhindsa, D. S. and Metcalfe, J. (1971). *In.* "2nd Int. Conf. Goat Breeding", Tours, France, pp. 367–372. ITOVIC, Paris.

INRA (1974). "Symposium sur la Traite Mécanique des Petits Ruminants", Millau, France, May 1973. *Ann. Zootechn.*, special issue, 1974.

INRA (1980). "zème Symposium sur la Traite Mécanique des Petits Ruminants", Alghero, Italy, May 1978. INRA, Paris.

INRA/ITOVIC (1975–1979). "1–5 Journées de la Recherche Ovine et Caprine". INRA, ITOVIC, Paris.

ITOVIC (1971). "2nd Int. Conf. Goat Breeding", Tours, France, July 1971, 391 pp. ITOVIC, Paris.

Jeannin, A., Lobry, M. and Robinet, A. (1971). Manuel des agents du conditionnement des cuirs et peaux en zone tropicale, 143 pp. IEMVT, Paris.

Joubert, D. M. (1973). *Z. Tierzücht. Züchtungsbiol.* **90**, 245–262.

Juárez, A. L. (1979). "VII Reunión Latinoamericana de Producción Animal Seminario de Producción Caprina en Medios Dificiles de America Latina", Panamá, 25 Septembre, 12 pp.

Kessler, J. (1979). *Schweiz. Landw. Mh.* **57**, 209–215.

Koller, J. (1979). *Schweiz. Landw. Mh.* **57**, 203–208.

Laor, M. (1971). *In* "2nd Int. Conf. Goat Breeding", Tours, France, pp. 114–116. ITOVIC, Paris.

Laor, M. (1979). Personal communication. Ministry of Agriculture, Tel-Aviv, Israel.

Larson, B. L. (1978). *J. Dairy Sci.* **61**, 1023–1029.

Lee, K. A., Mohd, W., Embong, B. W., Ng, Lh., Ng, F. C. and Phang, A. K. (1978). "Proceedings, Integration of Animals with Plantation Crops", Penang, Malaysia.

Le Jaouen, J.-Cl. (1976). L'évolution des charactéristiques de la production commerciale de lait de chèvre en France. 27th Annual Meeting, European Association for Animal Production, Zürich, 25.

Le Jaouen, J.-Cl. (1978). XX Int. Dairy Congr. Paris 8 EC.

Mackenzie, D. (1956). "Goat Husbandry" (2nd edn, 1965), 368 pp. Faber and Faber, London.

Martinez, J., Durand, G., Castillo, J. and Garcia, O. (1972). *Progr. Nac. Invest. Ovin Caprin. Min. Agric. Cría Bol. Inf. Venezuela*, No. 2, 24–32.

Maubecín, R. A. (1979). "VII Reunión Latinoamericana de Producción Animal Seminario Producción Caprina en Medios Dificiles de America Latina", Panamá, Sept. 1979.

Maule, J. P. (1966) *Anim. Breed. Abstr.* **34**, 153–158.

Moore, N. W. (1976). *Theriogenology* **6**, 638–639.

Moore, N. W. and Eppleston, J. (1979). *Aust. J. Agric. Res.* **30**, 973–981.

Moorhouse, S. R. and Batten, G. J. (1978). *N.Z. J. Agric.* **137** (12), 22–25.

Müller, D. (1977). *Dt. Kleintierzüchter* 1979 (15).

Murray, M., Morrison, W. I., Murray, P. K., Clifford, D. J. and Trail, J. C. M. (1979). *Wld Anim. Rev.* (FAO) No. 31, 2–12.

Nogueira, T. P. (1979). "VII Reunión Latinoamericana de Producción Caprina en Medios Dificiles de America Latina", Panamá, Sept. 1979.

Nolte, E. (1979). "VII Reunión Latinoamericana de Producción Caprina en Medios Dificiles de America Latina", Panamá, Sept. 1979.

Restall, B. J. (1976). *In* "2èmes Journées de la Recherche Ovine et Caprine" pp. 155–165. INRA, ITOVIC, Paris.

Robinet, A. H. (1967). *Rev. Elev. Méd. Vét. Pays Trop.* **20**, 129–186.

Rumich, B. (1968). The goats of South Korea: preliminary notes. FAO, Regional Office, Bangkok, Mimeograph, 7 pp.

Shelton, M. (1978). *J. Dairy Sci.* **61**, 994–1010.

Smith, I. D., Clarke, W. H. and Turner, H. N. (1973). *J. Aus. Inst. Agr. Sci.* **39**, 128–131.

Westergaard, J. M. (1972). *Wld Anim. Rev.* (FAO) No. 2, 49–52.

Winrock, (1976). "The Role of Sheep and Goats in Agricultural Development", 41 pp. Winrock International Center, Morrilton, Arkansas, USA.

Ziani, P. (1964). FAO Sem. For. (goat) 11. Cited by Bourbouze and Guessous (1977).

Conferences

1960, 1962: Seminars by the Food and Agricultural Organization of the United Nations (FAO) on Production and Utilization of Goat Milk in Mediterranean Countries. 1964: FAO Seminar on Goat Raising Policies in the Mediterranean and Near East Countries. 1964: First International Conference on Goat Breeding in London. 1969: FAO Expert Panel on utilization of sheep and goats' milk. 1971: Second International Conference on Goat Breeding in Tours, France (ITOVIC, 1971). 1972: Symposium on Milk Recording Practices for Sheep and Goats in Israel. 1973: First Symposium on Machine Milking of Small Ruminants in France (INRA, 1974). 1976: Workshop on the Role of Sheep and Goats in Agricultural Development, Winrock International Center, USA. 1977: Symposium on Goat Breeding in Mediterranean Countries, Spain (EAAP/CNEZ, 1977). 1979: Second Symposium on Machine Milking of Small Ruminants, Italy (INRA, 1980).

2. Wild Goats and their Domestication

I. L. Mason

A. Wild Goats

Taxonomically goats are placed in the tribe Caprini, family Bovidae, suborder Ruminantia, order Artiodactyla. The other members of the Caprini are the sheep (*Ovis*) and the two goat-like sheep or sheep-like goats, the bharal or blue sheep (*Pseudois*) and the Aoudad or Barbary sheep (*Ammotragus*). The so-called Rocky Mountain goat (*Oreamnos*) is in fact a goat-like antelope. Morphologically goats differ from sheep in the presence of beard and caudal scent glands in the male, and the absence of face glands and lachrymal pits in the skull and of foot glands, at least in the hind feet. In domestic breeds the easiest way to distinguish the two species is by the normal carriage of the tail—held up in goats and hanging down in sheep.

Goats are divided into two genera—*Capra* and *Hemitragus*. The latter is the tahr which has three species occurring in eastern Oman, in the central Himalayas and in the Nilgiri hills of southern India (see Fig. 1). The tahr has short stout horns and no beard (Fig. 3). The coat is long and shaggy and varies in colour from brown to fawn. Its chromosome number is 48 (Hsu and Benirschke, 1969), and it will not cross with the genus *Capra*, although abortions have been reported after matings between the two genera (Gray, 1972). Since it played no part in the ancestry of the domestic goat, it need not concern us further.

Goats of the genus *Capra* are divided into five species* (Ellerman and Morrison-Scott, 1951):

* But see note on p. 55.

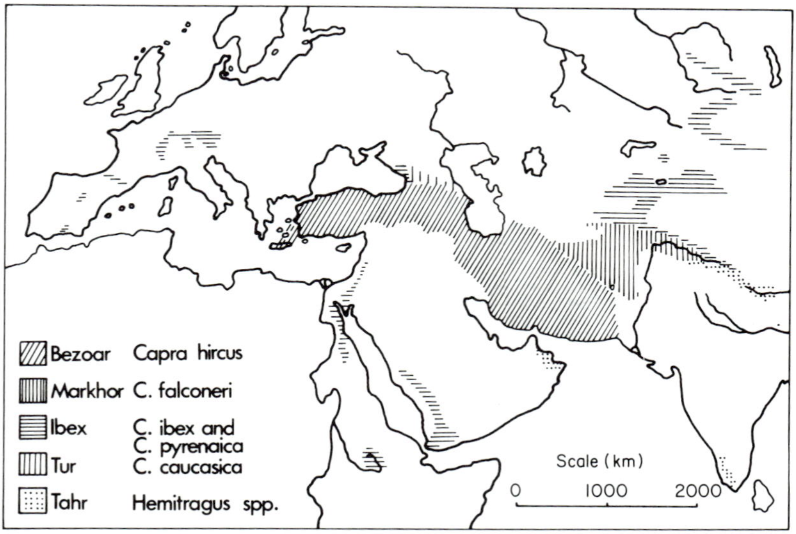

Fig. 1. Distribution of wild goats (Harris, 1962).

(1) *C. ibex*—the ibex—with subspecies in the Alps, the western Caucasus, Central Asia, the Near East and Ethiopia.
(2) *C. pyrenaica*—the Spanish ibex or Spanish wild goat.
(3) *C. caucasia*—the tur of the eastern Caucasus.
(4) *C. hircus*—the bezoar or pasan whose range extends from Greece to Pakistan.
(5) *C. falconeri*—the markhor of Afghanistan, Pakistan and neighbouring countries.

The first three have not been domesticated. The bezoar is the direct ancestor of the domestic goat and the markhor may have influenced some of the breeds of Central Asia. It was formerly claimed that a sixth species of wild goat, *C. prisca*, had existed in Central Europe and from it descended European breeds with twisted horns. The osteological remains on which this species was based have now been shown to be bones of domestic goats which had been wrongly dated (Herre, 1958; Epstein, 1971; Bökönyi, 1974).

All the species of *Capra* which have been so far examined (ibex, markhor, domestic goat) have shown the same number of chromosomes, namely 60, the basic number for the Bovidae (Schmitt and Ulbrich, 1968). Likewise all crosses which have been made between them have been fertile and, in so far as they have been tested, the offspring are also fertile (Gray, 1972) (see Table I).

Table I. Results of crossing between *Capra* species.

	Bezoar	Markhor	Ibex	Tur
Domestic	Fertile offspring	Fertile offspring	Fertile offspring	♀ offspring fertile ♂ offspring frequently fertile
Bezoar		Fertile offspring	No record	No record
Markhor			Offspring produced. Fertility not recorded	No record
Ibex				Offspring produced. Fertility not tested

On the basis of this interfertility, Kesper (1953) included all five in a single species, *C. hircus*, which he divided into three groups—the bezoar, ibex and markhor groups—and 19 subspecies. This classification was followed by Herre (1958). Nevertheless in view of the morphological differences between the five species listed, to reduce them to a single one seems unnecessarily pedantic and the system of Ellerman and Morrison-Scott (1951) is retained herein. However, the curious distribution of *C. ibex* (see Fig. 1) suggests that this may not be the last word on the subject.

1. *Capra ibex*

The ibex has long, sabre-shaped, untwisted horns. The front edge is rounded and shows regularly spaced cross-ridges (or knots) (Fig. 2). The Alpine ibex (*C. ibex ibex*) has a small beard and short ears. It was hunted to extinction in Switzerland (1800–1850), Germany and Austria (1720) in the eighteenth and nineteenth centuries and remained only in the Gran Paradiso park in the Aosta region of Italy. From there it has been successfully reintroduced into the other Alpine countries starting with Switzerland in 1911. The total population is estimated at 5000 in Whitehead (1972), of which about 2000 are in Italy and at 8000 in Grzimek (1972) of which 3500 are in Switzerland and the rest in Italy, France and Austria. The Caucasian ibex (*C. ibex severtzovi*) of the western Caucasus has horns typical of the ibexes but in other ways resembles the tur (Fig. 2). Schaller (1977) calls this subspecies the Kuban ibex, *C. i. caucasica*. The Siberian or Asiatic ibex (*C. ibex sibirica*) is the largest of the group, with massive horns and a long beard. Its range extends from the western Himalayas through the Tien Shan, Altai and Sayan

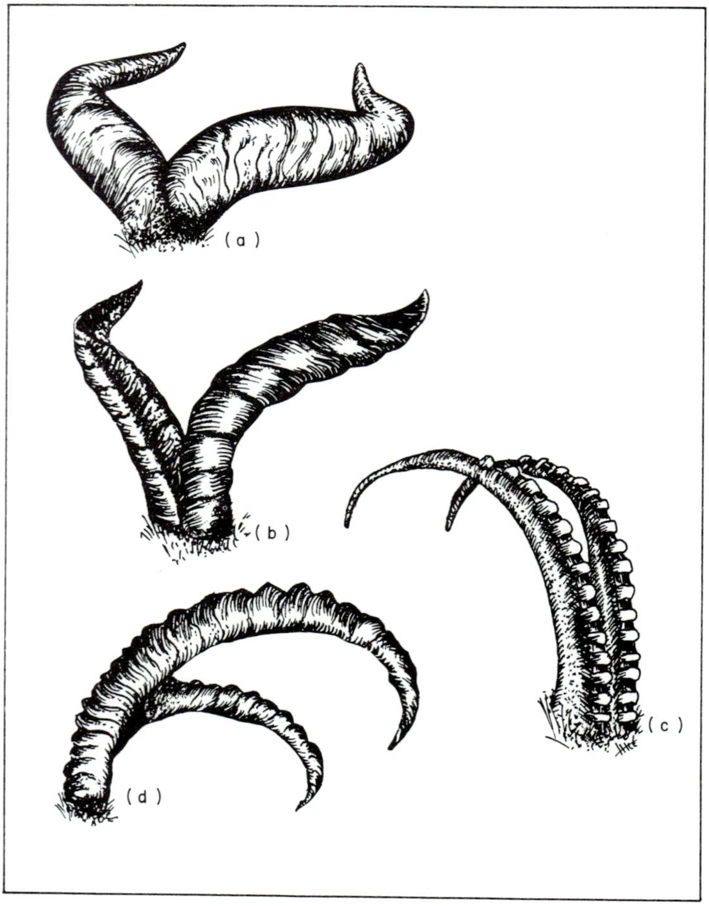

Fig. 2. Horn forms of *Capra* (after Harris, 1962). (a) Tur; (b) Spanish Ibex; (c) Nubian or Arabian Ibex; (d) Abyssinian Ibex.

mountains as far as Lake Baikal (see Fig. 1). It is protected in the USSR but suffers from natural predators.

The Nubian ibex (*C. ibex nubiana*) has long, slender, sharply curved horns, a long pointed beard and long ears. There are about 1500 in the rocky deserts of Judaea and the Negev in Israel. In Egypt and Sudan it is restricted to the hills along the Red Sea coast. Its present status in Sinai and Arabia is not known but it is endangered by hunting (Grzimek, 1972). It is sometimes subdivided into the Sinai or Syrian ibex and the South Arabian ibex. The Nubian ibex may have been domesticated by the Ancient Egyptians. There are certainly pictures of it being fed under control (along with addax, oryx and gazelle—Zeuner, 1963) but there is no proof that it was bred in captivity. The Abyssinian ibex is sometimes made a separate species but it is more usual to classify it as a fifth ibex subspecies, *C. ibex walie*. It differs from the Nubian in its shorter beard, shorter and thicker horns and convex facial profile. It is now restricted to the north and north-western escarpment of Semien between Adis Gay and Embarass. In 1975 only about 300 remained and it is clearly menaced with extinction (IUCN, 1976).

On the basis of its convex profile, Harris (1962) suggested that blood of the Abyssinian ibex may have entered into the domestic Nubian goat. As Epstein (1971) points out, however, when the Nubian goat has horns these do not have the knots which are characteristic of the ibex. Furthermore, the so-called "Nubian" goat does not in fact originate from Nubia, and certainly not from Ethiopia, but from much further east (see Chapter 3, p. 106).

Kesper (1953) attaches *nubiana* and *walie* to the bezoar group but Zeuner (1963) points out that they are both genuine ibexes for the horn cores are broad anteriorly while those of true goats are keeled.

2. *Capra pyrenaica*

The Spanish ibex somewhat resembles the tur in horn form but the horns have a sharp inner edge and the front surface is marked by knots like the ibex (Fig. 2). It was nearly extinct at the beginning of the century but by careful conservation its numbers have now risen to several thousand in the national reserve in Sierra de Gredos, Central Spain (*C. pyrenaica victoriae*). There are also small numbers in the south, especially in the Sierra Nevada (*C.p. hispanica*).

In the Pyrenees *C.p. pyrenaica* is menaced with extinction; only 20 animals remained in 1965 (IUCN, 1976). The Portuguese subspecies (*C.p. lusitanica*) became extinct at the end of the last century (about 1892).

3. *Capra caucasica*

The tur is a heavily built animal with a dull brown coat and massive horns which curve outward, backward and upward and are almost circular in cross-section (Fig. 3). It is confined to the eastern Caucasus. Schaller (1977) follows Heptner *et al.* (1966) in calling the tur *C. cylindricornis*. In Grzimek (1972) the tur is classed as a subspecies of ibex under the name *Capra ibex cylindricornis* on the grounds that it interbreeds with the neighbouring *C.i. severtzovi* of the eastern Caucasus producing a transitional population on the boundary of their distributions.

4. *Capra hircus*

The names bezoar and pasan (or pasang) both derive from the Persian and were originally applied to the Persian subspecies. Bezoar is a corruption (through Arabic) of the Persian *pād-zahr* meaning counter poison. The word was applied to the concretion found in the stomach of some ruminants and believed to be an antidote against poison. The wild goat was the best known source of this bezoar stone and took its name. *Pāzan* is the Persian for mountain goat.

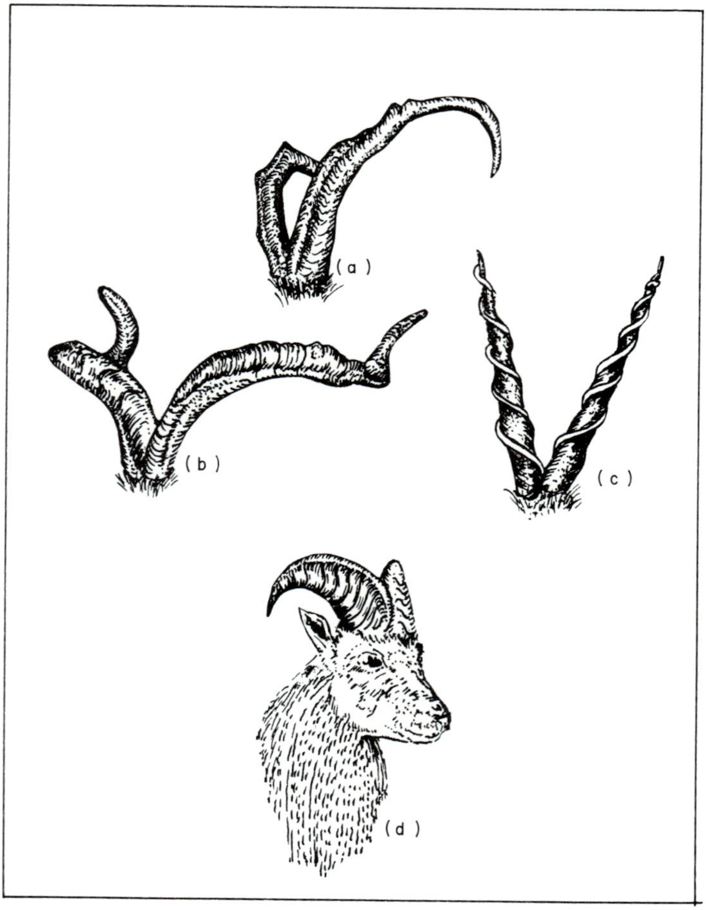

Fig. 3. Horn forms of *Capra* (after Harris, 1962). (a) Bezoar or Pasang; (b) Astor Markhor; (c) Suleman Markhor; (d) Tahr.

Capra hircus is divided (by Ellerman and Morrison-Scott, 1951) into two subspecies:

(a) *C.h. aegagrus*, the Persian wild goat of Iran and Turkey whose range extends into the southern Caucasus, southern Turkmenistan, western Afghanistan and some of the Greek islands.
(b) *C.h. blythi*, the Sind wild goat, of western Sind, Baluchistan and southern Afghanistan.

The distinguishing feature of this species is the long sabre- or scimitar-shaped horns of the male. These are laterally compressed in cross-section so that the anterior edge forms a sharp keel for some distance above the base. Further up the horn there are irregular knots on the front edge (Fig. 3). The horns of the female are shorter but of similar shape. Epstein (1971) describes the bezoar as follows:

> *Capra hircus aegagrus* is a relatively slender animal, with a shoulder height of up to 95 cm. The beard of the male is very long, occupying the whole width of the chin. The hair on the neck and shoulder is elongated in winter when a soft under-fur is developed in the colder parts of the animal's range. The general colour of the upper parts of the pelage is brownish grey in winter, reddish brown in summer. The under-parts and inner sides of the buttocks and thighs are either white or whitish. The face, a nucho-dorsal stripe, the tail, a collar on the neck expanding into a breastplate, the throat, chin, beard, front of limbs except the knees, and a flank stripe are all blackish brown, becoming nearly black on the beard, face and some other parts. The knees, the hind and inner surface of the fore-legs and corresponding parts of the hind-legs, including the hocks are white or whitish. A certain degree of individual variability is displayed in the extent of the black and white markings. Females and young males lack most of the dark markings.

C.h. blythi is smaller. The ground colour of the coat is lighter and the markings darker and more sharply defined. The horns are less curved, often nearly straight, and the knots are few or absent.

The races on the Greek islands, remnants of an earlier much wider distribution in Greece, are smaller still, averaging only 85 cm in wither height (Kesper, 1953). They have been given subspecific status under the names *C.h. cretica* for the agrimi of Crete and *C.h. pictus* for the wild goat of the island of Antimilos (Erimomilos) in the Cyclades. The agrimi was formerly present in the Ida and Lassithi mountains but is now restricted to the Levka Ori (white mountains) in the south-west. Here a national park has been formed to protect it. It has also been introduced into three islands off the north of Crete. In 1963 its numbers were estimated at about 300 in west Crete, 55 in Theodorou island, 20 in Agii Pantes island and 20 in Dia island. The Antimilos goat was termed *pictus* because of the sharpness of its colour markings. Its numbers were estimated at about 100 in 1963. The wild goat of Samothrace shows evidence of much mixture with domestic blood and that in Gioura in the northern Sporades is certainly a feral domestic goat (Schultze-Westrum, 1963).

5. *Capra falconeri*

The range of the markhor extends from Tajikistan through Kashmir and Afghanistan to the Baluchistan province of Pakistan. The name markhor derives from the Persian for "snake eater"; the kh is pronounced like the German ch or the ch in Scottish "loch". Its conspicuous feature is the huge spiral horns which are directed upwards. It also differs from the bezoar in its large size and well developed throat mane, or ruff. The horns are heteronymously twisted, i.e. the right-hand horn is screwed anti-clockwise and the left clockwise (Fig. 3). The form of the spiral varies from a very open corkscrew to that of an ordinary screw. In young animals there are sharp keels on both anterior and posterior edges; the front edge, however, becomes rounded with age. The horns measure up to 125 cm long in males and up to 35 cm in females. The markhor is a heavy thickset animal with a shoulder height of about 100 cm in males but much less in females. In summer the hair on body and legs is short; in winter it grows long and silky with little or no undercoat. Both sexes carry profuse beards. The general colour is grey in winter, reddish brown in summer. Belly and legs are creamy white. There is a dark stripe along the back and on the front legs from the knees and hocks to the fetlock. The tail is very short, dark brown in colour and fringed with coarse long black hairs. Markhors came into oestrus for about 2–3 weeks during December and January in the Himalayas and for about 6 weeks beginning early December in Baluchistan. Gestation length is about 160 days. Females can first kid at 19–20 months of age but do not generally do so until 3–3½ years. Young females produce single kids; older females usually produce twins (Roberts, 1969; Epstein, 1971).

The markhor has recently been studied in Afghanistan (FAO, 1977) and in Pakistan (Roberts, 1969; Schaller and Khan, 1975). The seven subspecies of *Capra falconeri* with their present distribution (see Fig. 4), as far as it is known, are as follows:

(a) *C.f. falconeri*. The Astor markhor is the typical race. It has very wide open spiral horns (Fig. 3) and is found in Azad Kashmir. The maximum horn length recorded is 150 cm. Most Astor markhor live on the steep slopes bordering the Indus and its tributaries upstream from Jalkot. These tributaries include the Astor, Gilgit, and Hunza rivers. They are under heavy pressure from hunting but are believed to be at least twice as numerous as the Kashmir markhor (Schaller and Khan, 1975).

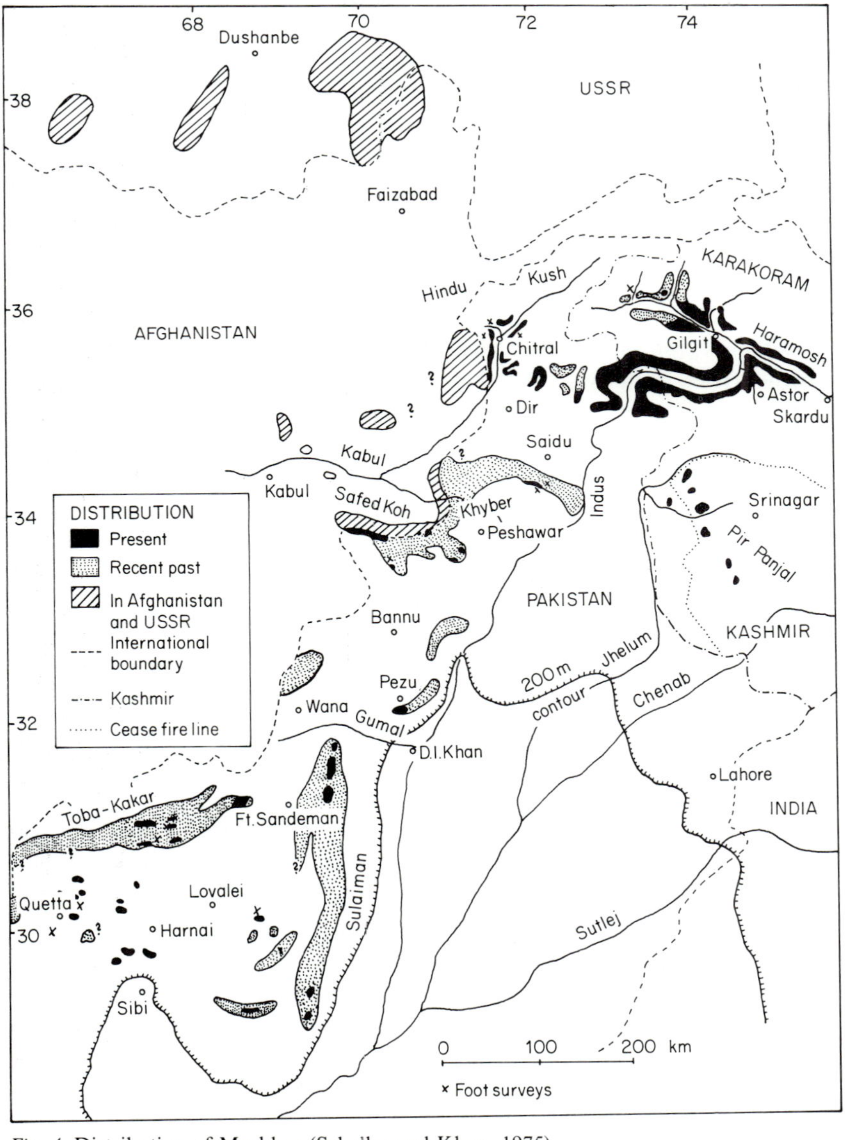

Fig. 4. Distribution of Markhor (Schaller and Khan, 1975).

(b) *C.f. cashmiriensis*. The Kashmir markhor has horns with a slightly closer spiral than those of the Astor markhor. Its typical location is the Pir Panjal range to the south of the valley of Kashmir, and the Shamsberi and Kaj-i-Nag ranges bordering the cease-fire line. Perhaps 250–300 animals are present in the eastern part of its range. In the western part of the range, in the Chitral, Dir and western Swat districts of Pakistan, their numbers are estimated at 1500. Just over the border in Afghanistan it is found in the north-east of the Hindu Kush mountains in the provinces of Konar and Laghman. Their numbers in this part of Nuristan are estimated at 600.

(c) *C.f. megaceros*. Horns are nearly straight with a closer spiral than those of the Kashmir markhor. Only remnants of the Kabul markhor exist scattered in several areas between Kabul and Jalalabad. It has been sighted in the Kohe Safi region of Kapisa province but numbers are very low. The Kabul markhor used to have a wide distribution from the Indus river to the Afghanistan border north of the Khyber Pass and then west to the Safed Koh range. It is now reduced to a few isolated pockets. Schaller and Khan (1975) estimate the total numbers of Kabul markhor at less than 1000.

(d) *C.f. jerdoni*. The Sulaiman markhor has vertical straight horns round which the two keels trace spiral ridges like the thread of a screw (Fig. 3). Numbers are now estimated at 1000 at the most. In Afghanistan it is still being successfully hunted in the mountains of Paktia province near the Pashtunistan border; its numbers are low. It was once numerous in the Sulaiman range west of the Indus, in the Tobar Kakar range along the border between Baluchistan and Afghanistan, and in some isolated hills north and west of Quetta.

(e) *C.f. heptneri*. Horns are thick and heavy with close corkscrew spirals. Heptner *et al.* (1961) estimate that there are about 1000 Badakhshan markhor in the USSR in the mountains of Tajikistan on the right bank of the Piandj river. In Afghanistan they occur in the region of Farkhar (Takhar province) and Darwaz—the north-western part of the Badakhshan province near the Soviet frontier. Estimates of numbers are very low—only 122–150.

(f) *C.f. ognevi* is found in Kugitang, Turkmenia. It has corkscrew horns which are less tightly coiled than *C.f. heptneri* but not so straight as those of *C.f.* megaceros.

(g) *C.f. chialtanensis*. See p. 49.

Ellerman and Morrison-Scott (1951) follow earlier authors in listing seven subspecies of *Capra falconeri*. However, Roberts (1969) points out that in Pakistan the subspecies fall into two distinct groups—larger in the north with more open spiral in the horns, smaller to the south with straighter horns and practically no chest ruff in winter. Schaller and Khan (1975) and Schaller (1977) go further and, because of the variation in horn form within subspecies and the overlapping between them, would reduce the number of subspecies to three. The straight-horned markhor would include *megaceros* and *jerdoni* (under the name of the former) and the flare-horned markhor would include *falconeri* and *cashmiriensis* (under the former name). *C.f. chialtanensis* is believed to belong to the species *C. hircus* (see p. 49) and a subspecific distinction between *heptneri* and *ognevi* "may not be justified".

All subspecies are in danger from hunting and *C.f. jerdoni* and *C.f. megaceros* are in the IUCN list of species threatened with extinction.

B. Domestication

The goat was probably the first ruminant to be domesticated. Bökönyi (1974) suggests that the principal reasons for this were that the wild goat was present in those regions of south-west Asia where agriculture was developing and the goat is an extremely hardy animal which could withstand the rigours of being "reduced to the state of domestication" as Zeuner (1963) so aptly puts it. The goat would also be more suitable than sheep in a forest environment. The species concerned was undoubtedly *Capra hircus aegagrus*. Its horns have the anterior keel of domestic goats and their scimitar shape is mirrored in that of many modern breeds. Those of the ibexes, the tur, and the markhor are quite different. The bezoar sometimes has a slight twist at the end of the horn and it is presumably by selection that this has given rise to the homonymously twisted horns of many modern breeds.

The time of domestication was before 7000 B.C. and the place was the slopes of the Zagros mountains on the borders of present day Iran and Iraq. The evidence for time and place is based on the identification of bones from archaeological sites. Much doubt remains because it is difficult to distinguish between wild and domestic forms. It is also difficult to distinguish between bones of sheep and those of goats so that many reports from excavations speak of "ovocaprids". Skulls and horns of sheep and goats are comparatively easily separated but the bones of the extremities are very similar. However, there are differences which can be detected if the sample is large enough (Boessneck *et al.*, 1964). For instance, Payne (1969) has demonstrated that when the mediolateral width of the distal metacarpal condyle is plotted against the anteroposterior width of its external trochlea, the points for sheep fall in a different cluster from those for goats.

The ibex can be distinguished from the true goat because the horns of the former lack the sharp anterior keel of the latter. The distinction between bezoar and domestic goat is naturally more difficult. It depends on finding remains in which the bones of one particular age predominate (due to selective slaughter of a domestic animal) or in finding the bones in sites which are outside the range of the wild species.

Because of these uncertainties there should be more "probablys" and "apparentlys" in what follows. However, while further discoveries may well place the date of first domestication earlier in time, they are unlikely, given the distribution of the wild goat in relation to early civilization, to move it far geographically.

1. Near East

The earliest osteological remains which can be identified with certainty as belonging to domestic goats were found at Tepe Ali Kosh, Deh Luren, Khuzestan, south-west Iran. They date from before 7000 B.C. Here they are already outside the habitat of the bezoar in the mountains and the identification of the remains as being from domesticated animals is based on the high proportion of bones from yearlings. However, according to Bökönyi (1976), the earliest domestic goats were found at Asiab on the Iranian plateau, about 8000 B.C. Goat remains precede those from sheep. Domestic goats from succeeding centuries have been recorded in other sites in the Near East. They were probably present at Çayönü in south-east Anatolia about 7000 B.C. They were certainly present at Jarmo in Iraqi Kurdistan which site is dated to either 6500 B.C. or the end of the sixth millennium. They were also present in Palestine (Beidha, El Khiam and probably Jericho) in the seventh millennium and, early in the sixth millennium, at Sialk on the Iranian plateau, at Belt Cave on the shore of the Caspian Sea (in Iran) at Hassuna in the upper Tigris valley and at Amouq on the gulf of Iskenderun (both now in Turkey) (Reid, 1969; Epstein, 1971). At Jericho and Belt Cave remains of goats precede those of sheep.

During the Neolithic period domestic goats, with scimitar horns like the wild bezoar, spread over a large part of the Old World. They were represented in vase paintings from Ancient Mesopotamia in the fourth millennium. The same small generalized type was illustrated at about the same time in the tomb paintings of Ancient Egypt. However, by the Hyksos period (1730–1570 B.C.) the sabre-horned goats appear to have been completely replaced by goats with spiral or corkscrew horns which had originally entered Egypt from the East about 3500 B.C. These were presumably derived from the straight-horned goats by selection—as mentioned, there is sometimes a twist at the end of the bezoar's horn. If the horns are not very long they may be sabre-shaped in the female and twisted in the male. Goats played an important role in the economy of Ancient Egypt. Their flesh was eaten (at least by the poorer classes) and their skins were used to carry water. Probably they were also milked. Selection affected the ears as well as the horns and long drooping ears became the fashion. These Ancient Egyptian screw-horn goats had fairly long legs and short hair and their colour was red, black, fawn, white or pied.

Goats with twisted horns are also depicted in Ancient Mesopotamia from the fourth millennium B.C. The best example is the statuette found at Ur of a goat with vertical corkscrew horns standing on its hind legs to browse on a tree. It has been wrongly called the "ram in the thicket". The twisted horn goats of Egypt had horns diverging at a much wider angle and often extending laterally from the skull. Both angle of divergence and degree of twist vary considerably.

2. Europe

Since Europe (except for Crete and possibly other Greek Islands) had no wild goats proper—it was the territory of various species and races of ibex—goats must have come in already domesticated from south-west Asia. Goats are common in the remains from the Neolithic Swiss lake-dwellings and here it is usually a scimitar-horned animal—the so called "turbary" goat. Mainly scimitar-horned goats were also excavated at the early Neolithic sites of north-east Yugoslavia and Hungary (Bökönyi, 1974). From the middle Neolithic goats with twisted horns became more frequent in central and eastern Europe. The so-called "copper" goat had twisted horns. In the Bronze Age this type became dominant in Austria and Germany but scimitar horns remained in Switzerland, Hungary and Scandinavia (Zeuner, 1963).

In Ancient Greece both goats with scimitar and with twisted horns are represented on coins and seals. During the Roman Empire hornless goats first appeared and the size of the animals generally increased. They decreased again up to the fourteenth century A. D. after which they again increased in size. Animals with scimitar horns were still found—mostly in the northern part of central Europe—a case of survival in border areas (Bökönyi, 1974). Both scimitar and twisted horns occur in modern European breeds but the moderately twisted horns predominate. True corkscrew horns are found, however, in the Agrigento breed of Sicily, and in the Ulokeros breed of Greece.

C. Origin of Screw-horned Breeds

The variation in horn form among goat breeds was studied by Herre and Röhrs (1955) and emphasized by Herre (1958). Epstein (1971) described the great variation in the horn shape of modern Iraqi goats. Generally the males have laterally directed horns with an open twist while the female horns may be scimitar-shaped or twisted. However, some animals have vertical screw-like horns. In fact, all intermediates are present between the original untwisted scimitar through the scimitar with tips turned inwards, the open homonymous twist and the spiral, to the tight corkscrew. It is thus easy to imagine how the corkscrew horn may have developed from the original bezoar shape by artificial selection.

Fig. 5. White Cashmere goats, males. *Left*: with (H. Epstein) homonymous horns; *centre*: with erect heteronymous horns; *right*: with scimitar horns.

It is not necessary to implicate the other wild species—the markhor—because of its twisted horns. Such involvement is unlikely because the screw-horn goats of Europe, Africa and the Near East have homonymously twisted horns while those of the markhor are heteronymous, i.e. the right-hand horn is twisted anti-clockwise. However, animals with heteronymous horns are present in many Asiatic domestic breeds in the area between the Caucasus and Mongolia. These include the Kurdi of northern Iraq, the down goats of Russian Central Asia and the Cashmere goat of Mongolia. Animals with heteronymous and with homonymous twists occur in the same herd (Epstein, 1969) (see Fig. 5).

Up to 1975 there was general agreement that the gene (or genes) for heteronymous twist was derived from the wild *Capra falconeri*. The heteronymous twist of the markhor is dominant to the homonymous twist of the domestic goat (Hilzheimer, 1933). Domestic goats also differ from the markhor in the shape of the cross-section of the horn. In the domestic goat the base of the horn has an anterior keel so that the cross-section is pear-shaped. The markhor horn has keels both fore and aft but the major

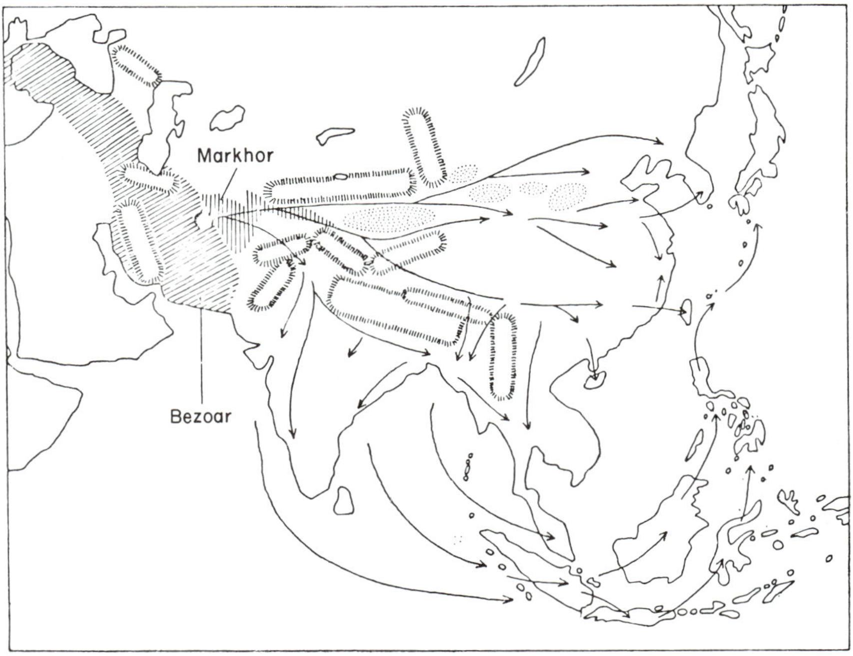

Fig. 6. Postulated dispersion route of domestic goats from domestication area (closed circle) to East and South Asia (Devendra and Nozawa, 1976).

keel is posterior. In this case it is the domestic character which is dominant (Hilzheimer, 1933). The heteronymous horns of domestic goats have a more vertical direction than those of the markhor and have a much more open screw. The same features, together with the anterior keel, occur in the horns of the subspecies *Capra falconeri chialtanensis* described by Lydekker (1913) from the Chiltan range near Quetta (Pakistan). It was suggested by Hilzheimer (1933), Roberts (1969) and others that this may in fact be a cross between markhor and domestic goats. Schaller and Khan (1975) and Schaller (1977) go further and state categorically that the so-called Chiltan markhor is in fact a variety of the wild goat, *Capra hircus*. In addition to the horn features mentioned above they point out that the Chiltan goat lacks a ruff and that in general its coat colour resembles that of the wild goat more than that of the markhor. Furthermore wild goat and Chiltan goat are said to occur in the same herds on several hills south of the Chiltan range. If this is confirmed it shows that heteronymous horns can exist as a variant within the wild species *C. hircus* and therefore it is not necessary to search further for the origin of this trait within domestic breeds.

Devendra and Nozawa (1976) have suggested the routes by which domesticated goats reached eastern from western Asia (see Fig. 6). One is the northern route via Afghanistan and Turkestan to Mongolia and northern China, the other is through the Khyber Pass to the Indian subcontinent and hence to south-east Asia. A sea route was also available.

D. Feral Goats

While the numbers of wild goats and ibexes are being systematically reduced by the spread of agriculture and of forestry, and by hunting, in many parts of the world the domestic goats have gone wild to form populations which are called "feral" to distinguish them from the true wild species. These populations have been particularly successful on islands and in mountainous districts and their success is undoubtedly due to that same hardiness which helped the original domestication. Feral goats have been reported for instance from the British Isles, New Zealand, Australia, the Canary Islands, some of the Aegean islands, Cyprus, Hawaii, Mexico, some of the Caribbean islands and some of the Japanese islands. The goats in the Pacific are descendants of those brought by western navigators, e.g. to the Ogasawara (Bonin) Islands of Japan by Admiral Perry in 1853, to Hawaii by Captain Cook in 1778 and by Captain Vancouver in 1792 (Bryan, 1954). The first wild goats in New Zealand descend from those left by Captain Cook in 1777.

The best recent descriptions come from the British Isles (Milner *et al.*, 1966; Whitehead, 1972) and from New Zealand (Rudge, 1976) and these will be quoted as examples of the origin, nature and fate of feral populations. In Britain the first specific references to "wild goats" date from the seventeenth century. Some of these may have been descendants of the goats which were pastured in forests—a practice which is reported as early as the fourteenth century. Others were probably strays from the large herds which were driven from the Lowlands of Scotland towards the north and from Ireland through Wales to England. The depopulation of the Scottish Highlands in the nineteenth century led to the abandonment of goat herds and there are more recent examples of domestic goat herds being set at liberty. In some cases goats were deliberately introduced as animals to be hunted, as attractions at a wild beauty spot or even to keep down adders or help herd sheep on the hills.

Fig. 7. Feral male goat. Highland Wildlife Park, Kingcraig, Scotland.

Most of the feral populations would derive from the Old English, Welsh, Irish or Scottish breeds, which differed slightly from each other but were all long-haired animals of variable colour and with long twisted horns, at least in the male. Werner (1977) suggests that it may be more useful to think of the original British breeds as only two—Upland and Lowland. The Upland goat would include the Irish, Welsh, Scottish Highland and Hebridean of other writers. It was also found in the hills of Northumberland and Cumberland in northern England. It was small, long-haired and variable in colour. Horns were scimitar-shaped or twisted. The Lowland goat would include the Scottish Lowland and the English breeds of other writers. It was larger with a short coat (though males were maned). Colours were variable but white was commoner and black rarer. Horns were variable but often strongly twisted.

The Swiss breeds were first introduced into Britain at the end of the nineteenth century and now have replaced the original British types in domestic herds. No doubt their blood also contributed to the more recent "wild" population. Whitehead (1972) gives the following description of the feral British goat (Fig. 7):

> Owing to their mixed origins there is no fixed type of colour pattern in the wild goats of Britain and they may vary from jet black to pure white, although most are a mixture of brown, grey, white or black. Some of the darker animals have small tufts of white hair on the forehead, whilst others may have a black or grey forequarter and a white rear quarter; silver-grey, brown, dun or yellow are among colour varieties and the nannies appear to be more prone to colour variation than the males. Most truly feral goats, especially the billies, develop long shaggy coats but one occasionally comes across a short-haired animal running wild.

The majority of these feral goats in Britain are found in north and north-west Scotland (some thousands), in the islands off the Scottish west coast (some thousands), in Kirkcudbrightshire, south-west Scotland (about 750), in the borders of England and Scotland (some hundreds), and in the mountains of north Wales (about 300). In Ireland there are up to a thousand in small herds in the coastal mountains and on islands. In recent years their numbers both in Britain and Ireland have been considerably reduced and many herds have been exterminated in the interests of afforestation. The hard winter of 1947 also took a heavy toll.

The adult males form their own small herds during most of the year and only join the females and young during the mating season. The establishment and recognition of a social order within the herd reduces the amount of fighting—except among males during the breeding season which is normally from early September to the end of October. The rut lasts for about a month; during this time mixed herds form, of all ages and both sexes. The young are born between the end of January and the end of March. Twins are not uncommon but triplets are rare. Mortality is high—especially due to cold weather and snowstorms. Kids may also be killed by foxes and by golden eagles. Feral males do not live longer than 12 or 13 years at the most and their weight varies between 50 and 65 kg.

In New Zealand goats are widespread on the lower parts of forested ranges in both North and South Islands (Rudge, 1976). They were at one time present on 22 off-shore islands where they were introduced from 1842 onwards. Systematic extermination (starting in 1914) has now reduced this number to 6–7. On the mainland goats were often introduced to control blackberry. Angoras were introduced as a commercial proposition. Most of the present feral populations consist of the common or scrub goat. They have thick faces, heavy limbs and rectangular bodies and are variously patterned in combinations of brown, white and black. However, traces of Angora ancestry remain in the herds of Hawkes Bay and North Auckland.

Toggenburg, Saanen and Alpine colours and markings in Taranaki and King Country indicate the influence of dairy breeds.

Because of the tremendous damage goats have caused to vegetation and to soil stability, control measures have been undertaken since the 1930s, first by the Department of Internal Affairs and, since 1956, by the Forest Service. However, goats are present in such large numbers throughout so much of New Zealand that there is no likelihood of their being exterminated on the mainland. In some places they are still used to control bush and their meat is a source of subsidiary income. The remaining island populations could be exterminated if necessary and indeed such is the policy for islands which are reserves of native flora and fauna.

The early Angora imports failed because of susceptibility to footrot and other diseases and the animals were abandoned. After 70–80 years in the bush they have become adapted to New Zealand conditions, but fleece weight has declined. Captured feral animals are being used to increase the hardiness of the experimental Angora herd belonging to the Department of Lands and Survey at Dargaville.

Feral goats are present in all the semi-arid scrublands of Australia and particularly in the States of Queensland, New South Wales, South Australia and Western Australia. They are derived from the goats introduced by early settlers in the outback—escapes from the dairy herds and releases from Angora flocks disbanded about the turn of the century. There are no reliable estimates of their numbers. They can supply additional income to graziers who may export them as meat or sell them for grading to Angora bucks (Mitchell, 1977).

In both New South Wales and Queensland they are being used as foundation stock for upgrading to Angoras for mohair production (Smith *et al.*, 1973). Some of these feral goats carry a fine undercoat resembling cashmere and presumably they descend from Cashmere goats introduced in the nineteenth century. About 50 000 feral goats are slaughtered each year yielding 10–22 kg carcasses according to age. Kidding rate is over 150%. Growth rate to weaning is about 1 kg per week (Restall, 1976).

References

Boessneck, J., Müller, H. H. and Teichert, M. (1964). *Kühn-Archiv.* **78**. 1–129.

Bökönyi, S. (1974). "History of Domestic Mammals in Central and Eastern Europe". Akadémiai Kiadò, Budapest.

Bökönyi, S. (1976). *Nature* **264**, 19–23.

Bryan, E. H. (1954). "The Hawaiian Chain". Bishop Museum Press, Honolulu. (Quoted by Devendra and Nozawa, 1976.)

Devendra, C. and Nozawa, K. (1976). *Z. Tierzücht. ZüchtBiol.* **93**, 101–120.

Ellerman, J. P. and Morrison-Scott, T. C. S. (1951). Checklist of Palaearctic and

Indian Mammals, 1758 to 1946. London; Trustees of the British Museum.
Epstein, H. (1969). "Domestic Animals of China". Commonwealth Agricultural Bureaux, Farnham Royal, England.
Epstein, H. (1971). "The Origin of the Domestic Animals of Africa", Vol. II. Africana Publishing Corporation, New York, London, Munich.
FAO (1977). Conservation and utilization of wildlife resources, Afghanistan: interim report. FAO, Rome.
Gray, A. P. (1972). "Mammalian Hybrids: A Checklist with Bibliography". Commonwealth Agricultural Bureaux, Farnham Royal, England.
Grzimek, B. (Ed.) (1972). "Grzimek's Animal life Encyclopedia", Vol. 13, "Mammals IV". Van Nostrand Reinhold, New York. Translated from "Grzimek's Tierleben, Enzyklopädie des Tierreichs". Band XIII, Säugetiere IV, 552 pp. Kindler, Zürich.
Harris, D. R. (1962). *Proc. Linn. Soc. Lond.* **173**, 79–91.
Heptner, W. G., Nasimowich, A. A. and Bannikow, A. G. (1966). "Die Säugetiere der Sowjetunion". Band I. Paarhufer und Unpaarhufer. G. Fischer, Jena. Translated from the Russian (1961).
Herre, W. (1958). *In* "Handbuch der Tierzüchtung" (Hammond, Johansson and Haring, eds), Vol I. pp. 1–58. Paul Parey, Hamburg and Berlin.
Herre, W. and Röhrs, M. (1955). *Der zoologischen Garten, Leipzig* **22**, 85–110.
Hilzheimer, M. (1933). *Arch. Tierernähr. Tierz.* **8**, 323–371.
Hsu, T. C. and Benirschke, K. (1969). "An Atlas of Mammalian Chromosomes", Vol. 3. Folio 141.
IUCN (1976). Red Data Book on Endangered Species.
Kesper, K. D. (1953). Dissertation, University of Kiel. (Quoted by Zeuner, 1963 and Epstein, 1971.)
Lydekker, R. (1913). Catalogue of the Ungulate Mammals in the British Museum (Natural History), Vol. I. London.
Milner, C., Goodier, R. and Crook, I. G. (1968). *Nature in Wales* **11**, 3–11.
Mitchell, T. D. (1977). *In* "Proceedings of a Symposium on Goats for Fibre and Meat Production", University of Sydney, 18 June 1977, pp. 25–28. Australian Mohair Research Foundation and Australian Society of Animal Production (New South Wales Branch).
Payne, S. (1969). *In* "The Domestication and Exploitation of Plants and Animals", Ucko and Dimbleby, eds), pp. 295–305. Duckworth, London.
Reed, C. A. (1969). *In* "The Domestication and Exploitation of Plants and Animals", (Ucko and Dimbleby, eds), pp. 361–380. Duckworth, London.
Restall, B. J. (1976). *In* "2èmes Journées de la Recherche Ovine et caprine", December, 1976, pp. 155–165. INRA, ITOVIC, Paris.
Roberts, T. (1969). *Säugetierk.* **34**, 238–249.
Rudge, M. R. (1976). *In* "The Value of Feral Mammals in New Zealand" (Whitaker and Rudge, eds). Information Series, Vol. I, pp. 15–21. New Zealand Department of Lands and Survey.
Schaller, G. B. (1977). "Mountain Monarchs: Wild Sheep and Goats of the Himalaya". University of Chicago Press, Chicago and London.
Schaller, G. B. and Khan, S. A. (1975). *Biol. Conserv.* **7**, 185–198.
Schmitt, J. and Ulbrich, F. (1968). *Säugetierk.* **33**, 180–186.
Schultze-Westrum, T. (1963). *Säugetierk. Mitt.* **11**, 145–182.
Smith, I. D., Clarke, W. H., and Turner, H. N. (1973). *J. Aust. Inst. Agric. Sci.* **39**, 128–131.

Werner, A. R. (1977). *The Ark* **4** (3), 79–82.
Whitehead, G. K. (1972). "The Wild Goats of Great Britain and Ireland". David and Charles, Newton Abbot.
Zeuner, F. E. (1963). "A History of Domesticated Animals". Hutchinson, London.

Note added in proof. The taxonomy of the genus *Capra* is very fluid. The most recent standard nomenclature (Corbet, 1978—a revision of Ellerman and Morrison-Scott, 1951—and Corbet and Hill, 1980) gives six species:

1. *Capra ibex*. The important change here is the removal of the subspecies Caucasian ibex (*C.i. severtzovi*) to be given specific rank as *C. caucasica*.
2. *C. pyrenaica*. No change.
3. *C. caucasica*. This name is now given to the west Caucasian tur or Kuban, formerly called *C. ibex severtzovi*.
4. *C. cylindricornis*. The east Caucasian tur which was earlier called *C. caucasica*.
5. *C. aegagrus*. The wild goat or bezoar. The name *C. hircus* is restricted to the domestic goat.
6. *C. falconeri*. No change.

References

Corbet, G. B. (1978). "The Mammals of the Palaearctic Region: a Taxonomic Review". British Museum (Natural History), London. Cornell University Press, Ithaca, N.Y.
Corbet, G. B. and Hill, J. E. (1980). "A World List of Mammalian Species". British Museum (Natural History), London.

3. Breeds

I. L. Mason

Since goats are characteristic of the warmer and less developed parts of the world, breeds in the traditional Western sense with breed society, breed standard and herdbook, are rare, indeed almost entirely confined to Europe and North America. Descriptions of breeds in other continents are very largely descriptions of populations. Thus for present purposes a breed may be defined as an interbreeding population of more-or-less similar animals living in a given geographical area. It is tempting to use a geographical basis and describe the goats of each country in turn. There are several objections to doing this. First, it would involve repeating almost verbatim recently published accounts, e.g. West Africa (Mason, 1951), Eastern and Southern Africa (Mason and Maule, 1960), China (Epstein, 1969), Africa (Epstein, 1971), India (1970), Pakistan (1970; Wahid, 1973), Nepal (Epstein, 1977). Readers who need a geographical account can refer to these books or to the less detailed but worldwide survey given by Börger (1961). Secondly, it would lead to duplication because several breeds occur in more than one country, for example the Angora and Cashmere and their relatives, and the derivatives of dairy breeds such as the Saanen, the Nubian and the Maltese.

The agriculturalist might prefer a classification based on products—milk, meat, fibre, skin. But all goats produce these products to a greater or less extent. Furthermore the major fleece breeds—the Angora and the Cashmere—can be sharply separated on morphological grounds, and although dairy goats may form a recognizable block in Europe and America, elsewhere they tend to be dual- or triple-purpose animals. To separate the predominantly dairy from the predominantly meat types would lead to a very confused classification.

Ideally a biological classification should be based on genetic relationship. In goats we have too little information on apparently neutral characters such as blood groups and other biochemical polymorphisms, to use them as a basis for classification. We are thus thrown back on the usual morphological breed characters of size, colour, coat, conformation, horns and ears. (Unlike sheep, goats show little or no variation in tail form.) These have the advantage of being easily seen features, which, for this very reason, have been selected by man. Their genetic plasticity has enabled artificial selection to produce striking changes. However, where there has been convergent selection for the same character in genetically distinct strains, or conversely the selection of a single strain into two morphologically distinct types, a classification based on present-day appearance might be a poor indication of earlier history.

It was shown in Chapter 2 that one of the main features which distinguishes the species and subspecies of wild goats is the shape and size of horns. Furthermore, being so readily preserved, horns form important clues in the study of the domestication and early history of the goat. As soon as pictorial representations appear on the archaeological and historical scene it is again these conspicuous anatomical features which define different types.

The wild bezoar has scimitar- (or sabre-) shaped horns and short erect ears and so therefore did the first domesticants. At an early date artificial selection produced twisted horns and lop ears. Markhor blood may have introduced the heteronymous twist in the horns. Classification on those features should therefore have some genetico-historical value and it will be attempted here. It can, however, serve only as an initial framework as there are so many mixed, variable and ill-described populations. Productive and geographical considerations have been introduced as extra criteria for making the final groupings—particularly in the case of fleece goats in which product and morphology are closely linked.

The main groups finally decided upon are as follows:

- A Short-eared goats with small or sabre horns or none.
- B Short-eared goats with twisted (*prisca*) horns.
- C Pashmina or Cashmere goats and their neighbours.
- D Angora goats.
- E Lop-eared horned goats.
- F Long-eared hornless dairy goats.

There is little recent information from the USSR so this country is very inadequately dealt with. As far as possible, breed names follow Mason (1969) but some unpublished new spellings have been used.

A. Short-eared Goats with Small or Sabre Horns or None

The typical animals of this group resemble the wild bezoar in having short hair, prick ears, straight facial profile, and sabre- or scimitar-shaped horns. However, they have been selected away from the original colour of the wild goat and the horns have often been reduced in size or eliminated. In this group it has been necessary, for the sake of convenience, to include some breeds with the beginning of a twist at the end of the horns (often difficult to see if the horns are very short) and others with horizontal rather than erect ears (e.g. Spanish). Some breeds have long coats (e.g. Appenzell, Norwegian, Valais Blackneck, Poitevin). Short-eared goats are the type most familiar in Europe but similar types have been described from Africa and southern Asia. Indeed Devendra (1973), remarking on the similarity between the goats of south-east Asia, East Africa and the West Indies, suggests that they may belong to a single racial group.

1. Europe

(a) *Alpine goats*

The various breeds of Swiss goat have been so successfully selected for milk yield that they have largely displaced the local varieties in central, western and northern Europe (except Norway). In Switzerland they are divided into seven breeds as described below and in Section B1. For their size and productivity see Tables I and II and for their distributions see Fig. 1. The first three breeds are usually selected for polledness and therefore suffer from the intersexuality which is an effect of the polling gene (see Chapter 4).

Table I. Swiss breeds—Herdbook animals.

	Recorded adults 1973		Breed standard			
	No.	Yield (kg)	Body weight (kg)		Withers height (cm)	
			Male	Female	Male	Female
Saanen	870	776	75	50	80–95	74–85
Toggenburg	503	728	65	45	75–85	70–80
Appenzell	113	652	65	45	70–80	65–75
Chamois Coloured	245	661	65	45	75–85	70–80
Verzasca			70	50	80–90	75–85
Grisons Striped	505	436	65	45	75–85	70–80
Valais Blackneck	46	691	65	45	75–85	70–80

Source: Kleinviehzüchter (1975), Switzerland (n.d.).

Table II. Swiss breeds—overall performance.

	1973[a] Total no.	%	No.[b] included	Gestation length (days)	No. born alive	Lactation length (days)	Milk yield (kg)
Saanen	16551	23·9	3 154–3 702	154·3	1·8	275	216
Toggenburg	7633	11·0	2 381–2 787	152·4	1·8	276	177
Appenzell	2484	3·6	586–703	153·8	1·7	264	183
Chamois Coloured	10825	15·6	1 061–2 513	153·0	1·5	263	173
Verzasca	10818	15·6	2 198–2 205		1·2	222	86
Grisons Striped	5725	8·3	2 419–2 695		1·2	229	112
Valais Blackneck	1898	2·8					
Nondescript	13272	19·2					

Source: [a] Kleinviehzüchter (1975).
[b] Ammann (1974).

(i) *Saanen*. The Saanen breed originated in the Saane valley in the Bernese Oberland and now covers most of the west and north-west of Switzerland. It has been exported to many countries of the world so that it is now the most widely distributed of the improved breeds. It is white, short-haired and hornless, the best milker in Switzerland and the breed of largest size.

In many other countries of Europe local goats have been graded up to the Saanen and new breeds formed which are named according to their origin and their appearance. Some examples follow (Mason, 1969). Figures on size and yield are given in Table III.

Belgium: *Campine* or *Belgian White*.
Bulgaria: *Bulgarian White Milk* goat.

Table III. Some Saanen derivatives—size and yield.

Breed	Sample	Body weight (kg)		Milk yield (kg)	Fat (%)	Source
		Male	Female			
French Saanen	Estimate	80–120	50–90	600–900	3·4	Quittet (1976)
German Improved White	Herdbook animals			1 000–1 200	3·6–4·0	Roesch (1973)
Israeli Saanen	Recorded mature animals (1965–70)	70–90	45–55	1 020		Laor (1973)
Banat White		50–70	40–48	470–500	3·8–4·4	Taftă (1975)

Czechoslovakia: *White Hornless Shorteared* goat.
France: *French Saanen*, derived from local goats graded up to the Swiss breed but with blood also from Germany, England and the Netherlands. The breed is widely distributed especially in the south-west (Isère, Drôme, Ardèche, Loire, Rhône) (Quittet, 1976; France, 1977).
Germany: *Weisse Deutsche Edelziege* = German Improved White goat. Local goats (especially in Hesse) were graded to Saanens which were first imported in 1892. The name was given in 1927 (Roesch, 1973).
Great Britain: *British Saanen*. Imports from Switzerland, chiefly in 1903 and 1922, were used to grade up the British goat. Purebred descendants of imported animals are kept in a separate *Saanen* herdbook.
Israel: The *Israeli Saanen* was formed by grading up the Mamber with Swiss and Dutch Saanens. The breed numbered 26000 in 1970 (Laor, 1973).
Netherlands: *Nederlandse witte geit* = Dutch White goat.
Poland: *Polish Improved White*, same origin as German Improved White.
Romania: *Albă de Banat* = Banat White. From a cross of the Carpathian with the Saanen and the German Improved White. Its withers height is only 69 cm (Taftă, 1975).
USSR: *Improved North Russian*, graded from the North Russian by imports from 1905 onwards.

(ii) *Toggenburg*. The *Toggenburger* originated in Obertoggenburg and Werdenberg in St Gallen canton in north-west Switzerland but it is now also present in the cantons of Schwyz and Lucerne. It is a little smaller than the Saanen and has a lower milk yield. It is brown to mouse-grey with white stripes on face and feet. The coat is rather longer than in the Saanen and some animals have long hair on back and thighs.

The Toggenburg has also been exported, though less extensively than the Saanen. It has given rise, for instance, to the *British Toggenburg*, *Dutch Toggenburg* and the *Thuringian* variety of the *Bunte Deutsche Edelziege*.

(iii) *Chamois Coloured*. The *Gemsfarbige Gebirgsziege* or *Chamoisée des Alpes* is a short-haired brown goat with black belly and lower legs and black stripes on the face and along the back. It includes the horned *Bündner* variety in the canton of Graubünden (Grisons) and the usually polled *Oberhasli-Brienz* in the Bernese Oberland and Freiburg canton. The Oberhasli-Brienz has a considerably higher yield than the Graubünden variety due, at least in part, to its better environment. The Swiss Chamois coloured breed has contributed to both the French Alpine and the Bunte Deutsche Edelziege.

The *Camosciata delle Alpi* of northern Italy is effectively the same breed. It is found in the Alpine provinces of Novara, Trento, Bolzano, and Udine.

It invariably carries scimitar-shaped horns of medium size. Age at first kidding is one year and litter size averages 1·85 (Italy, 1973).

For size and yield of Italian breeds see Table IV.

Table IV. Italian breeds—numbers, size and yield.

	Number (1971)	Body weight (kg) Male	Body weight (kg) Female	Withers height (cm) Male	Withers height (cm) Female	Milk yield (kg)	Litter size
Camosciata delle Alpi	19000	65	45	75–85	70–80	300–900	1·85
Alpina	5000					500–600	
Garganica	45000		40		65	100–150	1·2–1·3
Girgentana	9000	65	45	70+	70−	400	2·0
Maltese	42000		35–45		65–75	500–600	
Nondescript	939000						

Source: Italy (1973).

(iv) *Appenzell.* This breed is similar to the Saanen but is smaller and has medium to long hair. It is almost restricted to the two Appenzell cantons, thus forming an island in the Toggenburg area but there are also some in the canton of Zürich where they were at one time crossed with the Saanen to give the *Zürcher* variety.

The following two breeds are mountain breeds with long horns and their lower milk yield is related to their inferior environment.

(v) *Bündner Strahlenziege* or *Grisons Striped* is restricted to the canton of Graubünden. It is black with white face stripes, muzzle, escutcheon, and lower legs. It is horned, the male having scimitar horns turned out at the tip. The British Alpine is derived from it.

(vi) *Verzasca.* The Verzasca goat of Ticino (*Schwarze Tessiner*) is uniformly black with long scimitar horns. It is the hardiest of the Swiss breeds.

(vii) *Bunte Deutsche Edelziege.* Swiss goats were imported to Germany from 1887 on and used to grade up the local goats. In 1927 the coloured varieties of improved German goats were grouped into a single breed named the "Coloured German Improved Goat". (*"Edel"* means improved and signifies short hair and hornlessness in addition to high milk yield.) Formerly three varieties were recognized—the *Thüringer Waldziege* (or German Toggenburg) and the *Rehfarbige Deutsche Edelziege* (Fawn German Improved) in dark and light types. The dark variety was typical of Franconia and is derived from the Chamois Coloured of Switzerland. The pale variety has the pattern of the wild goat i.e. with pale face stripes and belly, and darker bands along the back and flank. It used to be typical of the Black Forest and Harz mountains. Now there is only one recognized breed with three colour variants. Its milk yield is less than that of the Weisse Deutsche Edelziege but the fat percentage is higher. Herdbook animals average 900–1200 kg milk and 3·6–4·2% fat. Also it has somewhat finer bone as befits a mountain animal (Roesch, 1973).

There is also a *Belgian Fawn* (*Chamoisée* or *Hertegeit*) and a *Polish Improved Fawn* with the same origin as the German breed.

(viii) *French Alpine*. This breed is found throughout the goat producing districts of France but particularly in the Pays de Loire—that is the departments of Indre, Cher, Loire-et-Cher, Indre-et-Loire. It is short-haired and may or may not carry horns, beard and wattles. All colours occur except white—white goats have recently been separated off from the Alpine breed as the French Saanen. Chamoisée is the commonest colour. Other common colour patterns are: white spots on a black or brown coat; black with or without pale legs; and *mantelée*—back and sides dark, neck and shoulders pale. (For size and yield of French breeds see Table V.) Although of Alpine origin and related, at least in colour, to the Swiss Chamois Coloured and Grison Striped, the Alpine is now as well adapted to the plains as to the mountains (Quittet, 1976; France, 1977).

The French Alpine is a recognized breed in the USA where the minimum weight is given as 55 kg and the breed is described as large, rangy and angular. Its colour varieties are listed as follows: *cou blanc*—white forequarters and blackish hindquarters with black or grey markings on the head. *Cou clair*—same as cou blanc but with forequarters tan or saffron. *Chamoisée–Cou noir*—the opposite of cou blanc with forequarters black and hindquarters white. *Sundgau*—black with Toggenburg markings. *Pied*—spotted (Shelton, 1978).

(ix) *British Alpine*. Alpine goats were first imported from Switzerland in 1903 and used to grade up the local British goats. The British Alpine is one of the results. It is black with Toggenburg markings; belly may be black or white. This colour relates them to the Bündner Strahlenziege and the Sundgau.

(x) *Italian Alpine*. The Alpine breed of Italy is similar to the Alpine breeds of Switzerland and France. As in France it is found away from the Alps and chiefly in the provinces of Brescia (Lombardy), Imperia (Liguria) and Lucca (Tuscany). Its colour is generally all black or all white but it may be grey or pied. Thin horns curve straight back but may be absent (Italy, 1973).

(b) *Other French Breeds*

(a) *Poitevin.* The third recognized breed in France (along with the Saanen and the Alpine) is the Poitevin (from Poitou). The breed is of medium size and rangy in conformation. Its colour varies from dark brown to black, being darker in the male. The belly, inner surface of the legs, and beneath the tail, are white. There are also two white spots or lines on the face which tend to disappear with age. Horns are usually absent. Facial profile is straight. In the female the coat is fine and short on the head, neck and legs but long on the back, hanging down like a cloak. In addition, the male has a longer mane on neck and shoulders. The chief area of distribution is in the departments of Deux-Sèvres, Vienne, Charente and Charente-Maritime. The performance of the three French breeds is similar. Milk yield is 600–900 kg with protein content 2·9–3·0% and fat content 3·4%. Prolificacy is 135% for first births and 180% for multipares (Quittet, 1976; France, 1977). (See also Table V.)

(b) *Common French.* This refers to the nondescript goats of France. Since some at least appear to have spiral horns they have been included in the *prisca* horned group and described in Section B2(a).

Table V. French breeds—numbers, size and yield.

	% of milk recorded goats	Body weight (kg)		Withers height (cm)		Milk yield (kg)
		Male	Female	Male	Female	
Alpine	70	80–100	60–90	90–100	70–80	600–900 (Alpine and Poitevin)
Poitevin	3	55–75	40–60	65–85	65–75	
Saanen	9	80–120	50–90			
Rove		65	50			150–450
Corsican			40–60			150–250

(c) *North European Goats*

Norway is the only country of northern Europe which has kept its original goats and turned them into milk goats with little foreign influence. There are only a few thousand goats in Denmark, Finland and Sweden but in Norway there are about 70000 (1975). Similar goats were common in the British Isles before the introduction of Alpine blood. There are records of Old English, Irish, Scottish and Welsh goats. These have all now been graded up to Swiss breeds or else have become feral (see Chapter 2, p. 50).

(i) *Norwegian.* The Norwegian goat is long-haired, short-eared and with sabre horns. In 1971 milk yield averaged 570 kg with 3·5% fat. There is no sharp division into breeds but four local varieties are usually distinguished. The *Telemark* has a rather fine build and is usually white and polled but it can also be horned. Body weight is 45–55 kg. It is a good milker. The *Døle* from Gudbrandsdal has many colours but is often blue-grey or brown. It weighs about 50 kg and is usually polled. It occurs over most of Østland. The *Vestland* is also rather variable but is normally of a heavy build, and usually horned. The commonest colour is dark brown. Males are long-haired but females have long hair only on the thighs. The *Nordland* (North Norwegian) has been much influenced by the Gudbrandsdal goat and is now very similar to it (Skjervold, 1973).

(ii) *North Russian.* This is the native breed of the European USSR north of a line from Smolensk to the Tatar Republic. It is a dairy breed and usually white.

(d) *Spanish Breeds*

(i) *Malaga.* The *Malagueña* breed is also called *Costeña* because it is found along the southern coast from Cadiz to Almería (although chiefly in Malaga province) and inland to Córdoba and Seville. Ears are fairly long and horizontal and facial profile is straight. Horns are sabre-shaped; polled males are sometimes used in the lowlands. The colour is usually light red but varies from blond to chestnut. Tassels are present. The coat is short except for the mid-dorsal line and the hind legs; the male has a top knot and beard.

Females are served first at 12–14 months of age. Although oestrus occurs throughout the year there is only one kidding per year. The average number of kids per birth is two but males for breeding are chosen from single births. This is a dairy breed; yield varies greatly according to the system of management but averages 400 litres per goat per year. Milk recording has recently started and a herdbook has been opened (Belinchón Valera *et al.*, 1977).

(ii) *Murcia-Granada.* The *Murciana* and *Granadina* used to be considered separate breeds but they differed only in colour, the Murcia being mahogany and the Granada black. However, there were in fact never more than 10% mahogany animals in Murcia and never fewer than 4–5% in Granada. The two colour variants are now combined in a single breed under the joint name Murciana-Granadina; most of them are in the provinces of Granada, Murcia, Alicante and Almería. From Granada the breed has spread into the whole of Andalucía and from Murcia into the whole of the Levante so that the breed is now encountered in nearly all the southern half of Spain.

Facial profile is straight and ears are small. Horns may or may not be present. The male has a beard and sometimes the female does too. Wattles are normally present.

The female can breed at any time of the year. Average litter size is two; the 20% of singles are compensated by a proportion of triplets and even quadruplets (which, however, are not considered desirable). Milk yield averages 450–500 litres in a lactation of 8–9 months. Selection is based on milk recording and a herdbook (Belinchón Valera *et al.*, 1977).

(iii) *Spanish Dual-purpose Breeds.* The goat populations of the Meseta, Pyrenees (*Pirenaica*), Badajoz and Caceres (*Cacereña*) cannot be called breeds. However, they have a certain amount in common—convex profile, sabre horns, large head, horizontal ears, colour usually black or red also brown or chestnut often with paler stripes on the face. The Pyrenean appears to represent the original type and the populations elsewhere have been modified by natural selection, cross-breeding and human selection for different colours. The whole group numbers over a million head of breeding females, chiefly in the south of the country.

Their potential for milk production may be as great as that of the recognized dairy breeds but they rarely receive treatment adequate to show it. When housed and fed, a yield of 300 litres can be obtained in addition to that taken by the kid. With the more usual extensive type of management, the surplus is only about 125 litres per year. The females can breed at any time of the year. Prolificacy is 125%. The young goats are normally sold at 30–45 days of age weighing 8–10 kg (Belinchón Valera *et al.*, 1977). For size and yield of Spanish goats see Table VIa.

Table VIa. Spanish breeds—numbers, size and yield.

	Number of adult females	Body weight (kg)		Milk yield (kg)	Litter size
		Male	Female		
Malagueña	200000			400	2
Murciana-Granadina	450000	70–75	45–55	450–500	2
Mesetas				125–300	1·25
Serrana	500000	80–100	40–50	80–90	1·1–1·25
Canaria	66000	60–70	30–35	350–400	usually 2

Sources: Belinchón Valera *et al.* (1977), Sánchez Belda (1977).

Table VIb. Spanish breeds—numbers (1970)

	Number	%
Serrana	1014000	39·5
Mesetas	635000	24·7
Murciana	235000	9·1
Granadina	233000	9·1
Malagueña	121000	4·7
Pirenaica	39000	1·5
Blanca Andaluza	34000	1·3
Blanca Celtiberica	31000	1·2
Not specified	219000	8·5

Source: Saraza Ortiz and Tejón Tejón (1973).

2. America

The vast majority of American goats are in Latin America and of these 75% are in Brazil, Mexico and Argentina. There are also considerable populations in Bolivia, Peru, Venezuela and Haiti. They are described here on a regional and linguistic basis—Creole, Criollo, Brazilian, USA—on the grounds that colonizers brought their own language and their own goats.

(a) *Creole Goats of the West Indies*

In the West Indies the three islands of Hispaniola, Jamaica and Cuba carry 95% of the goat population. The West Indian or Creole goat has short erect ears and short scimitar horns. Males commonly have beards; females rarely do. The coat is short but males have a mane of coarse hair over the shoulders. Black and brown are the commonest colours, particularly black which is often broken by white patches, especially on the belly. The conformation is compact. Males and females average 25 and 20 kg respectively in weight and 60 and 50 cm in height (Devendra and Chenost, 1973). However, litter size is high (average 2·33) and the average interval between kiddings is only 8 months (Guadeloupe) (Cognie *et al.*, 1973). These goats must have come originally from Europe or from West Africa but they do not appear to exhibit the achondroplastic dwarfism of the goats from the Guinean zone of West Africa.

(b) *Criollo*

Goats were first taken to America from Spain at the beginning of the sixteenth century. In Peru, for instance (Nolte, 1973), the first goats arrived in 1536. Today there are 4 million goats of which 60% are in the dry northern province of Piura. The Criollo breed is very variable. It derives from the Granada, Murcia and Malaga breeds of Spain and more recently has been influenced by the Anglo-Nubian. Milk yield in northern Peru is 1–1·5 litres per day in a lactation of 2–3 months. Meat comes from milk kids at 6–8 weeks old or from old culled animals.

In Venezuela there are 1·25 million goats; 85% of them are in three dry north-western provinces—Falcon, Lara, and Zulia. Most belong to the Criollo type. These have short hair and are of various colours but commonest are black, white and red and their combinations. They are small, weighing only 30–45 kg with withers heights of 75 cm for males and 65 cm for females. Milk yield is low—40–60 kg in 200 days, with 5·5% fat, 4·2% protein and 17·1% total solids. Meat production is also important. The Criollo goat is very hardy and a good grazing animal (Castillo, 1973).

In Mexico the origin of the Criollo is very similar to that in South America. At the beginning of the twentieth century, some Saanen, French Alpine, Toggenburg and Anglo-Nubian were imported and used for cross-breeding. Two different types may be distinguished (Mexico, 1971):

—in Central Mexico there is a smaller type with dairy characteristics and much influenced by Spanish breeds (Granadina and Murciana);

—in Northern Mexico the Criollo is of more varied appearance due to repeated crosses with Alpine, Toggenburg and Anglo-Nubian; it is larger in size with lower milk yield but higher carcass yield than the Central type.

There were 12·9 million goats in 1970, half of them in the northern zone of Mexico. Some production figures from Mexico and Venezuela are given in Table VII.

Table VII. Comparative productivity of Latin American goats (from Winrock, 1977).

Location	Breed	Mature weight		Slaughter		Lactation length (months)	Yield per day (kg)	Litter size
		males (kg)	females (kg)	weight (kg)	age (mo)			
North Mexico	Criollo[a]	45	35	6–9	1–1·2	5–6	0·5–0·8	1·6
	Exotic[b]	70	50	10	1	6–10	1–2	1·8
N.E. Brazil	Local[a]	35–45	30	25–35	12–14	—	—	—
	Bhuj[c]	70–80	40–50	30–40	8–12	—	—	—
N. Central Venezuela	Criollo[d]	40–50	30–35	20–25		6–7	0·2–0·3	—

Sources: [a] Observations and responses to questions by local producers.
[b] Production statistics summarized at the Centro de Cria Tlahualilo.
[c] de Andrade, T. W. and Eduardo, T. M. (1976). Possibilidades da Caprinocultura e Ovinocultura do Nordeste. Banco do Nordeste do Brasil, S. A. Fortaleza, Brasil.
[d] Garcia *et al.* and Castillo *et al.*, Estacion Experimental El Cuji, Barquisimeto (no precise references given).

(c) *Brazilian*

The local, *Crioulo* or national goat is equivalent to the Criollo of Spanish America but derives primarily from Portuguese rather than Spanish breeds. It varies considerably in size, coat and conformation. Ninety-two per cent of the 6·6 million Brazilian goats are in the nine north-eastern states and here the Crioulo goats predominate. Centuries of neglect have produced by natural selection, small goats which are hardy, prolific, undemanding and adapted to a harsh environment. Horns and ears are small and facial profile is straight. Coat colour is light, dark or pied, sometimes with a distinctive pattern. Body weight is 30–35 kg and withers height 60–65 cm. Meat and skin are the most important products; milk yield is only 0·5–1·0 litres per day. Among this general population of short-eared, short-haired goats, four colour types have been named and described as follows:

(i) *Marota*—white.

(ii) *Canindé*—black or dark brown with pale face stripe and belly and dark extremities.

(iii) *Moxotó*—named after the Moxotó valley in Pernambuco it is also found in Paraíba, Ceará, Piauí and Bahia. Colour is white or buff, with black back stripe, face stripe and belly; ears, lower legs, hooves and mucosae are also black.

(iv) *Repartida*—black forequarters, brown hindquarters, with black back-stripe.

More recently, various Oriental lop-eared breeds have been imported—especially the Gujarati from India (called Bhuj) and the Anglo-Nubian. Crossing these on to the Crioulo type has introduced the lop ears and increased body size and produced the present population which is called SRD (Sem Raça Definida = without defined breed) (Jardim, 1971, 1977; Mason, 1980).

(d) *USA*

In the south-western United States the term *Spanish* is used to distinguish range meat goats from Angoras and dairy breeds. Most are of the same origin as the Mexican Criollo but they may show traces of Nubian and Toggenburg blood. They are highly variable in appearance and performance.

Some of these goats lack external ears or have very small auriculae. On the west coast this type was selected and bred to top sires of dairy breeds in the 1920s and 1930s to produce a new dairy breed—the *American La Mancha*. It retains the vestigial ear character which is presumably controlled by a single gene (Shelton, 1978).

3. Africa

The division into small short-eared and large lop-eared types has been used already, with varying success, to describe the goats of Africa (Mason and Maule, 1960). Epstein (1971), on the other hand, picked out three special types, namely dwarf equatorial, Maltese, and Nubian, and grouped all the rest as "Savannah goats of Africa".

The goats of Africa south of the Sahara are predominantly short-eared. The long-lop-eared animals of the north and north-east are included in the Near East group (see p. 100). The other population of long-eared goats is in the south (see p. 103). Although Ethiopia has more goats than any other African country while Zaïre has more goats than any other species of farm animals, little reference is made to these countries here because their goats have been inadequately described.

Some weights and heights of African goats are shown in Table VIII.

Table VIII. Body size of some African goats.

	Body weight (kg)			Withers height (cm)		
	Male		Female	Male		Female
Short-eared						
West African Dwarf		20–25			40–50	
African Pygmy (USA)			25		45–50	
Southern Sudan		11–25			40–50	
Somali	30–50		25–30	72–76		65–68
Somali Arab			26			65
Small East African		25–30			60	
West African Long-legged		30		80–85		60–75
Maradi or Red Sokoto	25–30		23–28	62–65		60–62
Long-eared						
Benadir (Somalia)	43–45		32–36	72–78		70–72
Boer (South Africa)	90		60–75			
Nguni (Swaziland)		30–40				
Sudanese Nubian			27	71–76		
Shukria (Eritrea)		40–60			70–85	

Sources: Mason (1951); Mason and Maule (1960); Epstein (1971).

(a) *West African Dwarf*

In the Guinean zone of West Africa and in the forest and coastal zones of West Central Africa there is a type of goat which is called *Fouta Djallon* after its presumed centre of origin in Guinea or West African Dwarf to describe its appearance. Similar types are found in the west of Zaïre and Angola and as far south as Ovamboland in northern Namibia (Mason, 1951; Mason and Maule, 1960; Epstein, 1971).

The West African Dwarf has very short horns—so short that their twist is not conspicuous. The ears are always short and erect. The wild colour—agouti with black stripes and pale belly is common but all other colour and combinations are encountered, especially black, brown and pied. The coat is short but the male often has a ridge of longer hair along the spine; he also has a beard. Conformation is typical of the achondroplastic dwarf with plump body, disproportionately short legs and broad short head. This breed is very fertile. At the Nungua experimental station in Ghana, age at first kidding was 12 months, interval between kiddings $8\frac{1}{2}$ months and litter size 1·84 (Vohradský and Šáda, 1973).

The West African Dwarf from Cameroon has reached the USA via Germany. It is now used as an experimental animal under the name *African Pygmy*. A breed society unites the breeders of this animal. In the colony at the University of Oregon Medical School, Portland, most of the animals are solid colour with black extremities; some have a black dorsal stripe. The commonest ground colours are black, various shades of grey, and white, followed by caramel and buff. Almost all animals have a white spot on the forehead and white or light grey ears and muzzle; some have white spots elsewhere. Females normally give birth to 1–2 kids with an average birth weight of 1·8 kg; triplets are not uncommon and quadruplets occur. Sexual maturity is very early and pregnancy has occurred at 3 months of age (Hoversland *et al.*, 1973).

(b) *Southern Sudan*

The typical goat of the Nilotic region of southern Sudan is also dwarf but it is harmoniously proportioned and shows no signs of the achondroplasia of the West African Dwarf. In the male the horns are typically 5–8 cm long, erect or curving backwards and outwards with the tip turned slightly forwards; in the female horns are even shorter and may be absent. Ears are usually short and erect—but they may be of medium size and carried horizontally. The hair is very short but the male may have a mane and a small beard. All colours and mixtures of colours occur (Mason and Maule, 1960; Epstein, 1971).

Similar goats are present in the adjacent areas of north-east Zaïre but some have short legs like the West African Dwarf and some have long hair.

(c) *Somali*

Except in the Webi Shibeli valley where they have long ears, the goats of

Somalia are short-eared, short-horned and short-haired. A similar type is found in the Somali areas of Ethiopia (Ogaden) and of north-east Kenya where it is called *Galla*. The colour is always brilliant white, often spotted with brown or black around eyes, ears and nose, and sometimes with coloured patches on neck or shoulders. Horns are short and slender and may be spiral or crescentic in shape; in the female they are smaller and may be absent. Ears are small and either erect or directed forward. The male has a short beard. This is primarily a meat animal adapted to dry zones (Mason and Maule, 1960).

In Somalia there is also a small goat introduced from Arabia. Ears and horns are short and erect. The only short-eared goats reported across the Gulf of Aden are the *Thamood* goats of South Yemen which Purohit (1973) describes as short-haired and of white-grey colour. Weight of female is 16–18 kg and height 50–52 cm.

(d) *Small East African*

The goats of Kenya, Uganda and Tanzania belong to the general type which has been called *Small East African* (Mason and Maule, 1960). The same description seems to cover also the goats of Rwanda, Burundi, the *Bantu* goats of Malawi, Zimbabwe (Mashonaland) and Botswana and the *Landim* goat of Mozambique. (The Nguni goat of Zululand and Swaziland has flabby ears and is considered on p. 104) (Joubert, 1973).

Naturally there is considerable variation over so wide an area but nowhere do specific types or colour patterns seem to have been selected. All colours and patterns are encountered. Horns vary from very short (2·5 cm) to moderately long (25 cm). Their direction is backwards and upwards or lateral; many animals, especially females, have scimitar horns. Ears are always short and usually erect; sometimes they are horizontal. The hair is usually short but males often have a dorsal ridge of long hairs along the spine and less commonly they may have long hairs on back, flanks and shoulders. Beard and tassels may or may not be present.

Sexual maturity can be reached at 4 months of age but the twinning rate is only 5–15%. The chief merit of this goat is its extreme hardiness. It is used chiefly for meat (and skin) and is rarely milked. Although no distinct local varieties have been developed, two names have been used in Uganda for fleece types of individuals found in mixed herds. These are the *Mubende* and the *Kigezi*. The first refers to a skin type favoured by the leather trade which is obtained from a goat found mainly in the Mubende district of Uganda. It is nearly always pure black with short, fine hair. The males are maned. The other refers to a long-haired black goat in the Kigezi district of Western province. It is smaller, more compact and shorter in the leg than the Mubende. It is prized for its long black or grey hair which is used for clothing.

4. South and East Asia

In south-east Asia, southern India and southern China the indigenous goats are of the small, short-eared type already described for Africa. In south-west Asia and northern India they belong mostly to the screw-horned lop-eared type (see p. 100). Central Asia is the homeland of the screw-horned fleece goats (Cashmere). On the boundaries of these regions intermediate types occur and the towns may have a different population from the countryside.

(a) *Pakistan*

Information is from Pakistan (1970) and Wahid (1973). For size and performance of Pakistani breeds, see Tables IX, X and XI.

Table IX. Pakistani breeds—size and numbers.

	Numbers ('000)	Body weight (kg)		Withers height (cm)		Ear length (cm)	Horn length (cm)
		Male	Female	Male	Female		
Short-eared							
Barbari	235	21	20	76	66	5–7·5	10
Chapper	2520	26	23	75	64		6–10
Damani	374	22	20	69	61	→ 11·5	13–14
Kajli	234	30	25	80	75		15
Himalayan							
Gaddi	91	50	40	88	82	23–30	23 (♂) 18 (♀)
Kaghani	403	36	32	82	77	13–15	
Kel		28	23	81	77	15–18	20 (♂) 15 (♀)
Long-eared							
Beetal	807	36	30	91	82	→ 46	
Bikaneri	85	25	20	72	65	15–16·5	
Daira Din Panah	106	45	40	82	71	17–24	25–28
Kamori	159	60	50	90	85	20–25	13–15
Khurasani	88	30	25	71	64	18	
Lehri	828	33	31	90	72	→ 35	18–20
Sind Desi	282	27	23	81	74	15–28	

Source: Pakistan (1970).

Table X. Pakistani breeds—performance.

	Milk yield (kg)		Lactation length (days)	Fleece weight (kg)	Age at first kidding (months)	Kidding interval (months)	Kids per birth
	per day	per lactation					
Short-eared							
Barbari	0·9–1·3	100	100	—	14–18	7–8	usually 2
Chapper	0·7–0·9	90	120	0·7	15–18	7	often 2
Damani	1·0–1·4	107	100	0·7	12–16	7½	usually 2
Kajli	0·9–1·3	118	120	0·7–0·9	15–18	7	usually 1
Himalayan							
Gaddi	0·9–1·0	125	150	1·4–1·6	16–20	12	usually 1
Kaghani	0·5–0·7	68	140	0·9–1·1	14–18	12	often 2
Kel	0·5–0·7	90	150	0·7–0·9	14–18	12	usually 1
Long-eared							
Beetal	1·8–2·7	260	130	—	12–18	7–8	often 2
Bikaneri	0·7–0·9	73	100	0·7	15–20	6–7	often 2
Daira Din Panah	1·6–2·2	200	135	0·9–1·3	14–20	8	usually 2 or more
Kamori	1·8–2·2	225	120	0·2	15–18	6–7	usually 2
Khurasani	0·9–1·3	180	170	0·5–0·7	14–16	12	
Lehri	0·7–0·9	90	120	0·7–0·9	14–16	7–8	often 2
Sind Desi	0·9–1·8	140	110	0·5	15–18	6–7	2

Source: Pakistan (1970).

Table XI. Pakistani breeds—fleece characters.

	Fleece weight (g)	Hair length (cm)	% short fibres (5 cm and less)	% non-medullated fibres	Fibre diameter (mm)
Short-eared					
Chapper	600	9·2	8·6	19·0	77·0
Damani	585	9·2	8·9	11·6	72·3
Kajli	600	9·1	8·8	9·7	79·3
Himalayan					
Gaddi	900	11·8	4·6	20·2	80·1
Kaghani	800	10·2	6·7	10·1	80·2
Long-eared					
Bikaneri	500	8·5	8·8	8·0	74·4
Daira Din Panah	1074	9·4	7·8	9·9	81·2
Kamori	271	4·8	48·0	4·9	98·6
Khurasani	540	10·4	6·0	15·2	78·0
Lehri	700	9·3	6·2	18·5	73·8
Sind Desi	500	6·3	38·8	5·6	78·2

Source: Pakistan (1970).

(i) *Chapper*. Also known as *Kohistani* or *Jabli*, this breed is found purest in the mountains of Sind and Baluchistan but less pure specimens occur throughout Sind. Facial profile is usually straight; in some animals it is slightly concave. Ears are small to medium in size. Horns are usually straight, broad at the base and blunt at the tip. The hair is long and there is a long forelock. Colour is usually black or white but some animals are pied. Goats are shorn yearly; some have an undercoat of down (pashmina). This is primarily a meat breed; milk yield is low.

(ii) *Damani*. This breed is centred in the district of Dera Ismail Khan and is bred primarily for meat and milk. Colour is usually black with head, ears and lower legs of camel colour. It is a medium-sized breed with small pointed ears and curled pointed horns. Throat tassels are sometimes present.

(iii) *Kajli* or *Pahari*. Bred mainly for meat and hair the Kajli breed is found in Loralai, Dera Ghazi Khan, and other mountain tracts. It is a large breed covered with long white or brown hair except on the face. Colour may also be grey or black and white or light brown face stripes are present. Profile is straight. Beard may or may not be present. Horns are short and thin; they grow backwards and sometimes outwards. Ears are erect and pointed.

(b) *India*

Information is from India (1970). For size and performance of some Indian breeds see Table XII.

Table XII. Indian breeds—size and yield.

	Body weight (kg)		Withers height (cm)		Milk yield per day (kg)	Lactation length (days)
	Male	Female	Male	Female		
Short-eared						
Barbari	35–45	27–36	66–76	61–71	0·9–1·25	110
Bengal	14–15	8–13			0·9–2·8	
Long-eared						
Jamnapari	70–90	45–65	90–102	76–86	2·2 – 2·7	250
Beetal	65–85	45–60	91–99	76–83	1·8	180

Source: India (1970)

(i) *Barbari*. Small and short-haired with short prick ears and short erect horns, this breed is found in the towns of Punjab, Uttar Pradesh and Haryana states as well as in West Punjab, Pakistan. It probably derives from Berbera in Somalia. Colour is variable but white with red or tan spots is common.

These animals are usually stall fed. They are prolific breeders kidding every 6–7 months and giving 33% twins. (Sengar, 1976).

(ii) *Berari.* This black or dark brown breed of the Nagpur and Wardhar districts of Maharashtra and the Nimar district of Madhya Pradesh is reported to yield 0·63 kg milk per day.

(iii) *Surti.* The breed of south Gujarat (Surat) and north-west Maharashtra (Bombay and Nasik) is believed to be derived from a small Arabian milking goat. It resembles the Berari in having short hair, ears and horns. Milk yield is 1–2 kg per day. The *Malabari* or *Tellicherry* is a mixed population with a preponderance of *Surti* blood.

(iv) *Deccani.* This is the breed of the Deccan plateau in western Andhra Pradesh. Colour is usually black but may be black and white or red and white. It is of medium size and milk yield is 1·35–2·25 kg per day. The *Osmanabadi* of south Maharashtra is closely related; it may be long- or short-haired and ears may lop. Horns are 13–15 cm long.

(v) *Bengal.* This diminutive meat breed weighs less than 15 kg. Ears and coat are short. It has three colour varieties—black, brown and white. It can kid twice a year and produce 60% twins and 10% triplets (Sengar, 1976).

(vi) *Assam Hill* with long white hair, short ears and horns with a slight twist.

(c) *Nepal*

Information is from Epstein (1977). For size of Nepalese goats see Table XIII.

Table XIII. Nepal goats—size.

	Body weight (kg)		Withers height (cm)	
	Male	Female	Male	Female
Tarai	8–12		45–50	
Southern Hill	12–16		50–55	
Northern Hill	35–45	24–30	60–70	50–65
Sinhal			65–75	

Source: Epstein (1977).

(i) *Tarai.* The breed of the Tarai, the southern lowlands, is a small animal with short hair, short horns and straight or slightly concave facial profile.

However, ears are semi-pendulous or carried horizontally. Black and chamois are the commonest colours but some animals are white or pied. Kidding may occur throughout the year and some animals kid twice. Twinning rate is estimated at 40%.

(ii) *Southern Hill* goats of the Siwalik Range resemble the Tarai goats but are larger. Coat colour is very variable.

(d) *South-east Asia*

Although goats were known in India by 100 B.C. it is suggested that they did not reach south-east Asia until early in the first millennium A.D. and indeed Yamane (1943) thinks that they were not brought to Indonesia until after the Muslim invasion in the fourteenth century. From there they spread north, via the Philippines and eastern Taiwan, and reached the south-western islands of Japan in the fifteenth century (Nozawa, 1972). Apparently they never reached the Japanese mainland where there is no evidence of goat husbandry until a century ago.

(i) *Malaysia*. The indigenous breed in Malaysia is called *Kambing Katjang* (*Kambing* = goat and *Katjang* = bean) and the same type is found in Burma, Thailand, Indonesia, the Philippines and Taiwan. Devendra (1966) describes the Kambing Katjang from Malaysia as follows: the horns are short and scimitar-shaped; the ears are short and erect. The male always has a beard but this is less frequent in the female. Colour is usually black sometimes with a few white patches. Twinning is common.

(ii) *Indonesia*. The description by Rumich (1967a) of the *Kambing Kachang* of Indonesia is very similar but colour is more variable—any shade of brown, black or white, or a combination of the three. Age of puberty is about 6 months in females and 7 months in males. Age of first kidding is 12–13 months. Twins are the rule; triplets are common and quadruplets have been recorded. This breed is very hardy and widely distributed; it is used for meat (and skins).

(iii) *Philippines*. Rumich (1968a) distinguishes two types among the native goats—coarse-hair and fine-hair. The coarse-hair variety usually lacks horns; if present they are small (5–10 cm long). Ears are horizontal and males and old females have beards. Tassels are rare. Coat colour is commonly cream, beige or light brown with darker face stripes and dorsal line. The fine-hair variety is similar but hair and bone are fine. Short twisted horns are generally present in both sexes. Ears are horizontal or semi-erect. Colour is black or

Fig. 1. French Saanen, female (ITOVIC).

Fig. 2. French Alpine (Chamoisé) male (ITOVIC).

Fig. 3. French Alpine (Chamoisé) female (ITOVIC).

Fig. 4. Poitevin (ITOVIC).

Fig. 5. Norwegian Dairy goat (H. Skjervold).

Fig. 6. Créole goat in Haiti (I. L. Mason).

Fig. 7. Nilotic goat in Southern Sudan, female (CBABG).

Fig. 8. Galla goat in Northern Kenya, male (CBABG).

Fig. 9. Small East African goat in Tanzania (H. G. Hutchinson via CBAGB).

Fig. 10. Small East African goat in Malawi (CBABG).

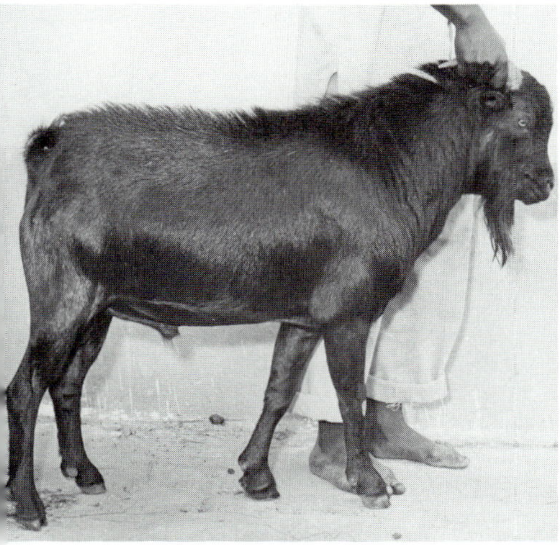

ig. 11. Mubende, male (CBABG).

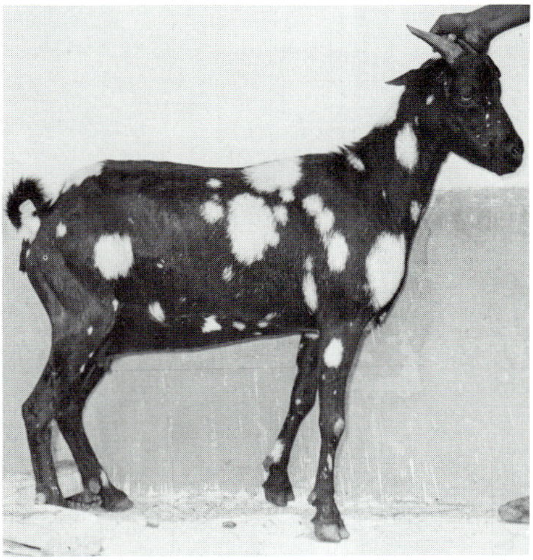

Fig. 12. Mubende, female (CBABG).

Fig. 13. Kigezi (CBABG).

Fig. 14. Barbari (R. M. Acharya).

Fig. 15. Surti (Malabari) male (K. L. Sahni).

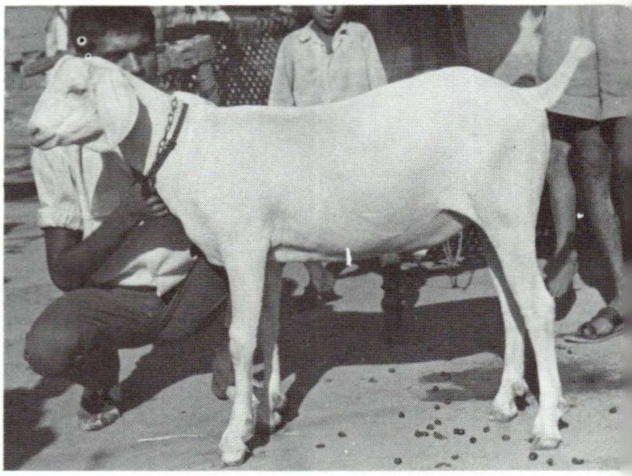

Fig. 16. Surti castrate (B. Rumich).

Fig. 17. Deccani (Osmanabadi) male.

Fig. 18. Black Bengal, female (R. M. Ācharya).

Fig. 19. Black Bengal, male (H. Guha via CBABG).

Fig. 20. White Bengal, female (B. Rumich).

ig. 21. South China goat—Kwantung, male H. Epstein).

Fig. 22. South China goat—Kwantung, female (H. Epstein).

g. 23. Kambing Kachang, Indonesia, female . Rumich).

Fig. 24. Korean Black, male (B. Rumich).

Fig. 25. Korean White, male (B. Rumich).

Fig. 26. Corsican goats (ITOVIC).

Fig. 27. Girgentana, female (A. M. Pilla).

Fig. 28. Head of Greek goat, male (N. Zervas).

Fig. 29. Head of Tuareg goat (Heck via Epstein).

Fig. 30. Cashmere goat, Inner Mongolia, female (H. Epstein).

Fig. 31. Orenburg, males (CBABG).

Fig. 32. Don, male (CBABG).

Fig. 33. Gaddi, female (B. Rumich).

Fig. 34. Kaghani, male (B. Rumich).

Fig. 35. Pashmina goats (R. M. Acharya).

Fig. 36. Changra pack train in Dhauli valley, Nepal (H. Epstein).

Fig. 37. Yearling Angora males (M. Shelton).

Fig. 38. Soviet Mohair goat, males (CBABG).

Fig. 39. Negev (left) and Syrian mountain (right) male goats (H. Epstein).

Fig. 40. Mamber, female (H. Epstein).

Fig. 41. Iraqi goats (I. L. Mason).

Fig. 42. Sudanese Nubian (CBABG).

Fig. 43. Boer, female (D. M. Joubert via CBABG).

Fig. 44. Long-eared Rhodesian goat, female (D. A. Robinson via CBABG).

Fig. 45. Jamnapari, female (D. S. Bhatnagar).

Fig. 46. Jamnapari, male (K. L. Sahni).

ig. 47. Beetal, female (D. S. Bhatnagar).

Fig. 48. Beetal, male (D. S. Bhatnagar).

Fig. 49. Gujarati (Zalawadi variety), female (B. Rumich).

Fig. 50. Gujarati (Bhuj) in Brazil, male (I. L. Mason).

Fig. 51. Gujarati (Cutchi) (R. M. Acharya).

Fig. 52. Gujarati (Mehsana) R. M. Acharya).

Fig. 53. Daira Din Panah (B. Rumich).

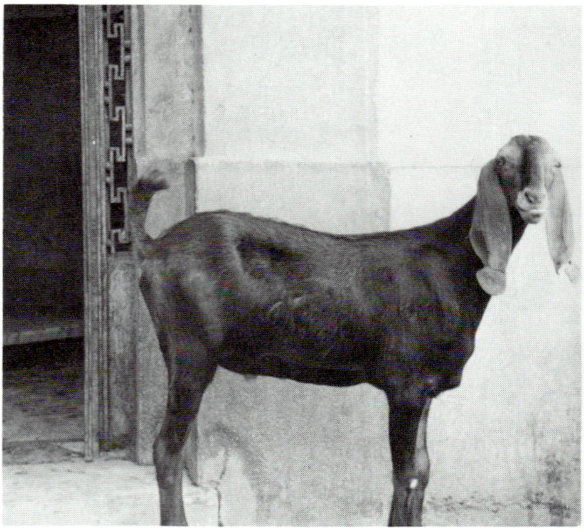

Fig. 54. Zaraibi (I. L. Mason).

Fig. 55. Damascus (Shami) (H. Epstein).

Fig. 56. Damascus (Shami), female (H. Epstein).

Fig. 57. Maltese, male (H. Epstein).

Fig. 58. Maltese, female (H. Epstein).

Figs 1–58. The author and publisher thank the people and institutions cited in brackets after each figure for permission to use the photographs. ITOVIC = Institut Technique d'Elevage Ovin et Caprin, Paris; CBABG = Commonwealth Bureau of Animal Breeding and Genetics, Edinburgh. Figures 9, 19, 43 and 44 were furnished by CBABG, but they were taken by the person mentioned. Figure 29 was taken by Mr Heck but given to the author by Dr Epstein.

brown with a white or white-spotted band. Intermediates are common as well as evidence of foreign blood.

(iv) The *Sarawak* goat is almost the same as the Kambing Kachang but smaller. Colour is brown, black, white or a combination of two or all three of these colours (Rumich, 1976b).

(v) *Thailand.* Brown with black dorsal and shoulder stripe is the commonest colour among the Kachang type of goat. Horns are always present but tassels never (Nozawa, 1974).

(vi) *Taiwan.* As elsewhere in south-east Asia the native goats are raised exclusively for meat. They have horns but not tassels. Cross-breeding with Saanen is common. The goats to the west of the central mountain chain are predominantly black and are believed to have descended from the goats of Kwangtung and Fukien brought to the islands by Chinese immigrants since the seventeenth century. The goats in the east of the island have a high frequency of the brown-with-black-stripes colour pattern. The brown goats are slightly larger than the black ones (Nozawa and Watanabe, 1969).

This original Kambing Kachang breed has been much influenced by the long-eared dairy breeds from India (particularly the Jamnapari) so that in Malaysia, Thailand and Indonesia the majority of the goats are now larger than the Kambing Kachang with a somewhat convex profile, longer, frequently lop, ears and horns sometimes twisted in a spiral (Devendra and Nozawa, 1976).

For the sizes of south-east Asian goats see Table XIV.

Table XIV. Size of goats in south-east and east Asia.

<table>
<tr><th></th><th colspan="2">Body weight (kg)</th><th colspan="2">Withers height (cm)</th><th>Source</th></tr>
<tr><th></th><th>Male</th><th>Female</th><th>Male</th><th>Female</th><th></th></tr>
<tr><td>Kambing Kachang</td><td></td><td></td><td></td><td></td><td></td></tr>
<tr><td>Malaysia</td><td>25</td><td>20</td><td>60–65</td><td>56</td><td>Devendra (1966)</td></tr>
<tr><td>Indonesia</td><td colspan="2">30</td><td colspan="2">50–60</td><td>Rumich (1967a)</td></tr>
<tr><td>Philippines</td><td></td><td></td><td>55–70</td><td>50–60</td><td>Rumich (1968a)</td></tr>
<tr><td>Sarawak</td><td colspan="2">15–18</td><td colspan="2">40–45</td><td>Rumich (1967b)</td></tr>
<tr><td>Thailand</td><td></td><td></td><td>57</td><td>52</td><td>Nozawa (1974)</td></tr>
<tr><td>Taiwan native</td><td></td><td></td><td>51</td><td>48</td><td>Nozawa and Watanabe (1969)</td></tr>
<tr><td>Korean Black</td><td>40–45</td><td></td><td></td><td></td><td>Rumich (1968b)</td></tr>
<tr><td>China</td><td></td><td></td><td></td><td></td><td></td></tr>
<tr><td>South China</td><td>30</td><td>25</td><td>55</td><td>50</td><td>Epstein (1969)</td></tr>
<tr><td>Ma T'ou</td><td colspan="2">31</td><td colspan="2">55</td><td>Epstein (1969)</td></tr>
<tr><td>Ma</td><td>42</td><td>77</td><td></td><td>64</td><td>Epstein (1969)</td></tr>
</table>

(e) *Southern and Eastern China*

The information is from Epstein (1969). For sizes see Table XIV.

(i) The *South China* goat which Epstein (1969) describes from Yunnan and Kwangtung is very similar to the Kambing Kachang except that it has disproportionately short legs. The head has a concave profile and short prick ears. Both sexes have a beard and strong twisted horns about 15–25 cm long. The coat is usually short and always black. Adult males develop a mane over the neck and shoulders. Twin births are the rule. As in south-east Asia crossing with Saanen and Jamnapari is common. This is a thickset muscular meat-type animal.

(ii) The goats of *Chekiang* in eastern China are slightly smaller but with more harmonious proportions. They are usually of an angular dairy type rather than the compact meat-type. The coat is nearly always white.

(iii) Further north in *Tanyang* country of Kiangsu there is a small meaty type which may be grey or black as well as white. It has a longer coat and may be polled.

(iv) The *Ma T'ou* is bred in Chushan, Chushih and adjacent districts of Hupeh. It is of medium size. Both sexes are polled. Ears are short and erect. Hair is white and may be long or short. The beard is large especially in long-haired males. Fecundity is high. Females usually kid twice a year. Twenty-two per cent of births give singles, 70% twins or triplets and 8% quadruplets with a total annual birth rate of 448 kids per 100 females. Lactating females yield up to 1·5 kg of milk per day.

(v) The *Ma* goat from Chengtu district of Szechwan is slightly larger than the Ma T'ou. Males have short thick horns; those of the females are more slender. Ears are straight and erect. The coat is usually short but there may be longer hair on flanks and thighs. Colour varies from pale to dark brown with black stripe on face and backline. This is primarily a meat breed. The mean birth rate is 2·1 kids per female.

(f) *Korea*

Goats were introduced from China into Korea during the early years of the 1st millennium A.D. In the south there are now three varieties: black, grey and white. The black is commonest; it is similar to the goat of western Taiwan. Ears are small and horizontal or semi-erect. Horns are of meduim size and slightly twisted. Facial profile is straight. Beard and forelock are present; they are more pronounced in the male.

The grey goats are much fewer in number and are slightly smaller in size. The white variety is being gradually absorbed by crossing with the Saanen (Rumich, 1968b; Watanabe, 1970).

B. Short-eared Goats with Twisted Horns

1. Europe

The wild goat's horns are curved like a sabre in one plane only. At an early stage they were selected by man into a loose spiral (starting with a twist at the tip) to give the so-called *prisca* horns (see Chapter 2, p. 36). Breeds with this shape of horn still survive, or have recently become extinct, in many parts of eastern and southern Europe. In most other parts of Europe they have been replaced by more productive Alpine breeds which retained the earlier horn shape.

(a) *Switzerland*

(i) *Valais Balckneck*. The *Col noir du Valais* or *Walliser Schwarzhalsziege* is restricted to the Swiss canton of Valais. It has long hair and long curved horns. The front half of the body is black and the hind part white.

A feral herd of the same phenotype grazed the Blithfield Hall park in Staffordshire, England, and is believed to be descended from Swiss imports in the fourteenth century. They are named *Bagot*, after the family owning them.

(b) *France*

(i) *Common French. La chèvre commune francaise* is the name used by Quittet (1976) for animals which cannot be included in one of the three recognized breeds in France—Alpine, Saanen and Poitevin. It is very variable in size, conformation, length and colour of coat, ear shape and size, and performance. However, it is very hardy and well adapted to local conditions. Quittet (1976) gives no information about its horns but it is included here because it contains traces of ancient breeds some of which (e.g. Valais Blackneck and Pyrenean) have spiral horns. Quittet lists the following among the extinct or nearly extinct breeds included now in his Common French: Mont d'Or, Drôme, Sundgau, Pyrenées, Béarn, Auvergne, Berrichon, Tarentais, Valais. It may be noted that other authors consider the Sundgau as the black variety of the Alpine (French), and the Tarentais as its Cou clair variety.

Among local breeds which still exist, Charlet and Le Jaouen (1977) mention the following.

(ii) The *Rove* of Provence and Haute Provence which is short-haired, black or mahogany and with spiral horns.

(iii) The *Pyrenean* (*la chèvre des Pyrenées*) between Catalonia and Languedoc, a long-haired variety varying in colour between black, grey and chestnut. It is thus different from the short-haired white Pirenaica of Spain.

(iv) The *Corsican* goat (*la chèvre Corse*) which exhibits a certain uniformity in spite of much crossing in the past. Its coat is long and of all colours and patterns but with a predominance of grey and chestnut. In spite of their small size females produce 150–250 kg of milk in 150–200 days under extensive husbandry systems (figures from official recording).

(c) *Spain*

(i) *Spanish Mountain*. Belinchón Valera *et al.* (1977) use the term *cabras serranas españoles* for two mountain breeds and their crosses, namely the *Blanca celtibérica* or *Serrana di Castilla* and the *Serrana andaluza*, each numbering about 250 000 breeding females. The former is purest in the mountains of Cuenca, Teruel, Guadalajara, Reino de Valencia, Albacete and Estremadura. The latter occurs around the Guadalquivir valley and its tributaries and in the Sierra Morena.

Sánchez Belda (1977b), on the other hand, uses the general term *Raza blanca española* to unite the Blanca celtibérica and the Blanca andaluza varieties and says that the 1974 census recorded only 23 610 of the first and 29 626 of the second. (Presumably this refers to purebred specimens only.)

Colour is usually white but there may be some darkening of head and legs. The coat is short but males have long coarse hair on the chest and on the front of the legs. The beard is well developed in the male but is not always present in the female. The profile is convex in the male, straight or slightly convex in the female. Horns are long and spirally twisted. Ears are large, horizontal or slightly pendent. Tassels are rare.

These animals are very hardy; they remain outside summer and winter with no supplementary feed. First kidding is at 18–20 months of age. Prolificacy is 110–125%. Males are castrated young and sold for meat at 4–5 months weighing 25–30 kg. Milk yield is low. The lambing season is late winter or early spring.

(ii) *Canary Island*. Sánchez Belda (1977a) describes *la raza caprina canaria* which exhibits some influence of African blood. It can be of any colour or combination of colours. Hair is generally short but may be long on the back of the thighs; also the male often has long hair along the backbone and sometimes on the forequarters. Facial profile is straight. Horns are usually of *prisca* type but may be of *aegagrus* and exceptionally of *falconeri* type. Ears are medium to large and drooping; they are often atrophic. Beard may or may not be present. First kidding is at 12–14 months and twins are the rule. This breed is very hardy and resistant. It is a good milker yielding 350–400 kg at 5% fat with a single milking daily. When housed the yield rises to 600 kg. A herdbook was started in 1976.

(d) *Portugal*

The goat population of Portugal is very heterogeneous but two breeds can be distinguished.

(i) *Serra da Estrela*. It is found in the mountains of the centre and north (hence *Serrana*) and numbers 45% of the total population. It has a slightly concave profile and usually carries horns, beard and tassels. Ears are small and horizontal. Live weight is 40–45 kg and height at withers 64 cm. The coat is long and silky and its colour is black, dark brown, grey or reddish. This is a milk breed giving 300–400 litres in 210–220 days.

(ii) *Charnequeira*. Larger than the Serrana with a weight of 45–50 kg and a height of 78 cm. The facial profile is convex; ears are medium-sized and erect, and horns are long and corkscrew—or lyre-shaped. The coat is short, usually brownish red but sometimes pied. This is a meat breed.

There are two local varieties of Charnequeira origin, the *Raiana* or *Serpentina* in Alentejo and the *Algarvia* in Algarve (Cabral Calheiros, 1976; Da Silva Lobo, 1973).

(e) *Italy*

Ninety per cent of goats in Italy belong to populations which may have been given geographical designations (e.g. *Murgese*, *Calabrese*, *Sarda*, *Romana*, *del Gran Sasso*, or Murge, Calabrian, Sardinian, Roman, Gran Sasso) but which are not sufficiently distinct and uniform to merit the name breed. In general the goats in the north are of the Alpine type (see p. 59) while those in the centre and south (including the islands) tend to be long-haired and often have twisted horns. Signs of Maltese blood are also present in the south. Ninety per cent of Italy's goats are in the south and in the islands—30% in Sardinia alone. However, two breeds stand out.

(i) *Garganica*. This is the breed of the Gargano peninsula in Apulia where 80% of its 45000 head are still found. It is of moderate size. The head is relatively small with long flattened horns which make a half turn in an open spiral. The coat consists of coarse, long, dark chestnut hairs. The kids are born with curly black hair and their skins make good furs. The colour begins to change at about 3 weeks of age. Milk yield and prolificacy are low (see Table IV) but the breed is extremely hardy.

(ii) *Girgentana*. The surprising feature of the breed from Agrigento (or Girgenti) province of Sicily is the closely twisted vertical corkscrew horns. They may be 50 cm long in males but are shorter in females. Apart from the horns the breed is similar to other local breeds. Facial profile is slightly convex. Topknot and beard are present. Ears are short and erect. The coat is coarse and rather long; it is white with brown (occasionally roan) spots on the ears, around the eyes and sometimes on the neck. Adult females give 90–95% of twin births and about 5% triplets. This breed is now also common in the Palermo province of Sicily and in Catanzaro (Calabria) (Italy, 1973).

(f) *East and South-east Europe*

(i) *Balkan*. The local breed of southern Yugoslavia, Albania, Greece and Bulgaria is called Balkan. Of the 4·6 million goats in Greece 80% are of the native breed. (The actual figures are 78·2% local, 14·3% crossbred, 7·5% Saanen. Maltese and Nubian have also been introduced.) Boyazoglu and Zervas (1977) describe the local breed of Greece as follows: The hair is usually long and coarse, in which case it is sheared, but there are also animals with short hair on all parts of the body. Colours are very diverse and include black, brown, grey, reddish and white and combinations of these in a variety of patterns. Nearly all have long curved horns. Ears are horizontal or slightly drooping.

Traditionally three types were distinguished varying in the shape of their horns: the *Vlach* in Epirus and Macedonia with horns directed backwards, the straight-horned variety of Central Greece and Euboea, and the type with curved horns in Attica and especially around Thebes. A fourth type of larger size and with a short coat has been described on some islands, e.g. Skopelos.

In both Greece and Albania varieties with corkscrew horns have been recorded. These are called *Capore* (Albania) and *Ulokeros* (Greece).

The live weight of Greek goats varies from 30 to 50 kg in females and 40 to 65 kg in males. Average milk yield in Greece in 1975 was 118 kg but it varied considerably according to breed and habitat—100 kg for local goats under range conditions, 115–120 kg under extensive conditions in the hills, and 155–160 kg for imported breeds and crosses under intensive conditions in the lowlands. The goat is also an important source of meat.

(ii) *Carpathian*. The Carpathian breed accounts for 75–80% of the goats of

Romania. It also occurs in southern Poland. In Romania it is smaller than the Banat White and much smaller than the pure Saanen. Adults weigh 37 kg and stand 63 cm high. It is very variable in colour, conformation and productivity. The majority of animals have *prisca* horns which are very large in the male. The dominant colour is grey of different shades; other colours are red, white, black and pied. The coat consists of hairs 7–10 cm long and a fine undercoat of down 2·5–3·5 cm long. In the district of Alba Iulia the fleece length exceeds 15 cm. Milk yield is 250–280 kg in 6–7 months with 4·8–5% fat. Prolificacy is 130–140% (Taftă, 1975).

In south-east Poland, in addition to the Carpathian, there are two lowland dairy breeds—the black *Kazimierz* between the Vistula and the San, and the pied or tricolor *Sandomierz* east of Radom.

2. Africa

From several short-eared African breeds which show signs of twisting in their horns two have been picked out here on the ground that the horns are somewhat longer and the twisting is more universal.

(a) *West Africa*

(i) *West African Longlegged.* The *Sudan* or *Sahel* goat of the savannahs lying between the tropical forest and the desert is difficult to fit into the present classification. It is similar to the West African Dwarf except for its size and its long, often twisted horns. It is bred by the Maure, Fulani and Tuareg in the area between Chad and Senegal. Facial profile is straight or slightly convex. Ears are usually short and horizontal, sometimes moderately long and pendulous. Horns are thin, flat and twisted in the male, sickle shaped in the female. Sometimes males have long twisted horns directed upwards. Beard and tassels are common. (For size see Table VIII.) Coat is short and fine but the male may have a forelock and tufts of long hair on the shoulders and the centre surface of the thigh. The commonest colours are white, black, and brown, usually pied or tricolor (Mason, 1951; Epstein, 1971).

(ii) *Maradi or Red Sokoto.* These names are used in Niger and Nigeria respectively for a breed famous for its skins which are used for the production of Morocco leather. It is intermediate between the West African Dwarf and the Sahel goat. It is commonest in Maradi and Tessoua in Niger and in Sokoto and Kano in Nigeria but its crosses extend east and west of these areas. Its population is estimated at 1 million purebreds and 2·5 million crossbreds. Colour is bright mahogany red and coat is short and sleek. It is a delicate animal which is kept under sedentary conditions and cannot undertake long walks to pasture. First kidding is at 12–14 months of age and females can produce twins twice a year but the average litter size is 1·5 giving three kids per female per year. It is a good meat animal but a poor milker. In Niger an improvement programme castrates inferior males (often merely because they are pied) and distributes Maradi males to the peripheral districts (Robinet, 1973; Haumesser, 1975).

C. Pashmina or Cashmere Goats and their Neighbours

This is the group whose horns are often heteronymously twisted (see Chapter 2, p. 48). Their ears are intermediate between short and drooping.

Because Europeans first encountered the fibre from the fine woolly undercoat (or down) of the Central Asian mountain goats in Kashmir, they named it "cashmere" (the old spelling for the country) and the goats from which it came "Cashmere goats". In fact the number of such goats in Kashmir itself is small; Kashmir was the route by which the fibre reached the European market from more inaccessible areas. The common goat of Kashmir, namely the Kel breed, does not produce down. Since the Cashmere goat is distributed in all the mountainous areas of Central Asia, a better name would be *Central Asian Down* goat or *Pashmina* goat. Pashm is the Persian word for wool and pashmina means woollen.

Pashmina goats are found at altitudes of 3000–5000 m in the highlands of Central Asia. This includes: (1) The Tibetan plateau (Tibet, Tsinghai and north-east Kashmir); (2) The Tien Shan of Sinkiang (China) and Kirgizia (USSR); (3) The mountains of Mongolia. Down producing goats are also found in Iran, Afghanistan and in European Russia. The three major producers of cashmere are China, Mongolia and Iran in that order.

(a) *China and Mongolia*

(i) *Cashmere Goats.* The typical Cashmere goat is a medium-sized animal with a withers height of 60–70 cm and a live weight of about 60 kg in males and 30–40 kg in females. The head is small and fine with a straight or slightly concave profile. The ears are fairly short; they may be erect or directed laterally (semi-lop). Sometimes the external ear is absent. Both sexes have a beard and strong flattened horns. In the female the horns are 20–35 cm long and usually homonymously twisted; more rarely they are sickle-shaped or straight. In the male the horns are 40–50 cm long and are either homonymously or heteronymously twisted. They are directed upwards and backwards, or upwards and outwards, sometimes curving in again so that they cross. The body (i.e. excluding face and legs) is covered with a heavy outer coat of coarse, straight medullated hairs which are 4–20 cm long and averages 14·5–16·5 mm in diameter. Weight of cashmere varies between 100 and 200

g. The typical colour is white but the outercoat may also be yellow, brown, grey, blue, black or pied and the undercoat may also be coloured (Epstein, 1969).

(ii) *Chungwei.* In Chungwei and adjacent countries of the Ningsia Hui Autonomous Region is another breed which also produces pashmina. It is also called the *Tan* goat because the pelt of the kid resembles that of the Tan lamb. The Chungwei is a small goat weighing 45–50 kg (male) or 28–36 kg (female). The head is small with a straight profile and slender tubular ears carried laterally. Both sexes have large homonymously twisted horns. The body is covered with an outer coat of lustrous white hair 15–20 cm long and an undercoat of short white woolly fibres which are combed out and sold as cashmere. The chief product is the pelt of the kid.

(b) *Iran and Iraq*

(i) *Morghose.* The pashmina of Iran is obtained from the Morghose (or Markhoz) breed of Kurdestan. It is white or pale brown. The down is coarser than that from China with a diameter of 17·5–19·5 mm. Milk yield is low (80–85 litres) (Sattari, 1973, 1977).

(ii) The *Kurdi* breed of Iraqi Kurdistan is similar if not identical (Epstein, 1971). Facial profile is straight or slightly concave. Ears are directed laterally with a forwards and downwards inclination; they vary from 5 to 14 cm in length. The horns of the male reach 50 cm in length; they are heteronymously twisted and vary from an open corkscrew to a tight vertical screw. Adult males weigh 55–60 kg. The fleece is 25–30 cm long, usually yellowish white with brown head, ears, neck, shoulders and shanks.

(iii) *Raini.* This is the other fleece breed in Iran, found in the Raini region of Kirman. It has long hair which is white, black or yellow. Weight is 30–40 kg and height is 70–75 cm in males and 65–70 cm in females. Ears are of medium size and horns are directed backwards. The fleece weighs 1–1·5 kg. It is a good milker giving 1·5–2 litres per day for 4–6 months.

(c) *Afghanistan*

(i) *Vatani.* Cashmere is produced by the *Native Black* or Vatani goats which are found in Farah, Badakshan and elsewhere where the grazing is too poor and the country too rough for sheep production. They make up 80% of the goat population. The fleece is usually black but may be grey, white or brown. Horns average 25 cm in length; they curl backwards and outwards. Ears are 10 cm long. This breed is also called *Kabuli*, *Kandahari* or *Tajiki*. In eleven samples the undercoat averaged 16·6 mm in diameter and 6·9 cm in length. This fineness is intermediate between Chinese and Iranian cashmere.

(ii) *Asmari*. The second distinct Afghan breed is the Asmari. This is larger than the Vatani and with longer fleece. It is bicoloured—white face and body, black neck and shoulders. (Demirüren, 1958).

(d) *USSR*

(i) *Kirgiz*. In Kirgizia the goats are almost equally divided between three colour groups—white, black, and brown or grey. Light or dark markings may also be present. They are large goats, males averaging 73 cm at withers, but have a compact, robust appearance. Horns are usually heavy and fairly short. There are varieties of the Kirgiz in: (1) Uzbekistan; (2) Kazakhstan and (3) South-east Bashkiria and Chkalov province south of the Urals. The goats of group (3) are usually black and constitute the *Orenburg* breed.

(ii) *Don*. The *Pridonski* goat in the basin of the river Don and the lower Volga valley is usually black but it may be pied, grey, or white. It has been used as an improver breed on down goats in many parts of the USSR. One of its most important offshoots is the Altai Mountain breed.

According to the breed census of 1 January 1969 there were 50000 pure-bred Orenburgs, 7000 pure and 34000 grade Dons and 93000 grade Altai Mountain goats in the USSR (Vinogradova, 1970).

(e) *Pakistan*

Among the 14 breeds described by Pakistan (1970) and Wahid (1973) are three from the northern mountains. For numbers, size and performance, see Tables IX, X, XI.

(i) *Gaddi* is found in Kiwai, Paras, Shangran of the Kaghan valley, Hazara and also in Azad Kashmir (Chinar, Chiker and Suddum Gali). It is a large animal with massive head, long ears and medium-length twisted horns. Hair is long and the usual colour is black but it may be white or grey. Hair and milk are the main products. (It does not appear to produce pashmina.)

(ii) *Kaghani* also found in the Kaghan valley, Hazara, as well as Swat State, Azad Kashmir, and the northern borders of Punjab. Ears are of medium size. Facial profile is straight. Horns are thick, broad at the base and extend upwards, backwards and then upwards. About 50% are black, 40% grey and 10% white. The chief products are milk, hair and pashmina.

(iii) *Kel* (or *Kail*). The Kel or Kashmiri of Kashmir is a medium-sized breed with a straight profile. Ears are of medium size. Medium length spiral-horns are present in both sexes. Colour is usually black, grey or white. Milk and meat are the chief products.

(f) *India*

Recent accounts from India (India, 1970) are incomplete and confusing because they do not always match up with accounts from Pakistan. This section, then, is largely a plea for further investigation and clarification.

(i) *White Himalayan.* This goat is used for transport and for production of hair. It is called *Chamba*, *Gaddi* or *Kashmiri* according to its habitat which extends through the Kangra and Kulu valleys, Chamba, Sirmur and Simla in Himachal Pradesh to parts of Jammu and Kashmir.

(ii) *Chegu* goats from Spiti, Yaksar and Kashmir are similar to the Gaddi but smaller; they live at higher altitudes. They have lop ears and spiral horns and may be grey or red although usually white. They yield pashmina as well as milk and meat.

(iii) *Pashmina.* The "diminutive" pashmina goat of Tibet is found in large numbers in Gilgit and Ladakh districts of Kashmir and also in the Lahul and Spiti valleys of Himachal Pradesh. Since the advent of Tibetan refugees they number about 180000 in Ladakh, mainly in the Changthong area at altitudes of 3700–4300 m. The yield of pashmina is 21–56 g per year.

(g) *Nepal*

In addition to the small Tarai breed of the lowlands and the Southern Hill goat (see p. 76). Epstein (1977) describes three mountain breeds. Size increases as one goes northwards.

(i) *Northern Hill* goat. The profile is straight and ears are pendulous or semi-pendulous. Horns are of moderate length and homonymously twisted. The coat is hairy and long, black, white or grey, the grey being the heterozygote. White is commonest in the Mahabharat Range and black close to the Himalayas. Females are milked and, in the west, castrated males are used as pack animals.

(ii) *Sinhal.* This is the goat of the high mountains. It is larger than the Northern Hill goat and is covered with long, coarse hair. Colour may be white, cream with darker head and legs, brown, black, grey, or pied. Yield of hair, which is devoid of pashmina, is about 1 kg (females) or 1·5 kg (males). Some females are milked.

(iii) *Changra or Bhotia.* The Cashmere goat from Tibet is found only in the Mustang district, which projects into Tibet. This is the typical pashmina goat described on p. 95. The name Changra derives from the Kangra valley in Himachal Pradesh.

D. Angora Goats

The Angora or mohair goat is a lop-eared spiral-horned breed whose fleece has evolved in a manner parallel to that of the Merino sheep, namely by an increase in the number of the secondary fibres and a reduction in the diameter and medullation of the primary fibres. It comes originally from the dry plateau of Central Anatolia and takes its name from Ankara. It numbers about one-quarter of the goats in Turkey.

The facial profile is straight or slightly concave. Horns are short to medium in length with a homonymous twist. Ears are of medium length and pendulous or semi-pendulous. The fleece is pure white. In an intensively managed experimental flock in Turkey (Aritürk and Yalçin, 1970), body weight of adult females at mating averaged 30–40 kg. Pregnancy rate was 83%, twinning rate 1% and 95% of kids survived to weaning. Greasy fleece weight was 2·46 kg, compared with an average of 1·7 for the whole country. The total number of fibres per mm^2 was 18–20 and the ratio of secondaries to primaries was 9·0–9·3. Staple length averaged 16 cm and fibre diameter 25–39 μm depending on age. Kemp fibres amounted to 2–3%. Crimp number was 7–8.

Angora goats were first exported to South Africa in 1838 with further imports up to 1880. A grading up programme has produced a large population on the dry Karoo in the south-central districts of Cape Province. About half the goats in Lesotho are Angora or its grades; some of these are black (Joubert, 1973).

Angora were imported into the USA from Turkey during 1849–76 and from South Africa in 1925–26. They were tried in many states but they are now almost confined to the Edwards Plateau in Texas. In a study by Basset and Engdahl (1969) of five Texas flocks of mixed age females, body weight averaged 27 kg, grease fleece weight 2·5 kg (every 6 months), clean yield 78% and fibre diameter 35 μm.

In the 1930s grading up of local coarse-haired goats in the Central Asian republics of the USSR with American Angoras produced the *Soviet Mohair* breed.

Table XV. Angora goat populations.

	1971[a] (in millions)	Other sources (in millions)
Turkey	4·1	5·6[b]
USA	2·2	1·4[c]
South Africa	0·9	1·6[d]
Lesotho	0·7	0·4[d]
Other		0·3[e]
TOTAL	7·9	

Source: [a] Commonwealth Secretariat (1973).
[b] Aritürk and Yalçin (1970).
[c] Shelton (1978) personal communication.
[d] Joubert (1973).
[e] USSR—Vinogradova (1970).

In spite of the different numbers of animals in the three chief producing countries, each produces about the same weight of mohair.

E. Lop-eared Horned Goats

The second main type of goat in Africa and southern Asia is larger than the short-eared type. It has long lop ears and frequently long spiral horns.

1. Near East and Africa

(a) *Syrian type*

The typical breed of the countries lying between Egypt and Iraq is the long-haired, black breed known as Baladi in Egypt and Jordan, Mamber or Syrian Mountain in Syria and Israel, Iraqi in Iraq and Anatolian Black in Turkey. Mason (1969) classifies these as the Syrian type of goat. For sizes of Near Eastern and North African goats see Table XVI.

(i) *Mamber.* Facial profile is straight or moderately convex. Ears are long and pendulous—25–32 cm long and 11 cm wide. The male has heavy twisted horns often describing more than two complete turns over a length of 80 cm. In the female the horns are about 27 cm long, sickle-shaped or slightly twisted with a backward sweep; polled females are rare. The prevailing colour is black, occasionally with white markings on forehead and lower legs; some animals are brown or grey. Both sexes have beards (Epstein, 1971).

In Israel the Mamber goat weighs about 60 kg (male) or 40 kg (female).

Table XVI. Sizes of Near Eastern and North African goats.

	Body weight (kg)		Withers height (cm)		Source
	Male	Female	Male	Female	
Mamber (Syria)	60–70	35	85	73	Epstein (1971)
Mamber (Israel)	60	40			Israel (1962)
Negev	55	30	68	61	Epstein (1971)
Negev	40	20			Israel (1962)
Libyan	40–60	30–35	50–65		Epstein (1971)
Moroccan	50–65				Epstein (1971)
Hejazi	22	19	→ 60		Epstein (1971)
Sudanese Nubian	50–70		70–75		Epstein (1971)

Starting at 2 years of age females produce about 130 kids for every 100 births and yield about 150 litres of milk including that sucked by the young (Israel, 1962).

(ii) *Negev*. In southern Israel the goat is much smaller and is adapted to desert conditions. Ears are also smaller—15–19 cm long. Milk production is only 60 litres.

(iii) *Iraqi*. In Iraq the goat is similar to the Syrian Mountain but ears and horns are not so long. It usually has white marks on the ears and often also on head and legs. Very occasionally pied or all white or grey individuals occur.

(iv) The *Cyprus* goat is usually red to black, horned, and long-haired. It is also called the Cyprus free-range goat to distinguish it from the tethered dairy goats which have a mixed origin from local, Maltese and Damascus breeds.

(v) The *Anatolian Black* or *Turkish Native* is called *Kil-Keçi* in Turkish meaning "hairy goat". It is usually black but may be brown, grey of pied.

(vi) The *Native Egyptian* or *Baladi* goat is a small animal with a straight or slightly convex facial profile and small curved horns or none. Ears vary in size and position but the long lop is common. Tassels are common but there is no beard. Black is the prevailing colour but brown, white and pied animals are common. In the University of Alexandria herd 26% produced singles, 46% twins, 25% triplets, and 4% quadruplets (Tantawy and Ahmed, 1960).

(vii) *Libyan.* The same type extends into Libya. Males have a slightly or moderately convex profile and large spiral horns. The female has a straight profile and shorter twisted horns. Ears are large and usually pendulous. Tassels occur occasionally. The hair is long and both sexes are bearded. Colour is brown, black, white or grey, usually in combination. Dark goats have a white blaze on the face and speckled ears (Epstein, 1971).

(b) *The Berber Goat of North Africa*

The Libyan goat is already showing modifications from the original Syrian type. Further west the divergencies are greater and the influence of short-eared, short-haired goats is clear. In Algeria and Tunisia the local goats are black or brown with hair and ears of medium length. The *Moroccan* goat is small. Profile is convex and ears are pendulous or horizontal. Horns project backwards with only a slight twist. Tassels are common. Hair is of medium length and black or brown in colour (Epstein, 1971).

(c) *Hejazi*

From Hejaz and the west of Nejd, Epstein (1971) described a small goat which shows some features of the small short-eared goats of Africa. Horns are of medium length and with a scimitar or sickle shape. Ears are long and slender and wavy and directed laterally. Hair is long and black.

The black hairy goat described by Purohit (1973) in Yemen sounds very similar to the Hejazi.

(d) *Sudanese Nubian*

In Sudan the so-called *Nubian* goat forms the bulk of the goat population in the riverain and urban areas north of the twelfth parallel. The Sudanese Nubian goat is of medium size. Its profile is straight or convex and horns are nearly always present. The coat is black and long. It is a good milker.

The *Shukria* goat of Eritrea is similar. Both appear nearer to the Syrian Mountain than to the classical Nubian type.

(e) *North-east Africa*

(i) *Sudanese Desert.* The typical goat of the desert and semi-desert areas of Northern Sudan is called the Sudanese Desert. It is of moderate size. Both sexes have horns; those of the male are long and twisted and project laterally. Ears are of moderate size and pendulous. Hair is short. The male, and sometimes the female, have a well developed beard and a profuse bushy mane.

Colour is light grey, often with brown or black markings. Milk yield is low (Mason and Maule, 1960).

Recent studies on sedentary flocks in Southern Darfur (Wilson, 1976) have demonstrated a high productivity of 2·1 kids born per breeding female per year. Growth rate to 6 months was 67 g per day.

(ii) *Benadir* breed of southern Somalia. Its habitat is the Webi Shilbeli valley, i.e. roughly the southern third of Somalia north of the river Juba. It is larger and better muscled than the short-eared breeds in the rest of the country. Colours are various but red or black spotted are common and plain white is rare. The coat is usually short and smooth but it is sometimes long and coarse. Ears are large and pendulous with turned up tips; they average 16 cm in length. Horns are more or less robust; they point backwards and run parallel. This breed is divided into three varieties: *Bimal*, *Garre* (or (*Gherra*), and *Tuni* which differ in their tribal ownership, geographical distribution, size, and colour (Mason and Maule, 1960).

(f) *Southern Africa*

(i) *Boer* goat of South Africa. The Boer goat derives principally from the goat of the Hottentots. These animals were observed as early as 1661 but the name Boer was used only later to distinguish them from the Angoras imported in the nineteenth century. Three types were recognized in the original unimproved Boer population: a common goat with short hair, usually white with brown spots, a larger long-haired type and a multicoloured polled type with evidence of exotic dairy blood. During the 1920s farmers in the eastern Cape Province selected for improved meat conformation and in 1959 a breed society for the *Improved Boer* goat was established. The breed standard prefers a white goat with reddish brown head and neck. The horns are prominent and the ears broad and drooping. Coat is smooth, and short to medium in length with no woolly undercoat. Castrates can reach 100 kg without supplementary feeding (Joubert, 1973).

(ii) *Other Lop-eared Goats* in Southern Africa. The goats of Southern Africa do not fit easily into the short-eared or long-eared categories. There are many mixed flocks and many intermediates.

Goats are common in the *Gwanda-Tuli* area of Rhodesia. As in the *Tswana* goats of Botswana, ears are usually long and pendulous with turned-up tips but they may be horizontal. Horns are short and erect. Colours include white, black, brown and spotted or pied. Hair is usually short but it may be long on the hindquarters.

The *Pafuri* of Mozambique is found only on the Limpopo, north-west of

Gaza in the Suldo Save province. It is also called the Boer goat, indicating its origin. It has a convex profile, well developed horns and drooping ears.

The *Nguni* goats of Swaziland and Zululand have twisted horns and flabby ears of medium length. They are intermediate between the Small East African and the lop-eared type. Colour is variable.

The *Damara* or *Herero* goats of Namibia have horns of medium length, a straight or convex profile and long wide drooping ears. The coat is short and usually white, red and white or brown and white, sometimes red or grey. (Joubert, 1973: Mason and Maule, 1960).

2. India and Pakistan

(a) *India*

The following information on Indian breeds is taken from India (1970). For size and performance see Table XII.

(i) *Jamnapari.* This is the largest of the Indian breeds; selected males can stand 127 cm high and females 107 cm. It is tall and leggy with prominent convex facial profile and very long pendulous ears folded up at the tip. The horns are short and flat. Coat is short except for the thighs which have long, thick hair. The commonest colour is white with chestnut or light brown patches but the pied areas may be tan or black. The home tract of the breed is the Etawah district and between the rivers Yamuna and Chambal in Uttar Pradesh State. From there it has spread to many other parts of the dry northern region of India. It has been taken to south-east Asia (in Indonesia it is known as Etawah) and to South America to increase milk yield. Selected females can give 360–540 kg of milk at 3·5% fat in a lactation of 250 days. Only 10% kid twice a year; the twinning rate is only 10% (Sengar, 1976).

(ii) *Beetal.* The Beetal breed of Punjab (both east and west) was derived from the Jamnapari. It is smaller and the male has long twisted horns. The usual colour is red or tan with white spots but it may be white, grey, red or black. In India only 20% kid twice a year and the twinning rate is 30% (Sengar, 1976).

(iii) *Gujarati.* The breeds of south Rajasthan, north Gujarat, and neighbouring parts of Madhya Pradesh are similar although they go under a number of different local names—Bhungri, Cutchi, Kathiawari, Marwari, Mehsana, Zalawadi. They usually have long black hair, long white or speckled ears, long corkscrew horns and a convex profile. They are triple purpose producing milk, meat and hair and are believed to be derived from a cross of Jamnapari and Himalayan goats. Milk yield is 0·75–1·25 kg per day.

(iv) The *Bikaneri* breed of Rajasthan has spread into the Bhawalpur, Multan and Sahiwal districts of Pakistan where it is described as medium-sized with medium-size ears and thin horns. Colour is usually black but may be pied. Milk and hair production are low (see Tables IX–XI.).

(b) *Pakistan*

The following information on Pakistan goats is taken from Pakistan (1970) and Wahid (1973). For size and performance see Tables IX–XI.

(i) *Daira Din Panah* (or Dera Deen Panah). This breed from the Muzzafargah district, Punjab, is bred for milk, hair, meat and skins. It is a heavy breed with convex profile and thick spiral horns. Ears are long and hanging. Beard is present and sometimes throat tassels. The body is covered with long black hair.

(ii) The *Kamori* is a large dairy breed with massive head and convex profile. Horns are small and ears are long. Common colours are reddish brown, white, black and pied. Its area of distribution is central Sind (Hyderabad, Nawabshah, Dadu and Larkana).

(iii) The *Khurasani* is found in Quetta, Qila Saifullah, Loralai, Zhob and Chagi in Baluchistan and over the borders into Afghanistan and Iran. It is bred for milk and meat. It is of medium size with a straight, or sometimes slightly convex, profile. Beard and throat tassels may be present. Ears are of medium size. In the male horns are large with one or two spirals upwards, backwards, outwards and upwards. Colour is usually black but may be white or grey. There is a small amount of pashmina under the long hair.

(iv) The *Lehri* breed of Kachi district, Sibi and other parts of Baluchistan is bred for hair and meat. It is a heavy breed with straight or slightly convex profile. Ears are very long, broad and pendulous. Horns are fairly short and broad and sometimes spirally curved. Colour is usually black but may be white or grey.

(v) The *Sind Desi* is the common breed of the Dadu, Shikarpur, Sukkur and Nawabshah districts of Sind. It is believed to be derived from a cross of Kamori and Chapper. The facial profile is convex. Ears are long. Colour is commonly black but black and white and grey also occur. Products are milk and meat. Hair is of medium length, occasionally long.

F. Long-eared Hornless Dairy Goats

These are all town goats or at least goats kept under intensive, not range, conditions.

(a) *Nubian*

The Nubian is not so much a specific breed as a concept. It should be leggy, short-haired, polled, with a pronounced Roman nose (convex profile), and a high milk yield. Black or red is the characteristic colour. Nubia is the region which includes northern Sudan and southern Egypt but it is difficult to find Nubian goats there.

(i) *Zaraibi.* Many older authors (1893–1930) describe the *Zaraibi* (or *Theban*) goat of upper Egypt (see Epstein, 1971). These descriptions fit the Nubian type but include sometimes short thick horns. The predominating colours are red, brown and pied. There are no recent descriptions of this goat from Egypt and it now appears to be rare. In fact riverine Nubia has been almost eliminated by the formation of Lake Nasser.

(ii) The *Sudanese Nubian* is Nubian only in name. Its chief breeding area is the south of the territory of ancient Nubia but its appearance relates it to the Syrian Mountain (see p. 100). In fact the description of the ideal Nubian type has many features in common with the lop-eared goat of the Near East from which it could theoretically be derived by selecting for polledness, short hair and milk.

(iii) The *Damascus* or *Shami* breed of Syria is close to the Nubian type but it is often horned and its coat is long. It is a tall goat standing about 75 cm at withers. Live weight is 60–80 kg for males and 40–60 kg for females. The facial profile is strongly convex and ears are 25–30 cm long and pendulous. In the male the horns are spreading and twisted; in the female they are sickle-shaped. Tassels are generally present on the throat. Colour is usually red or brown, also pied or grey.

The Damascus goat is a good milker. The herd at Terbol Research Station, Lebanon, averaged 392 kg milk in 238 days and gave 53% twins and 6% triplets (Choueiri, 1973). The government herd in Cyprus averaged 230 kg up to weaning (75 days) and 250 kg thereafter (200 days). Kidding rate was 1·7–1·8 kids per birth (Cyprus, 1977).

(iv) The *Kilis* goat of south-eastern Turkey is derived from crossing the Damascus on to the Anatolian Black. In the University of Ankara herd (Tuncel, n.d.) milk yield was 327 kg and kids per birth 1·42.

(v) The *Mzabite* goat of southern Algeria may also be of the Nubian type. It is red in colour and a good milker.

(vi) The *Mishri* (Egyptian) goat of Yemen Democratic Republic is polled, long-haired, brown or pied, and yields 1–1·5 litres per day (Purohit, 1973).

(vii) *Anglo-Nubian.* However, the best known example of the Nubian type is the breed called *Nubian* in America which is entirely derived from the Anglo-Nubian of Great Britain. This breed in turn was formed by grading up native English goats with lop-eared goats from the East. The breeds used were apparently the Zaraibi, the Jamnapari and a third breed identified only by the name of its place of origin—Chitral in the extreme north of Pakistan.

The Anglo-Nubian is a tall animal with a sleek coat, lop ears and a Roman nose. Colours are various but preference is given to black and tan or reddish brown with or without black or black and white markings. The average female weighs 64 kg or more. Milk yield is good but lower than that of the Alpine breeds. Little trace of the English ancestry remains. Horns, if present, are small and curved downwards.

(b) *Maltese*

Goats on the island of Malta have been selected intensively for milk production. They are light slender animals. According to Epstein (1971) adult males from Tripolitania weigh 45–50 kg and females 22–28 kg. Corresponding withers heights are 77–81 cm and 59–63 cm. For Maltese goats in Italy (1973) weight is given as 34–45 kg and height 65–73 cm. The facial profile is normally straight but it may be slightly convex in the male. Beard is absent but throat tassels are common. Ears are horizontal, semi-pendulous or pendulous; they are 14·5–16·5 cm long and 6·5–7·5 cm wide. Horns are normally absent; if present they are short and thin, sickle-shaped or curved. Coat is either short or long. Predominating colours are red and brown but black, white, grey and pied also occur. In Italy the hair is always long and usually yellowish white, often with black head, neck and ears.

The origin of this breed is not known, but it probably derives from an early union of stock from Italy and from North Africa. It has more recently been exported to other parts of the Mediterranean—particularly Italy, Greece, Libya and the Maghreb countries.

References

Ammann, P. (1974). *Schr. Schweiz. Ver. Tierzucht.* **46**, 65–68.

Aritürk, E. and Yalçin, B. C. (1970). Genetic and environmental aspects of goat production. Mimeographed report, University of Ankara.

Basset, J. W. and Engdahl, G. B. (1969). Texas Agric. Expt. Sta. Progress Report 2642, pp. 27–29.
Belinchón Valera, P., Baro Shakery, E., Cruz Sagrado, J. and Martínez Delicado, J. (1977). *In* "Symposium on Goat Breeding in Mediterranean Countries", Spain, October 1977, pp. 21–30.
Bode, A. M. and Brooks, H. (1976). "Australian Angora Goat Husbandry" (3rd edn). Published by the authors.
Börger, H. (1961). *In* "Handbuch der Tierzüchtung" (J. Hammond, I. Johansson, and F. L. Haring, eds), Vol. III, Part 2, pp. 304–323. Paul Parey, Hamburg and Berlin.
Boyazoglu, J. G. and Zervas, N. (1977). *Elev. Bovin-Ovin-Caprin* **67**, 66–75.
Cabral Calheiros, F. (1976). Caprinos. Situacão e perspectiva. Direccão-Geral dos Serviços Pecuários, [Lisbon].
Castillo, M. J. (1973). *In* "2nd Int. Conf. Goat Breeding", Tours, 1971, pp. 163–168. ITOVIC, Paris.
Charlet, P. and Le Jaouen, J. C. (1977). *Options Méditerranéennes* **35**, 45–54.
Choueiri, E. (1973). *In* "III World Conf. Anim. Prod." Pre-Conf. Vol. I, Short Contrib. p. 1 (b)1.
Cognie, Y., Houix, Y. and Logeay, B. (1973). *In* "2nd Int. Conf. Goat Breeding", Tours, 1971, pp. 345–350. ITOVIC, Paris.
Commonwealth Secretariat (1973). *Wool Intelligence* **26**, 702–707.
Cyprus (1977). "Akhelia-Orites" Goat Breeding Unit—brief description. Mimeograph. Department of Agriculture, Nicosia.
Da Silva Lobo, J. M. (1973). *In* "2nd Int. Conf. Goat Breeding", Tours, 1971, pp. 151–154. ITOVIC, Paris.
Demirüren, A. (1958). Report to the Government of Afghanistan on sheep and goat improvement. Appendix II. FAO Report No. 853. Rome.
Devendra, C. (1966). *Malays. Agric. J.* **45**, 268–274.
Devendra, C. (1973). In "2nd Int. Conf. Goat Breeding", Tours, 1971, pp. 47–54. ITOVIC, Paris.
Devendra, C. and Chenost, M. (1973). *Z. Tierzücht. ZüchtBiol.* **90**, 83–93.
Devendra, C. and Nozawa, K. (1975). *Z. Tierzücht. ZüchtBiol.* **93**, 101–120.
Epstein, H. (1969). "Domestic Animals of China". Commonwealth Agricultural Bureaux, Farnham Royal, Bucks, England.
Epstein, H. (1971). "The Origin of the Domestic Animals of Africa". Vol. II. Africana Publishing Corporation, New York, London, Munich.
Epstein, H. (1977). "Domestic Animals of Nepal". Holmes and Meier Publishers, New York and London.
France (1977). *Bull. Elevage Français* **9**, 2–19.
Haumesser, J. B. (1975). *Rev. Elev. Med. Vet. Pays Trop.* **28**, 225–233.
Hoversland, A. S., Dhindsa, D. S. and Metcalfe, J. (1973). *In* "2nd Int. Conf. Goat Breeding", Tours, 1971, pp. 367–371. ITOVIC, Paris.
India (1970). "The Wealth of India. A Dictionary of Indian Raw Materials and Industrial Products. Raw Materials", Vol. VI. Supplement: Livestock (including poultry). Publications and Information Directorate, CSIR, New Delhi.
Israel (1962). Breeding dairy goats in Israel. Ministry of Agriculture and "Asisa" Israel Goat Breeders' Association.
Italy (1973). Indagine nazionale su alcuni aspetti degli allevamenti e delle produzioni caprine. Ministero dell' Agricoltura e delle Foreste: Associazione Nazionale della Pastorizia, Rome.

Jardim, W. R. (1973). *In* "2nd Int. Conf. Goat Breeding", Tours, 1971, pp. 39–45. ITOVIC, Paris.
Jardim, W. R. (1977). "Criação de Caprinos". Nobel, São Paulo. 2nd ed.
Joubert, D. M. (1973). *Z Tierzücht. ZüchtBiol.* **90**, 245–262.
Kleinviehzüchter (1975). *Kleinviehzüchter* **23**, 936–938.
Laor, M. (1973). *In* "2nd Int. Conf. Goat Breeding", Tours, 1971, pp. 114–116. ITOVIC, Paris.
Mason, I. L. (1951). "The Classification of West African Livestock". Commonwealth Agricultural Bureaux, Farnham Royal, Bucks, England.
Mason, I. L. (1969). "A World Dictionary of Livestock Breeds, Types and Varieties". Commonwealth Agricultural Bureaux, Farnham Royal, Bucks, England.
Mason, I. L. (1980). *Wld. Anim. Rev.* (FAO) **34**, 23–28.
Mason, I. L. and Maule, J. P. (1960). "The Indigenous Livestock of Eastern and Southern Africa". Commonwealth Agricultural Bureaux, Farnham Royal, Bucks, England.
Mexico (1971). La Ganaderia Caprina (Importante Recurso Ganadero) Banco Nacional Agropecuario, S.A., Mexico, D.F. 281 pp.
Nolte, E. (1973). *In* "2nd Int. Conf. Goat Breeding", Tours, 1971, pp. 147–150. ITOVIC, Paris.
Nozawa, K. (1972). *Wld Rev. Anim. Prod.* **8**, 18–26.
Nozawa, K. (1974). *Rep. Soc. Res. Native Livestock* **6**, 113–121, 184–185
Nozawa, K. and Watanabe, S. (1969). *Rep. Soc. Res. Native Livestock.* **3**, 92–105, 160–162.
Pakistan (1970). Scheme on investigation and research on Pakistan animal hair. Final report. Wool Test House, Department of Agriculture, Marketing and Grading, Government of Pakistan, Karachi.
Purohit, K. (1973). Interim report on animal husbandry. Special Fund Project for Hadhramaut valley and northern areas of People's Democratic Republic of Yemen. Mimeograph. FAO, Rome.
Quittet, E., Ed. (1976). "La Chèvre. Guide de l'Eleveur". La Maison Rustique, Paris.
Robinet, A.-H. (1973). *In* "2nd Int. Conf. Goat Breeding", Tours, 1971, pp. 127–140. ITOVIC, Paris.
Roesch, C. (1973). *In* "2nd Int. Conf. Goat Breeding", Tours, 1971, pp. 25–28. ITOVIC, Paris.
Rumich, B. (1967a). The goats of Indonesia. FAO Regional Office, Bangkok.
Rumich, B. (1967b). The goats of Sarawak. FAO Regional Office, Bangkok.
Rumich, B. (1968a). The goats in the Philippines: preliminary notes. FAO Regional Office, Bangkok.
Rumich, B. (1968b). The goats of South Korea: preliminary notes. FAO Regional Office, Bangkok.
Sanchéz Belda, A. (1977a). *In* "Symposium on Goat Breeding in Mediterranean Countries", Spain, October 1977, pp. 48–50.
Sanchéz Belda, A. (1977b). *In* "Symposium on Goat Breeding in Mediterranean Countries", Spain, October 1977, pp. 51–53.
Saraza Ortiz, R. and Tejón Tejón, D. (1973). *In* "2nd Int. Conf. Goat Breeding", Tours, 1971, pp. 55–62. ITOVIC, Paris.
Sattari, M. (1973). *In* "2nd Int. Conf. Goat Breeding", Tours, 1971, pp. 105–113. ITOVIC, Paris.
Sattari, M. (1977). *In* "Symposium on Goat Breeding in Mediterranean Countries",

Malaga-Granada-Murcia (Spain), pp. 43–44. EAAP/CNEZ, Madrid.
Sengar, O. P. S. (1976). Sheep and goat production systems in India. Paper for conference on "Role of Sheep and Goats in Agricultural Development". Winrock International Livestock Research and Training Center, Morrilton, Arkansas, USA.
Shelton, M. (1978). *J. Dairy Sci.* **61**, 994–1010.
Skjervold, H. (1973). "Husdyravel". Landbruksforlaget, Oslo.
Switzerland (n.d.). Goat breeding in Switzerland.
Taftă, V. (1975). "Creşterla Caprelor". Editura "Ceres", Bucharest.
Tantawy, A. O. and Ahmed, I. A. (1960). *Emp. J. Exp. Agric.* **28**, 74–82.
Tuncel, E. (n.d.) Some crossing experiments for developing a new dairy goat in Turkey. Mimeograph. University of Ankara.
Vinogradova, V. M. (1970). *Ovtsevodstvo* **16**, (12) 1–4.
Vohradský, F. and Šáda, I. (1973). *Sborn. Vys. Školy Zeměd. Praze* **6**, 161–172.
Wahid, A. (1973). Draft monograph on goat production in Pakistan, with particular reference to goat controversy. Karachi University, mimeograph.
Watanabe, S. (1970). *Rep. Res. Group Native Farm Anim. East. Asia* **4**, 82–90, 127–128.
Wilson, R. T. (1976). *Trop. Anim. Hlth Prod.* **8**, 221–223.
Winrock (1977). "The Role of Sheep and Goats in Agricultural Development". A state of the arts study. Winrock International Livestock Research and Training Center, Morrilton, Arkansas, USA.
Yamane (1943). [Animal husbandry of the East Indies] Yoken-do, Tokyo (in Japanese). Quoted by Devendra and Nozawa (1976).

4. Genetics: Breeding Plans

G. Ricordeau

A. Formal Genetics: Mendelian Inheritance

1. Karyology

(a) *Normal Karyotype*

The goat (*Capra hircus*) has $2n = 60$ chromosomes. Apart from the Y-heterosome, which is the smallest and the only metacentric chromosome, all are acrocentric, the 29 pairs of autosomes being of regularly decreasing size (Fig. 1). The X-chromosome is one of the largest and is distinguished from the autosomes by a small appendix, which is often difficult to see but can be made to appear on a preparation by changing the focus (Evans, 1965; Hulot and Lauvergne, 1967; Basrur and Stoltz, 1967; Hansen, 1973). Caprine chromosomes have been studied using banding techniques and G band homology between cattle, goat and domestic sheep has been demonstrated (Evans *et al.*, 1973; Schnell and Czaker, 1974; Franck *et al.*, 1977; Henderson and Bruère, 1979).

(b) *Chromosome Anomalies*

(i) *Intersexual XX/XY* (blood chimaera or freemartin). Intersexual goats occur with a mosaic of XX and XY cells (Padeh *et al.*, 1965; Ilbery and Williams, 1967; Hamerton *et al.*, 1969; Soller *et al.*, 1969). They may be horned or polled and are impossible to distinguish from intersexuality linked with the polled homozygous genotype, because the genital tract is also

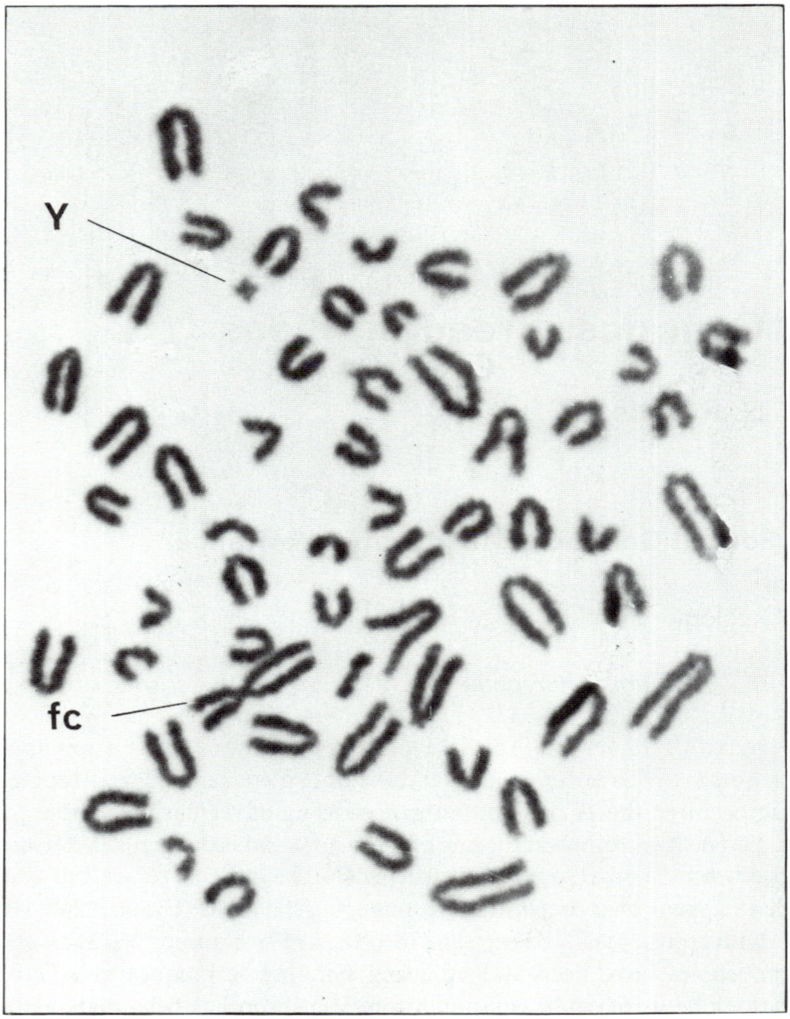

Fig. 1. Karyotype with centric fusion, fc (F. Hulot).

masculinized. The observations of Ilbery and Soller do not prove free-martinism since they show no vascular anastomosis *in utero*, but they do correspond to some typical XX and XY cell mosaics, while practically all cases of intersexuality linked with absence of horns do not show XY cells (Basrur and Coubrough, 1964; de Grouchy *et al.*, 1965). Other cases of intersex in a horned goat have been detected (Divekar, 1953; Kaikini and Puranik, 1964).

(ii) *Centric Fusion or Robertsonian Type Translocation.* The chromosome system most frequently found is 59 including a new submetacentric chromosome, probably the result of the translocation of a small acrocentric to a large acrocentric (Soller *et al.*, 1966, 1969; Hulot, 1969; Padeh *et al.*, 1971; Sohrab *et al.*, 1973). This anomaly was transmitted as a dominant character, and therefore in a normal population one should find heterozygotes with 59 chromosomes and homozygotes with 58 chromosomes. Out of 30 progeny of a heterozygous buck, Popescu (1972) found 20 heterozygotes (14 males out of 19 and 6 females out of 11) but no homozygotes. The bucks with this anomaly have a lower sperm production than normal bucks and does with the anomaly have a significantly lower prolificacy than normal does (Ricordeau, 1972b).

(c) *Goat–Sheep Hybrids*

Many experiments in Bulgaria, the USA and England have demonstrated that fertilization of goat eggs by ram spermatozoa is obtained consistently, while the reverse is more rare, but that in the former case, embryonic development does not continue beyond 2 months (Hancock, 1965; Bunch *et al.*, 1976; Moore and Eppleston, 1976). McGovern (1969) discusses the various theories put forward to explain the immunological barriers to crossbreeding. In a study when either ewes and goat were inseminated into the uterus with semen of the other species, or ova were transferred to the oviducts of recipient females of other species, Eppleston and Moore (1977) showed clearly that goat ova can be fertilized by ram sperm in the tracts of either species but this is not the case with sheep ova: it may be that the reproductive tract of the ewe does not provide an environment suitable for the retention of fertilizing capacity of goat sperm.

Several crosses have produced fertile hybrids. Bratanov *et al.* (1961, 1969, 1972) obtained F_1 offspring ($2n = 57$ chromosomes) by mating goats ($2n = 60$) with rams ($2n = 54$) and backcross offspring ($2n = 55$) by mating F_1 females ($2n = 57$) with rams ($2n = 54$). Ilbery *et al.* (1967) confirmed the

Sire × Dam		Offspring	Autosomes	Sex chromosomes
Barbados ram × $2n = 54$	goat $2n = 60$	F_1 ($2n = 57$)	3 metacentrics 52 acrocentrics	2 large acrocentric Xs
Barbados ram × $2n = 54$	F1 $2n = 57$	B.C ($2n = 55$)	5 metacentrics 48 acrocentrics	2 large acrocentric Xs
		B.C ($2n = 55$)	5 metacentrics 48 acrocentrics	1 acrocentric X and 1 small Y

karyotype of F_1 fetuses. More recently, Bunch *et al.* (1976) checked the results of Bratanov and his team: a Spanish breed goat ($2n = 60$) mated with a Barbados ram ($2n = 54$) gave two offspring, one of which was a female ($2n = 57$) polled like its dam and with the Barbados colour type. The back-cross (F_1 dam × Barbados ram) also gave two offspring ($2n = 55$) with horns but resembling the Barbados as to colour and development.

Some authors have compared the karyology and biochemistry of hybrids between the goat (*Capra hircus*) and the Aoudad or Barbary sheep (*Ammotragus lervia*: $2n = 58$), and between the goat and the bharal (*Pseudois nayaur*: $2n = 54$) which are considered as animals having sheep–goat characters. The goat × aoudad hybrid had $2n = 59$ chromosomes and the goat × bharal hybrid had $2n = 57$ chromosomes but was almost unviable (see Bunch *et al.*, 1976, 1977, 1978).

2. Mendelian Inheritance of External Traits

(a) *Horns*

Polledness is due to a dominant autosomal gene *P* (Asdell and Crew, 1925; Lush, 1926; Ricordeau and Lauvergne, 1967: Table II). So we have three genotypes but two phenotypes: the homozygotes *PP* and heterozygotes *Pp* which are polled, and the recessive homozygotes *pp* which are horned.

(i) *Phenotype Distinction of* PP *and* Pp *Polled Genotypes.* They may be distinguished by observation of the two boney knobs on the skull. In *PP* males, the protuberances are rounded, well separated and without scurs, whereas in *Pp* they take the form of two beans in the shape of a V pointed forwards, with scurs of 2 or 3 cm in many cases. Distinction is possible at 3 months but more accurate at 5–6 months (Ricordeau, 1972a; Fig. 2).

(ii) *Horn Development and Types.* Horns are not always visible at birth. In the Alpine goat, they appear by the age of 28 days at the latest. They appear earlier in males and in kids with a high birth weight. Therefore it is necessary to wait a month to decide whether kids have horns or not. Hair whorls on each horn bud almost always indicate the future appearance of horns (Ricordeau and Bouillon, 1969).

In the domestic breeds, horns are predominantly of the homonymous type, the right horn forming a right-handed spiral and the left forming a left-handed spiral. Epstein (1951) reports the existence of a heteronymous type (right horn forming a left-handed spiral and vice versa) in the wild *Capra falconeri* (see also Chapter 2). Sometimes four or six frontal horns are found, the extra horns being more or less floating (polycerism; Sionneau, 1964).

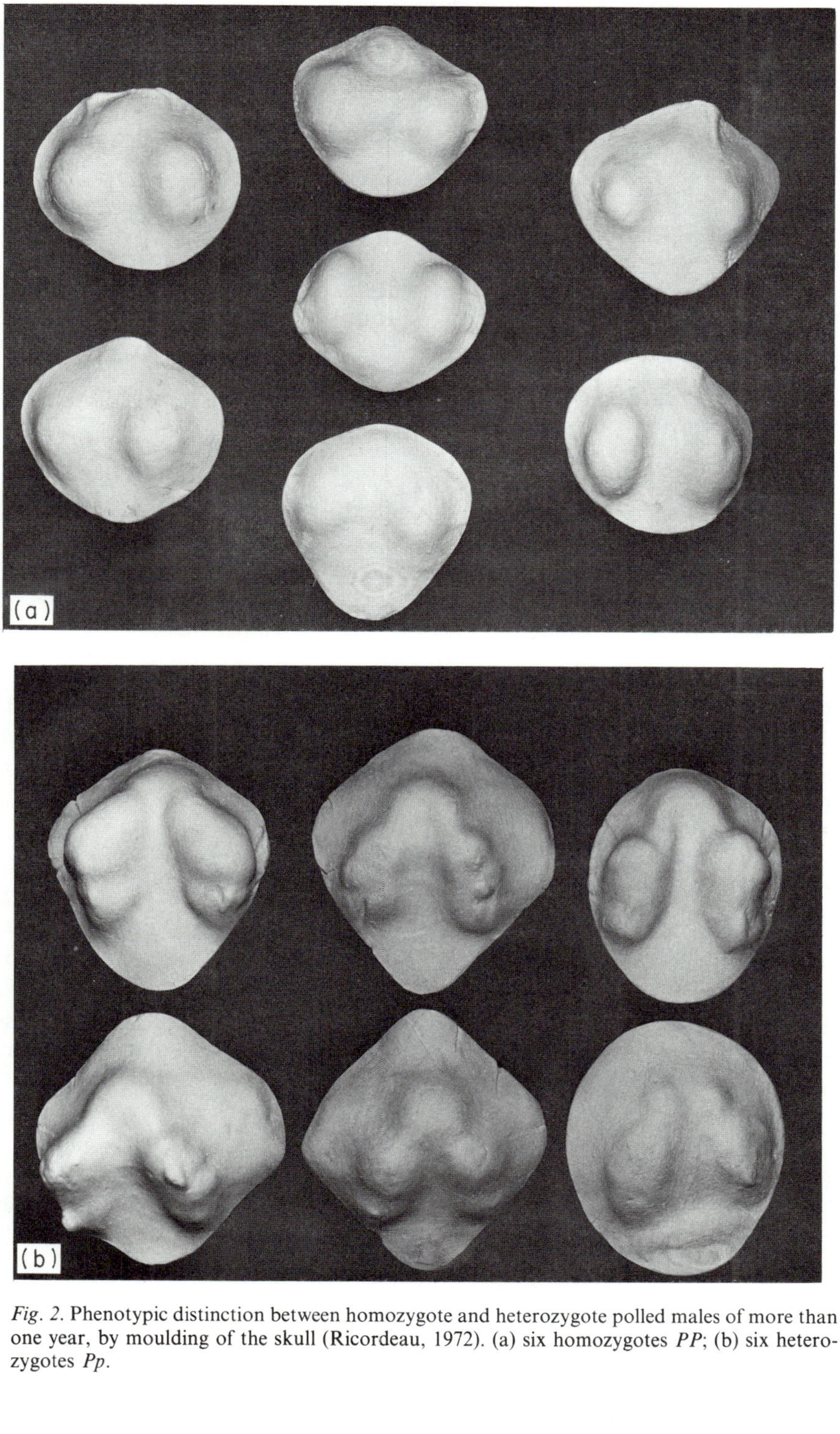

Fig. 2. Phenotypic distinction between homozygote and heterozygote polled males of more than one year, by moulding of the skull (Ricordeau, 1972). (a) six homozygotes *PP*; (b) six heterozygotes *Pp*.

Ovarectomy has no apparent effect upon horn formation, whereas castration of males stops it. Therefore the male sex hormones (androgens) are major determinants of the "potential" rate of horn formation.

(b) *Wattles or Tassels*

Wattles are due to a dominant autosomal gene (*W*) with complete penetrance but variable expressivity (one or two wattles on the neck or under the ears, long or very thin, etc...). The frequency of the gene for absence of wattles (*w*) is given in Table I. Does with wattles (*WW* or *Ww*) are 7% more prolific

Table I. Frequency of gene for absence of wattles q^w.

Breed (country)	q^w	Source
French Saanen	0·81	Ricordeau (1967)
French Poitevine	0·72	
French Alpine	0·64	
Japanese Native	0·8–0·98	Nozawa (1970)
Japanese Saanen	0·83	
Spanish, USA	0·80	Shelton (1978)
Angora	near 1·00	

than does without (*ww*). The difference in prolificacy is particularly marked between polled does with wattles—double heterozygous for the most part—and horned does without wattles—double homozygous (Ricordeau, 1967; Laor, 1971).

(c) *Beard*

According to Asdell and Smith (1928), this character is dominant and linked with sex. It is difficult to check this theory because the beard does not develop until puberty, but after the age at which surplus males are slaughtered. Beardless males have a small tuft of hair under the chin but this beard is not comparable with the large growth of bearded goats having the dominant gene.

(d) *Absence of Ears*

In a genetic study, Paredes (1952) observed three types of morphologically distinct ears in the Granada and Murcia breeds: wide and normal-length ears, short ears and no ears. Crossbreeding with short-eared Spanish goats produced the American La Mancha breed, with very short ears (elf or gopher ears). Wassin (1928) showed that the absence of ears was a recessive character.

(e) *Colour Inheritance*

Many findings have been published on this subject but it is difficult to draw general conclusions and therefore we shall confine ourselves to the few essential results that seem the most reliable.

The Saanen white colour is due to a gene-inhibiting coloration, dominant over the other colours (Lush, 1926; Asdell and Smith, 1928). The gene has been given the symbol *I* and more recently *R* to represent the roan pattern (Ricordeau and Lauvergne, 1971). In the pure Saanen homozygote (*RR*), the action of the gene which gives the white colour seems to be total. Conversely, in heterozygotes (*Rr*) produced by crossing Saanen with Chamoisée, some white animals are found which resemble the dominant parent and some of roan colour diluted to varying extents. The backcross between crossbred females (Saanen × Chamoisée) and Chamoisée males gives 48·5% chamoisé kids and 51·5% white or roan kids. In the roans, the colour becomes darker from birth to 1 year of age. Berge (1966), who has reviewed findings on the way in which colour pattern and types of pigmentation are transmitted, gives the following order of dominance: white > red > fawn or chamoisé > black. In the Toggenburg type, the wild or mouflon type (i.e. dark-coloured back and flanks and light belly) is dominant over the reverse type (badgerface). Searle (1968) proposes three alleles at the Agouti locus (agouti wild type, black and tan, recessive black) but notes that by contrast in certain breeds, a dominant black seems to be present. In observing the coloration of Corsican goats and of offspring of matings between Alpine Chamoisée and Poitevine, Lauvergne (1977 and 1978) suggests seven alleles:

A : Wild type;
A^{r} : red;
A^{g} : dark grey;
A^{b} : fawn or chamoisé;
A^{t} : black and tan with white belly;
A^{tn} : black and tan with black belly;
a : black;
B : brown (probably).

In Angora goats, Nel (1975) notes black or red fibres on the head, ears and legs. His observations point to a change of colour with age but do not provide any basis for genetic analysis.

3. Hereditary Defects

(a) *Intersexuality or Masculinization Linked to Polled Trait*

In all the populations in which polled males and females have been selected, it is usual to find intersexuals, an excess of males and a deficit of females (Tables II and III). Since intersexuality occurs only in polled animals, a link

Table II. Test of hypothesis: allele *P* (polled) is a simple dominant autosomal (Ricordeau and Lauvergne, 1967).

Mating type			Horned			Polled			Horned		
No.	Sire × dam	*N*	male	female	h	male	female	h	Observed (%)	Expected (%)	χ^2
1	*PP* × [*P*]	366	0	0	0	251	87	28	0	0	0
2	*Pp* × [*P*]	655	74	71	0	274	202	34	22·1	25·0	NS
3	*pp* × [*P*]	233	59	74	0	51	49	0	57·1	50·0	NS
4	*PP* × *pp*	164	0	0	0	81	81	2	0	0	0
5	*Pp* × *pp*	255	60	53	0	76	63	3	44·3	50·0	NS
6	*pp* × *pp*	64	32	32	0	0	0	0	100·0	100·0	0
3 + 5		488	119	127	0	127	112		50·4	50·0	

N = number of offspring.
h = hermaphrodites.
χ^2 test with the supposition that all the polled phenotype dams [*P*] are genotype *Pp*.

Table III. Sex distribution of offspring from polled parents.

Mating type		Number of	% observed			
No.	Sire × dam	offspring	Male	Female	Hermaphrodite	Source
1	*PP* × [*P*]	50	48·0	28·0	24·0	(1)
		140	61·4	20·0	18·6	(2)
		404	67·1	24·0	8·9	(3)
	Expected %		50·0	25·0	25·0	
2	*Pp* × [*P*]	676	51·8	38·3	9·9	(1)
		2 094	54·4	38·9	6·7	(4)
		648	52·2	39·9	7·9	(2)
		1 366	55·8	39·9	4·3	(3)
	Expected %		50·0	37·5	12·5	

Hermaphrodites are *PP* females and part of the males are genetic females (pseudo-males) which are infertiles due to testicular hypoplasia. Thus, the abnormal sex ratio found in those two matings (excess of males over females plus hermaphrodites) is not an independent phenomenon, but simply a trivial consequence of some extreme pseudo-hermaphrodites being mistakenly classified as males.

[*P*] = polled phenotype considered as genotype *Pp*.

(1) Eaton (1945).
(2) Soller and Angel (1964).
(3) Ricordeau and Lauvergne (1967).
(4) Haugen (1960).

between the two characters was assumed fairly early on. The first genetic theory was put forward in 1944 by Asdell, but it was not until 1962–1970 that all the associated anomalies were explained, particularly through the determination of the genetic sex of intersexuals and of the two distinct groups of males (Hulot and Basrur, 1969). The situation at present may be summed up as follows: polledness is due to a dominant autosomal gene *P*; this gene

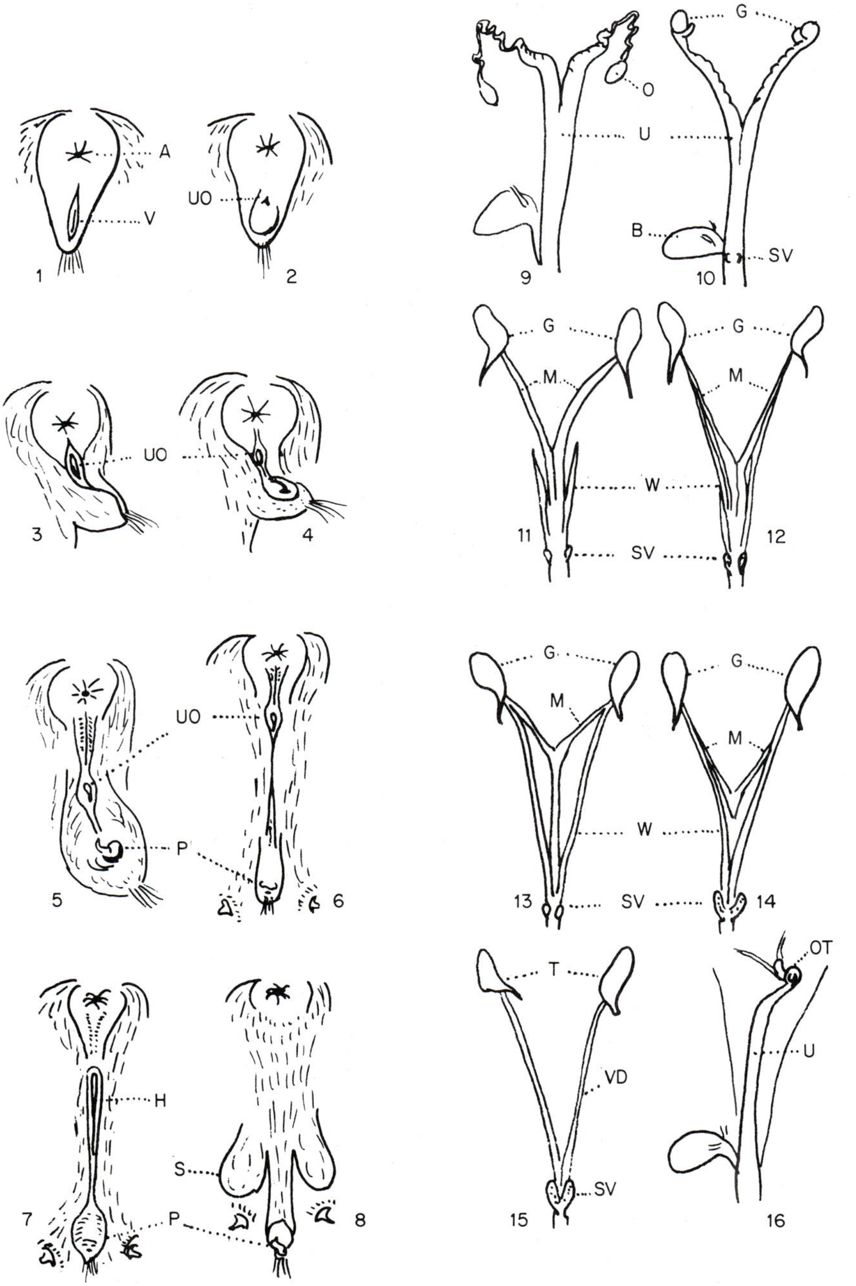

Fig. 3. Types of hermaphrodites or masculinized females (Eaton, 1943). UO = bulbous vulva (urinary orifice); P = penis like clitoris; H = hypospadia; S = scrotum; O = ovary; U = uterus; SV = seminal vesicles; G = gonad; M = Müllerian ducts; W = Wolffian ducts; T = testis; OT = ovotestis; VD = vas deferens.

also has a recessive masculinizing effect with complete penetrance in females and incomplete penetrance in males (after Lauvergne, 1969 and Boyajean, 1969).

(i) *Genetic Females* (*XX*). In the homozygous state, the *P* gene causes masculinization of all *PP* females, which may be malformed to various degrees and leads to pseudohermaphrodism.

a. External genitalia (Eaton, 1943; Fig. 3)

(1) Seemingly normal female. The vulva projected slightly.

(2) Slightly bulbous vulva. The vulvar cleft was represented by only a short slit or simply by a rounded opening.

(3) Pronounced projecting bulbous vulva. Usually, there was a slit in the dorsal surface which represented the vulvar cleft.

(4) Projecting bulbous vulva with prominent projecting clitoris. Resembling somewhat the glans penis of males.

(5) Penile clitoris. Large projecting bulb with long prominent raphe, projection from bulb resembling glans penis with urethral process present. Urethral orifice about midway between anal opening and projecting glans.

(6) Shortened distinct penis. Usually terminated in a bulbous structure which did or did not have a urethral orifice. If the latter was absent, the bulb was swollen from the accumulation of urine and frequently broke open. A long raphe is usually present.

(7) Hypospadia. A penis with the ventral surface open for a part or the whole of its length, through which the urine seeped nearly constantly. Usually a raphe anterior and another posterior to the open portion. Bulbous end of penis frequently present.

(8) Seemingly normal male (pseudomales). Penis complete. Testes and scrotum usually present. In some of the cases examined, the penis was normal, but one or both testes were in the abdominal cavity (cryptorchid).

Types (1) to (4) are characteristically female or slightly deviating from the normal female type; penis and scrotum are absent.

Types (5) to (8) are of male type with a raphe along part of the "penis"; scrotum and testes are sometimes present. In many cases, these females have a little beard at puberty (personal observation of the author).

This classification presented a series beginning with a female type and gradually grading into the male type. In a similar manner internal genitalia may be classified from female to male type.

b. Internal genitalia

(9) Normal female type with ovaries, oviducts and uterus.

(10) Nearly normal type of uterus, but with testes present and very small seminal vesicles.
(11) Uterine horns and remnants of the Wolffian ducts, small seminal vesicles.
(12) Both Müllerian and Wolffian ducts present and closely parallel.
(13) Müllerian and Wolffian ducts separated.
(14) Wolffian ducts and seminal vesicles well developed, lower portion of Müllerian ducts degenerating.
(15) Seemingly normal male with vasa deferentia and seminal vesicles.
(16) Unicorn uterus and ovotestis.

Of three pseudomales of type (8), Basrur *et al.* (Basrur and Coubrough, 1964; Basrur and Kanagawa, 1969) observed meiotically active cells in some of the seminiferous tubules of their testes; however, they did not exclude the possibility of these animals being XX/XY chimaeras (Short, 1971). Recently, Sulimovici *et al.* (1978) observed that the histological and hormonal situations in the hypoplastic XX male goat resembled those found in Klinefelter and XX human males: their testes showed various degrees of degeneration, in one case a "sperm pocket" full of spermatozoa was observed (sperm cells with heads separated from tails in the semen), and in the other two cases, semen was devoid of sperm cells.

As all the genetic female homozygote polled PP *are infertile, all the reproducing females are either heterozygotes polled* (Pp) *or horned* (pp).

(ii) *Genetic Males* (*XY*). Of the three possible genotypes, *PP*, *Pp* and *pp*, all horned *pp* and polled *Pp* males seem to be fertile (Table IV). Conversely,

Table IV. Distribution of 540 young males 6 to 12 months of age, observed in farms, by type of mating of the parents (Ricordeau *et al.*, 1972a).

	Type of mating of parents		
Type of young buck	Both parents horned	1 parent polled 1 parent horned	Both parents polled
Polled	3 fertile	122 fertile 1 sterile	101 fertile (genetic males) 25 sterile (genetic males) 13 pseudo-males (genetic females of caryotype XX)
Horned	129 fertile	118 fertile	28 fertile

more than half of the polled *PP* males are partly or completely sterile through unilateral or bilateral obstruction of the epididymis, their testicles being of normal size and consistency. The description given by Soller *et al.* (1969) corresponds to males with symptoms of epididymal sperm granuloma. They had granulomatous swellings in the heads of the epididymis. One section contained a yellowish cheese-like mass of spermatozoa which could not reach the tail of the epididymis since the tubules of the head were apparently closed. The tubules were distended with varying degrees of calciferous degeneration and atrophy of testicular tissue. In all the cases observed, the epididymal tails were soft, small and void of spermatozoa. Their peripheral blood testosterone concentration is apparently normal regarding sex and season, and their spermatogenetic activity is related to the testicular weight (Corteel *et al.*, 1969). Males with unilateral obstruction are fertile and may remain so throughout their life, but sometimes, the obstruction develops on the other side the following year (Weber, 1969, 1977).

(iii) *Other Effects of* P *Gene.* The action of the P gene is not limited to embryonic development alone, in masculinizing all PP females and making more than half of the PP males sterile. The females heterozygous for the polled gene are 6–7% more prolific than horned females and PP males 7–8% more prolific than Pp and pp bucks (Soller and Kempenich, 1964; Ricordeau, 1969; Hancock and Louca, 1975). However, those effects may be a result of a selection for polledness because, for example, the hyper-prolificacy of females Pp over pp seems to decrease when the selection increases the frequency of gene p in the population (Ricordeau and Bouillon, 1978).

Masculinization also manifests itself by a maternal effect (E^{m}) of the P

Table V. Mode of action of gene P according to genetic sex and genotypes.

Genotypes		
PP (polled homozygote)	Pp (polled heterozygote)	pp (horned)
Genetic females (Karyotype XX)		
All sterile $c = 1$	All fertile	All fertile
sterile females γ_3	Advantage of prolificacy (litter size)	
hermaphrodites γ_2	of polled Pp goats *vs* horned goats	
pseudo-males with	($t = 1{\cdot}044$ to $1{\cdot}069$)	
testicular hypoplasia γ_1		
	Favourable maternal effect of polled Pp mothers on the weight of their kids from 1 to 7 months	
Genetic males (Karyotype XY)		
$(1 - b)$ fertile males	All fertile	All fertile
b males with obstructed epididymal ducts	males partly or completely sterile	
	($a = 0$)	
{ unilateral blocking β_1		
{ bilateral blocking β_2		
Coefficient of hyperprolificacy of fertile homozygous males PP *vs* polled Pp and horned pp males ($r = 1{\cdot}080$)		

gene on the weight of breeding does. After accounting for non-genetic effects, there are no differences in weight between horned and polled kids, but the weight at 1 month of age of those born of *Pp* polled dams is 5–6% higher than that of young from horned dams (Ricordeau *et al.*, 1972b).

(iv) *Genetic Parameters.* A detailed study of the various effects of the *P* gene, by genotype and sex, shows that at least ten parameters have to be used whose significance is shown in Tables V and VI. Three are penetrance

Table VI. Estimation of parameters which characterize the mode of action of the gene polled *P* (Lauvergne, 1969; Ricordeau, 1972).

Definition of parameters	Coefficients[a]		Authors
Penetrance of male infertility in			Soller *et al.* (1963)
male heterozygote *Pp*	*a*	0	Ricordeau *et al.* (1972a)
Penetrance of epididymal stenosis			Schönherr (1954), Brandsch
in male homozygotes *PP*	*b*	0·91	(1959), Soller *et al.* (1963)
Penetrance of sterility in female			Brandsch (1959), Ricordeau
homozygotes *PP*	*c*	1	and Lauvergne (1967)
Expressivity of unilateral blocking		0·64	Schönherr (1954)
in male *PP* with epididymal stenosis	β_1	0·43	Weber (1969)
Expressivity of bilateral blocking		0·36	Schönherr (1954)
in male *PP* with epididymal stenosis	β_2	0·57	Weber (1969)
Percentage of male homozygotes		0·33	Schönherr (1954)
PP completely steriles	$b.\beta_2$	0·79	Soller *et al.* (1963)
		0·52	Weber (1969)
		max 0·50	Soller *et al.* (1969)
		0·55	Ricordeau *et al.* (1972a)
Expressivity of pseudo-male type		0·43	Buechi (1957)
in female homozygotes *PP*	γ_1	0·56	Brandsch (1959)
		0·66	Ricordeau and Lauvergne (1967)
Expressivity of hermaphroditic type		0·57	Buechi (1957)
in female homozygotes *PP*	γ_2	0·44	Brandsch (1959)
		0·34	Ricordeau and Lauvergne (1967)
		0·50	Hancock and Louca (1975)
Expressivity of female sterile type		0	Brandsch (1959)
in the female homozygotes *PP*	γ_3	0	Buechi (1957)
		ϵ	Ricordeau and Lauvergne (1967)
Coefficient of overprolificacy of homozygotes males *PP vs* heterozygotes *Pp* and horned males *pp*	*r*	1·080	Ricordeau (1969)
Coefficient of overprolificacy of		1·062	Soller and Kempenich (1964)
heterozygotes females *Pp vs* horned	*t*	1·069	Ricordeau (1969)
females		1·044	Hancock and Louca (1975)
		1·060	Ricordeau and Bouillon (1978)
Maternal effect: kids born from heterozygotes polled mothers have a higher weight than kids born from horned mothers	E^m	+5·5% at 1 month	Ricordeau *et al.* (1972b)

[a]*Coefficients:* a, b, c, β_1, β_2, γ_1, γ_2, γ_3, are percentages varying between 0 and 1
$\beta_1 + \beta_2 = 1 \quad \gamma_1 + \gamma_2 + \gamma_3 = 1$

parameters: *a* and *b* for epididymal disorders in heterozygotes and polled male homozygotes respectively, and *c* for sterility in polled female homozygotes. Two are coefficients of differential selection: *r* to take account of the higher prolificacy of *PP* males compared with *Pp* and *pp* males, and *t* to take account of the higher prolificacy of *Pp* females compared with *pp* females. Five are expressivities: β_1 for the unilateral form (fertility maintained) and β_2 for the bilateral form (with sterility) of epididymal stenosis in *PP* polled homozygous males; and γ_1, γ_2, and γ_3 are the expressivities of the various forms of masculinization occurring in *PP* homozygous females (Table VI). The situation is somewhat simplified by the fact that epididymal anomalies do not appear to exist in *Pp* heterozygous males ($a = 0$) and that sterility seems to be complete in *PP* females ($c = 1$).

(v) *Appearance of the Polled Mutant in Horned Populations.* On this subject, references are rare or debatable, and the findings of Bilgemre and Düzgünes (1952) are exemplary. In a herd of Angora goats in Turkey consisting of 140 does and four bucks—all horned—a polled buck was born and widely used by insemination on horned does. Examination of the horns on the 169 kids produced showed that he was a *Pp* heterozygote. According to the authors, the mutation must have been produced in the dam of the buck because his sire continued to produce nothing but horned young after the birth of the polled kid. The occurrence of an intersexual among the young from matings between the polled buck and 12 of his polled daughters shows clearly that intersexuality is genetically linked with the *PP* homozygous genotype, regardless of breed.

(vi) *Practical Conclusion.* In selecting polled animals, it is impossible to obtain a fertile and homozygous polled line. On the other hand, this selection has a very negative influence on the quantity of milk produced in reducing the efficiency of selection: calculations of Gaillard and Kropf (1978) show that horned breeds respond better to selection than hornless ones to an extent of around 40 per 100.

(vii) *Mechanisms Affecting the Sexual Differentiation of Homozygotes* PP *Females.* Three hypotheses are proposed.

(1) According to Hamerton (1968), the H-Y structural gene resides on the X-chromosome while the Y-chromosome carries its activator gene: XY males, or pseudo-males, with an autosomal testis-determining mechanism, might simply have developed an autosomal mimic of this switching action normally carried on by the Y-linked gene. From an evolutionary point of view, the X linkage of the H-Y gene is plausible, the Y-linked activating gene and an autosomal repressing gene controlling its expression. According to this

alternative, testicular organization in the absence of the Y is caused either by an autosomal mimic of the Y-linked activator, or by a defective mutation of the autosomal repressor (Hamerton, 1968, 1969; Short, 1971; Ohno, 1979).

(2) The H-Y structural gene is Y linked, but how then is the masculinization of the females explained? There is now a body of evidence to indicate that it is not the presence of the Y-chromosome *per se* that determines differentiation of the testes, but, in a stricter sense, the presence of H-Y antigen (it is a cell surface component that is present in males of all mammalian species tested, including human). Wachtel *et al.* (1978) confirmed that H-Y antigen is present in the blood cells of normal male goats but not in the blood cells of normal female goats; on two polled XX intersex goats, they found that they were H-Y antigen positive, although the degree of expression of the antigen was somewhat less than that of normal male goats. This finding is consistent with the view that there may be a cluster of testis-determining H-Y genes on the Y-chromosome and that a subcritical part of the Y-chromosome had become translocated to the autosome carrying the gene *P*, so that it had no sterile effect on normal ovarian development in the heterozygous condition but caused partial or complete sex reversal of the gonad when present in double dose in the homozygous *PP* female (Wachtel *et al.*, 1978; Short, 1979).

(3) The third hypothesis presupposes an autosomal recessive gene mutation similar to the autosomal dominant gene mutation described in the mouse which causes genetic females to develop as phenotypic males, with small testes which, in the adult, are devoid of germ cells (Cattanach *et al.*, 1971). For a long time it was thought that the XX male goat was a unique example of an autosomal recessive mode of sex determination but recently evidence has been obtained for a similar condition in our own species: 80 cases of XX men have now been described; they all have small, azoospermic testes and tend to be shorter in stature than normal men (cf. Short, 1979).

(b) *Dwarfism*

There are several types of dwarfism. In the dwarf goats of Africa two different hereditary forms occur: pituitary hypoplasia and achondroplasia (Epstein, 1953).

Pituitary hypoplasia is defined as a general arrest of growth and development of all organs and systems of the body because of a hereditary hypofunctioning of the anterior lobe of the pituitary (Engelbach, 1932). This leads to an individual of short stature, but normally developed and proportionately built. A typical example is the slender-bodied Sudan goat. (see Fig. 7, p. 79).

Achondroplasia is defined by a marked disproportion between the length of the trunk, which is nearly that of an ordinary goat, and the length of the legs which are stunted. The head looks enormous. Achondroplastic goats, as typified by the dwarf goat of the Cameroons, are distinguished by the abnormality of cartilage bone formation arising in fetal life and resulting in a deficient growth of the long bones, causing dwarfism with a long body but short limbs.

In discussing the origin of the dwarf goat commonly found in Central and West Africa, Forde (1934) points out that this type has a much higher immunity to fly-borne diseases than the large brown goat of the African savannas. A similar dwarf goat occurs in Arabia, particularly in the province of Hejaz. The goats are of remarkably uniform type, and the general conformation shows the harmonious proportions of the pituitary dwarf (Epstein, 1946).

These types are not always as unmistakably manifest as in the goats of the Cameroons and the Sudan. Many of the dwarf goat breeds of Africa represent transitional types. If African dwarf goats are crossed with Alpine dairy goats, the F_1 crossbreds exhibit heterosis in terms of weight and milk production (Blanks, 1971; Blanks and Bogart, 1974; Cumlivski, 1976). This is not true, however, of the Puerto Rico Saanen × local breed cross (Table VII). African dwarf goats have been used regularly since 1963 as a laboratory animal in the USA (Hoversland *et al.*, 1965; Ragan *et al.*, 1966; Mehren, 1973). The observations on the small size breeds of goats are very diverse: Zambia (Quatermain, 1975); Tibetan dwarf, Chungwei, South China goat, etc. (Epstein, 1969); Kambing Katjang in Malaysia, Korean native (Devendra and Nozawa, 1976); Creole goats in Guyana and West Indies (Devendra, 1975; Devendra and Chenost, 1973).

(c) *Hereditary Goitre*

This is a dominant character, studied in 130 goat families in Switzerland (Gyger, 1925). In South Africa, several cases of goitre in Angora goats, and also in newborn kids, are discussed by Bath *et al.* (1978).

(d) *Brachygnathia*

Upper brachygnathia (parrot mouth) has been observed in Nubian, Etawah and Katjang goats of Java. It is a recessive anomaly (Fisher, 1955). Lower brachygnathia (undershot jaw) is also recessive, sometimes linked with the absence of a limb (Kobozieff, 1943; Machens, 1914).

(e) *Anomalies in Members or Joints*

Rheumatoid arthritis was reported in the Toggenburg breed in 1969 and in the polled Swiss breeds (Stunzi *et al.*, 1964). It is a chronic inflammation

Table VII. Body weight and milk yield of dwarf, normal and crossbred goats.

	Weight at 6 months (kg)			
	Male	Female		Source
Normal (Alpine and Nubian	21·8	17·2		
3/4 Normal, 1/2 Pygmy	22·3	20·9		
F_1 (Pygmy × Normal)	18·3	15·5		Blanks (1971)
1/4 Normal, 3/4 Pygmy	15·9	13·4		
African Pygmy	10·8	10·4		

	Weight (kg)	Milk yield in 210 days (kg)	Fat (kg)	Fat (%)	
Saanen (Czechoslovakia)	56·5	596	22·1	37·1	
F_1 (Saanen × WAD)	43·5	517	19·4	37·6	Cumlivski (1976)
W. African dwarf	19·5	241	9·1	37·8	

	Weight at 2 years (kg)	Body length (cm)	Chest girth (cm)	
Korean Native	23·8	54·4	68·5	
F_1 (Saanen × K. Native)	33·3	62·2	75·6	
3/4 Saanen, 1/4 K.N.	37·0	64·8	75·5	Park *et al.* (1977)
7/8 Saanen, 1/8 K.N.	39·8	67·4	76·4	
Saanen	46·2	70·1	80·8	

	Weight at birth kg	Weight at 6 months kg	Weight at 12 months kg	
Kambing Katjang (KK)	1·5	10·8	22·2	Devendra and
F_1 (Jamnapari × KK)	2·8	24·1	40·0	Nozawa (1976),
F_1 (Anglo-Nubian × KK)	2·5	20·0	35·6	Malaysia

of the carpal joints appearing mainly in the synovial capsules, the tendon sheaths and the tendons themselves, while the joint cartilage is scarcely affected. Histologically, this periarthritis is characterized by a proliferation of mesenchymal cells and a deposit of plasma cells and lymphocytes and by a focalized fibrinous swelling and calcification of the collagenous connective tissue. The disease seems to be hereditary. Animals are not generally affected before 10 months of age.

Abnormal growth of hooves forces the animals to walk "on their knees", starting at the age of a few months or even at a later age. It does not appear to be caused by malnutrition but is more likely to be a hereditary character comparable with a similar recessive gene found in bovines in South Africa (Boyazoglu *et al.*, 1969).

Inherited metacarpal gait in kids was studied in Saanen. It only affected males which were born co-twin to females. All affected males would stand, but walked on the metacarpus and died within a few weeks of birth. This defect seems due to a single, recessive, sex linked, semi-lethal gene (Bazargani, 1975).

We have referred to these three anomalies separately, but we do not know whether there is a link between them.

(f) *Absence of Hair or Recessive Atrichosis*

The animals are born hairless and remain so during the early months of their lives whereafter the normal coat is gradually acquired (Kislovsky, 1937).

(g) *Anomaly in Number of Teeth*

This is mainly a matter of too few incisors (Sionneau, 1964).

(h) *Supernumerary Teats*

The mode of inheritance is not clear, but a multifactorial inheritance can be presumed. The Japanese natives have supernumerary teats, whereas the Saanen introduced in Japan do not (Nozawa, 1970). The presence of a double or a fused teat in a strain of milk goats is analysed by Cunningham (1932).

(i) *Blind Halves of the Udder*

The gland tissue is capable of secreting milk, but the primary milk ducts are blind, preventing the drainage of milk from the mammary gland. This condition is due to an inherited defect in the gland (Turner and Berousek, 1942).

(j) *Nervous Goats*

If these goats are suddenly surprised or frightened they become quite rigid. They recover the use of their muscles in the front end of the body first. Often when these goats recover from fright they regain control of their forelegs enough to start running away, but the two hind legs will drag as if the goat were still quite stiff from the loin downwards although fully in control of the muscles of its front legs, head and neck. After such a fright a goat cannot usually be frightened again, until it has had at least 20 or 30 minutes rest. Lush (1930) thinks that this is a simple recessive character, similar to that found in other species, like congenital palsy of the guinea pig or waltzing mice, rats and rabbits (Cole and Ibsen, 1920; Cole and Steele, 1922). The character is similar to the myotonia discussed by Sionneau (1964).

(k) *Congenital Afibrinogenaemia*

Carriers in the Saanen breed suffer from serious haemorrhagic diathesis with bleeding into the joints. The blood plasma contains no fibrinogen and will not clot. This anomaly is due to an error of metabolism, and thought to be of incomplete dominance: heterozygotes seem to have a low fibrinogen count (Breukink *et al.*, 1972).

(l) *Congenital Dropsy*

An unusual obstetric case was described by Walser (1963). The goat was from a herd in which 11 kids with fetal dropsy were born, all sired by one buck. Two of the dams had dropsy kids in two consecutive pregnancies and two were full sisters (Gall, personal communication). These observations suggest a simple recessive character.

(m) *Cancerous Skin Lesions*

The Boer goat is essentially white with red or brown head and neck, but exposed skin—below the tail and surrounding both anus and female genitalia—is adequately pigmented (Joubert, 1973). Cancerous skin lesions occur in Boer goats lacking pigment around the labia vulvae in a low-lying tropical region with, *inter alia*, high infra-red radiation (Hofmeyr *et al.*, 1965).

(n) *Length of Sexual Cycle*

Instead of the 21-day cycle, goats may have short oestrus cycles averaging 6 days (Phillips *et al.*, 1943; Parer, 1963; Lyngset, 1964; Corteel, 1966; Sahni and Roy, 1967; Ricordeau and Bouillon, 1975; Skjevdal, 1975). These short cycles are most frequent early in the sexual season and predominate in certain progeny, suggesting a partly hereditary determinism. They can be compared with the "buck effect" (Shelton, 1960b, 1961; Shelton and Morrow, 1965; Ott *et al.*, 1979) which is natural synchronization of oestrus 6–13 days after bucks are introduced into the herd just before the beginning of the sexual season. They are probably due to an inadequate hormonal equilibrium because aborting Angora goats seem to have a higher proportion of short cycles than normal does.

4. Special Anomalies in Angora Goats

(a) *Genetic Abortion* (*gestational failure syndrome*)

In Angora goats of South Africa, abortions which are not caused by infections or nutrient deficiencies occur frequently. They occur between 30 and 140 days of gestation and most frequently between 100 and 110 days. In Texas, the large and strong goats which produce relatively small quantities of fairly coarse hair rarely abort. But in the herds of the smaller South African type selected for fine hair, abortion rate may be as high as 50%. Goats that had aborted exhibited abnormally short oestrus cycles and cystic corpora lutea (Rensburg, 1964, 1971b) which reduced their fertility, but Pretorius (1973) reported that their sexual activity did not appear to be different from that of normal does: the reproductive season was 94 days long, i.e. 4–7 oestrus cycles of 4–72 days with an average of 20·2 days. Strangely enough, kids born of does who had aborted before were heavier, had finer coats at birth and produced more hair than average. This anomaly is thought to be hereditary (Heerden, 1963, 1964). According to Rensburg (1971a, b), it can be explained by a "metabolic competition" which is the result of endocrine deficiency (inadequate rate of secretion of steroids by the adrenal glands) and a metabolism of fetuses oriented towards the growth of hair, with a high genetic potential for this character because of intensive selection and inbreeding. Low energy diet had a highly significant effect on the incidence of abortions among confirmed aborters, but not among normal does. Abortions following a low level of nutrition were characterized by expulsion of apparently normal fetuses (Wentzell *et al.*, 1974). The stress of underfeeding and, above all, abrupt changes to unfavourable environmental conditions contribute to abortion (Rensburg, 1971a; Wentzell *et al.*, 1975). Selection against disposed animals is desirable but difficult, because not all abortions are identified (Westhuysen and Wentzell, 1971). Joining does with vasectomized bucks before the beginning of the mating season improved fertility or reduced the number of visible abortions and increased the kidding percentage, although kid mortality within 5 days of birth was higher. It is therefore evident that the pre-mating association with vasectomized bucks stimulates ovarian activity and improves conditions for full-term pregnancy but not sufficiently to ensure the birth of a vigorous kid. However, this system increases the number of offspring which are genetically carriers of this abnormality (Lamont, 1964). This anomaly does not seem to exist in Angora herds in Turkey, Madagascar, India, Albania and the USSR, although, in a genetic experiment carried out in Turkey, Bilgemre and Düzgünes (1952) observed 23 abortions out of 132 does conceiving after insemination. For Shelton and Groff (1974), the poor reproductive efficiency in Texas Angora goats is largely explained by the following factors: failure of oestrus and ovulation due to underdevelopment of the female, lack of strength and mating vigour in breeding males due to malnutrition, abortion and death losses of kids (see also, Westhuysen, 1979).

(b) *Cryptorchism* (ridglings)

This anomaly has mainly been found in the Angora breed, in the USA and South Africa. It is thought to be due to a recessive gene which inhibits the descent of the testis. Its expressivity is variable since cryptorchism may be unilateral or bilateral, with absence of scrotum (Lush *et al.*, 1930). Males with only one testis are often fertile. Warwick (1961) has shown that selection, for or against cryptorchism, is effective. In South Africa, Skinner *et al.* (1972) found 2% cryptorchids in 8000 males whereas intersexuals were very rare. The cryptorchids were of the XY male karyotype, confirming that it is not the same anomaly as intersexuality.

(c) *Fetal Dwarfism*

Dwarf kids, mostly males, are born with very little capacity for life. An excessive metabolic rate and poor heat tolerance seem to cause malnutrition of the fetus. Dissection shows inadequate development of the pituitary, thyroid and adrenal glands. The anomaly seems to be hereditary, but occurs only when the dams are in an unfavourable environment, particularly in great heat and towards the end of gestation (Skinner *et al.*, 1971).

(d) *Hair Cover of Face and Belly*

The degree to which the head is covered with hair—a highly hereditary character (Table XX)—has a significant and unfavourable influence on live weight and fertility (Shelton, 1960a). At 2 years of age, live weight of does with hair covering their face was 55 kg and fecundity rate 16%, whereas the figures for does with no hair covering the face were 61 kg and 78% respectively. This is attributed to lower vigour associated with restricted vision (see Chapter 12).

5. Polymorphism of Genetic Markers

(a) *Blood Groups*

The search for natural antibodies in goats has so far been fruitless. Blood groups in sheep, however, have been determined with the aid of prepared reagents. Schmid and Suzuki (1971) have identified 18 phenogroups of the B system and shown that the Coloured German improved and the Saanen goat differ sharply from one another in the frequencies of the 40 factors detected. Kunz (1974) and Schmid *et al.* (1975) have described the frequency of the 45 blood factors in five Swiss breeds (Appenzell, Grisons Striped, Toggenburg, Verzasca and Valais Blackneck). Nguyen (1979) gives the number of phenogroups actually observed in five systems of blood groups similar to the systems B, C, M, R, F 30 of ovine blood groups (Table VIII) and discusses the homology between sheep, cattle and goat blood groups.

Table VIII. Blood factors detected in goats with the aid of sheep antigens on 232 goats of Saanen, Alpine Chamoisée and Poitevine breeds (Nguyen, 1979).

Genetic System	Number of blood groups factors in red cells	Minimum number of phenogroups
A	0	—
B	13	34
C	4	9
M	2	3
R	1	2
F30	1	2

Phenogroup = combination of factors which are transmitted as one genetic unit and probably controlled by closely linked loci.

(b) *Blood Proteins* (*Table IX*)

The transferrin, haemoglobin, albumin and serum amylase systems are polymorphous but their variability is low. Two to four transferrin variants are found but in most of the populations studied, variant A—the fastest—is predominant. Two variants are detectable for the haemoglobin system (apart from fetal Hb and anaemic Hb), type A being the fast and type B the slow fraction. In breeds of European origin and in the South African Angora and Boer breeds, variant A is more frequent. In the Niger breeds, however,

Table IX. Polymorphism of blood proteins.

Breed	Transferrin q^A	Transferrin q^B	Transferrin q^C and q^D	Haemoglobin q^A	Serum Albumin q^{Fast}	Source
Angora	0·80	0·19	⩽0·01	0·94	0·15	(1)
Boer goat	0·70	0·30	0	0·91	0·08	(1)
Italian	0·84	0·16	—		0·34	(2)
5 Swiss breeds	1·00	0	—			(3)
Fr. Saanen	0·97	0·03	—	0·99	no polym.	(4)
Fr. Saanen				0·99		(5)
Fr. Alpine				0·97		(5)
Fr. Poitevine				0·94		(5)
Granada	0·64	0·34	0·02	0·88		(6)
Toggenburg	1·00	0	—		0·17	(7)
Bulgarian	0·77	0·23	—		0·24	(7)
Swiss Saanen					0	(8)
German Alpine					0·29	(8)
Italian Alpine					0·17	(8)
Angora					0·47	(8)
Japanese					0·96	(8)
Korean	0·32–0·76	0·20–0·67	<0·02			(9)
Alpine	0·89	0·11		0·88		(10)
Beetal	0·44	0·56		0·92		(10)
Anglo-Nubian	0·30	0·70		0·92		(10)
Norway	1·00	0		1·00	no polym.	(11)
Barbari				0·93		(12)
Jamnapari				0·94		(12)
Algeria				3 phenotyp.		(13)
				4 phenotyp.		(14)

Breed	Haemoglobin q^N	Haemoglobin q^S	Haemoglobin q^F	Haemoglobin q^{S1}	Serum amylase q^A	Source
Red Sokoto	0·50	0·33	0·17	—		(15)
Kano Brown	0·47	0·40	0·13	—		(15)
Sahel	0·55	0·23	0·19	0·03		(15)
Angora					0·95	(16)
5 Swiss breeds					no polym.	(3)

(1) Osterhoff and Ward-Cox (1972).
(2) Salerno *et al.* (1970).
(3) Kunz (1974), Crottaz (1975).
(4) Nguyen *et al.* (1975).
(5) Ricordeau and Grosclaude (1970).
(6) Garzon *et al.* (1976).
(7) Tjankov (1972).
(8) Watanabe and Suzuki (1967).
(9) Watanabe and Suzuki (1973).
(10) Goel and Nair (1976).
(11) Efremov and Braend (1965).
(12) Joshi *et al.* (1975).
(13) Cabannes and Serain (1955).
(14) Garrick and Charlton (1969).
(15) Enyenihi (1974).
(16) Fechter and Pretorius (1970).

Enyenihi (1974) has found four types of Hb—F (fast), N (normal), S (slow) and S^1 (extra-slow)—but the last only in the Sahel breed. The rate of occurrence suggests that there is a selective advantage for AA homozygotes or AS heterozygotes over the FF and SS homozygotes which have disappeared even in Sokoto goats. In 129 Alpine, Beetal and crossbred goats, Gopinathan and Nair (1976) observed that females Hb AA were significantly younger at first kidding than females of other Hb phenotypes. There are also two detectable variants of the serum albumin and amylase systems, the most frequent being the fast amylase variant and the slow albumin variant, except in local Japanese breeds. For the serum β globulin there are two variants (Millsson and Pattison, 1961). Genetic variations at 27 blood-protein loci were examined recently for about 300 individuals from eight island populations of the Okinawa meat goats and from the Japanese Saanen. Significant differences of the gene frequencies at the LAP (plasma leucin aminopeptidase), Alp (plasma alkaline phosphatase) and PA-3 (plasma prealbumin 3) loci were observed. Genetic distance and migration rate were estimated (Nozawa *et al.*, 1978).

(c) *Milk and Lactoserum Proteins*

These, too, can be differentiated by electrophoretic migration. Lactoserum basically contains two main proteins: α lactalbumin and β lactoglobulin. Research by Sen and Chaudhuri (1962) on 23 Indian breeds produced no evidence of any polymorphism but Schmidt and Ebner (1972) and Boulanger (1977) identified two variants of α lactalbumin and β lactoglobulin polymorphism. Boulanger (1977) has made a thorough genetic analysis of whole casein. It consists of four major proteins called α_{S1}, α_{S2}, β and κ accounting for 5, 25, 50 and 20% respectively. Casein α_{S1} shows three variants (*r, l, ls*) whereas casein α_{S2} has two (*R* and *L*) alleles, $\alpha_{S1}{}^{l}$ and $\alpha_{S2}{}^{R}$ being the most frequent (Table X). Electropherogrammes of goat whole casein are much more complex than those of cow casein because of the superposition of certain caseins: α_{S1}, and α_{S2}, β and κ in alkaline gels, α_{S1} and κ in acid gels.

Table X. Gene frequency (q) of alleles for casein α_{S1} and α_{S2} (Boulanger, 1977).

Genetic type (herd)	$q\alpha_{S1}$			$q\alpha_{S2}$	
	l	*r*	*ls*	*R*	*L*
Saanen × Alpine (B)	0·97	0·03	—	0·91	0·09
Saanen × Alpine (M)	0·87	0·11	0·02	0·93	0·07
Saanen (BR)	0·86	0·06	0·08	0·88	0·12

The amino acid sequence of goat α lactalbumin is very similar to the predominant bovine B form, differing in only 12 positions (Macgillivray *et al.*, 1979). The amino acid sequence of β lactoglobulin is given by Phillips and Jenness (1965) and Preaux *et al.* (1979).

(d) *Absence of β Carotene*

Only vitamin A—and no β carotene—is found in goats' milk (and colostrum), whereas this liposoluble pigment is always present in cows' milk. This makes it possible to detect whether cows' milk has been mixed with goats' milk. This is still possible in cheese because the carotene concentration in the cheese does not change very much during the ripening process (Kuzdzal—Savoie and Kuzdzal, 1959). The method is of no use, however, if the adulteration is done with skimmed cows' milk. In this case the protein fractions may be compared by electrophoresis, taking advantage of the fact that casein α_{S1} of cows' milk has greater mobility than the fastest fraction of goats' milk casein (Aschaffenburg and Dance, 1968; Pierre and Portmann, 1970; Portmann and Pierre, 1971).

(e) *Glutathione Polymorphism* (GSH)

Like sheep, goats can be divided into two types according to the GSH concentration in their red blood cells (Table XI). This concentration is thought to be hereditary, and in man GSH deficiency is inherited as an autosomal recessive character (Prins and Loos, 1969). It may be noted that the major role of glutathione would be as a donor of γ glutamyl.

Table XI. Concentration of the red blood GSH (Agar *et al.* 1974).

Breeds	GSH low type		GSH high type		% of low type
	n	mg	*n*	mg	
Saanen	22	52	229	91	8·8
Angora	15	55	43	78	25·9
British Alpine	3	55	38	111	7·3

Low type: less than 60 mg/100 ml red blood cells.
High type: more than 60 mg/100 ml.

(f) *Serum Alkaline Phosphatase*

In 290 Norwegian does aged 8–10 months, Lode (1970) found two peaks at 0·16 and 20 to 73 units respectively; the distributions indicate that the high type enzyme is a simple dominant (A). In a population of Japanese Saanen,

Tsunoda *et al.* (1976) also observed two phosphatase phenotypes which are probably controlled by a pair of genes: Alp. F dominant and Alp. O recessive.

(g) *Nucleoside Phosphorylase* (NP)

Nucleoside phosphorylase plays an important part in red cell metabolism. In sheep, NP activity is controlled by a pair of allelic genes—NP^H and NP^L. Goats have high enzyme activities (NP^H), but no polymorphism has been detected, although only a small number of animals was tested (Tucker and Young, 1976).

(h) *Phosphoglucomutase*

No polymorphism was found in Boer goats. In Angora goats, three different phenotypes F, S and FS are observed. It is an autosomal locus with two alleles $PGM_3{}^F$ and $PGM_3{}^S$ with frequencies of 0·62 and 0·38 respectively (Pretorius *et al.*, 1976).

(i) *Antigen Factors*

In some goats G1 and G2 antigenic substances were secreted in their saliva and milk, but in others they were not; so goat saliva and milk may be classified into two types, secretor and non-secretor (Suzuki *et al.*, 1960). Japanese native goats had very high frequency of the appearance of dominant genes which control G1 and G2 antigen (Suzuki *et al.*, 1967). In blood sera, Watanabe *et al.* (1965) observed four new cellular antigen factors (Ch1, Ch2, Ch3, Ch4 antigen): these antigenic factors differ from the two agglutinogen types G1 and G2.

Electrophoretic studies showed that the seminal plasma consisted of six different antigenic components. These antigens might constitute the sperm coating antigens of this species (Prasad *et al.*, 1978), p. 136.

(j) *Histocompatibility*

Analysis of the results of typing and graft exchanges on dwarf and Saanen goats, and the segregation in the F_1 of 4 antigenic specificities, are in accordance with Mendelian codominant inheritance (Dam *et al.*, 1976). The major histocompatibility complex (MHC) seems to consist of at least three distinct loci: SD1, SD2 and LD (Dam *et al.*, 1979). Since recent studies indicate that the MHC not only serves as a marker system, but that its products play a role in immune response regulation itself, knowledge about the MHC in goats might be an important tool in the control of diseases.

B. Genetic Improvement

Before determining selection methods it is necessary to define the selection characters, their variability, their heritability and the genetic correlations between them, and to have some idea of their economic importance. In the case of dairy breeds, the goals of selection may be summed up as follows:

—to improve the fertility of the herd by removing any cases of sterility linked with polledness, reduce the percentage of does barren the first year, increase the length of the breeding season and maintain prolificacy at a sufficiently high level;
—to increase production per animal, particularly from first lactation at one year;
—to maintain or improve the cheesemaking qualities of the milk;
—to improve suitability for milking, particularly by machine;
—to obtain strong animals capable of adjusting to various environments.

1. Genetic Aspects of Reproductive Performance

(a) *Breeding Season and Prolificacy*

In Alpine and Saanen goats in France, the average date of the first oestrus in the season varies from June to September. This character (period of first oestrus) is repeatable and heritable (Ricordeau *et al.*, 1979). In young does, the breeding season starts 1–2 months later and 15–20% of them will not conceive in the first year. By selecting on the basis of the date of the first oestrus, therefore, it should be possible to reduce the percentage of barren yearling goats. Similarly, breeds differ in length of breeding season. The Angora goats of South Africa have a short season of 94 days or 4 to 7 oestrus, between April and July (Pretorius, 1973). In Texas, Spanish does, run continuously with males, showed a complete absence of ovulation in March, April and May, and did not kid twice per year apart from one kidding in October/November and then in May (Shelton and Spiller, 1977). In the same farming conditions, Sengar (1976) showed that Jamnapari does were definitely more seasonal than Beetal, Barbari and particularly Black Bengal does which kid throughout the year and have a high prolificacy (Tables XII and XIII). Observations on 12 080 Deshi goats (Patna, India) revealed heats in any season, but the maximum number were recorded in October and June, the minimum in March (Mishra and Biswas, 1966). In the tropical parts of Venezuela (10°09′ lat. N.), Nubian, French Alpine, Saanen and Toggenburg does imported from the USA have a main breeding season from

July to December (maximum in August–September) with a second peak of activity from March to June (Gonzalez *et al.*, 1974): the maximum fertility is obtained in August (77%) and the lowest in November–December (43%) with a maximum prolificacy in July and August. In the Niger (also 10° lat. N.), the Maradi (or Red Sokoto) goat is sexually active throughout the year, but the interval between kiddings is still 327 days and only 12% of intervals are less than 8 months (Haumesser, 1975); in this breed, the age at first kidding is 435 days (Adu *et al.*, 1979). The interval between kiddings in the Creole goat in Guadeloupe is only 237 days (Devendra and Chenost, 1973).

Table XII. Reproductive performance. Percentage of does kidding once or twice a year and producing singles, twins and triplets (Sengar, 1976).

		Jamnapari	Beetal	Barbari	Black Bengal
Number does kidded		49	55	63	61
once (%)		92	82	52	46
twice (%)		8	18	48	54
and producing	singles (%)	88·7	69·2	66·7	30·8
	twins (%)	11·3	29·2	33·3	59·6
	triplets (%)	0	2·6	0	9·6

Table XIII. Breed differences in litter size.

Breed	Country	Litter size	*n*	Source
Anglo-Nubian		2·1		Asdell (1929)
Toggenburg Saanen	USA	1·8		Eaton and Simmons (1939)
Maltaise-Damas	Cyprus	1·85	300	Maule (1949)
Saanen	Germany	1·83		Loertscher (1959)
Banat	Romania	2·00		Ciolca *et al.* (1959)
Baladi	Egypt	2·01	284	Tantawy and Ahmed (1960)
Kilis	Turkey	1·27		Yarkin and Eker (1961)
Kambing-Kartjang Anglo-Nubian	Malaya	1·65 1·43		Devendra (1962)
Saanen French Alpine	France	1·68 1·84	4 774 531	Ricordeau (1963)
Saanen	Israel	1·95	1 697	Soller and Kempenich (1964)

Table XIII. (cont.)

Breed	Country	Litter size	*n*	Source
Syria	Israel	1·44	199	Epstein and Herz (1964)
Neguev		1·38	29	
Appenzell		1·27	14	
Saanen		1·90	2 066	
Damas		1·76	21	
Anglo-Nubian		1·75	20	
German Alpine		1·65	17	
Maltese		1·59	17	
Anglo-Nubian × local	Mauritius	2·45	30	Delaitre (1965)
Anglo-Nubian		2·29		
Beetal	India	1·90	2 050	Amble *et al.* (1964)
Black Bengal	India	2·04	703	Moulick *et al.* (1966)
Ma T'ou	China	2·24		Epstein (1969)
Murcia	Spain	1·97	95	Belinchon and Marques (1971)
Angora	USA	1·15	246	Shelton and Stewart (1973)
Creole	Guadeloupe	1·91[a]	307	Cognie *et al.* (1971)
		2.11[a]		Devendra and Chenost (1973)
Nubian	Venezuela	1·52	498	Gonzalez *et al.* (1974)
Alpine		1·47	165	
Toggenburg		1·39	68	
Saanen		1·33	76	
Red Sokoto	Nigeria	1·47	850	Haumesser (1975), Adu *et al.* (1979)
Jamnapari	India	1·11[a]	49	Sengar (1976)
Beetal		1·32[a]	55	
Barbari		1·33[a]	63	
Black Bengal		1·79[a]	61	
Local	Norway	1·40		Steine (1976)
Anglo-Nubian	Mexico	1·92	111	Montaldo *et al.* (1978)
French Alpine		1·70	230	
Saanen		1·68	139	
Granada		1·65	129	
Toggenburg		1·62	115	

[a] Goats with more than 1 kidding per year.

The heritability of litter size is very low ($h^2 = 0{\cdot}08$) and similar to that obtained in sheep (Ricordeau *et al.*, 1979). Prolificacy of several breeds is given in Table XIII. Black Bengal and Baladi are prolific breeds, but the most productive breed could well be the Central Chinese Ma T'ou, which regularly has two kiddings a year, with 70% twins and triplets, producing annually 4·5 kids per doe (Epstein, 1969).

(b) *Length of Gestation*

Table XIV shows distinct differences between breeds—143 to 153 days. The period of gestation is longer for larger goats. However, as with cattle and sheep, this length of time depends more on the genotype of the fetus (direct genetic effect) than on the genotype of the dam (maternal genetic effect): Table XV.

Table XIV. Length of gestation: differences between breeds.

Breeds	Country	Average (days)	Source
Black Bengal	Bangladesh	143·0	Ali *et al.* (1973)
Black Bengal	India	144·3	Gupta *et al.* (1964)
Local	Brazil	145·0	Santiago (1946)
Local dwarf	Japan	145·6	Surgie and Soma (1970)
Barbari	India	146·0	Slater and Bhatia (1935)
Local dwarf	S. Africa	146·5	Wilson (1957)
Local F_1 (A. Nubian × local) 3/4 A. Nubian, 1/4 local)	Brazil	146·5 146·6 148·0	Jardin *et al.* (1965)
Anglo-Nubian	Brazil	148·0	Santiago (1946)
Nubian	Venezuela	148·7	Gonzalez (1977)
Local	Philippines	148·1	Arriola (1936)
Local	Mexico	148·2	Carrera and Hernández (1971)
Angora	USA	149·2 (139–156)	Shelton (1960b, 1961)
Angora Saanen British Alpine Toggenburg	S. Africa	149·4 (143–153) 149·9 149·0 150·5	Rensburg (1971b) Hofmeyr (1962) Hofmeyr (1962) Hofmeyr (1962)

Table XIV. (cont.)

Breeds	Country	Average (days)	Source
Toggenburg	Venezuela	149·0	Gonzalez (1977)
Boer	S. Africa	150·0	Skinner (1971)
Saanen Criollo	Venezuela	149·7 150·3	Gonzalez (1977)
Jamnapari	India	150·0	Slater and Bhatia (1935)
Anglo-Nubian		150·0	Asdell (1929)
Schwarzwald	Germany	150·8	Hinterthur (1933)
Norwegian	Norway	150·0 (140–162)	Lyngset (1964)
Saanen	Great Britain	150·0 (151, 150 and 149 with 1, 2 or 3 kids)	Peaker (1978)
Toggenburg Saanen	Brazil	151·0 152·0	Santiago (1946)
French Alpine	Venezuela	152·2	Gonzalez (1977)
Alpine	France	152·0 (142–162)	Ricordeau (1973)
Murciana	Spain	152·8	Saraza Ortiz (1955)

See also Devendra and Burns (1970, p. 83), Asdell (1929).

Table XV. Direct and maternal genetic effects on gestation length: transplantation of embryos in goats of 3 to 5 years, kidding 1, 2 or 3 kids (Surgie and Soma, 1970).

Dam	Saanen (days)	Native Japanese (days)
Kid		
Saanen	150·8	149·1
N. Japanese	146·1	145·6

Live weight: Saanen = 50–60 kg; dwarf Japanese = 25–30 kg.

(c) *Generation Interval as Component of Annual Genetic Gain*

The length of the generation interval has an important effect on the efficiency by which a trait in a population can be improved by selection. It is defined as the average age of parents when their offspring are born, or as the average

age difference between parents and selected offspring. The mean generation intervals are estimated separately for the four paths by which genes are transmitted from one generation to the next:

from sire to son;
from dam to son;
from sire to daughter;
from dam to daughter.

In France, the mean generation interval is 3·4 years in Saanen and 3·6 in "Alpine Chamoisée". In these two breeds, the dam–son interval is longest and the sire–daughter is the shortest of the different paths (Table XVI).

Table XVI. Estimates of generation intervals in years.

Japanese Saanen	3·4		Shotake (1971)
British Alpine	2·8		Ikin and Herron (1978)
French Breeds			Ricordeau *et al.* (1979)
Alpine Chamoisée	3·6	(3·3, 2·9, 4·5, 3·6)	
Saanen	3·4	(3·1, 2·9, 4·1, 3·6)	
		(sire–son, sire–daughter, dam–son, dam–daughter)	

2. Genetic Aspects of Milk Yield and Composition

(a) *Differences between Herds; Heritability of Differences*

The phenotypic differences between herds are very large. For example, if we group herds in six classes according to the average level of production after correction for lactation number, we observe:

(i) lactations are considerably shorter in low level herds but the protein and fat contents are not dependent on herd level;

(ii) the increase in milk production from first to second lactation rises from 24 kg in class 1 to 196 kg in class 6, whereas the increase beyond the second lactation differs little from one class to the next (Table XVII).

The heritability of herd differences, based on the intra-sire regression of the daughters on the herd average, is 0·08 for milk yield, 0·10 for fat yield and 0·40 for fat percentage (Rønningen, 1967).

(b) *Heritability of Individual Performance* (Table XVIII)

The h^2 for milk yield is of the same order as that of milk composition. According to Prakesh *et al.* (1971), the h^2 for milk yield corrected for lactation length and kidding season is highest in the first lactation.

Table XVII. Milk performance by herd class and lactation number: 1360 herds in 1977 (Ricordeau and Sigwald, 1977).

Herd class	*N*	Milk (kg)	Days lact.	PC	FC	Lact. number
1	444	223	164	28·3	31·9	
2	2 500	299	207	28·9	32·6	
3	6 414	399	227	29·1	31·7	1st
4	7 116	500	238	29·3	31·2	
5	3 074	612	259	29·7	31·6	
6	749	764	262	29·7	30·2	
1	353	279	177	28·3	31·4	
2	2 062	391	230	29·2	33·0	
3	4 937	525	251	29·2	31·7	2nd
4	5 447	658	264	29·3	31·6	
5	3 013	799	276	29·6	31·8	
6	702	1 006	282	29·3	30·0	
1	944	310	185	28·9	33·4	
2	4 255	426	235	29·3	34·4	
3	11 168	565	256	29·3	32·5	$\geqslant 3$
4	11 336	700	267	29·5	32·2	
5	5 894	852	278	29·6	32·3	
6	1 771	1 042	289	29·6	30·8	

6 Herd classes comprising 5, 15, 30, 30, 15 and 5% of the herds, respectively.
N = number of goats.
Protein content (PC) and fat content (FC) in g/kg of milk.

The casein content (or coagulable protein content) is highly heritable, but very low variable ($h^2 = 0{\cdot}65$–$0{\cdot}90$; coefficient of variation 2–3% with mean of 70–71%: Ricordeau and Bouillon, 1971c; Sigwald *et al.*, 1979). Flavour of milk is relatively variable (CV 21%) and as heritable as milk yield ($h^2 = 0{\cdot}25$: Rønningen, 1965a): a significant difference has been found between Saanen imported to Norway and the native breed which produced milk with a higher flavour intensity score and a higher free fatty acid content (Bakke *et al.*, 1977; Steine, 1978; Skjevdal, 1979).

(c) *Repeatability*

Between different lactations of the same goat, repeatability of milk or fat yield is as high as that of fat or protein content (Table XIX). Protein content is very useful for milk recording:* while fat content shows considerable variation from one milking to the next and from day to day, protein content varies very little (Carta *et al.*, 1975).

* For accuracy of determinations of composition of individual milk samples, see Grappin and Jeunet (1979).

Table XVIII, Heritability of traits of primary interest.

Milk yield per lactation	0·36(1); 0·33(2); 0·25(3); 0·32(4); 0·64(5)
Fat yield	0·30(1); 0·33(2); 0·49(5)
Protein yield	0·59(5)
Fat percentage (FC)	0·32(1); 0·29(2); 0·62(5)
Protein percentage (PC)	0·59(5)
Coagulable protein content (CPC)	0·66 and 0·90(6); 0·65(12)
Lactose percentage	0·38(7)
Flavour intensity in milk	0·27(7)
Milking time at machine milking	0·67(8)
Age at first kidding	0·54(3); 0·55(5)
Interval 1st and 2nd kidding	0·15(3)
Live weight at 7 months	0·70(5)
Live weight at kidding	0·29(1); 0·55(5)
Litter size	0·07(5); 0·24(9); 0·09(10); 0·15(11); 0·08(13)
Birth weight	0·01(17)
Angora	
Body weight	0·50(14)
Fleece weight	0·20(14); 0·15 and 0·26(16)
Fleece yield	0·48(14)
Fiber diameter	0·11(14)
Face cover	0·31(14); 0·38(15); 0·53(14)

(1) Rønningen (1967) (2500 lactations, 230 sires).
(2) Steine (1976).
(3) Singh *et al.* (1970) (1236 lactations, 110 sires).
(4) Prakash *et al.* (1971) (4912 lactations, 114 sires).
(5) Ricordeau *et al.* (1979) (1350 first lactations; 126 sires; 100 days: station).
(6) Ricordeau and Bouillon (1971c) (626 goats, 55 sires, 2 herds).
(7) Rønningen (1965a, b), Skjevdal (1979).
(8) Bouillon (1975).
(9) Tantawy and Ahmed (1960) (82 dam–daughter pairs).
(10) Moulick *et al.* (1966) (83 dam–daughter pairs).
(11) Amble *et al.* (1964).
(12) Sigwald *et al.* (1979) (432 lactations, 45 sires).
(13) Adu *et al.* (1979) (50 dam–daughter pairs).
(14) Shelton and Bassett (1970), Shelton (1960a) (453 parent–offspring pairs).
(15) Sincer (1967).
(16) Sincer (1963).
(17) Moulick and Syrstad (1970) (284 does, 20 sires).

Table XIX. Repeatability of milk performance estimated by the correlation between two successive lactations.

Variables	Correlations
Milk yield	0·40(1); 0·63(2); 0·57(3); 0·78(4); 0·18(5)
Fat yield	0·61(2); 0·51(3)
Protein %	0·75(4)
Fat %	0·71(2); 0·64(3); 0·85(4)
Lactose %	0·38(1)
Kidding interval	0·05(5)

(1) Rønningen (1964, 1965a, b).
(2) Rønningen (1967).
(3) Steine (1976).
(4) Ricordeau and Bouillon (1967).
(5) Kumar *et al.* (1962).

(d) *Phenotypic and Genetic Correlations*

Table XXa shows that the phenotypic correlations are mainly genetic, and therefore it is more interesting to discuss the genetic correlations only (Table XXb, Fig. 4).

Milk yield has a negative correlation with fat percentage (−0·20 to −0·45), protein percentage (−0·50) and coagulable protein content (−0·14), and a positive correlation with lactose percentage. The correlation between protein yield and fat content is −0·28 and between protein yield and protein content −0·19. Rønningen (1965) found a positive genetic correlation between milk yield and flavour intensity (0·77), but the results from the selection experiment of Steine (1978) tend to give a negative genetic correlation between these two traits. Protein content has a positive correlation with coagulable protein (+0·43) and fat content (+0·62).

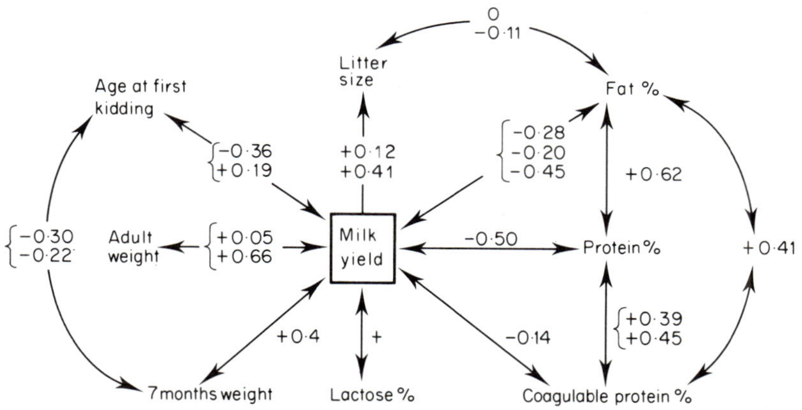

Fig. 4. Genetic correlations.

Milk yield is in positive or negative correlation with age at kidding (+0·19 in Alpine of 13–14 months at first kidding; −0·36 in Beetal goats 2 years old) and in positive correlation with weight at 7 months, so that selection for precocity is not incompatible with selection for milk production. The correlations between milk yield and live weight are zero or positive (0·05 and 0·46). We note also a positive correlation between milk production (total lactation or daily maximum yield) and litter size (+0·12 to 0·48) and a zero or negative correlation between fat percentage and litter size (0 and −0·29).

Table XX. Phenotypic and genetic correlations.

(a) *Phenotypic correlations*	
Milk yield–lactation period	+0·70(7); +0·66(9); etc...
Milk yield–fat %	−0·18(10); −0·31(13); −0·25(19)
Milk yield–protein %	−0·39(10); −0·39(13)
Milk yield–litter size	+0·16(6); +0·18(10); +0·09(19); >0(4 and 5)
Milk yield–weight at mating	+0·23(2)
Milk yield–weight at kidding	+0·31(6); 0·32(10); +0·06(19)
Flavour of milk–fat %	−0·04 to −0·35(18)
Flavour of milk–protein %	−0·28 to −0·35(18)
Weight at mating–ovulation rate	+0·58(15)
Body weight–litter size	+·026(4); −0·07(10); +0·41(15)
Body weight–face cover	−0·35(14)
Body weight–fleece weight	+0·14(14)
FCM–body measurements	+0·59 to 0·79(3)
(b) *Genetic correlations*	
Milk yield–fat %	−0·20(7); −0·45(10); −0·28(19) −0·31 to −0·47; between months test (19)
Milk yield–protein %	−0·49(10)
Fat %–protein %	+0·62(10)
Protein yield–fat %	−0·28(10)
Protein yield–protein %	−0·19(10)
Fat %–flavour of milk	−0·28(11)
Milk yield–flavour of milk	+0·77(11); realized r_g = negative (18)
Milk yield–lactose %	>0(11)
Coagulable protein %–milk yield	−0·14(8)
Coagulable protein %–protein %	+0·43(8); +0·39(16)
Coagulable protein %–fat %	+0·41(8)
Milk yield–milking time	+0·37(1)
Milk yield–age at first kidding	+0·19(10); −0·36(17)
Milk yield–weight at kidding	+0·66; 1st kidding (10); +0·05(19)
Milk yield–weight at 7 months	+0·48(10)
Milk yield–litter size	+0·41(10); +0·12(19)
Maximum daily yield–litter size	+0·23(19)
Fat %–litter size	−0·11(10); 0(19)
Protein %–litter size	+0·20(10)
Body weight–face cover	0(14)
Body weight–clean fleece weight	−0·31(14)
Body weight–fibre diameter	0(14)

(1) Bouillon and Ricordeau (1975b).
(2) Fehr and Sauvant (1975).
(3) Gall *et al.* (1967, 1971) (multiple regression r^2).
(4) Hayden *et al.* (1979).
(5) Lee *et al.* (1974).
(6) Montaldo *et al.* (1978).
(7) Prakash *et al.* (1971).
(8) Ricordeau and Bouillon (1971a).
(9) Ricordeau and Sigwald (1977).
(10) Ricordeau *et al.* (1979).
(11) Rønningen (1965).
(12) Rønningen (1974).
(13) Sauvant *et al.* (1973).
(14) Shelton and Basset (1970) (Angora).
(15) Shelton and Stewart (1973) (Angora).
(16) Sigwald *et al.* (1979).
(17) Singh *et al.* (1970).
(18) Skjevdal (1979).
(19) Steine (1976).

(e) *Response to Selection for Milk Yield and Composition*

The expected change per year is estimated by the common formulae:
Direct responses for character x

$$\frac{R_x}{\text{year}} = \frac{i\, h^2\, \sigma_x}{L}$$

Correlated or indirect response upon the character y in selecting directly upon the character x

$$\frac{CR_{y,x}}{\text{year}} = \frac{i\, h_x\, h_y\, r_G\, \sigma_y}{L}$$

where h^2_x and h^2_y = heritabilities of x and y; σ_x and σ_y = phenotypic standard deviation of x and y; r_G = genetic correlation between x and y; i = selection differential expressed in standard deviation units; L = generation interval.

Results of calculations are given in Table XXI. Direct selection for milk yield increases the quantities of milk, protein and fat, and also body weight, but it reduces protein and fat contents. Selection for protein yield reduces fat and protein contents but maintains the protein/fat ratio, whereas selection for fat yield reduces protein content. As regards cheese weight, the best response is obtained by selecting for milk yield (+2·72 kg), protein yield (+2·56) and fat yield (+2·46); however, this approach reduces the "useful dry matter" (fat + protein) or the cheese yield. A selection index could be designed to maximize milk quantity and keep milk composition unchanged, but that always implies reducing maximum genetic gain in cheese weight (Bouillon and Ricordeau, 1975a). In practice, we adopt selection on the basis of protein yield, with a minimum protein content (29·5 g/kg) at the level of dams of sires, in order to avoid bucks that improve milk yield but have an inadequate index for protein or fat content.

(f) *Effects of Inbreeding*

In Turkey, in a study on crossbred goats (3/4 Saanen, 1/4 Kilis), Tuncel (1976) found that increasing the rate of inbreeding (up to 20% on average in the herd) had a significant depressive effect on the quantity of milk per lactation (regression of 7·6 kg of milk) and the quantity of fat, but not on fat content or the length of the milking period. In the USA, in a study of 42 Alpine does, Ormiston and Touchberry (1959) found a curvilinear relation between milk production and inbreeding: standardized production (at 2 years and in 240 days) fell from 598 to 498 kg if inbreeding increased from 5 to 25% (a regression of 5 kg) and then increased to 622 kg if inbreeding increased further from 25 to 45%.

Shotake (1971) has investigated the relationship between inbreeding co-

Table XXI. Expected change per generation in a character y when selecting for a character x (Bouillon and Ricordeau, 1975a; Ricordeau *et al.*, 1979).

Selected character (x)	Direct and indirect response on character y						
	Milk yield (kg)	Fat yield (kg)	Protein yield (kg)	Fat (‰)	Protein (‰)	Cheese yield[a] (%)	Weight at first kidding (kg)
Milk yield	+29·8	+0·724	+0·749	−0·90	−0·60	+8·3	+1·7
Fat yield	+22·2	+0·732	+0·616	+0·12	−0·21	+7·5	+1·5
Protein yield	+27·0	+0·726	+0·754	−0·53	−0·23	+7·8	+1·7
Fat (‰)	−13·1	+0·55	−0·215	+1·95	+0·75	−2·8	−0·3
Protein (‰)	−14·0	−0·158	−1·45	+1·20	+1·18	−5·3	+0·3

[a] Additional cheese yield in 100 days, estimated by the following equations of Ricordeau and Mocquot (1967) (1), Portman *et al.* (1968) (2):

$$\text{kg of cheese per 100 kg milk} = -3{\cdot}54 + 0{\cdot}15 \text{ fat } ‰ + 0{\cdot}45 \text{ protein } ‰ \quad (1)$$
$$\text{kg of cheese per 100 kg milk} = +1{\cdot}47 + 0{\cdot}09 \text{ fat } ‰ + 0{\cdot}33 \text{ protein } ‰ \quad (2)$$

Characters considered = performance in the first 100 days of lactation of goats kidding at 13 months in station.

efficient, population size and mortality for the breeding population of Saanen in Japan: he observed an increase of inbreeding but no significant variation in mortality. To identify possible causes of reproductive difficulties and general lack of viability amongst British Alpine in Australia, a genetic analysis was carried out by Ikin and Herron (1978): total inbreeding was 28% and it was clear that without importations and with inbreeding increasing at about 2% per year, breeding difficulties will continue. Pure breeds are often closed populations which may have a rather small effective size, in which it is necessary to avoid a high rate of inbreeding. Chevalet and de Rochambeau (1979) have proposed general and practical solutions: (1) define reproduction groups of equal size and not split into different flocks, then minimal ancestry information may be used to split a flock into groups; (ii) choose a circulation scheme between groups so that males of a given group successively mate the females of all other groups; these schemes are found to be the best ones, if the number of groups is more than 10; (iii) reproduction groups must be large enough to give birth to the necessary males, as the number of reproductive males remains the most important factor. This method was applied to the Poitevine breed with 11 groups of reproduction.

3. Genetic Aspects of Milking Characters

(a) *Milking Time*

Ease of machine milking may be expressed in terms of milking time measuring milk flow rate. Because of the close relationship between the yield of milk and the rate of flow, the milking time is usually corrected for milk yield in order to obtain a better basis for comparison of milking at different periods (Naito *et al.*, 1965). Corrected milking time has a h^2 of 0·67 and is in positive genetic correlation of 0·37 with milk yield.* To account for the three variables, milk yield, protein content and milking time, Bouillon and Ricordeau (1975b) calculated a selection index for maximizing the gain in cheese weight by changing, to an acceptable degree, milk composition and ease of milking ($I = Q - 0{\cdot}10\ t + 11.\ PC - C^{te}$; in which Q = kg of milk in 100 days, t = corrected milking time in seconds, PC = protein content in g per kg milk). Dams of sires should be tested for milking time before progeny testing.

(b) *Selection for One Milking per Day*

In Alpine goats, milking only once a day (1 x) instead of twice a day (2 x) throughout lactation reduced milk production by 35%; if done after a 2-month lactation (2 x.1 x), the drop is only 20% (in 100 days of lactation 2 x = 275, 1 x = 163 and 2 x.1 x = 253 kg). The distribution of quantities produced with a single milking per day is very asymmetrical: many goats are not sufficiently stimulated for normal milking, whereas some young does respond very well. Heritability of milk yield for the group of superior level is high, and therefore genetic improvement should be possible by selection (Mocquot and Auran, 1974; Mocquot and Guillimin, 1975, 1978; Mocquot, 1978).

4. Progeny-testing of Bucks

In most goat populations, the replacement rate is 0·25. If the genetic level of herds is to be improved, both males and females have to be selected. Male selection is based on the test of young bucks in station or alternatively in farms, with daughters produced in different herds.

*With regard to machine milking of goats, it is useful to consult recent studies on equipment and factors affecting milking performance, material, udder preparation, stripping, udder conformation, etc: Ricordeau and Labussière (1970), Cumlivski (1978), Darracq *et al.* (1978), Le Mens *et al.* (1978a, b), Sharav and Carasso (1978), Horak and Kasing (1970). See also Chapter 11.

(a) *The Norwegian "Buck Circles"*

Each circle consists of at least 500 goats in 5 herds and 5 young males under test (at least 1 year old) which pass through each herd during the mating season. Before the final index at the end of the first lactations, a provisional index is calculated on part lactations, so that the bucks can be used for élite matings when they are $2\frac{1}{2}$ years old and the generation interval is reduced (Steine, 1975a, b). The dam index is calculated on the basis of milk production and udder conformation is scored. Artificial insemination with frozen semen is limited to élite sires at least 3 years old tested in farms. The system is discussed in more detail in Chapter 15.

(b) *France*

(i) *Progeny Testing Station.* This was instituted in 1965 at a time when milk recording and artificial insemination were limited. There are three stages (Fig. 5).

(1) *Ancestry selection.* The males selected for pretesting have to come from planned "élite matings". They are preferably horned (bucks from two polled parents are excluded). It is desirable to take animals born in November and December, so that the test inseminations with frozen semen can begin the next year, when they are 6–7 months old.

(2) *Individual control* in three pretesting centres, of 100 young males, where growth, sexual behaviour, sperm production and rate of sperm survival after deep freezing and defreezing are controlled (Corteel and de Montigny, 1975; Corteel, 1976). Male weight at 7 months is in positive correlation with sperm production (repeatable character with a CV of 31%) and with the average weight of their daughters at 7 months (progeny index): their selection for weight at 7 months could therefore improve their reproductive qualities and increase the average weight of their daughters at first mating (Table XXII).

TableXXII. Correlation between progeny tests of sire and son and between performance test of the sire and his progeny index.

Correlation between		r	Source
Milk yield	Sire's index and son's index	0·42	Ricordeau and Bouillon (1971b)
	dam's index and son's index	0·25	
7 months weight	Sire's index and son's index	0·36	Ricordeau and Bouillon (1971d)
	Sire's weight and average of his daughters	0·34	
	Sire's weight and his own sperm production	0·41	Bouillon (1971)

The sire's index and son's index are estimated on 28 and 15 daughters respectively.

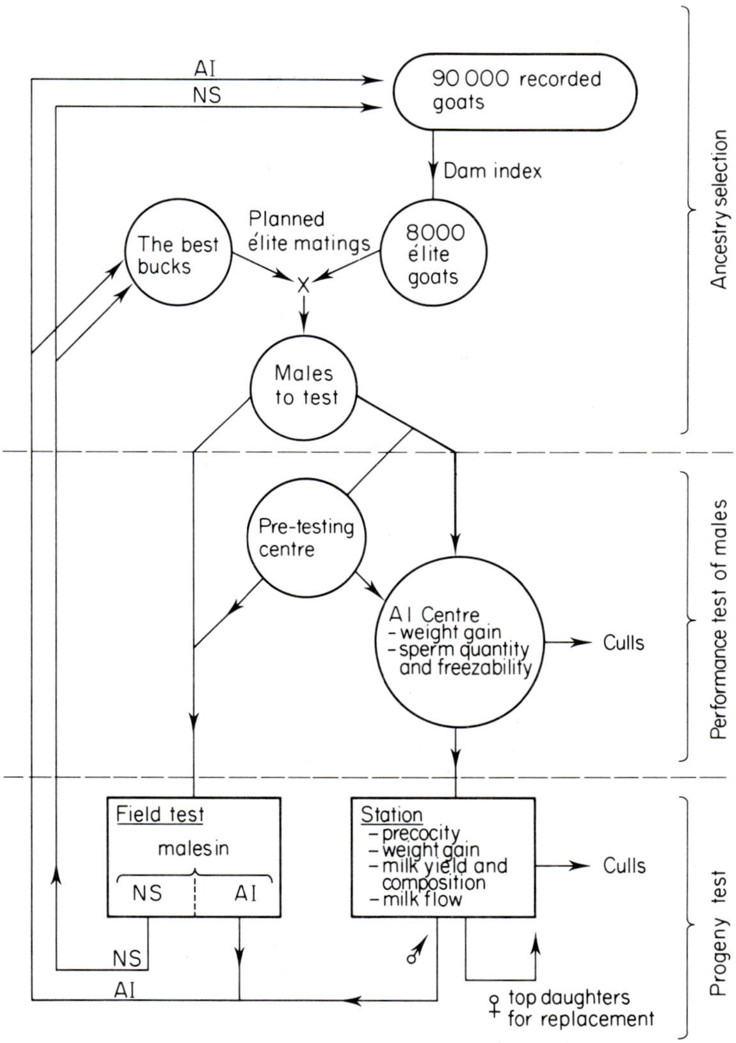

Fig. 5. Selection scheme of goats in France. AI = artificial insemination; NS = natural service.

(3) *Progeny testing.* Frozen semen of the 20 best males is sent to the station in July for inseminating 250 adult does in the basic herd, in order to obtain 10–15 daughters per sire. The daughters are born at the station and kept there up to the end of their first lactation, the best then being selected as herd replacements. The traits controlled are growth (weight

taken montly from 1 to 8 months and after each kidding), quantity of milk and composition in protein and fat (every 14 days), coagulable protein, milking time and shape of teats. The first provisional indices are calculated in June on the basis of the first 100 days of lactation, before the insemination season starts. With h^2 of 0·64 and the present station capacity of 200 daughters, the equations of Razungles (1974) show that the maximum genetic gain with the minimum confidence interval is obtained with 7–9 daughters per sire.

(ii) *Progeny Testing in Farms.* This is developing due to increasing milk recording: 1000 goats in 1961, 27 000 in 1971 and 86 000 in 1979 (Ricordeau *et al.*, 1979, Table XXIII). With 20 000 first lactations, it is theoretically possible to test annually 200–300 young males on 30 to 40 daughters, but at the present time we obtain only less than 100 bucks with a correct estimation of their breeding value. The following priorities are defined in the present programme: to keep the semen of the improver sires for milk recorded herds, increase the number of pretested sires to have 60 young bucks for field progeny-testing by AI, obtain a valid index for 60 sires used for natural mating in realizing systematic connections between herds with AI reference sires, and to plan élite matings with dams having the best indices in order to produce enough young males with the required pedigree.

Table XXIII. Performance recording in France (Ricordeau *et al.*, 1979).

	1963	1973	1979
Number of recorded goats	2 300	36 700	86 400
Number of herds	160	820	1 660
Average size of herds	15	45	52
Milk yield (1st lactation)	413	440	475
in kg (all lactations)	533	562	595

Repartition of goats: Alpine: 73%; Saanen: 11%; Poitevine: 1%; "croisées": 15%.

5. Estimation of Breeding Values

The first indices for sires (Grönbeck, 1955) were based on a comparison between average production of daughters ($\bar{D}$) and that of their dams ($\bar{M}$): $I = 2\ (\bar{D} - \bar{M})$. But the actual indices are based on the "relative genetic value" method of Johansson and Robertson which takes differences between herds into account and estimates the relative production by comparison with the contemporary daughters in the same herd (Norwegian method) or in the same level of herd (French method).

(a) *Norwegian method* (Rønningen, 1967; Steine, 1977, personal communication)

(i) *Sire Index*

$$I = \frac{[2b\,(D - C) + 0{\cdot}1\,(C - R) + R]\,100}{R}$$

in which D = corrected average production of daughters in first lactation; C = herd average of contemporary yearling goats in the same herds but born of other sires; R = breed average production in the same area; 0·1 = heritability of genetic differences between herds in milk yield; b = regression coefficient of production of future daughters of the sire on his first daughters, based on heritability of milk production per lactation (h^2) and number of recorded daughters (n).

$$b = \frac{0{\cdot}25\,n\,h^2}{1 + 0{\cdot}25\,(n - 1)\,h^2}$$

For each herd i, the difference d_i between average production of daughters and of contemporary does is weighted with the factor w_i which takes into account the number n'_i of contemporary does, so that

$$(D - C) = \frac{\sum w_i\,d_i}{\sum w_i}\text{; in which } w_i = \frac{n_i n'_i}{n_i + n'_i}$$

With this method, the daughters have to be distributed among several herds and there have to be at least five contemporary daughters per herd (see also, Rønningen, 1974).

(ii) *Dam Index*

$$I = 100\frac{[0{\cdot}8\,h_n^2\,(D - C) + 0{\cdot}08\,(\mathrm{C} - \mathrm{R})]}{R} + (0{\cdot}5 - 0{\cdot}4\;h_n^2)(I_p + I_m - 200) + 100$$

in which $h^2 = 0{\cdot}3$, $r = 0{\cdot}6$; $h_n^2 = h^2$ for n observations of FCM yield $= \dfrac{\mathrm{n\,h}^2}{1 + (n - 1)\,r}$; C = herd average; G = production of the goat in FCM; R = Breed average; I_P = sire's index and I_M = dam's index.

The best 5% of the does are regarded as élite goats and are eligible for planned matings with the best of the progeny tested bucks. Practical details of the Norwegian breeding scheme are discussed in Chapter 15.

(b) *The INRA French Method*

This method, designed originally for dairy cattle, was adapted to dairy goats and applied from 1979 onwards (Poutous *et al.*, 1980). The production x_i of a goat is corrected for environmental (herd level, lactation number, month of kidding, age at first kidding, herd level × lactation number interaction) or genetic influences (for example, in the first lactation, one-half of the index of the dam is deducted to counterbalance a possible selection effect). These environmental correction factors are estimated together by successive iterations taking into account data of the previous year and those of the new lactations. The corrected production, CP, thus expresses the genetic value of the goat and can be used to compute the sire and dam indexes. Different values are given to the variables, depending on their precision and the lactation number. The computation programs of these indexes are very complicated and require many iterations: the formulae presented here are simplified and are only meant to give an indication.

(i) *Sire Index*

$$I = 2\left[\text{Ref} + \frac{\sum[p(CP - \text{Ref})]}{\sum p\left(1 + \frac{0{\cdot}3}{NE}\right) + \left(\frac{4}{h^2} - 1\right)}\right] - I_{BM}$$

where Ref = half the mean of the indexes of sires whose first daughter kidded the same year as the first daughter of the sire to be indexed; CP = corrected production of the sire's daughters, including all the lactations; $\sum$ = Sum of the weighted factors of the considered lactations; NE = number of herds with daughters in first lactation; and I_{BM} = mean index of the last three series of young indexed sires. This index is constructed so that every year 50% of the young sires have a positive index.

(ii) *Dam index*

$$I = \left[I_A + \frac{\sum p\, h^2}{1 + (\sum p - 1)\, h^2}\,(\overline{CP} - I_A)\right] - I_{BM}$$

where I_A = half the sum of the two parents indices; $\overline{CP}$ = corrected performance of the goat, or average of corrected differences due to environmental factors; I_{BM} = mean index of goats in second lactation on the two preceding years; $\sum p$ = sum of all lactations considered.

The information from all lactations is taken into consideration with the following parameters (heritability, repeatability):

milk yield and protein yield: $h^2 = 0{\cdot}30$; $r = 0{\cdot}50$
protein percentage : $h^2 = 0{\cdot}40$; $r = 0{\cdot}60$

(c) *Projection Factors for Lactation Records*

The information from all lactations is taken into consideration with the between records of different lengths. The milk yield is estimated between using the cumulative yield Y_n, after n days in milk, and with a yield of LP on the last sample day.

(i) *INRA Method* (Mocquot, 1978, 1979, personal communication)
Corrected yield for partial lactation:

$$C.\ PL = 0{\cdot}90\left(\frac{385 \,.\, Y_n}{n+80}\right) + LP\, a^{(n-b)/c}$$

where $a = 0{\cdot}81$; $b = 85$; $c = 12{\cdot}5$ in first lactation;
$a = 0{\cdot}78$; $b = 63$; $c = 12{\cdot}5$ in other lactations.

Corrected yield for total lactation:

$$C.\ TL = \text{total yield} \times \frac{385}{n+80} \times 1{\cdot}3$$

(ii) *DHI Method* (Wiggans *et al.*, 1979)

$$Y_{305\,\text{days}} = Y_n + [\text{factor} \times LP(305 - n)]$$

when production on the last sample day is not available, the 305-day yield is predicted with a factor k for cumulative yield:

$$Y_{305} = Y_n[1 + k(305 - n)/n]$$

Wiggans *et al.* give the last sample day factors for two groups of breeds, four ages at kidding, two seasons and two levels of herd production.

(d) *Validity of Indices*

In theory the accuracy of an index can be measured by the correlation between the true breeding value of the sire and his index ($r = \sqrt{\tfrac{1}{2}\beta}$). In practice, use is made of the regression of the son's index on the sire's index or the correlation between them. When the breeding value is based on the average of the sire's daughters, the equation of Searle (1964) is used

$$E(r) = \tfrac{1}{4}\beta\beta'$$

in which

$$\beta = \frac{2nh^2}{4 + (n-1)h^2};\ \beta' = \frac{2n'h^2}{4 + (n'-1)h^2};$$

n and n' = the number of daughters used for calculating the indices of the sire and the son respectively.

For in-farm indices, allowance is made for the coefficients of determination (CD) for the indices of the sire and the son (Mocquot *et al.*, 1970)

$$E(r) = \tfrac{1}{2}\sqrt{CD_{sire} . CD_{son}}$$

$$E(b) = \tfrac{1}{2} CD_{son}$$

If the population of sires is selected, we use the Cochran equation (1951).

6. Organization of Selection

(a) *Systems with AI or "Bucks Circles" for Progeny Testing*

This consists essentially of the following steps:

—selection of the best dams on the basis of the "dam index";
—planned matings of superior dams with the very best progeny-tested bucks, in order to breed potential sires of the next generation;
—rational use of the best proven males on dams selected for the production of future breeding stock.

In order to be able to exert a high selection pressure on dams and to use proven sires, but at the same time to test young bucks, it is necessary to expand milk recording, monitor the mating system (accuracy of paternity) and institute a test programme on a sufficiently large scale. For calculating regularly and rapidly the partial and final indices of sires and the indices of dams likely to produce bucks, all the available data should be centralized and processed by computer. Since all these operations are interdependent and need to be programmed in advance, a high degree of coordination needs to be established between artificial insemination centres, milk recording and data processing bodies and the officials responsible for selecting sires.

(b) *Group Breeding Schemes*

Since in extensive goat breeding it is very difficult to realize planned matings and progeny-test with the aid of AI, breeders may adopt the new cooperative group breeding schemes developed recently in New Zealand for sheep. They vary in size, objectives, organization and success, but there are three common features:

—cooperation among breeders in running a jointly owned central or nuclear breeding unit to produce their replacement breeding males;
—a two-way flow of tested and selected stock, of selected males from the

nucleus to cooperators, and of highly selected females from the cooperators' farms to the nuclear unit;

—selection for economically important traits.

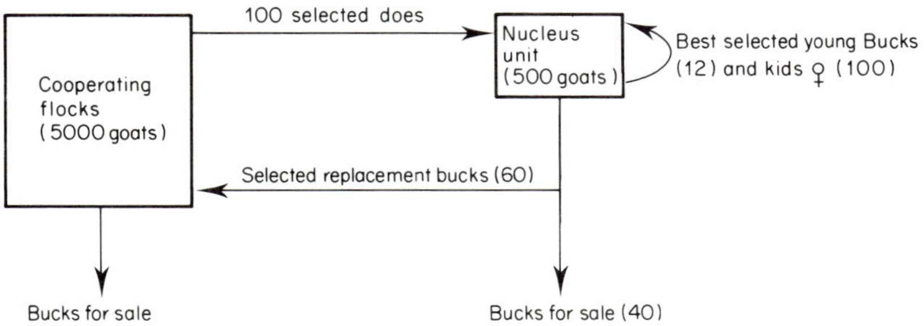

Fig. 6. Example of goat group-breeding scheme. Assuming 5000 goats on the cooperating farms, about 500 goats will be required in the nuclear unit. These will be taken in gradually over the first 5 years to build up the total. After 4–5 years, the annual replacement in the nucleus will be about 100 selected recorded goats (used for 1–2 years) from the cooperators and 100 kids ♀ (used for 3–4 years) from the nuclear unit. Bucks will be available to the cooperators, and for sale, after 2 years of operation.

The general form of a group breeding scheme is shown in Fig. 6. A group of breeders record their animals on their own farms and select their best performing females and send them to the jointly-owned breeding unit. Here, after further recording, the best young males and females are retained for breeding in the nucleus. Other selected young males from the nucleus will be used in the cooperating farms as replacement sires. This cycle of recording and selection is continued over the years. The nucleus is always kept open to the highly productive selected females from the cooperating farms so there is a two-way flow of stock and genes. The maximum rate of gain is achieved when the nucleus has 5 to 10% of the total number of animals in the cooperating farms, when 40 to 50% of the nucleus female replacements come from the contributing flocks and when all nucleus-born young goats not needed for replacements in the nucleus are used in the contributing herds (Rae, 1977). There are several advantages to a breeder in belonging to a group.

(i) Because of their size and potential, a breeding group will have a broader spectrum than any individual breeder, a common set of breeding goals, of recording systems and breeding methods and the costs of improvement will be less than for each breeder separately.

(ii) Breeding stock from the different herds will be compared under common conditions in the nuclear unit. Differences in genetic merit between strains from different herds can be measured and used in selection. There will also be an incentive for each breeder to do well in these comparisons, and a stimulus to pay more attention to his recording and selection.

(iii) The breeding group can act as a focus for technical aid to breed improvement and may be able to exercise more influence in the market than the individual breeder.

(iv) In theory group breeding should be able to obtain a greater rate of improvement (15 to 25%) than an individual breeder: goats are recorded in practical farm conditions; the records are taken for measurable traits of economic value; there can be intense selection on the cooperating farms by screening large numbers of females, thus, the best individuals can be concentrated in the nuclear unit and the nuclear unit itself can also be selected more intensely; detailed recording, special tests or matings and further selection can be done in the nuclear unit; replacement sires for the cooperating farms come from the nuclear unit, so the improvements are quickly spread throughout the group.

(v) Inbreeding problems which may arise with small breeding units can be avoided. Continuity in breeding objectives and selections over the years could be ensured, giving a steady improvement in economically important characteristics.

7. Improving Productivity of Goats by Crossbreeding

The introduction of specialized breeds for milk or meat production may be a rapid solution for increasing productivity (Keeping, 1951; Sanfiorenzo, 1962; Jardin *et al.*, 1965; Garcia *et al.*, 1972; Geerts, 1975; Eker *et al.*, 1976; Mishra *et al.*, 1976; Tuncel *et al.*, 1976; Boyazoglu and Zervas, 1977). The results concerning the milk performances are often favourable in F_1, but they may be less so in F_2 (Sanfiorenzo, 1957, 1962; Table XXIV). On the

Table XXIV. Crossbreeding between local and specialized breeds. Comparison of exotic breeds.

Type	*N*	Milk yield (kg)	Days[a]	FC[b]	Source
Local	(146)	157	236		(1)
Barbados	(20)	381	276		
F_1 (Barbados × Local)	(138)	247	294		
F_2 (Barbados × Local)	(39)	175	290		
F_1 (Saanen × Barbados)	(11)	374	314		
F_1 (Saanen × Local)	(14)	281	347		
F_1 (Saanen × Kilis)	(170)	710	293	40·0	(3)
3/4 Saanen, 1/4 Kilis	(178)	717	295	37·4	
Saanen	(235)	424	240	41·6	(4)
Native Korea	(55)	91	173	57·2	
F_1 (Saanen × Native Korea)	(33)	288	214	53·8	
3/4 Saanen, 1/4 Native Korea	(50)	355	236	43·0	
7/8 Saanen, 1/8 Native Korea	(20)	373	234	42·4	

Table XXIV. (cont.)

Type	*N*	Milk yield (kg)	Days[a]				Source
Native Cyprus[c]		264					
Maltese[c]		345					
Maltese[d]		785					(7)
Local Malaya[e]	(11)	95					
F_1 (A. Nubian × Malaya)	(15)	291					(6)
3/4 A. Nubian, 1/4 Malaya	(12)	236					
Saanen Mexico	(139)	508	267				
Toggenburg	(115)	498	264				
French Alpine	(230)	462	252				
Granadina	(129)	387	252				(8)
Anglo-Nubian	(111)	344	236				
Saanen Venezuela	(37)	295	277				
Toggenburg	(32)	282	283				(9)
French Alpine	(53)	232	264				
Nubian	(189)	154	224				
Saanen USA	(1 769)	979					
Toggenburg	(2 282)	921					(10)
Alpine	(3 444)	970					
Nubian	(2 624)	817					
La Mancha	(884)	835					
					Mortality of kids (%)	(*N*)	
Native of Puerto Rico	(62)	188			15·8	(66)	
F_1 (Saanen × Native PR)	(17)	285			16·6	(92)	(2)
3/4 Saanen, 1/4 Native PR	(29)	285			11·2	(50)	
Saanen	(23)	292			13·9	(38)	
				Litter size	Age 1st kidding (days)	Kidding interval (days)	
Beetal		183	189	1·7	526	308	
F_1 (Alpine × Beetal)		313	239	1·6	461	357	(5)
Alpine		315	217	1·5	542	382	

[a] Days = lactation length.
[b] FC = fat content in g/kg.
[c] Grazing only.
[d] Grazing + 2 lb concentrate.
[e] Average of first and second lactation.

(1) Sanfiorenzo (1957).
(2) Sanfiorenzo (1962).
(3) Eker *et al.* (1976).
(4) Lee *et al.* (1974).
(5) Mishra *et al.* (1976).
(6) Mahmud and Devendra (1966).
(7) Maule (1949).
(8) Montaldo *et al.* (1978).
(9) Garcia *et al.* (1972).
(10) Dickinson and King (1977).

other hand, we have very few objective data on viability and reproduction (principally the length of the breeding season, post-partum fertility at different periods of the year, variability of the litter size). Before considering the introduction of a new breed in a local population, one should estimate the rate of progress which can be expected in the local breed, with an efficient breeding programme, including the participation of the maximum number of breeders and goats, even if this is only possible with simplified recording procedures. If crossbreeding really appears to be the best solution for increasing the variation of selection traits (milk yield, length of the breeding season, litter size, growth rate or fleshiness, adult size), a decision as to whether females, males or both should be introduced must be made. In order to do so it is necessary to compare the local breed and the exotic (or specialized) breed experimentally with their two reciprocal crosses, and to estimate the genetic components (genetic, maternal and heterosis effects) of all the characters of economic importance (Dickerson, 1969):

♂♂ local breed × ♀♀ local breed
♂♂ new breed × ♀♀ new breed
♂♂ local breed × ♀♀ new breed
♂♂ new breed × ♀♀ local breed

8. Disease Resistance

(a) *Climatic Sensitivity*

There were suggestions of breed differences in resistance to different climatic conditions. Appleman and Delouche (1958) found that the limit of heat tolerance lay between 35 and 40°C. Although Saanen goats have given high yields of milk in several areas, there are reports of their sensitivity to strong sunlight and the need to protect them from the sun, keep them indoors and manage them very carefully (Maule, 1966). Observations of Bianca and Kunz (1978) show that the Toggenburg appeared to be relatively cold tolerant (−5°C) owing to the high thermal insulation of their hair coat, and also heat tolerant (40°C), possibly due to their low normal heart rate (87 beats/min compared with 94 in the Saanen and Alpine Chamoisée). In South Africa, severe losses among Angora does during periodic cold climate spells are a serious limiting factor in the mohair industry. Physiological and endocrinological reactions to cold stress (0° to 5°C, wetting at 30 min intervals) were investigated by Wentzell *et al.* (1979): pregnant mature animals showed a better resistance as compared to young barren animals; goats which survived the 48 h experimental period maintained a relatively high heart rate (140 to 160 beats/min), normal body temperature and relatively high plasma thyroxine concentration. In the Damascus breed, genotype environment interactions for post-weaning growth is not of importance and should not constitute a handicap to selection (Louca and Hancock, 1977).

The black bedouin goat seems to have a genetically determined ability to withstand prolonged water deprivation (see p. 524).

(b) *Resistance to Parasitism*

Bucks were selected for resistance to stomach worms (*Haemonchus contortus*). The survivors were bred with untested females, and the survival rate of the progeny was very high, 71 *vs* 33% for controls. Untested daughters of resistant males bred to resistant bucks gave progeny with 83 survival *vs* 31% for controls. This result supposes the existence of polygenes with possibly true dominance for resistance (Warwick *et al.*, 1949).

Dwarf goat breeds indigenous in tse-tse areas are innately more resistant to trypanosomiasis than exotic breeds. In Kenya, indigenous East African and Galla goats were tolerant to experimental infection with *Trypanosoma congolense* while Saanen goats suffered severely and had a high mortality rate; tolerance was intermediate in Saanen × Galla crossbreeds (Griffin and Allonby, 1979).

References

Adu, I. F., Buvanendran, V. and Lakpini, C. A. M. (1979). *J. Agric. Sci.* **93**, 563–566.
Agar, N. S., Gruca, M. and Harley, J. D. (1974). *Anim. Blood Groups Biochem. Genet.* **5**, 63–64.
Ali, S. Z., Hoque, M. M. and Hasnath, M. A. (1973). *Indian Vet. J.* **50**, 438–440.
Amble, V. N., Khandekar, N. C. and Garg, J. N. (1964). *I.C.A.R. Res. Ser.* **38**, New Delhi.
Appleman, R. D. and Delouche, J. C. (1958). *J. Anim. Sci.* **17**, 326–335.
Arriola, G. G. (1936). *Philippine Agric.* **25**, 11–29.
Aschaffenburg, R. and Dance, J. D. (1968). *J. Dairy Res.* **35**, 383–384.
Asdell, S. A. (1929). *J. Agric. Sci.* **19**, 382–396.
Asdell, S. A. (1944). *Science* **99**, 124.
Asdell, S. A. and Crew, F. A. E. (1925). *J. Genet.* **15**, 367–374.
Asdell, S. A. and Smith, A. D. (1928). *J. Hered.* **19**, 425–430.
Bakke, H., Steine, T. and Eggun, A. (1977). *Acta Agric. Scand.* **27**, 245–249.
Basrur, P. K. and Coubrough, R. I. (1964). *Cytogenetics* **3**, 414–426.
Basrur, P. K. and Kanagawa, H. (1969). *Ann. Genet. Sel. Anim.* **1**, 349–378.
Basrur, P. K. and Stoltz, D. R. (1967). *J. Hered.* **58**, 261–262.
Bath, G. F., Wentzell, D. and Van Tonder, E. M. (1978). *The Angora* **21**, 63–65.
Bazargani, T. T. (1975). *J. Vet. Faculty, Univ. Teheran* **31**, 82–86.
Belinchon, P. and Marques, F. (1971). *In* "2nd Int. Conf. Goat Breeding", Tours, France, pp. 339–343; ITOVIC, Paris.

Berge, S. (1966). *Meld. Norges Landbruk* **45**, 1–24.
Bianca, W. and Kunz, P. (1978). *Livestock Prod. Sci.* **5**, 57–69.
Bilgemre, K. and Düzgünes, O. (1952). *Z. Tierzücht. ZüchtBiol.* **60**, 282–284.
Blanks, G. A. (1971). Ph.D. Thesis, Oregon State University.
Blanks, G. A. and Bogart, R. (1974). *J. Anim. Sci.* **38**, 1319 (Abstr).
Boulanger, Anne (1977). Thèse Doctorat, University of Paris.
Bouillon, J. (1971). *In* "2nd Int. Conf. Goat Breeding" Tours, France, pp. 319–321. ITOVIC, Paris.
Bouillon, J. (1975). *In* "Ières Journées de la Recherche Ovine et Caprine", pp. 133–136. INRA, ITOVIC, Paris.
Bouillon, J. and Ricordeau, G. (1975a). *In* "Ières Journées de la Recherche Ovine et Caprine", pp. 124–132. INRA, ITOVIC, Paris.
Bouillon, J. and Ricordeau, G. (1975b). *In* "Ières Journées de la Recherche Ovine et Caprine", pp. 149–155. INRA, ITOVIC, Paris.
Boyajean, D. (1969). Bibliographie signalétique. *Ann. Génét. Sél. Anim.* **1**, 447–463.
Boyazoglu, J. G., Meyer, E. and De Boom, H. P. A. (1969). *Ann. Génét. Sél. Anim.* **1**, 180–181.
Boyazoglu, J. G. and Zervas, N. (1977). *L'Elevage* **67**, 66–75.
Brandsch, H. (1959). *Arch. Geflügelz. Kleintierk.* **8**, 310–362.
Bratanov, K. and Dikov, V. (1961). *Proc. IVth Int. Cong. Anim. Reprod., The Hague*, **IV**, 744–751.
Bratanov, K., Dikov, V. and Somlev, B. (1969). *Genet. Selekts. Sofia* **2**, 363–368.
Bratanov, K. T., Somlev, B. P. and Tsoncheva, V. E. (1972). *C.R. Acad. Sci. Bulgare* **25** (8), 1109–1111.
Breukink, H. J., Hart, H. C., Arkel, C. V., Velden, N. A. and Watering, C. C. (1972). *Zentralbl. Veterinärmed.* **19**, 661–676.
Buechi, H. F. (1957). *Z. Tierzücht. ZüchtBiol.* **69**, 30–90.
Bunch, T. D., Foote, W. C. and Spillett, J. J. (1976). *Theriogenology*, **6**, 379–385.
Bunch, T. D., Rogers, A. and Foote, W. C. (1977). *J. Hered.* **68**, 210–212.
Bunch, T. D., Nadler, C. F. and Simmons, L. (1978). *J. Hered.* **69**, 316–320.
Cabannes, R. and Serain, C. (1955). *C.R. Soc. Biol.* **149**, 1193.
Carrera, C. and Hernández, E. (1971). *XII°* Infme. Invest. Esc. Agric. Ganad. Inst. Technol. Estud. Sup. Monterey, 1969–1970. 162.
Carta, R., Casu, S., Flamant, J. C. and Ricordeau, G. (1975). *Ann. Génét. Sél. Anim.* 7, 293–298.
Cattanach, B. M., Pollard, C. E. and Hawkes, S. G. (1971). *Cytogenetics* **10**, 318–337.
Chevalet, C. and De Rochambeau, H. (1979). *In* "V Journées de la Recherche Ovine et Caprine", pp. 282–292, INRA, ITOVIC, Paris.
Ciolca, N., Stafanesco, C., Stoianovici, V., Raico, L. and Schneider, A. (1959). *Lucrarile Sti. Inst. Crecetari Zootech.* **17**, 407–423.
Cochran, W. G. (1951). *Proc. II Berkeley Symp. Math. Stat. Prob.* 445–470.
Cognie, Y., Houix, Y. and Logeay, B. (1971). *In* "2nd Int. Conf. Goat Breeding", Tours, France, pp. 345–350. ITOVIC, Paris.
Cole, L. J. and Ibsen, H. (1920). *Am. Nat.* **54**, 130–151.
Cole, L. J. and Steele, D. (1922). *J. Hered.* **13**, 290–294.
Corteel, J. M. (1966). *Bull. Tech. Inf.* **210**, 463–471.
Corteel, J. M. (1976). *Ann. Zootech.* **25**, 567–571.
Corteel, J. M. and Montigny, G. (1975). *In* "Ières Journées de la Recherche Ovine et Caprine", pp. 56–63. INRA, ITOVIC, Paris.

Corteel, J. M., Hulot, F., Courot, M., Attal, J. and Philippon, A. (1969). *Ann. Génét. Sél. Anim.* **1**, 341–348.

Crottaz, M. (1975). Thèse Vét. Bern.

Cumlivski, B. (1976). *27th Ann. Meet. Europ. Ass. Anim. Prod. Zurich*, 1–5.

Cumlivski, B. (1978) *In* "2 ème Symposium sur la Traite Méchanique des Petits Ruminants", Alghero, Italy, pp. 164–167. ITOVIC, Paris.

Cunningham, O. C. (1932). *New Mexico. St. Rept* **43**.

Dam, R. H., Werkhoven, C. B., Donk, J. A. and Goudswaard, J. (1976). *J. Immunogenet.* **3**, 237–244.

Dam, R. H., Amaro, J., Kooten, P. J. S., Donk, J. A. and Goudswaard, J. (1979). *Anim. Blood Groups Biochem. Génét.* **10**, 121–124.

Darracq, J., Le Mens, P. and Perrot, C. (1968). *In* "2ème Symposium sur la Traite Mécanique des Petits Ruminants", Alghero, Italy, pp. 324–345 ITOVIC, Paris.

Delaitre, C. (1965). Cited by Devendra and Burns, 1970.

Devendra, C. (1962). *Malay. Agric. J.* **43**, 265–280.

Devendra, C. (1975). *Z. Tierzücht. ZüchtsBiol.* **92**, 305–317.

Devendra, C. and Burns, M. (1970). "Goat Production in the Tropics", Technical Communication No. 19, CBABG. Commonwealth Agricultural Bureaux, Farnham Royal, Bucks, England.

Devendra, C. and Chenost, M. (1973). *Z. Tierzücht. ZüchtsBiol.* **90**, 83–93.

Devendra, C. and Nozawa, K. (1976). *Z. Tierzücht. ZüchtsBiol.* **93**, 101–120.

Dickerson, G. (1969). *Anim. Breed. Abstr.* **37**, 191–202.

Dickinson, F. N. and King, G. (1977). *Ann. Meet. Am. Dairy. Sci. Ass.* Iowa.

Divekar, K. V. (1953). *Indian Vet. J.* **29**, 538–540.

Eaton, O. N. (1943). *Am. J. Vet. Res.* **4**, 333–343.

Eaton, O. N. (1945). *Genetics* **30**, 51–61.

Eaton, O. N. and Simmons, V. L. (1939). *J. Hered.* **30**, 261–266.

Eker, M., Tuncel, E., Askin, Y. and Yener, M. (1976). *Ank. Univ. Zir. Fak. Yilligi.* **26**, 451–462.

Efremov, G. and Braend, M. (1965). *Proc. 9th Eur. Conf. Anim. Blood Groups*, Prague, 1964.

Engelbach, W. (1932). "Endocrine Medicine". Baltimore.

Enyenihi, U. K. (1974). *Res. Vet. Sci.* **17**, 300–303.

Eppleston, J. and Moore, N. W. (1977). *Theriogenology* **8**, 165.

Epstein, H. (1946). *J. Hered.* **11**, 345–352.

Epstein, H. (1951). *J. Hered.* **62**, 81–84.

Epstein, H. (1953). *East Afric. Agric. J.* **18**, 123–132.

Epstein, H. (1969). "Domestic Animals of China". Commonwealth Agricultural Bureaux, Farnham Royal, Bucks, England.

Epstein, H. and Herz, A. (1964). *J. Agric. Sci.* **62**, 237–244.

Evans, H. J. (1965). *Exp. Cell. Res.* **38**, 511–516.

Evans, H. J., Buckland, R. A. and Summer, A. T. (1973). *Chromosoma* **42**, 382–402.

Fechter, H. and Pretorius, G. (1970). *Anim. Blood Groups Biochem. Genet.* **1**, 63.

Fehr, P. M. and Sauvant, D. (1975). *In* "Ières Journées de la Recherche Ovine et Caprine", pp. 73–84. INRA, ITOVIC, Paris.

Fisher, H. (1955). *Hemera Zoa* **62**, 17–27.

Forde, C. (1934). Cited by Epstein (1953).

Franck, M., Froget, J. and Nain, M. C. (1977). *Ann. Génét. Sél. Anim.* **9**, 535 (Abstr).

Gaillard, C. and Kropf, W. (1978). *Schr. Schweiz. Verein. Tierz.* **54**, 20–31.

Gall, C. (1971). *In* "2nd Int. Conf. Goat Breeding", Tours, France, pp. 287–294. ITOVIC, Paris.
Gall, C., Frahm, K., Graf, F., Meyer, J. and Osterkorn, K. (1967). *In* "Arbeit. Inst. Tierzucht. Vererb. Konst. For. Univ. München" **7**, 38–50.
Garcia, O., Castillo, J. and Gado, C., (1972). *Agron. Tropical* **22**, 239.
Garrick, M. D. and Charlton, J. P. (1969). *Biochem. Genet.* **3**, 392–402.
Garzon, G. E. R., Zarazaga, B. I., Vallejo, V. M. and Rodero, F. A. (1976). *Arch. Zootech.* **25**, 147–170.
Geerts, H. A. (1975). *Dairy Goat. J.* **53**, 9.
Goel, K. C. and Nair, P. G. (1976). "Proc. 2nd Workshop all India Coord. Res. Project Goat Breeding". NDRI, Karnal. Central Sheep and Wool Res. Inst., Avikanagar, India.
Gonzalez, S. C. (1977). "Symposium on Goat Breeding in Mediterranean Countries". Malaga, F.E.Z., pp. 315–321.
Gonzalez, S. C., Garcia, O. and Castillo, O. (1974). *Cienc. Veter. Maracaïbo* **4**, 223–248.
Gopinathan, N. and Nair, P. G. (1976). "Proc. 2nd Workshop All India Coord. Res. Proj. Goat Breeding", NDRI, Karnal. Central Sheep and Wool Res. Inst., Avikangar, India.
Grappin, R. and Jeunet, R. (1979). *Le Lait* **587**, 345–360.
Griffin, L. and Allonby, E. W. (1979). *Vet. Parasit.* **5**, 97–105.
Gronbeck, H. (1955). *Arch. Geflüg. Kleintierk.* **4**, 265–404.
Grouchy, J. de, Lauvergne, J. J. and Ricordeau, G. (1965). *C.R. Acad. Sci.* **260**, 2932–2935.
Gupta, S., Som, T. L. and Bhattacharya. (1964). *Indian Vet. J.* **41**, 668–672.
Gyger, E. (1925). Thèse Vet, 21 pp. Berne.
Hamerton, J. L. (1968). *Nature, Lond.* **219**, 910–913.
Hamerton, J. L., Dickson, J. M., Pollard, C. E., Grieves, S. A. and Short, R. V. (1969). *J. Reprod. Fert.* (Suppl. 7), 25–71.
Hancock, J. L. (1965). Agric. Res. Council, A.B.R.O. Report, Jan. 6–11.
Hancock, J. and Louca, A. (1975). *Anim. Prod.* **21**, 227–231.
Hansen, K. M. (1973). *Hereditas* **75**, 119–130.
Haugen, E. (1960). *Meld. Norges Landbruk.* **39**, 1–33.
Haumesser, J. B. (1975). *Rev. Elev. Méd. Vét. Pays Trop.* **28**, 225–234.
Hayden, T. J., Thomas, C. R. and Forsyth, I. A. (1979). *J. Dairy Sci.* **62**, 53–57.
Heerden, K. M. (1961). *Proc. IVth Int. Congr. Anim. Prod. Den Haag* **3**, 586–589.
Heerden, K. M. (1964). *J.S. Afr. Vet. Med. Ass.* **35**, 19–20.
Henderson, L. M. and Bruere, A. N. (1979). *Can. J. Genet. Cytol.* **21**, 1–8.
Hinterthur, E. (1933). *Züchtungsk.* **8**, 55–62.
Hofmeyr, H. S. (1962). M. Sc. (Agr.) Thesis, University of Pretoria.
Hofmeyr, H. S., Joubert, D. M., Badenhorst, F. J. G. and Steyn, G. J. (1965). *Proc. S. Afr. Soc. Anim. Prod.* **4**, 191–195.
Horák, F. and Kasing, J. (1970). *Zivoc. Vyroba* **15**, 205–216.
Hoversland, A. S., Parer, J. T. and Metcalfe, J. (1965). *Fed. Proceed.* **24**, 705.
Hulot, F. (1969). *Ann. Génét. Sél. Anim.* **1**, 175 (Abstr).
Hulot, F. and Basrur, P. K. (1969). *Ann. Génét. Sél. Anim.* **1**, 383–390.
Hulot, F. and Lauvergne, J. J. (1967). *Ann. Génét.* **10**, 86–97.
Ikin, T. L. and Herron, N. D. (1978). *Proc. Aust. Soc. Anim. Prod.* **12**, 200.
Ilbery, P. L. T. and Williams, D. (1967). *Cytogenetics* **6**, 276–286.
Ilbery, P. L. T., Alexander, G. and Williams, D. (1967). *Austr. J. Biol. Sci.* **20**, 1245–1247.

Jardin, W. R., Silocira Filho, S., Peixoto, A. M. and Spallini, A. (1965). *Rev Agricultura* **40** (3), 131–147.
Joshi, S. C., Rawat, J. S. and Pandey, M. D. (1975). *Current Sci.* **44**, 673.
Joubert, D. M. (1973). *Z. Tierzücht. ZüchtBiol.* **90**, 245–262.
Kaikini, A. S. and Puranik, J. K. (1964). *Indian Vet. J.* **41**, 815–819.
Keeping, G. S. (1951). *Malay. Agric. J.* **34**, 32–39.
Kislovsky, D. (1937). *J. Hered.* **28**, 264–266.
Kobozieff, N. (1943). *In* "Précis de Génétique Appliquée à la Médecine Vétérinare". Vigot.
Kumar, S., Singh, G. and Kumar, A. N. (1962). *Indian J. Dairy Sci.* **15**, 101–104.
Kunz, H. (1974). Diss. Med. Bern. 56 pp.
Kuzdzal-Savoie, S. and Kuzdzal, W. (1959). *Ann. Technol.* **2**, 131–144.
Lamont, J. L. (1964). *S. Afric. J. Afric. Sci.* **7**, 305–310.
Laor, M. (1971). Unpublished results.
Lauvergne, J. J. (1969). *Ann. Génét. Sél. Anim.* **1**, 403–412.
Lauvergne, J. J. (1977). *Xth Int. Pigment Cell. Conf.* Boston, October (Abstr. 69).
Lauvergne, J. J. (1978). *Ann. Génét. Sél. Anim.* **10**, 181–189.
Lee, K. W., Chai, K. S., Tak, T. Y. and Sul, D. S. (1974). Res. Rep. Office Rural Dev., Suweon Livestock, 16, 7–14.
Le Mens, P., Toussaint, G. and Le Jaouen, J. C. (1978a). *In* "2ème Symposium sur la Traite Mécanique des Petits Ruminants", Alghero, Italy, ITOVIC, Paris.
Le Mens, P., Lequenne, D. and Toussaint, G. (1978b). *In* "Zème Symposium sur la Traite Mécanique des Petit Ruminants", Alghero, Italy, pp. 253–262. ITOVIC, Paris.
Lode, T. (1970). *Acta Vet. Scand.* **11**, 181–185.
Loertscher, (1959). *Arch. Klaus Stift. Vererb. Forsch.* **33**.
Louca, A. and Hancock, J. (1977). *J. Anim. Sci.* **44**, 927–931.
Lush, J. L. (1926). *J. Hered.* **17**, 73–91.
Lush, J. L. (1930). *J. Hered.* **21**, 242–247.
Lush, J. L., Jones, J. M. and Dameron, W. H. (1930). *Texas Agric. Exp. St. Bull.* **407**.
Lyngset, O. (1964). *Nord. Veterinaermedicin.* **16**, 833–845.
Macgillivray, R. T. A., Brew, K. and Barnes, K. (1979). *Arch. Biochem. Biophys.* **197**, 404–414.
Machens, A. (1914). *Tierärtzl. Rund.* **20**, 423.
Mahmud, A. B. and Devendra, C. (1966). *Exp. Agric.* **2**, 211–216.
Maule, J. P. (1949). *Emp. J. Exp. Agric.* **17**, 37–47.
Maule, J. P. (1966). *Anim. Breed. Abstr.* **34**, 153–158.
McGovern, P. T. (1969). *Anim. Breed. Abstr.* **37**, 1–10.
Mehren, M. J. (1973). *Diss. Abstr.* 3401–B.
Millson, G. C. and Pattison, J. H. (1961). *Vet. Rec.* **73**, 256.
Mishra, H. R. and Biswas, S. C. (1966). *Indian J. Dairy Sci.* **19**, 132–134.
Mishra, R. R., Bhatnagar, D. S. and Sundaresan, D. (1976). *Indian J. Dairy Sci.* **29**, 235–237.
Mocquot, J. C. (1978). *In* "2éme Symposium sur la Traite Mécanique des Petits Ruminants", Alghero, Italy, pp. 175–201. ITOVIC, Paris
Mocquot, J. C. (1978). *Ann. Génét. Sél. Anim.* **10**, 359–376.
Mocquot, J. C. and Auran, T. (1974). *Ann. Génét. Sél. Anim.* **6**, 463–476.
Mocquot, J. C. and Guillimin, P. (1975). *In* "Ières Journées de la Recherche Ovine et Caprine", p. 10. INRA, ITOVIC, Paris.
Mocquot, J. C., Flamant, J. C. and Poutous, M. (1970). *Ann. Génét. Sél. Anim.* **2**, 53–63.

Montaldo, H., Juárez, A., Forat, M., Berruecos, J. M. and Villarreal, M. (1978). *Abstr. 70th Ann. Meet. ASAS, Mich. Stat. Univ.*
Moore, N. W. and Eppleston, J. (1976). *Proc. 4th. Int. Symp. Reprod. Evol.*, Canberra, Aust. Acad. Sci.
Moulick, S. K. and Syrstad, O. (1970). *J. Agric. Sci.* **74**, 409–414.
Moulick, S. K., Guha, H., Gupta, S., Mitra, D. K. and Bhattacharya, S. (1966). *Indian J. Vet. Sci.* **36**, 154–163.
Naito, M., Shoda, Y., Kobayashi, H. and Fukushima, Y. (1965). *Jap. J. Zootech. Sci.* **36**, 170–176.
Nel, J. E. (1975). *Agroanimalia* **7**, 77–80.
Nguyen, T. C. (1979). Thèse doctorat. Etat, Paris.
Nguyen, T. C., Boulanger, A. and Raynaud, C. (1975). *In* "Iéres Journées de la Recherche Ovine et Caprine", pp. 109–118. INRA, ITOVIC, Paris.
Nozawa, K. (1970). *Jap. J. Genet.* **45**, 45–57.
Nozawa, K., Shinjo, A. and Stotake, T. (1978). *Z. Tierzücht. ZüchtsBiol.* **95**, 60–77.
Ohno, S. (1979). "Monographs on Endocrinology". Springer, Berlin.
Ormiston, E. E. and Touchberry, R. W. (1959). *J. Dairy Sci.* **42**, 924.
Osterhoff, D. R. and Ward-Cox, I. S. (1972). *Proc. 12th Eur. Conf. Anim. Blood Groups Biochem. Polymorphism*, Budapest, 1972, pp. 579–582.
Ott, R. S., Nelson, D. R., Lock, T. F. and Hixon, J. E. (1979). *J. Anim. Sci.* **49** (Suppl.), 449 (Abstr.).
Padeh, B., Wysoki, M., Ayalon, N. and Soller, M. (1965). *Israel J. Med. Sci.* **1**, 1008–1212.
Padeh, B., Wysoki, M. and Soller, M. (1971). *Cytogenetics* **10**, 61–69.
Paredes, J. R. (1952). *Arch. Zootechnia, Cordoba* **1**, 24–39.
Parer, J. T. (1963). *Am. J. Vet. Res.* **24**, 1223–1226.
Park, C. S., Lee, K. W. and Sul, D. S. (1977). *Res. Rep. Rural Rev. Livestock* **19**, 49–52.
Peaker, M. (1978). *Br. Vet J.* **134**, 379–383.
Phillips, N. and Jenness, R. (1965). *Biochem. Biophys. Res. Commun.* **21**, 16–20.
Phillips, R. W., Simmons, V. S. and Schott, R. G. (1943). *Am. J. Vet. Res.* **4**, 360–367.
Pierre, A. and Portmann, A. (1970). *Ann. Technol. Agric.* **19**, 107–130, 177–179.
Popescu, C. P. (1972). *Ann. Génét. Sél. Anim.* **4**, 355–361.
Portmann, A. and Pierre, A. (1971). *Ann. Fals. Exp. Chim.* **696**, 159–176.
Portmann, A., Pierre, A. and Vedrenne, P. (1968). *Rev. Laitière Fr.* **251**, 97–101.
Poutous, M., Briend, M., Calomiti, S., Doan, D., Felgines, C. and Steier, G. (1980). Polycopié INRA Départ. Génétique. Jouy en Josas.
Prasad, S. K., Pandey, R. S. and Nair, P. G. (1978). *Indian J. Exp. Biol.* **16**, 1193–1194.
Prakesh, C., Acharya, R. M. and Dhillon, J. S. (1971). *Indian J. Anim. Sci.* **41**, 356–360.
Preaux, G. (1979). *Hoppe Seyler's Z. Physiol. Chem.* **360**, 1595–1604.
Pretorius, P. S. (1973). *Agroanimalia* **5**, 55–58.
Pretorius, A. M. G., Schmid, D. O., Osterhoff, D. R. and Albert, E. D. (1976). *J. Immunogenet.* **3**, 291–296.
Prins, H. K. and Loos, J. A. (1969). *In* "Biochemical Methods in Red Cell Genetics", pp. 115–137. Academic Press, New York and London.
Quartermain, A. R. (1975). *Proc. III World Conf. Anim. Prod*, 102–106.
Rae, A. L. (1977). *Proc. N.Z. Soc. Anim. Prod.* **37**, 206–212.
Ragan, H. A., Hortsman, V. G., McClellan, R. O. and Bustad, L. K. (1966). *Am. J. Vet. Res.* **27**, 161–165.

Razungles, J. (1974). *Ann. Génét. Sél. Anim.* **6**, 415–426.
Rensburg, S. J. (1964). *Proc. Vth Int. Conf. Anim. Reprod., Trento* **3**, 375–378.
Rensburg, S. J. (1971a). *J.S. Afr. Vet. Med. Ass.* **42**, 305–308.
Rensburg, S. J. (1971b). *Onderstepoort J. Vet. Res.* **38**, 1–62.
Ricordeau, G. (1963). *Bull. Tech. Inf.* **179**, 189–204.
Ricordeau, G. (1967). *Ann. Zootech.* **16**, 263–270.
Ricordeau, G. (1969). *Ann. Génét. Sél. Anim.* **1**, 391–395.
Ricordeau, G. (1972a). *Ann. Génét. Sél. Anim.* **4**, 469–475.
Ricordeau, G. (1972b). *Ann. Génét. Sél. Anim.* **4**, 593–598.
Ricordeau, G. (1973). Annual Report (Polycopié INRA).
Ricordeau, G. and Bouillon, J. (1967). *Bull Tech. Inf.* **218**, 303–310.
Ricordeau, G. and Bouillon, J. (1969). *Ann. Génét. Sél. Anim.* **1**, 397–401.
Ricordeau, G. and Bouillon, J. (1971a). *In* "2nd Int. Conf. Goat Breeding", Tours, France, pp. 271–274. ITOVIC, Paris.
Ricordeau, G. and Bouillon, J. (1971b). *In* "2nd Int. Conf. Goat Breeding", Tours, France, pp. 275–277. ITOVIC, Paris.
Ricordeau, G. and Bouillon, J. (1971c). *In* "2nd Int. Conf. Goat Breeding", Tours, France, pp. 283–286. ITOVIC, Paris.
Ricordeau, G. and Bouillon, J. (1971d). *In* "2nd Int. Conf. Goat Breeding", Tours, France, pp. 351–353. ITOVIC, Paris.
Ricordeau, G. and Bouillon, J. (1975). *In* "Ières Journées de la Recherche Ovine et Caprine", pp. 48–55. INRA, ITOVIC, Paris.
Ricordeau, G. and Bouillon, J. (1978). Mémoire A. Pittet, Sept. 78.
Ricordeau, G. and Grosclaude, F. (1970). Annual Report, INRA.
Ricordeau, G. and Labussière, J. (1970). *Ann. Zootech.* **19**, 37–43.
Ricordeau, G. and Lauvergne, J. J. (1967). *Ann. Zootech.* **16**, 323–334.
Ricordeau, G. and Lauvergne, J. J. (1971). *Ann. Génét. Sél. Anim.* **3**, 425–432.
Ricordeau, G. and Mocquot, G. (1967). *Ann. Zootech.* **16**, 155–181.
Ricordeau, G. and Sigwald, J. P. (1977). *In* "Symposium on Goat Breeding in Mediterranean Countries", pp. 70–79. F.E.Z. Malaga.
Ricordeau, G., Bouillon, J. and Hulot, F. (1972a). *Ann. Génét. Sél. Anim.* **4**, 537–542.
Ricordeau, G., Poujardieu, B. and Bouillon, J. (1972b). *Ann. Génét. Sél. Anim.* **4**, 29–34.
Ricordeau, G., Bouillon, J., Sanchez, F., Mocquot, J. C. and Lajous, A. (1979). *In* "V Journées de la Recherche Ovine et Caprine", pp. 403–426. INRA, ITOVIC, Paris.
Rønningen, K. (1964). *Meld. Norges, Landbruk.* **43**, (18) 19p.
Rønningen, K. (1965a). *Acta Agric. Scand.* **15**, 301–342.
Rønningen, K. (1965b). *Meld. Norges Landbruk.* **44**, 19p.
Rønningen, K. (1967). *Meld. Norges Landbruk.* **46**, 17p.
Rønningen, K. (1974). *Swedish J. Agric. Res.* **4**, 61–69.
Sahni, K. L. and Roy, A. (1967). *Indian J. Vet. Sci. Anim. Husb.* **37**, 269–276.
Salerno, A., Montemurro, N. and L'Afflitto, A. (1970). *Proc. XIth Europ. Conf. Anim. Blood Groups Biochem. Polymorphism*, 517–520. Warsaw, 1968.
Sanfiorenzo, J. H. (1957). *Agric. Exp. St. Bull.* **139**.
Sanfiorenzo, J. H. (1962). *J. Agric. Univ. Puerto Rico,* **46**, 205–212.
Santiago, A. A. (1946). *Anim. Breed. Abstr.* **15**, 259.
Saraja Ortiz (1955). *In* "Enciclopedia de la Leche". Agenjo Cecilia C., Espagne, Calpe, S.A.
Sauvant, D., Fehr, P. M., Rodolphe, F., Tomassone, R. and Delage, J. (1973). *Ann. Biol. Anim. Bioch. Biophys.* **13**, 107–129.

Schnell, W. and Czaker, R. (1974). *Cytogenet. Cell. Genet.* **13**, 246–255.
Schmid, D. O., Odermatt, K. and Kunz, H. (1975). *Schweiz. Arch. Tierheilk.* **117**, 31–43.
Schmid, D. O. and Suzuki, S. (1972). *Zbl. Vet. Med. B* **18**, 604–609.
Schmidt, D. V. and Ebner, M. E. (1972). *Biochim. Biophys. Acta.* **263**, 714–720.
Schönherr, S. (1954). Vet. Med. Diss. Freie Universität, Berlin.
Searle, S. R. (1964). *J. Dairy Sci.* **47**, 414–420.
Searle, A. G. (1968). *In* "Comparative Genetics of Coat Colour in Mammals". Logos Press.
Sen, A. and Chaudhuri, S. (1962). *Nature* **195**, 286–287.
Sengar, O. P. S. (1976). Int. Conf. Winrock, USA, 1–22 (polycopy).
Sengar, O. P. S. (1976). *In* "The Role of Sheep and Goats in Agricultural Development", pp. 26–28. Winrock International Livestock Research and Training Center, Morrilton, Arkansas, USA.
Sharav, E. and Carasso, J. (1978). *In* "2ème Symp. Int. sur la Traite Mécanique des Petits Ruminants", Alghero, Italy, ITOVIC, Paris.
Shelton, M. (1960a). *J. Anim. Sci.* **19**, 302–308.
Shelton, M. (1960b). *J. Anim. Sci.* **19**, 368–375.
Shelton, M. (1961). *Texas Agric. St.* PR 2189.
Shelton, M. (1978). *J. Dairy Sci.* **61**, 994–1010.
Shelton, M. and Bassett, J. W. (1970). *Texas Agric. Exp. Stat.* PR2570, 38–41.
Shelton, M. and Groff, J. L. (1974). *Texas Agric. Exp. Stat. Bull.* **1136**.
Shelton, M. and Morrow, T. (1965). *Texas Agric. Exp. Stat.* PR2340.
Shelton, M. and Spiller, D. (1977). *Texas Agric. Exp. Stat.* PR3445.
Shelton, M. and Stewart, J. R. (1973). *Texas Agric. Exp. Stat. Bull.* **3187**, 28–31.
Short, R. V. (1971). "Proc. Int. Symposium on the Genetics of the Spermatozoon", Edinburgh, pp. 16–20, 325–345.
Short, R. V. (1979). *Br. Med. Bull.* **35**, 121–127.
Shotake, T. (1971). *Jap. J. Zootech. Sci.* **42**, 409–415.
Sigwald, J. P., Bouillon, J., Dusartel, S. and Ricordeau, G. (1979). Polycop., INRA, ITOVIC.
Sincer, N. (1963). *Lalahan Zootek. Arast. Enst. Derg. Ankara* **3**, 26–39.
Sincer, N. (1967). *Lalahan Zootek. Arast. Enst. Derg. Ankara* **7**, 32–37.
Singh, R. N., Acharya, R. M. and Biswas, D. K. (1970). *Acta Agric. Scand.* **20**, 10–14.
Sionneau, G. (1964). *Thèse Dr. Vet. Alfort.* 58 pp.
Skinner, J. D. (1971). *Meat Industry* **20**, 25–28.
Skinner, J. D., Rensburg, S. J. and Badenhorst, J. F. G. (1971). *S. Afr. J. Anim. Sci.* **1**, 69–71.
Skinner, J. D., Heerden, H. A. H. and Goris, E. J. (1972). *S. Afr. J. Anim. Sci.* **2**, 93–95.
Skjevdal, T. (1975). *Sau og Geit* **28**, 72.
Skjevdal, T. (1979). *Liv. Prod. Sci.* **6**, 397–405.
Slater, A. E. and Bhatia, S. S. (1935). *Allahabad Farmer* **9**, 156–159.
Sohrab, M., McGovern, P. T. and Hancock, J. L. (1973). *Res. Ind. Vet. Sci.* **15**, 77–81.
Soller, M., Laor, M., Barnea, R., Weiss, Y. and Ayalon, N. (1963). *J. Hered.* **54**, 237–240.
Soller, M. and Angel, H. (1964). *J. Hered.* **55**, 139–142.
Soller, M. and Kempenich, O. (1964). *J. Hered.* **55**, 301–304.
Soller, M., Wysoki, M. and Padeh, B. (1966). *Cytogenetics* **5**, 88–93.
Soller, M., Padeh, B., Wysoki, M. and Ayalon, N. (1969). *Cytogenetics* **8**, 51–67.

Steine, T. S. (1975a). *Meld. Norges Landbruk* **369**, 30p.
Steine, T. S. (1975b). *Meld. Norges Landbruk* **385**, 25p.
Steine, T. S. (1976). *Meld. Norges Landbruk* **398**, 19p.
Steine, T. S. (1978). 29th Ann. Meet. Europ. Ass. Prod. Stockholm.
Stünzi, H., Büchi, H. F., Le Roy, R. L. and Leemann, W. (1964). *Schweiz. Arch. Tierheilk.* **106** (12), 778–788.
Sulimovici, S., Weissenberg, R., Lunenfeld, B., Padeh, B. and Soller, M. (1978). *Clin. Genet.* **13**, 397–404.
Surgie, T. and Soma, T. (1970). *Bull. Na. Just. Anim. Industry* **23**.
Suzuki, S., Watanabe, S. and Iwase, N. (1960). *Jap. J. Zootech. Sci.* **31**, 144–147.
Suzuki, S., Hayashida, S., Yamouchi, C., Nozawa, K., Tanaka, K. Watanabe, S., Nishinakagawa, H. and Shotake, T. (1967). *Jap. J. Zootech. Sci.* **38**, 443–452.
Tantawy, A. C. and Ahmed, I. A. (1960). *Emp. J. Exp. Agric.* **28**, 72–82.
Tjankov, S. (1972). *XIIth Eur. Conf. Anim. Blood Groups Bioch. Polymorphism,* 575–578.
Tsunoda, K., Watanabe, S. and Suzuki, S. (1976). *Jap. J. Zootech. Sci.* **47**, 665–671.
Tucker, E. M. and Young, J. D. (1976). *Anim. Blood Groups Biochem. Genet.* **7**, 109–117.
Tuncel, E. (1976). *27th Ann. Meet. Eur. Ass. Anim. Prod.* Zurich. 1–5.
Tuncel, E., Elicin, A. and Riza, R. (1976). *Ank. Univ. Zir. Fak. Yilligi* **26**, 93–106.
Turner, C. W. and Berousek, E. R. (1942). *J. Dairy Sci.* **25**, 549–555.
Wachtel, S. S., Basrur, P. and Koo, G. C. (1978). *Cell* **15**, 279–281.
Walser, K. (1963). *Berlin. München. Tierärztl. Wschr.* **76**, 206–208.
Warwick, B. L. (1961). *J. Anim. Sci.* **20**, 10–14.
Warwick, B. L., Berry, R. O., Turk, R. D. and Morgan, C. O. (1949). *J. Anim. Sci.* **8**, 609–610 (Abstr).
Wassin, B. (1928). *Z. Indukt. Abst. Vererbungsl.* **49**, 95–104.
Watanabe, S. and Suzuki, S. (1967). *Jap. J. Zootech. Sci.* **38**, 487–494.
Watanabe, S. and Suzuki, S. (1973). *Anim. Blood Groups Biochem. Genet.* **4**, 23–26.
Watanabe, S., Kanemaki, M. and Suzuki, S. (1965). *Jap. J. Zootech. Sci.* **36**, 293–299.
Weber, W. (1969). *Ann. Génét. Sél. Anim.* **1**, 379–382.
Weber, W. (1977). *Schweiz. Arch. Tierkeilk* **119**, 201–204.
Wentzell, D., Morgenthal, J. C., Niekerk, C. H. and Roelofse, C. S. (1974). *Agroanimalia* **6**, 129–132.
Wentzell, D., Morgenthal, J. C. and Niekerk, C. H. (1975). *Agroanimalia* **7**, 35–40.
Wentzell, D., Viljoen, K. S. and Botha, L. J. J. (1979). *Agroanimalia* **11**, 19–22.
Westhuysen, J. M. (1979). *The Angora* **21**, 7–13.
Westhuysen, J. M. and Wentzell, D. (1971). *S. Afr. J. Anim. Sci.* **1**, 101–102.
Wiggans, G. R., VanVleck, L. D. and Dickinson, F. N. (1979). *J. Dairy Sci.* **62**, 797–801.
Wilson, P. N. (1957). *East African Agric. J.* **23**, 138–147.
Yarkin, I. and Eker, M. (1961). *Yb Fac. Univ. Ankara* 143–152.

5. Collection, Processing and Artificial Insemination of Goat Semen

J. M. Corteel

Introduction

The contribution of artificial insemination to the genetic make-up of highly productive strains and breeds of the bovine species has been such in the past 40 years that artificial breeding is now considered a desirable tool for genetic improvement in other species. In goats, the goal remains the same but, unlike the bovine species, the sperm supply from the male is limited when compared with the need for sperm in the female to ensure a high level of fertility. Moreover, the dilution and storage techniques used for bull sperm cannot be transposed to goat sperm. Research was necessary to provide a better knowledge of goat sperm and to develop specific techniques for its preservation. In this chapter an analysis of the sperm supply from the male will precede a description of the difficulties of *in vitro* storage of goat sperm cells and of ways round these difficulties. Taking into account the many situations where artificial insemination can be helpful, especially when maximum size progenies are desired for proven males, procedures to reach the desired goals will be suggested.

A. Semen Collection

1. The Start of the Sperm Crop

The postnatal development of the male reproductive tract has received some attention (Yao and Eaton, 1954; Onuma and Nishikawa, 1955; Skinner, 1970, 1974) and it is known that sperm cells begin to appear in the ejaculate of Boer goats at 157 ± 10 days of age. Then, the volume (0·17 ml), sperm concentration (0·09 × 10^9 spermatozoa per ml) and wave motion of the ejaculate (0·7 out of a possible 5·0 value) are low, but only 9 weeks later, wave motion has reached an average of 4·1 and only 4 more weeks are needed to obtain nearly 2 × 10^9 spermatozoa in 1·0 ml ejaculates. At this stage, the animals are only 8 months old and produce good quality semen in practical quantities (Louw and Joubert, 1964). The same amount of spermatozoa is obtained 6 weeks earlier from correctly fed Alpine bucks but there are extreme inter-breed differences in the age at which sexual maturity is acquired: from 4 to 8 months or from 1 to 4 years of age, depending on whether animals belong to early or late maturing breeds (Lall, 1947; Meklenburcev, 1949; Galeon, 1951; Nadal-Grau, 1955). The late sexual maturity of the Damascus Breed (Elwishy and Elsawaf, 1971; Elwishy *et al.*, 1971) is ruled by a dominant gene (Epstein and Herz, 1964).

2. Semen Collection Techniques and Related Problems

Semen may be ejaculated into an artificial vagina or following electrical stimulation of the genital tract (Fewson and Fisher, 1959; Bonadonna, 1938; Dziuk *et al.*, 1954; O'Brien, 1966; Austin *et al.*, 1968; Hafs and Williams, 1963). Then it is collected into a special vial attached to one end of the vagina or held by the operator. Electro-ejaculated semen contains a greater amount of seminal plasma than semen collected with the artificial vagina (Austin *et al.*, 1968). However, recent advances in anatomical (Richter, 1959; Hassa, 1966; Calislar, 1966; Wrobel, 1969, 1970, 1971; Starflinger, 1972) and physiological knowledge (Becket *et al.*, 1972, 1975; Igboeli, 1974) of the male reproductive tract may lead to electro-ejaculatory techniques more adapted to the specific anatomy and physiology of the male goat, i.e. enabling a normal sperm to seminal plasma ratio in the ejaculate. Semen collection with the artifical vagina depends on the libido of the animals.

Some male goats never display any breeding behaviour at all and their semen cannot be collected with the artificial vagina. The proportion of these (8·7%) was measured in a randomized group of 46 Alpine and Saanen goats in which lack of libido was obvious when 5 to 6 months old (Leboeuf, 1977).

In males with a seasonal sexual activity, although displaying a normal sexual behaviour in the breeding season, mountings and copulations cease to occur in some or all animals for several weeks or months in the non-breeding season (Shank, 1972). In those that continue to mount and copulate beyond the normal breeding season, a significant increase in the reaction times is always reported (Corteel, 1977). During the breeding season, libido may be severely depressed by zinc deficiencies (Neathery *et al.*, 1973).

In spite of the seasonality of mating behaviour, semen may be obtained all year round with the artificial vagina from such seasonal breeds as Alpine and Poitevine. But this was shown with males displaying normal sexual behaviour, trained to mount a teaser doe and serve the artificial vagina at an early age (5 months), with semen collections taking place twice a week on fixed days and hours (Fig. 1). Continuous training of the animals seems to

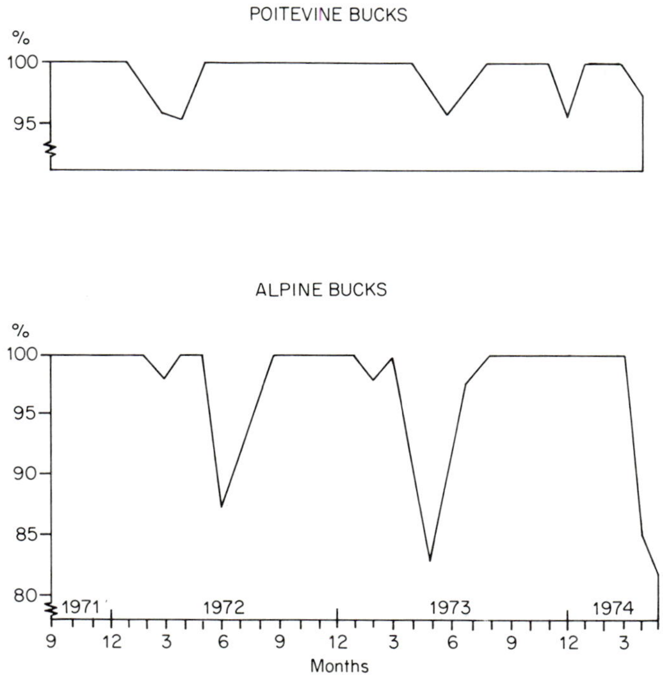

Fig. 1. Monthly variations in the sexual efficiency of male goats (% successes in all attempted semen collections). Five males per breed, aged 8 months at the beginning of observations.

maintain the mounting and ejaculatory activities in the animals. These activities may then be considered as reflex activities. Adequately trained animals may even continue to mount and "ejaculate" for months following castration (Hart and Jones, 1975).

Seasonality cannot be a limiting factor of libido in non-seasonal breeders such as the Japanese native goat (Hiroe and Tomisuka, 1965; Mohri *et al.*, 1970), the Red Sokoto of Nigeria (Hart and Jones, 1975) or the Barbari (Sahni and Roy, 1967).

3. Factors influencing the Size and Sperm Content of the Ejaculate

(a) *Effect of Season*

In seasonal breeders, the volume of the ejaculate is high in the breeding season and decreases in the non-breeding season. Sperm concentration follows the opposite trend (Eaton and Simmons, 1952; Shukla and Bhattacharya, 1952; Sharma *et al.*, 1957; Vinha, 1975). In these earlier studies, the time intervals between semen collections were 1, 2 and 4 weeks and did not allow for a precise appraisal of the magnitude of the phenomenon; this is most conspicuous in ejaculates collected twice a week from Poitevine

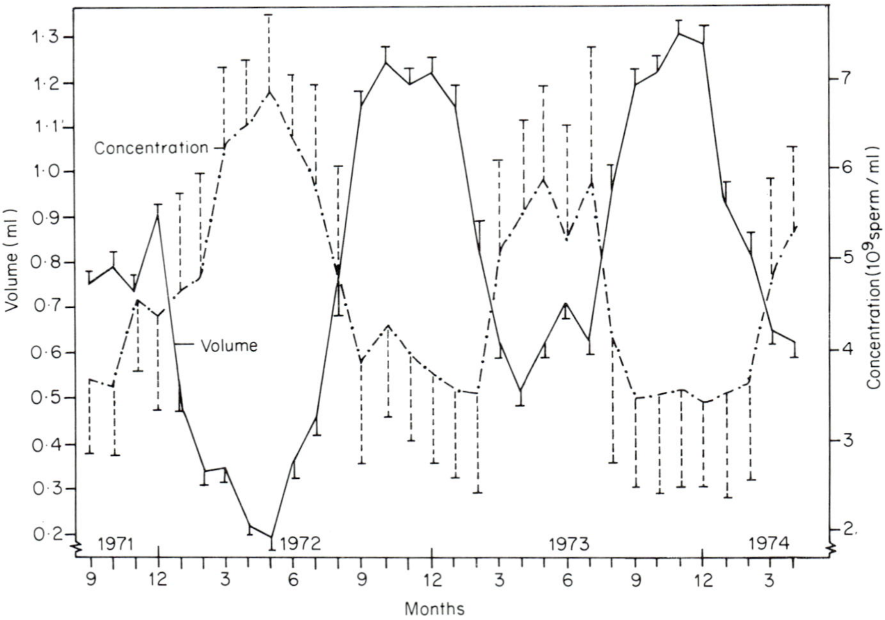

Fig. 2. Seasonal variations in volume and sperm concentration of ejaculates (m ± s.d.). Same poitevine males as in Fig. 1.

males (Fig. 2). No seasonal variations in numbers of sperm per ejaculate were reported except in the Poitevine breed and in the young Alpine male. The lack of such variations in the adult Alpine male was due to the lengthy time intervals between collections.

On an average yearly basis, seminal plasma accounts for about 70% of the total volume of the ejaculate. Therefore, when the volume increases or decreases, the changes of volume are mainly due to changes in quantities of fluids secreted by the epididymis and principally by the accessory glands. Both epididymal and accessory gland secretions are androgen-dependent (Orgebin-Crist *et al.*, 1975; Mann, 1964) but androgen receptors in the target organs need a preliminary preparation in which prolactin is involved. In turn, prolactin secretion is associated with daily light of long duration (spring and summer days in the northern hemisphere). This information was obtained from studies in the rat and the sheep (Hafiez *et al.*, 1972; Ravault *et al.*, 1977) but may well apply to the male goat in which blood plasma

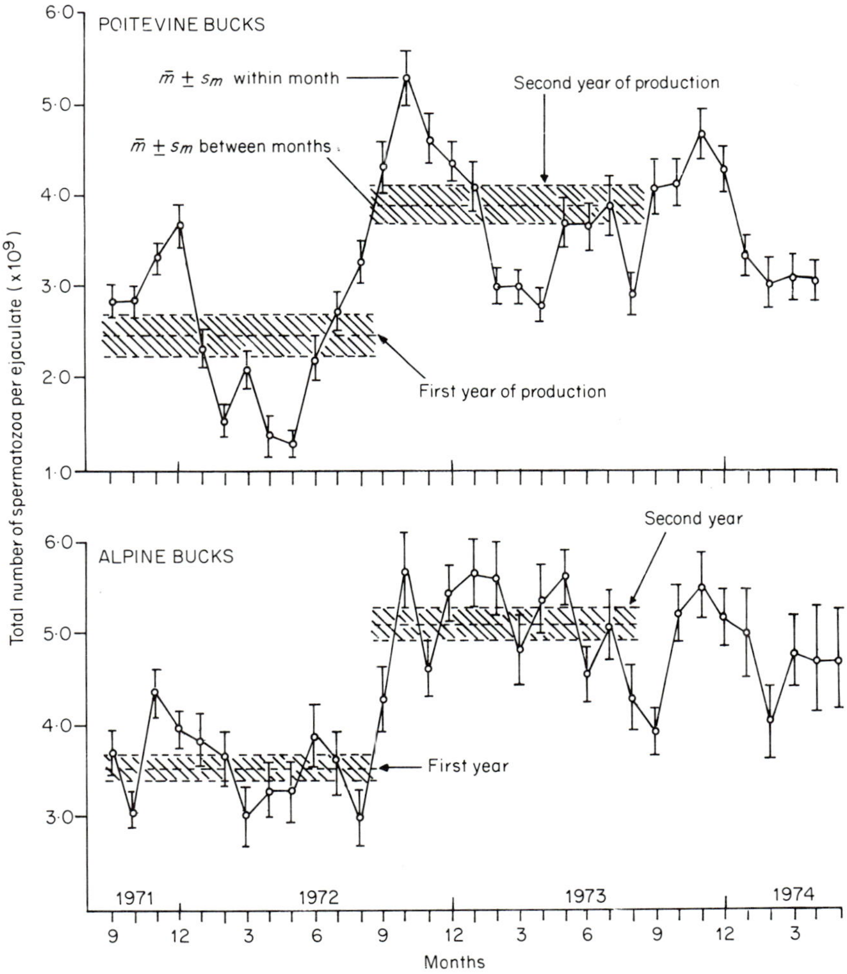

Fig. 3. Total numbers of spermatozoa per ejaculate. Same males as in Fig. 1.

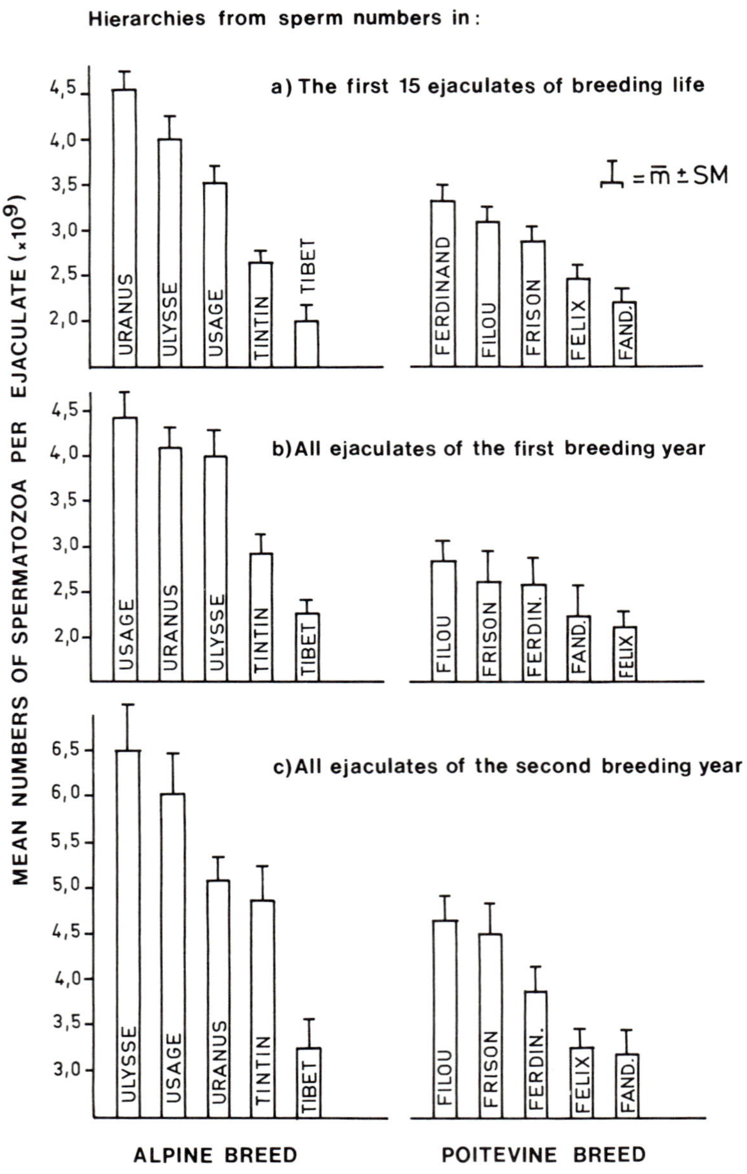

Fig. 4. Male goats of two breeds ranked according to sperm production.

prolactin levels rise in the spring, culminate in summer and drastically decline in autumn (Buttle, 1974), while the abrupt rise in androgen takes place in the middle of the summer to culminate in autumn and decline in the middle of winter (Saumande and Rouger, 1972; Muduuli *et al.*, 1979).

(b) *Age, Breed and Individual Differences*

In early maturing breeds, sperm numbers per ejaculate in the first breeding year represent only 60% of those collected in the second breeding year which are representative of adult production (Fig. 3). Seminal attributes (ejaculate volume, sperm concentration and total numbers of sperm per ejaculate) differ according to breed (Corteel, 1975c; Tewari *et al.*, 1968; Vinha, 1975). Some of these differences are great, e.g. the Boer goat may ejaculate three times as many spermatozoa as the native Zambian goat (Igboeli, 1974). Differences are also observed in the amount of sperm ejaculated by animals of the same age and breed. Animals may be ranked early in life according to these differences (mean of 15 ejaculates over an 8-week period). The resulting hierarchy remains fairly unchanged in the first two breeding years (Fig. 4).

(c) *Other factors*

The amounts of collected semen and sperm are decreased by hyperthermia (Yokoki and Ogasa, 1977), high ambient temperature (Hiroe and Tomizuka, 1966; Masaki and Masuda, 1968), associated with high relative humidity and rainfall (Mukherjee *et al.*, 1953), low levels of energy intake (Hiroe and Tomizuka, 1965), poor quality forage (Arbeiter, 1963) and lack of particular nutrients (Hildebrandt, 1967; Haertel, 1967; Neathery *et al.*, 1973). In spite of preliminary approaches, precise information is needed as to the most appropriate frequency of semen collection to obtain optimal sperm quantity and quality.

Sperm resorption is weak in the epididymis (Zankl, 1970) and epididymal sperm reserves available daily for ejaculation are readily depleted (62% of the potential daily sperm collection is obtained in the first two ejaculates (Fielden and Barker, 1964). Therefore most of the testicular sperm output could be collected with time intervals of 24 hours between collections, minimizing sperm losses in the urine or by masturbation; conversely, it is not worth ejaculating a male goat more than twice during the same day. In adult Alpine males ejaculated once a day during the breeding season, weekly production amounted to 25 billion spermatozoa per animal (Corteel, 1968).

4. Sources of Variation in Sperm Quality

The notion of sperm quality is rarely applied to a single cell but most commonly refers to that of a given population of cells, in most instances to those belonging to the same ejaculate. It is measured *in vitro* by the viability of the cell population reflected by the percentage of sperm cells with a progressive motility, at a given time or according to its evolution with time. It is also tentatively appraised in terms of motility, of enzyme release by the cells, on morphological grounds and various sophisticated tests which have not yet been used to assess goat sperm quality. In our opinion goat sperm quality has to be tested *in vivo*, because sperm quality is only truly accounted fòr by sperm fertility.

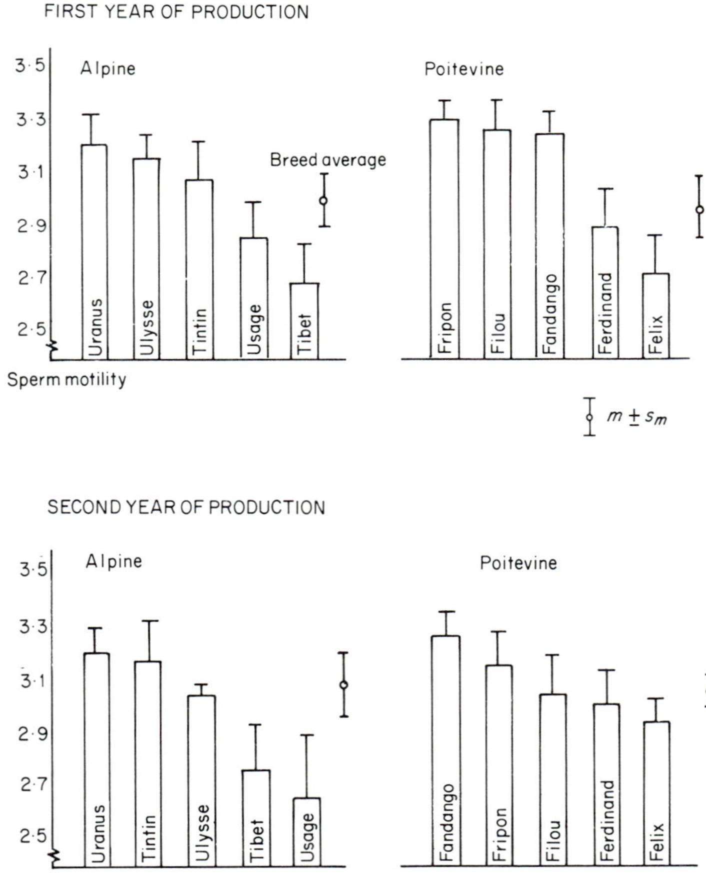

Fig. 5. Individual differences in sperm motility after thawing. Same males as in previous figures.

During the breeding season, semen motility (wave motion of freshly collected and undiluted semen) is high. Then, an average ejaculate contains 85 to 95% morphologically normal cells and the progressive motility of these cells (speed of travel of moving cells) is also the highest of the year. Sperm motility values in diluted frozen ejaculates after thawing differ from one male to another. As with numbers of sperm per ejaculate, the individual differences appear early in life with practically unchanged animal ranks all through the first two breeding years (Fig. 5). The percentages of motile sperm following dilution and cooling and after freeze-thawing are the lowest in sperm cells diluted, cooled and deep-frozen together with seminal plasma (Corteel, 1977).

During the non-breeding season (Phillips *et al.*, 1943; Yao and Eaton, 1954; Dziuk *et al.*, 1954) or following exposure of the males to high ambient temperatures (Masaki and Masuda, 1968), there is an increase in the incidence of morphologically abnormal sperm. On the basis of what is known in other species (Srivastava and Williams, 1971), we presume that this impairs fertility. Semen motility may or may not be decreased, depending on the breed, but sperm motility is severely depressed in diluted cooled and in diluted frozen thawed ejaculates (Fig. 6). This depressed sperm motility is associated with a severely decreased fertilizing ability (Corteel, 1976a). Impaired motility has been observed in each of 4 consecutive years in the same males and for the same length of time (3 months), but any time from March to September according to the year (Corteel, 1977). In seasonal male goats of the same breed, reared in the same environment, the lowest yearly peripheral blood testosterone levels were also observed for 3 consecutive months within the non-breeding season and also at a time varying from year to year in the same animals (Rouger, 1974). This suggests that goat sperm motility is in some way androgen-dependent. Altogether the seasonal variations in motility appraised in defined conditions are closely associated with the corresponding seasonal variations in fertility.

The percentages of total sperm cells that show a progressive motility in cell populations diluted together with the seminal plasma during the non-breeding season appear to be the highest of the year. However, they are only less depressed by the usual technology than those of the breeding season. This was recently demonstrated in Alpine spermatozoa washed before processing, for which percentages of motile sperm upon thawing were low in the non-breeding season and high in the breeding season, both breeding and non-breeding season levels being higher than those previously observed for unwashed spermatozoa (Rimbert, 1980). This, again, is related to interactions between sperm cells and seminal plasma *in vitro* and this will now be discussed.

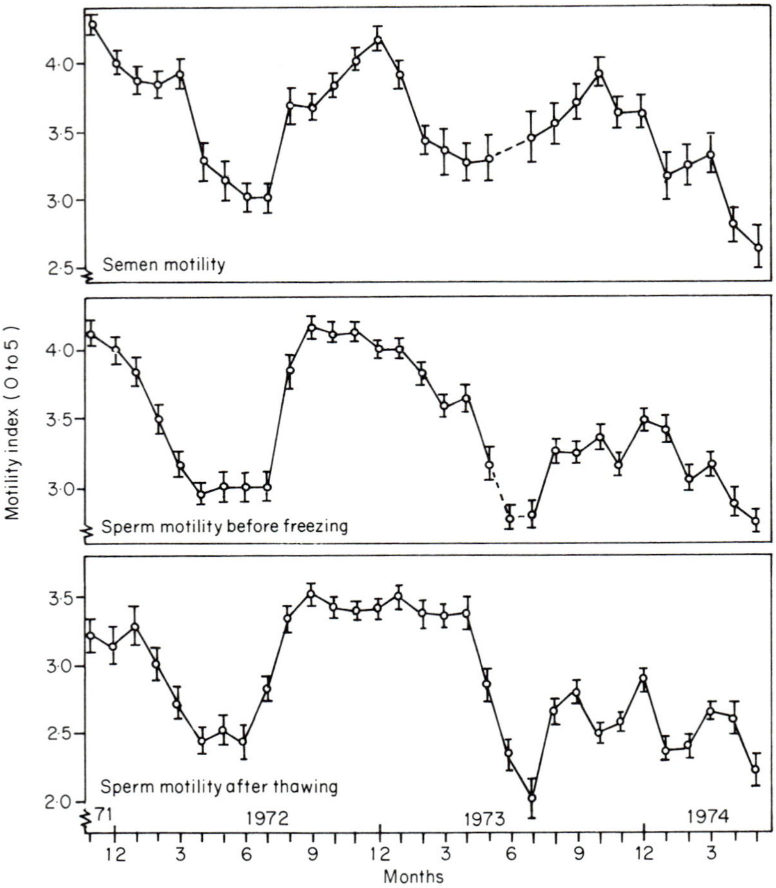

Fig. 6. Monthly variations in semen and sperm motilities (m ± sm). Same alpine males as in Fig. 1, but 10 months old at the beginning of observations.

B. Preservation of Sperm Cells

Undiluted semen can be used when donors and recipients are reared in the same large units. But then it is necessary to operate quickly in order to prevent sperm exhaustion and semen drying. This is why in most cases diluted stored semen is used.

1. Preservation Media

To our knowledge, there is not a single comprehensive study on the effects of goat semen preservation media on post artificial insemination fertility.

The merits and limits of these media can only be appraised in a very general manner. Dilutors containing egg yolk have given rise to very irregular fertility results, kidding rates after a single insemination ranging from 5% to 85%, and the improved fertility results reported were never repeated (Corteel, 1973). The use of dilutors prepared with cows' whole or skimmed milk never resulted in low fertility. Saline media have not been tested to a sufficient extent.

The very poor fertility results obtained by inseminating goat semen preserved in media containing egg yolk may be attributed to high concentrations of lysolecithins in the environment of the sperm cells. These high concentrations are toxic to spermatozoa (Corteel, 1980). Lysolecithins result from the hydrolysis of egg yolk lecithins to lysolecithins and fatty acids, the hydrolysis being triggered by an enzyme secreted in large quantities by the bulbourethral glands of the male goat. This enzyme was first called Egg-yolk Coagulation Enzyme (Roy, 1957). The hydrolysis depends on certain conditions, i.e. calcium availability (Roy, 1957), pH, temperature, seminal plasma concentration (Iritani and Nishikawa, 1961) and season of semen production (Iritani and Nishikawa, 1964). In some instances, fertility results were extremely good and obviously the hydrolysis did not take place or was considerably reduced. Whether goat breed differences exist in the quantitative secretion of the Egg-Yolk-Coagulating Enzyme by the Cowper's gland is unknown. Egg yolk composition may also be such that the hydrolysis of lecithins is prevented or delayed or reduced. An interesting observation must be reported here. Semen samples were diluted in yolk buffers prepared with eggs of New Hampshire, Sussex, White Leghorn and Brown Leghorn hens. The diluted semen was stored at 6°C and examined daily. Spermatozoal motility and survival time were greatest when New Hampshire eggs were used, followed by Sussex, White Leghorn and Brown Leghorn in decreasing order. Further semen samples were diluted and yolk from New Hampshire or Brown Leghorn eggs used as a buffer. With the former, 85% of 205 goats kidded *vs* 74% of 199 with the latter. The differences were greater the longer the storage time; the percentages were 75 and 60 respectively after storage for 48 hours (Aehnelt and Brockman, 1955). As long as the interactions between egg yolk and goat semen are not more precisely known and knowledgeably controlled, the author will not advocate the use of egg yolk in goat semen dilutors.

Current work in the author's laboratory has not included studies on dilutors, but cows' skimmed milk containing one centimole of glucose has been used for years with fairly steady fertility results. When semen is deep frozen, cryoprotective agents are used: usually glycerol (Corteel, 1977) and in some instances high concentrations of sugars (Nagase, 1966).

2. Storage in the Liquid and Frozen States

(a) *Liquid Stored Semen*

As a general rule, when extended semen is stored at temperatures above 0°C, sperm cell survival is of short duration (a few days) in most dilutors (Dauzier, 1956; Tewari *et al.*, 1968; Sahni and Roy, 1969; John and Raja, 1973; Hiroe and Udatsu, 1975) and sperm fertilizing ability cannot be preserved at a high level for more than a few hours (Dauzier, 1965, 1966b). Since sperm cell survival lasts longer than sperm fertilizing capacity, the former cannot be used to assess the latter.

In breeds where male fertility is greatly reduced during the non-breeding season, liquid stored semen cannot be relied upon to inseminate anoestrus goats induced to ovulate by the application of hormonal treatments (Corteel, 1975a) or any other manipulations unless the males are themselves manipulated.

(b) *Long Term Storage: Freezing of Goat Sperm Cells*

Good fertility results have been reported after insemination with deep frozen goat semen but in most instances these results have never been repeated. The only long-term effort to develop the use of deep-frozen thawed semen was that of a German worker (Bonfert, 1965) who failed to reach an average kidding percentage of 50 or more after one or more inseminations during a single oestrus period. Spermatozoa frozen together with the seminal plasma did not stand storage for more than 2 months. After that time, revival rates upon thawing constantly decreased (Corteel, 1975b). Fertility rates better than 50% were obtained only after removal of most of the seminal plasma from the environment of the sperm cells by washing them through dilution and centrifugation before cooling, freezing and thawing (Corteel, 1974; Fougner, 1974; Corteel, 1975b, 1976b). Washing the sperm cells before freezing has improved storage ability and post artificial insemination fertility but its efficiency is likely to differ according to the conditions in which it is used. We showed recently that the washing procedure only slightly improved the freezability and fertility of spermatozoa from ejaculates of very large volumes (> 1·9 ml). Saanen males have ejaculates of greater volumes than Alpine males, mostly in the breeding season. There is no seasonal variation in the percentage of motile sperm upon thawing in the ejaculates of Saanen males. This lack of variation is connected to an incomplete improvement of sperm cell freezability in the breeding season when ejaculates of high volumes are produced. On the other hand, washing has in the Alpine ejaculates completely reversed the awkward "seasonal trend" that was observed in unwashed deep-frozen spermatozoa (Rimbert, 1980). Although non-breeding season seminal plasma is more detrimental to sperm motility than breeding season seminal plasma (Nunes, 1980), it is more readily and efficiently removed by the washing procedure because then the volumes of the ejaculates are low.

Since the washing procedure complicates the task of the artificial insemination centres, it would be more satisfactory to have a better understanding of the mechanisms involved in the deterioration of the sperm cell freezability by seminal plasma and to prevent this deterioration biochemically. A low molecular weight protein of the Cowper's gland secretions is responsible for a drastic decrease of epididymal sperm motility and survival in a milk glucose medium at 37°C. This decrease in motility and survival is prevented and/or delayed by a constituent of the vesicular secretions. The control of sperm motility and survival by the non-epididymal compartment of seminal plasma is being studied in relation with sperm freezability and fertility. This work will hopefully, lead to more simple deep freezing techniques for ejaculated goat sperm (Nunes *et al.*, 1981).

C. Artificial Insemination (AI)

1. Methods of AI

(a) *Undiluted fresh semen* can be used where donors and recipients are reared in the same large unit. One advantage is that only simple equipment is required: artificial vaginas, semen collecting vials and micropipettes for semen deposition into the uterus through the cervix or into the cervix. In both cases the latter has to be made conspicuous by parting the vaginal walls with a speculum and illuminating it, with the help of the sunlight if possible, otherwise with some other light source (a head light or a flash light). A disadvantage of this method is that semen quality can only be visually appraised by the turbulence of semen seen through the collecting vial. This turbulence is nothing but semen motility which is a poor indicator of semen quality. Another disadvantage is that it is necessary to operate quickly in order to prevent sperm exhaustion and semen drying.

(b) *Liquid stored semen* allows more time between semen collection and artificial insemination. A rehydrated skimmed cow's milk dilutor (Table I) preserves sperm fertilizing ability at a practical level for about 12 hours. This method allows the appraisal of sperm motility which is a better indicator of sperm quality than semen motility and short distance shipping of semen. It requires more equipment than the previous method (a microscope, glass ware, high quality water, skimmed milk powder) and it cannot be used to inseminate anoestrus does induced to ovulate when sperm motility and fertilizing capacity are decreased in the non-breeding season, but this, of course, only applies to breeds with severe seasonal variations of sexual activity. It is surprising that this method is not widely used for the insemination of cyclic goats in those circumstances where the equipment for sperm deep-freezing is not available.

Table I. Skimmed cow's milk dilutor for goat sperm storage in the liquid form.

Skimmed cow's milk powder containing less than 1% butterfat	100 g
D. Glucose	0·194 g
Distilled water	100 ml
Following thorough mixing, heating at 85°C for 10 minutes and cooling down to room temperature antibiotics are added:	
Penicillin (Na salt)	50 000 IU
Dihydrostreptomycine (sulphate form)	0·050 g
(Quantities for 100 ml of the above dilutor).	

(c) *The use of frozen stored semen* has failed to develop because of the unavailability of efficient freezing techniques for goat sperm. The technical difficulties may now be overcome by sophisticated methods (washing of sperm prior to dilution, cooling and freezing). Simpler methods will result from a better understanding of the interactions between goat sperm cells and goat seminal plasma *in vitro*. Freezing of cells and use of frozen cells are unfortunately limited to countries with high and medium levels of technology.

2. Effect of AI Conditions on Post-AI Fertility

(*a*) *Time of Insemination from Onset of Oestrus*

Fertility in goats, inseminated at different time intervals from the onset of oestrus observed and reported by the farmer, indicates that insemination must be done at the beginning of the heat period (Table II). This is in accordance with time of ovulation (Harrison, 1948; Lopyrine, 1953; Gonzalez-Stagnaro, 1977; Rao and Bhattacharya, 1980), duration of sperm transport in the female genital tract (Ajello, 1958; Lopyrine, 1953) and time of survival of male and female gametes (Lopyrine, 1953). When goats are inseminated on a fixed time basis, i.e. following the application of hormonal treatments to induce and/or synchronize oestrus and ovulation, insemination time(s) from end of treatment must be gauged for each treatment in different breeds and physiological conditions (Corteel, 1975a).

(b) *Site of Semen Deposition in the Female*

Fertility also depends on the site of insemination in the female genital tract. The highest fertility is always obtained when the inseminate is deposited in the uterus through the vagina and the cervix, the lowest when released into the vagina. Following insemination in the cervix, fertility rates stand at an intermediate level (Fougner, 1976; Dauzier, 1966b) (Table III).

(c) *Number of Inseminated Spermatozoa*

Beyond 100 million total or 65 million motile sperm cells in a single inseminate, fertility no longer varies with numbers of deposited sperms; but insemination of smaller numbers leads to a decrease in fertility (Georgiev, 1962). However, motile sperm number requirements are likely to differ according to the efficiency of the technique used for the preservation of sperm cell motility (Corteel, 1977) and the site of semen deposition. When oestrus is induced and synchronized by long-lasting progestogen treatment, sperm transport in the female genital tract is impaired. This has to be counterbalanced by inseminating double numbers of total spermatozoa, if deep-frozen thawed semen is used.

Table II. Fertility of untreated goats inseminated with liquid stored semen at different intervals from onset of oestrus.

Hours from first appraisal of oestrus behaviour to insemination	Percentage of kidding does	Number of does
0–12	70[a]	742
12–24	63[a]	646
24	47[a]	248

[a]Statistically different at the 1% level
Dauzier (1966)

Table III. Fertility rates of untreated goats inseminated at different sites with frozen stored sperm cells withdrawn from seminal plasma before dilution.

Vagina	Cervix	Uterus	References
15 %	60%	71%	Fougner (1976)
40%	69%	—	Corteel (1975)

(d) *Size of the Inseminate*

It is obvious that semen deposited in the vagina has, mostly when frozen, less chance of ensuring fertilization. When semen is deposited in the cervix, backflow of semen into the vagina may be avoided with the use of inseminates of small sizes ($< 0{\cdot}2$ ml) provided they contain enough spermatozoa to ensure a high level of fertility. This is possible when undiluted semen is inseminated (Table IV).

Table IV. Fertility of untreated goats inseminated with various volumes of crude and unextended semen containing different numbers of spermatozoa.

Inseminate volume (ml)	Sperm contents ($\times 10^6$)	Kidding (%)	*N* Females
0·2	725	81·4	147
0·1	447	89·9	100
0·05	203	92·4	106
0·025	83	90·3	123

Polovceva and Fomenko (1936)

(e) *Fertility of the inseminate*

Attempts to relate traits of *in vitro* sperm survival to sperm fertility in different species have given rise to an endless list of papers (Salisbury *et al.*, 1978). Percentages of motile sperm and sperm progressive motility are the most commonly used parameters to appraise semen quality. But the correlation between these parameters and fertility is usually very low, unless percentages of motile sperm and progressive motility are estimated after sperm cells have been submitted to drastic conditions. The percentages of motile sperm and sperm motility of breeding seasonal male goat sperm cells that have been washed, diluted, cooled, deep frozen, thawed and incubated are related with fertility. But this has been shown only for breeding season sperm cells and when semen is deposited into the cervix (Fig. 7). In the non-breeding season, variations in sperm fertility are best accounted for by variations in sperm motility after dilution and cooling, as previously indicated.

Conclusions

Collection of semen following electrical stimulation of the genital tract cannot be advocated at present because it leads to ejaculation of semen contain-

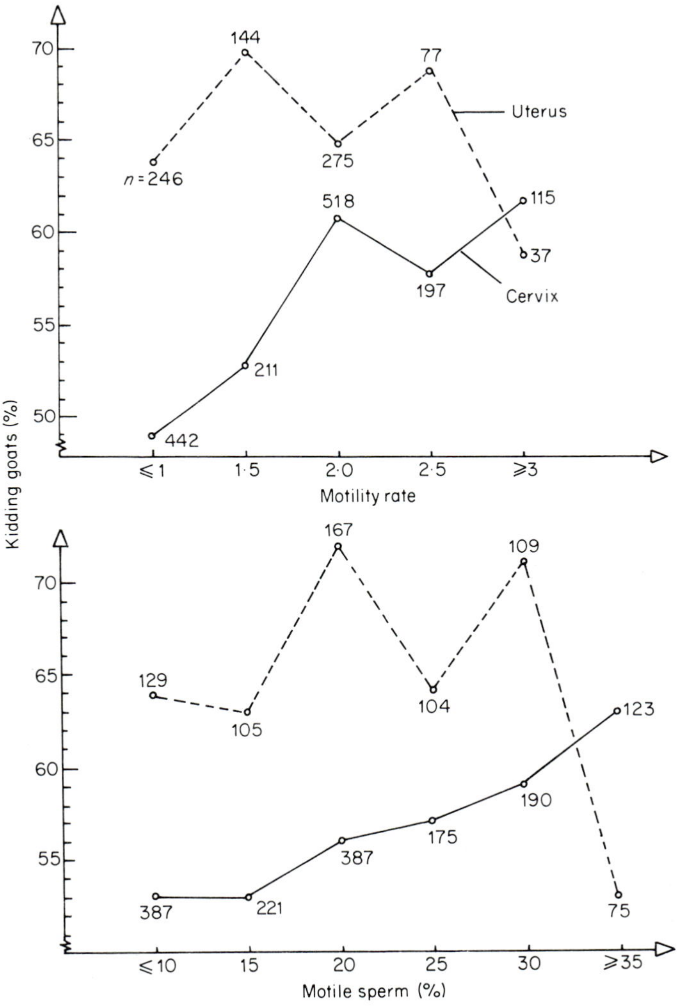

Fig. 7. Influence of motility and percentage of motile spermatozoa after 2 hours incubation at 37 C following freeze-thawing on fertility of goats inseminated in the uterus or the cervix (washed frozen-thawed breeding season ejaculates) (Rimbert, 1980).

ing more seminal plasma than semen collected with the artifical vagina. Semen may be collected by use of the artificial vagina without difficulty, provided males without libido are discarded and males with seasonal sexual activity are well trained for collection, starting from a young age and continuing all the year around. There is no definite age at which a young male may be used to produce semen for artifical insemination. This age must be investigated in each breed under its own climatic and managerial conditions keeping in mind the decisive role played by the genotype on age at puberty.

Males differ in their ability to produce semen both quantitatively and qualitatively. These individual differences appear early in the breeding life of the animals and are maintained in adulthood. One way of securing efficient AI males is to discard the poor donors when they are young and before they are progeny-tested. Some good young donors will not remain so all their life long because of exposure to high ambient temperatures, diseases and accidents. This is another good reason to get rid of the poor donors to begin with; otherwise AI centres could become crowded with poor and impaired donors. The policy we suggest also increases the chances of spreading the useful genes evidenced by the progeny test.

High ambient temperatures depress semen production. More should be known about the mechanisms involved in these depressive effects in order to counteract them efficiently. The need for more research is emphasized by the fact that the world goat population is mainly located in the tropics.

One of the greatest handicaps of goat AI is probably the lack of simple tests to predict sperm fertility before it is inseminated.

Collection, processing and insemination of goat semen are not easy tasks. They require basic knowledge and skills that still have to be perfected by continued research efforts. These efforts are worthwhile because they may contribute to the improvement of goat populations.

References

Aehnelt, E. and Brockman, P. (1965). *Dtsch. Tierärztl. Wschr.* **62**, *Suppl., Fortpfl. Zuchthyg. Haustierbesam.* **5** (6), 69–72.

Ajello, P. (1958) *Zootech. Vet.* **13**, 50–53.

Arbeiter, K. (1963). *Zuchthyg.* **7**, 349–362.

Austin, J. W., Leidy, R. B., Krise, G. M. and Hupp, E. W. (1968). *J. Appl. Physiol.* **24**, 369–372.

Becket, S. D., Hudson, R. S., Walker, D. F., Vachon, R. I. and Reynolds, T. M. (1972). *Biol. Reprod.* **7**, 359–364.

Becket, S. D., Purohit, R. C. and Reynolds, T. M. (1975). *Biol. Reprod.* **12**, 289–292.

Bonadonna, T. (1938). *Clin. Vet.* **16**, 54–63.

Bonfert, A. (1965). *Tierzüchter* **17**, 154–155.

Buttle, H. L. (1974). *J. Reprod. Fert.* **37**, 95–99.

Calislar, T. (1966). *Ankara Univ. Vet. Fak. Derg.* **13**, 173–188.

Corteel, J. M. (1968). Revue "La Chèvre", Special No., pp. 24.

Corteel, J. M. (1973). *World Rev. Anim. Prod.* **IX**, (1), 73–99.

Corteel, J. M. (1974). *Ann. Biol. Anim. Bioch. Biophys.* **14**, 741–745.

Corteel, J. M. (1975a). *Ann. Biol. Anim. Bioch. Biophys.* **15**, 353–363.

Corteel, J. M. (1975b). *Ann. Biol. Anim. Bioch. Biophys.* **15**, 525–528.

Corteel, J. M. (1976b). "2èmes Journées de la Recherche Ovine et Caprine", Ed. 2, pp. 4–17. INRA, ITOVIC, Paris.

Corteel, J. M. (1976a). *Ann. Zootech.* **25**, 567–571.
Corteel, J. M. (1976b) "2èmes Journées de la Recherche Ovine et Caprine", Ed. 2, pp. 283–287. INRA, ITOVIC, Paris.
Corteel, J. M. (1977). "Proc. Symposium on Management of Reproduction in Sheep and Goats", Madison (Wisc., USA), July 24–25, pp. 41–57.
Corteel, J. M. (1980). *Reprod. Nutr. Develop.* **20**, 115–127.
Dauzier, L. (1956). "Proc. IIIrd Int. Congr. Anim. Reprod. Artif. Insem". Cambridge, **III**, pp. 12–14.
Dauzier, L. (1966a). Personal communication.
Dauzier, L. (1966b). "International Encyclopedia of Veterinary Medicine" (Dalling *et al.*, eds), pp. 267–271. W. Green and Son,
Dziuk, P. J., Graham, E. F., Donker, J. D., Marion, G. B. and Petersen, W. E. (1954). *Vet. Med.* **49**, 455–458.
Eaton, O. N. and Simmons, V. L. (1952). *Am. J. Vet. Res.* **13**, 537–544.
Elwishy, A. B. and Elsawaf, S. A. (1971). *Indian J. Anim. Sci.* **41**, 350–356.
Elwishy, A. B., Elsawaf, S. A., Elmikkawi, F. and Omar, A. A. (1971). *Indian J. Anim. Sci.* **41**, 562–569.
Epstein, H. and Herz, A. (1964). *J. Agric. Sci.* **62**, 237–244.
Fewson, D. and Fischer, A. (1959). *Zuchthyg. Fortpflstör. Besam. Haust.* **3**, 132–134.
Fielden, E. D. and Barker, C. A. V. (1964). "Proc. Vth Int. Congr. Anim. Reprod. Artif. Insem.", Trento, pp. 488–497.
Fougner, J. A. (1974). *Norsk. Vet.* **86**, 551–553.
Fougner, J. A. (1976). "Proc. VIIIth Int. Congr. Anim. Reprod. Artif. Insem.", Krakow, IV, pp. 987–990.
Galeon, F. C. (1951). *Phillip. Agric.* **34**, 230–235.
Georgiev, G. (1962). "Izvestija, Etudes sur la Biologie et la Pathologie de la Reproduction", **2**, 95–100.
Gonzalez-Stagnaro, C. (1977). Project "Biol. de la Reprod. en las Cabras", No. 42, VIth ALPA Meeting, La Habana, Cuba.
Haertel, M. (1967). *Vet. Med. Diss.* Hannover, pp. 49.
Hafiez, A. A., Lloyd, C. W. and Bartke, A. (1972). *J. Endocr.* **52**, 327–332.
Hafs, H. D., and Williams, J. A. (1963). *J. Anim. Sci.* **22**, 859–860.
Harrison, R. J. (1948). *J. Anat.* **82**, 21–48.
Hart, B. L. and Jones, T.O.A.C. (1975). *Horm. Behav.* **6**, 247–258.
Hassa, O. (1966). *Ankara Univ. Vet. Fak. Derg.* **13**, 97–125.
Hildebrant, J. (1967). *Vet. Med. Diss.* Hannover, pp. 55.
Hiroe, K. and Tomizuka, T. (1960). *Jap. J. Anim. Reprod.* **6**, 28–30.
Hiroe, K. and Tomizuka, T. (1965). *Bull. Nat. Inst. Anim. Ind.* (*Chiba*) **8**, 17–24.
Hiroe, K. and Tomizuka, T. (1966). *Bull. Nat. Inst. Anim. Ind.* (*Chiba*) **9**, 27–38.
Hiroe, K. and Udatsu, S. (1975). *Bull. Fac. Agric. Miyazaki Univ.* **22**, 163–169.
Igboeli, G. (1974). *E. Afric. Agric. J.* **40**, 132–137.
Iritani, A. and Nishikawa, Y. (1961). "Proceedings for the Silver Jubilee". Lab. Anim. Husb., Coll. of Agric., Kyoto University, Japan.
Iritani, A. and Nishikawa, Y. (1964). *Jap. J. Anim. Reprod.* **10**, 52–56.
John, K. and Raja, C. K. S. V. (1973). *Kerala J. Vet. Sci.* **4**, 90–93.
Lall, H. K. (1947). *Ind. Farm.* **8**, 322–327.
Leboeuf, B. (1977). Unpublished data.
Lopyrine, A. I. (1953). *Goat Reprod. Physiol. Selkhozguiz Ed.* (*Moscow*), 163–177.
Louw, D. F. J. and Joubert, D. M. (1964). *S. Afr. J. Agric. Sci.* **7**, 509–520.

Mann, T. (1964). "The Biochemistry of Semen and of the Male Reproductive Tract", pp. 493. Methuen, London; J. Wiley, New York.
Masaki, J. and Masuda, H. (1968). "Proc. VIth Int. Congr. Anim. Reprod. Artif. Insem.", Paris, I, pp. 301–303.
Meklenburcev, R. N. (1949). *Zool. Z.* (*Mosk*), **28**, 482–483.
Mohri, H., Hasegawa, S. and Masaki, J. (1970). *Biol. Reprod.* **2**, 352–355.
Muduuli, D. S., Sanford, L. M., Palmer, W. M. and Howland, B. E. (1979). *J. Anim. Sci.* **49**, 543–553.
Mukherjee, D. P., Roy, A. and Bhattacharya, P. (1953). *Indian J. Vet. Sci.* **23**, 1–17.
Nadal-Grau, R. A. (1955). PhD. Thesis, Ohio State University, pp. 94.
Nagase, H. (1966). *Jap. Agric. Res. Quart.* **1**, 10–19.
Neathery, M. W., Miller, W. J., Blackmon, D. M., Pate, F. M. and Gentry, R. P., (1973). *J. Dairy Sci.* **56**, 98–105.
Nunes, J. (1980). *D.E.A. Physiol. Reprod. Univ. Paris* **VI**.
Nunes, J., Combarnous, Y., Baril, G. and Corteel, J. M. (1981). Unpublished data.
O'brien, C. A. (1966). *J. Anim. Sci.* **25**, 928.
Onuma, H. and Nishikawa, Y. (1955). *Bull. Nat. Inst. Agric. Jap.* **10**, 365–375.
Orgebin-Christ, M. C., Danzo, B. J. and Davies, J. (1975). "Handbook of Physiology", Sect. 7, Vol. V, pp. 319–328. Academic Press, New York and London.
Phillips, R. W., Schott, R. G., Eaton, O. N. and Simmons, V. L. (1943). *Cornell. Vet.* **33**, 227–235.
Poloceva, F. and Fomenko, M. (1936). *Uspek. Zootechnik. Nk.* (*Moscow*) **3**, 51–65.
Prasad, S. P., Roy, A. and Pandey, M. D. (1970). *Agr. Univ. J. Res. Sci.* **19**, 61–66.
Rao, V. H. and Bhattacharya, N. K. (1980). *J. Reprod. Fert.* **58**, 67–69.
Ravault, J. P., Courot, M., Garnier, D., Pelletier, J. and Terqui, M. (1977). *Biol. Reprod.* **17**, 192–197.
Richter, W. R. (1959). *Am. J. Vet. Res.* **20**, 603–606.
Rimbert, M. (1980). *Mém. fin d'études E.N.S.F.A.*, Rennes.
Rouger, Y. (1974). Thèse Doct. Sci. Nat., CNRS AO 10, 316, pp. 197.
Roy, A. (1957). *Nature*, 179–318.
Sahni, K. L. and Roy, A. (1967). *Indian J. Vet. Sci.* **37**, 269.
Sahni, K. L. and Roy, A. (1969). *Ind. J. Anim. Sci.* **39**, 1–14.
Salisbury, G. W., Vandemark, N. L. and Lodge, J. R. (1978). 'Physiology of Reproduction and Artificial Insemination of Cattle" (G. W. Salisbury, ed). W. H. Freeman and Co, San Francisco.
Saumande, J. and Rouger, Y. (1972). *C. R. Acad. Sci., Paris*, Ser. D, 274, 89–92.
Shank, C. C. (1972). *Z. Tierpsychol.* **30**, 488–528.
Sharma, G. P., Suri, K. R. and Vali, K. N. (1957). *Punjab. Univ. Zool. Res. Bull.* **1**, 217.
Shukla, D. D. and Bhattacharya, P. (1952). *Indian J. Vet. Sci.* **22**, 179–190.
Skinner, J. D. (1970). *Agroanim.* **2**, 177–180.
Skinner, J. D. (1974). *Agric. Res., Pretoria* (1974). p. 56, ABA 43, 4628.
Srivastava, P. N. and Williams, W. L. (1971). *In* "Control of Human Fertility", pp. 73–87. Proc. XVth Nobel Symp. Sweden.
Starflinger, F. (1972). *Zbl. Vet. Med. Cl.* 161–188.
Tewari, S. B., Sharma, R. P. and Roy, A. (1968). *Indian J. Vet. Sci.* **38**, 607–615.
Vinha, N. A. (1975). *Arch. Esc. Vet. Univ. Fed. Minas Gerais* **27**, 23–28.
Wrobel, K. H. (1969). *Berl. Münch. Tier. Wschr.* **24**, 473–477.
Wrobel, K. H. (1970). *Zbl. Vet. Med.* (A) **17**, 634–643.
Wrobel, K. H. (1971). *Zbl. Vet. Med.* (A) **18**, 250–263.

Yao, T. S. and Eaton, O. N. (1954). *Ann. J. Anat.* **95**, 401–432.
Yokoki, Y. and Ogasa, A. (1977). *Jap. J. Anim. Reprod.* **23**, 93–98.
Zankl, H. (1970). *Zuchthyg.* **5**, 32–36.

6. Nutrition and Feeding of Goats: Application to Temperate Climatic Conditions

P. Morand-Fehr

Although the number of goats in the world is very large, their nutrition and feeding up to now have been little studied. Because of the lack of information, some experts have applied results from sheep or cows to goats, generally without any modification. Others have tended to exaggerate the originality of the goat by considering it very different from other ruminants. From experimental results of the last 10 years, it is now possible to obtain a more objective view of the problem.

The most recent data on feeding behaviour, feed intake, digestibility and goat nutrient requirements will first be reported in this chapter. The effects of the diet on reproductive performance, milk yield and composition will then be discussed, before studying the composition of the diets and the establishment of feeding programmes in the relatively intensive production conditions of Western Europe with an oceanic temperate climate. Specific questions related to tropical conditions are discussed in Chapters 7, 16 and 17. The feeding of young goats is not included here, but in Chapter 8 on growth.

A. Nutritional Basis of Goat Feeding

1. Feeding Behaviour

In comparison with other ruminants, the feeding behaviour of goats has some typical features that must be taken into account for rearing goats on grazing ranges or on pasture, and for determining diet composition and feeding conditions.

(a) *Importance of Feeding Behaviour*

Even when the goat lives in favourable feeding conditions, all its activities are governed by food searching and consumption.

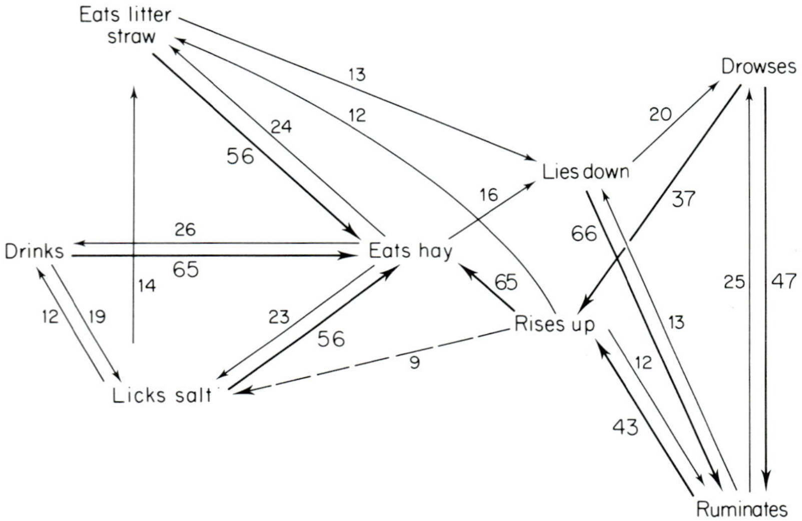

Fig. 1. Overall behaviour of a goat from 7 a.m. to 7 p.m. The arrows indicate changes from behavioural stage 1 to stage 2 and the figures near to the arrows indicate the percentage likelihood for a goat in behavioural stage 1 to pass to stage 2.

Figure 1 shows the main daily activities of six non-lactating Alpine goats, kept on straw litter and given lucerne *hay ad libitum*, water and salt lick-stone. The different behavioural phases and changes from one phase to another were observed twice a week over 12 weeks with 12 hours of daylight each day.

During the daytime, there seemed to be two kinds of activity: one in relation to rest and the other to feeding. In most cases, goats stood up for feeding, in particular hay. Ingestion of hay was interrupted for taking straw, water or salt. In more than half of the cases, the goats consumed hay again after eating straw, water or salt. Thus, a meal was ingested in different fractions which seem to be necessary for balancing fibre, water and salt intake.

Discontinuing hay intake marked the end of a meal in only 21 out of

100 observations, often revealed by the animal lying down. Frequently, after lying down, the goats began to ruminate. During resting, rumination and somnolence were closely related. Figure 1 clearly shows the importance of eating activity and its relationship to resting.

(b) *Main Characteristics of Goat Feeding Behaviour*

The average time goats spent consuming long hay, chopped hay, ground and pelleted hay and maize silage was 6 to 7, 5, 2·5 and 3·3 hours, respectively (Geoffroy, 1974; Fehr, 1971). This is a little more than sheep would spend at the same level of intake.

The most important meals are observed just after the feed is supplied. However, in the case of two forage supplies per day, minor meals between main meals must be taken into account; they are more important in goats than in sheep. The goats exhibiting the highest forage intake levels are generally those consuming the most during the minor meals. Martin *et al.* (1974) observed that when equal forage quantities are given twice a day, at 7 a.m. and 2 p.m., goats ingest more in the afternoon than in the morning. The duration of ingestion is variable from one animal to the other: 7–15 seconds per g hay (Martin *et al.*, 1974) and according to the physical form of the forage: 6·2 and 4·4 min/g D.M./kg $W^{0.75}$ ($W^{0.75}$ = metabolic weight) with chopped rye grass hay and silage, respectively (Geoffroy, 1974). This is generally shorter than in sheep (Table I), and the difference is more marked with long hay. The number of feed takings is generally higher in goats than in sheep, but their duration is shorter.

Table I. Comparison between the feed intakes of goats and sheep (chopped rye grass hay diet) (Geoffroy, 1974).

	No. of meals	Length of meals min.	Feed intake duration (min/g D.M./kg $W^{0.75}$)	Duration of feed intake in the day time (%)
5 Poitevine non-lactating goats 5 years old	9·2	39·0	6·0	22·6
5 Texel castrated rams 2 years old	5·4	67·8	4·5	20·7

D.M. = dry matter.

Goats ruminate mainly during the night. The cycle of rumination lasts 63 seconds on average in goats and 52 seconds in sheep (Geoffroy, 1974; Ruckebush and Bost, 1963). Rumination time seems to be little affected by the kind of forage except by forage pellets which reduce it strongly (Fehr, 1971).

(c) *Choice of Plants or Plant Fractions*

On range, on pasture or at the trough, goats seem to be very careful in choosing plant species or varieties, or morphological fractions. This is more evident when quantities available are large, forage quality poor and competition between animals limited. This selective feeding behaviour also exists in other ruminants, but it seems to be particularly well developed in goats.

When the feed allowance increases, not only is the intake of feed enhanced but its nutritive value is improved (Fig. 2). As shown in Table II, forage

Table II. Feeding value of lucerne hay: offered, refused and consumed (six non-lactating Alpine goats) (Morand-Fehr, *et al.*, 1977).

	Offered (O)	Refused (R)	Significance of differences (O–R)	Consumed (calculated values) (C)
Energy value of lucerne hay (F.U./kg D.M.)	0·54	0·40	***	0·59
% leaves	46	24	***	55
% stems	54	76	***	45
Energy value of leaves (F.U./kg D.M.)	0·74	0·69	*	0·75
Energy value of stems (F.U./kg D.M.)	0·38	0·31	**	0·40

F.U. = feed units
***$p \leqslant 0{\cdot}001$
**$p \leqslant 0{\cdot}01$
*$p \leqslant 0{\cdot}05$

chosen to eat has a better nutritive value than forage offered. The proportion of leaves is higher and the proportion of stems lower in ingested lucerne hay than in that offered. Likewise, ingested stems and leaves tend to have a slightly higher nutritive value than the stems and leaves offered. Thus, the higher the forage allowance, the more the goat can choose and consume the most nutritive fractions of the forage offered. However, this selection is different from one animal to another. The proportion of leaves and stems in lucerne hay refusals varies between goats (Morand-Fehr and Hervieu, 1977).

In intensive feeding conditions, the economic importance of goat forage refusals is often underestimated. Indeed, if the goat-keeper allows many forage refusals and consequently large wastage, he is managing his forage resources badly. If he accepts few refusals, feed intake level and performance of the goats will decrease. It is therefore necessary to develop a strategy according to forage availability and quality. In particular, with forages of

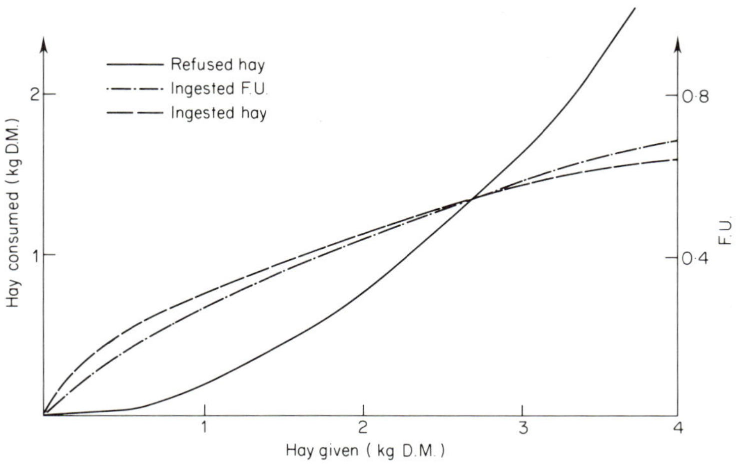

Fig. 2. Variation in hay consumption according to hay supply in adult Alpine goats (moderate quality lucerne hay) (Morand-Fehr and Hervieu, unpublished).

poor quality it is advisable to accept a higher proportion of refusals. As mentioned in Section C2, this question must be decided upon according to the quantities of concentrate given.

(d) *Goat Behaviour in Different Feeding Conditions*

(i) *At the Trough.* The goat frequently begins by sampling all forages offered before selecting the fractions it prefers to eat. The ingestion rate is very fast at the beginning but after some time, varying between goats, it decreases; the choice then becomes more selective and thereafter the feed intake stops.

(ii) *On Pasture.* It is frequently admitted that the goat uses pastures poorly. Indeed, the goat leaves many refusals on the pasture, sometimes more than 50% of available forage if no measures are taken. Nevertheless, it is possible to make (Masson and Decaen, personal communication) goats use the pasture better by enlarging it each day with an electric fence by a distance equivalent to their ingestion capacity, and by maintaining a high animal density to obtain greater competition between animals.

When goats arrive on pasture, they quickly begin to graze, but this intensive grazing does not last long. The choice of plants becomes more and more selective and the rate of intake decreases progressively. Goats will discontinue grazing at any minor disturbance like rain or the appearance of an unknown person.

Periods of grazing essentially depend on daylight duration and on temperature. Generally, the goat grazes in the morning and evening, and rests during the middle of the day. The higher the day temperature, the more the goat tends to graze early in the morning and late in the evening. When days are very hot, the goat may change its grazing periods and essentially graze during the night.

(iii) *On Range*. The goat's selective feeding behaviour is an advantage on the range, particularly in areas of sparse plant supply. In fact, the goat can stay in areas which cannot be used by other ruminants.

The behaviour of goats in dry areas is described in Chapter 7. Let us, however, mention here that goats on range tend to scatter more than sheep. Their feeding behaviour and feed intake level depend not only on the plant availability, but also on the availability of comfortable resting areas.

2. Feed Intake

(a) *Level of Feed Intake and its Variability*

The feed intake level of the goat depends on production capacity, physiological stage, age and breed of the animals. Very different values have been reported by various authors. Devendra (1971) indicated that in tropical areas, meat or dual purpose type goats consume less than 3 kg D.M./100 kg L.W./day. Dairy goats of European breeds, under intensive production conditions, can reach very high intake levels of up to 5–8 kg D.M./100 kg L.W./day (Mackenzie, 1967), i.e. higher levels of intake than in cows.

(b) *Factors affecting Level of Feed Intake*

Sauvant and Morand-Fehr (1978a) analysed the factors affecting feed intake capacity of the goat during late gestation and during lactation in two studies.

(i) *During Gestation*. A total of 5500 daily intake records in 130 Alpine goats (fed lucerne or pasture hay and concentrates) during the last 6 weeks of gestation show that the voluntary feed intake of the goat may vary from 0·82 to 2·13 kg D.M./day. The intake capacity during late gestation varies little, but it decreases markedly if it is related to live weight or metabolic weight ($W^{0·75}$) because of the increasing weight during this period (8 or 9 kg on average in Alpine goats).

The capacity for forage intake during late gestation also depends on age, increasing by about 50% between the age of 1 and 2 years and by 10% between the age of 2 and 3 years.

On the other hand, there is no relation at constant age between intake capacity during gestation and litter size. There is, therefore, no need to adjust feeding to the number of fetuses in the litter.

(ii) *During Lactation.* A total of 28 400 feed intake records were made during lactation in the same goats as in the previous study. Their milk production level varied from 0·78 to 6·26 kg/day. The results show that the dairy goat is indeed able to reach the feed intake levels mentioned by Mackenzie (1967). Table III indicates that during lactation, the intake reached a maximum of about 7 kg D.M./100 kg L.W., i.e. 180 g/kg metabolic weight ($W^{0.75}$).

Figure 3 shows the variation of feed intake capacity with stage of lactation. Generally, it increases by about 40% during the first weeks and the maximum

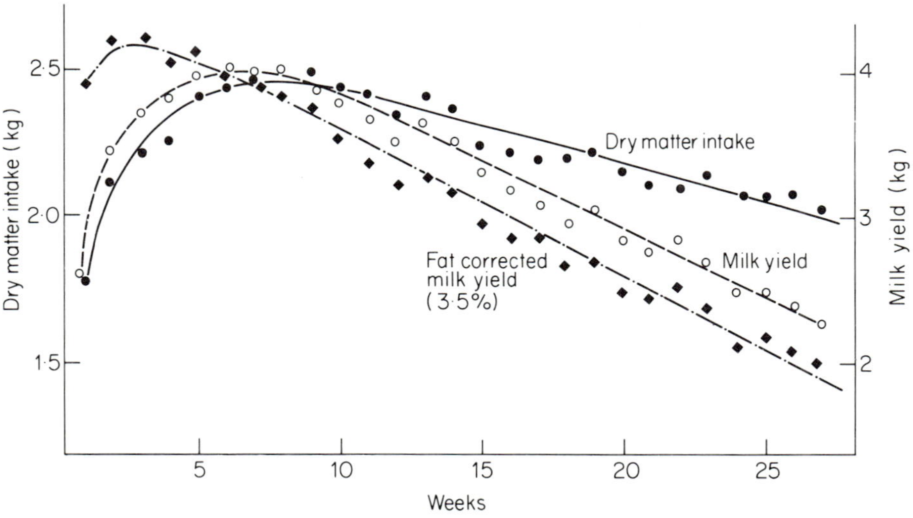

Fig. 3. Variations in the voluntary feed intake (D.M.) and in the milk yield of goats (72 one- to eight-year-old Alpine goats per week on average) (Sauvant and Morand-Fehr, 1978).

is reached between weeks 6 and 10 of lactation. It depends on the level of milk production and live weight, more markedly in mid-lactation than in early lactation. The relation of feed intake and milk produced is closer to uncorrected quantity of milk than to 3·5% fat corrected milk, particularly during early lactation.

(iii) *Prediction of Feed Intake Level.* Some authors have proposed regression equations to predict the feed intake level from measurable parameters (Opstvedt, 1969; Kondos, 1972; Skjevdal, 1974) and with the aim of formu-

Table III. Average value and variation of the voluntary feed intake in lactating goats[a] (D. Sauvant and P. Morand-Fehr, unpublished)

	Average	Standard deviation	Minimum	Maximum
D.M. intake (kg/day)	2·24	0·41	1·04	3·60
D.M. intake (kg/100 kg L.W.)	3·94	0·76	1·63	6·81
D.M. intake (g/kg $W^{0.75}$)	108·12	20·05	47·14	181·18

L.W. = Live weight.
[a] 2030 weekly average values from Alpine goats 1 to 8 years old.

Table IV. Prediction of feed intake in goats.

Prediction equations for daily dry matter intake	
1. During the whole lactation: basal diet with pasture hay or lucerne hay (Sauvant and Morand-Fehr, 1977a)	
T.D.M. = 315·8 M.Y. + 10·0 L.W. + 632	$r = 0.738$ (1)
2. During the whole lactation, basal diet with maize silage (de Simiane, 1977, unpublished)	
T.D.M. = 287 M.Y. + 6 L.W. + 642	$r = 0.606$ (2)
3. In mid-lactation (week 10–30), basal diets with pasture hay or lucerne hay (Sauvant and Morand-Fehr, 1977b)	
T.D.M. = 423·2 F.C.M. + 27·8 M.W. + 6·57 F + 440 N.W.G.	$r = 0.914$ (3)
B.F.I. = − 1271 B.C.I. + 537 F.C.M. + 48·3 M.W. + 159 N.W.G. + 142	$r = 0.954$ (4)
4. During the first 20 weeks of lactation, basal diets with hay, grass silage swedes and potatoes (Skjevdal, 1974)	
T.D.M. = 970 + 16 L.W.	$r = 0.51$ (5)

T.D.M. = Total dry matter (g/day).
M.Y. = Milk yield (kg/day).
F.C.M. = Fat corrected milk (3·5% fat) (kg/day).
L.W. = Live weight (kg).
M.W. = Metabolic weight (kg).
F. = Percentage of forage in the diet.
B.F.I. = Intake of forage (g/kg).
B.C.I. = Intake of concentrate (g/kg).
N.W.G. = Net weight gain (kg per month) from the equation:
N.W.G. = L.W. − 3·7 T.D.M.

lating feeding programmes in goat herds. Sauvant and Morand-Fehr (1978a) suggested a prediction equation for the whole lactation involving live weight and milk production and another one for mid-lactation (8–30 weeks of lactation) taking into account the weight gain and the proportion of forages in the diet (Table IV). The same authors also supplied an equation for predicting hay intake during mid-lactation. The coefficients of multiple correlation of the two latter equations indicate that in mid-lactation, the voluntary feed intake in goats can be predicted with sufficient accuracy. The first application of these equations shows that they are less utilizable in late lactation because of a slight bias.

3. Digestibility

The anatomy of the digestive tract of goats is similar to that of other ruminants. Sometimes the rumen of goats is very large probably because of the type of diet rather than because of a species difference. The different gastric segments in goats and in other ruminants develop at the same rate (Gihad and Morad, 1977).

El Hag (1976) compared the apparent digestibility and the metabolic utilization by goats and sheep of a rich forage (berseem), and of two poor forages (high crude fibre and low protein content). With poor forages in the goat, rumen volatile fatty acid and gas productions were higher, ammonia production lower and nitrogen retention was higher than in sheep. With berseem these differences between goat and sheep did not appear or were very slight. It may be concluded that goats use cell wall carbohydrates and ruminal ammonia of poor quality forage better than sheep. In addition, synthesis of microbial proteins is improved. These particularities of the goat are much more difficult to see with good or medium quality forages. This interpretation agrees with the results of Ridges and Singleton (1962) and Gray (1947) who observed that the proportion of cell wall carbohydrates disappearing in the rumen is larger in goats than in sheep in which this digestion takes place mainly in the large intestine.

With good and medium quality forage, grown in intensive or semi-intensive conditions, the apparent digestibility coefficients of dry matter, organic matter and crude protein are very similar in goats and sheep, but the results on cellulose digestibility are inconsistent (Table Va) (Baumgardt *et al.*, 1964; Jones *et al.*, 1972; Chenost, 1972; Wilson, 1977; Geoffroy, 1974; Huguet and Dulphy, 1976, unpublished; Blanchart *et al.*, 1977).

Organic matter of poor quality forage, particularly of tropical forages (Table Vb), does not seem to be better digested by goats than by sheep; sometimes protein is slightly better digested and in most cases crude fibre seems to be better digested by goats (Hossain, 1960, 1961; Mia *et al.*, 1960 a, b; Jang and Majumdar, 1962; Gihad, 1976; El Hag, 1976; Sharma and Rajora, 1977; Devendra, 1978).

Blanchart *et al.* (1977) showed that the apparent digestibility coefficients in male goats can be estimated from sheep coefficients accurately enough for organic matter and protein but less accurately for crude fibre. These results confirm the close relationship between organic matter and protein digestibilities in goats and in sheep whereas fibre digestibility in the two species is less similar.

Since results such as those of Blanchart *et al.* (1977) are still too scarce to determine accurately the digestibility of feeds by goats as compared to that by other ruminants, we are obliged to apply to goats the digestibility

Table V. Comparison of digestibility in goats and sheep: (a) High and medium value forages.

Type of feed	Species—No. of experimental animals	D.M. intake (g/kg $W^{0\cdot75}$)	Apparent digestibility coefficient: Dry matter	Organic matter	Crude protein	Crude fibre	References
Alfalfa—brome— grass hay (various stages of maturity)	4 castrated Saanen male goats (8 months)		62·1		75·6	64·9	Baumgardt *et al.* (1964)
	4 Holstein × Jersey bulls (8 months)		62·3		73·0	65·4	
	4 castrated Saanen male goats (15 months)		60·5		72·7	61·8	
	4 Jersey and Holstein bulls (15–18 months)		61·9		72·7	64·7	
3 Silages: lucerne, timothy, maize, harvested early and late	2 Toggenburg female goats (10 months)	39·6	64·0		62·4	63·1	Jones *et al.* (1972)
	2 yearling Cheviot ewes	54·3	64·2		55·8	65·4	
Lucerne hay + 2 silages: lucerne timothy and maize harvested early and late	4 Toggenburg female goats (6 months)	58·5	60·0		51·6	61·9	Jones *et al.* (1972)
	6 Cheviot × Suffolk ewe lambs	70–8	60·4		51·9	62·3	
4 Pelleted lucerne and fescue	Alpine male goats (2 years)	88·4		59·7	57·7	53·9	Chenost (1972a)
	Sheep (2 years)	86·7		59·3	59·8	51·4	
3 Pangola grass (*Digitaria decumbens*)	Alpine male goats (2 years)	32·9		66·3	68·5	70·4	Chenost (1972a)
	Sheep (2 years)	57·5		68·9	71·0	72·8	

5 Diets with rye grass	5 dry Poitevine goats						
hay with or without melasse	(5 years)	56·1		63·1			Geoffroy (1974)
urea and rye grass silage	5 castrated Texel rams						
	(2 years)	60·2		60·7			Geoffroy (1974)
19 Green forages	6 adult Alpine lactating goats	73·4	70·5	71·6	71·7	63·9	Huguet and Dulphy
7 Silages	6 castrated Texel rams	62·5	71·0	70·4	71·7	66·6	(1976)
15 Different diets	5–7 castrated Alpine male goats	58		71·6	69·4	59·5	Blanchart *et al.* (1977)
with hay or/and silage, with or	5·7 castrated Texel rams	46		69·4	67·9	53·7	
without concentrate							
			Apparent digestibility coefficient				
		D.M. intake (g/kg $W^{0.75}$)	Organic matter	Crude protein	N.D.F.[a]	A.D.F.[b]	
Medicago sativa	4 young wether goats	47·4	74·3	80·9	46·9	38·1	Wilson (1977)
(lucerne) (good quality)	10 young Merinos ewes (42 kg)	47·4	73·2	82·0	47·2	39·9	

[a] Neutral detergent fibre.
[b] Acid detergent fibre.

Table V. (b) Low quality forages.

Type of feed	Species—No. of experimental animals	D.M. intake (g/kg $W^{0.75}$)	Apparent digestibility coefficient: Dry matter	Organic matter	Crude protein	Crude fibre	References
Green spear grass (*Andropogon contortus*)	2 goats		59·7	64	66·4	66·9	Jang and Majumdar (1972)
	3 sheep		59·9	62·6	64·9	64·3	
	3 cattle		53·5	56·4	49·5	61·6	
Hay of natural grass with *Hyparrhenia*	12 small East African male goats	40·5	53·9	58·5	43·4	60·2	Gihad (1976)
	12 Dorper sheep	35·0	53·6	58·6	44·3	56·5	
Berseem hay (*Medicago sativa*) and lokl grass (*Dianthum annulatum*) dry desert grasses (*Hummra*)	4 Desert male goats (28–36 kg)		58·2	60·7	71·7	66·1	El Hag (1976)
	4 Sudan Desert rams (2 years, 26–36 kg)		57·1	60·1	69·9	63·9	
Dry grass (*Apluda aristata* and *Themada quadvivalvis* mixture)	4 growing female goats (Jamnapari)	41·1	53·6	54·9	22·8	63·9	Sharma and Rajora (1977)
	4 growing ewes (Sonadi)	50·9	48·8	49·8	18·7	57·4	
	4 growing cows (Gir)	62·4	36·6	40·3	4·3	53·4	
			Apparent digestibility coefficient: Organic matter	Crude protein	N.D.F.[a]	A.D.F.[b]	
Leaves of *Casuarina cristala*	4 feral wether goats (30 kg)	48·4	52·2	32·3	23·1	12·0	Wilson (1977)
	10 Merinos wethers (42 kg)	50·8	46·5	21·1	14·0	3·8	
Leaves of *Acacia pendula*	4 feral wether goats (30 kg)	57·4	58·8	68·8	31·8	20·7	Wilson (1977)
	10 Merinos wethers (42 kg)	57·0	56·6	63·2	27·9	14·2	

[a] Neutral detergent fibre.
[b] Acid detergent fibre.

coefficients observed in sheep. This is probably satisfactory for organic matter and protein but less so for crude fibre, especially in the case of tropical poor quality forages.

However, as most experiments have been carried out on male goats or on non-lactating female goats, digestibility in lactating goats and its variation according to milk yield (or feed intake) is completely unknown at the present time.

4. Requirements and Recommended Dietary Allowances

The information available at present essentially concerns energy requirements and, to a lesser extent, protein requirements of adult goats. Unfortunately, most experiments were carried out with too few goats, sometimes in precarious experimental conditions and with very different types of goats, such as dwarf goats or Angora goats, which may partly account for the variability of the results.

For a long time (French, 1944) goats were frequently treated like sheep for maintenance requirements and like cows for milk production. Now, on the basis of experimental results obtained from a great number of observations it will be possible to determine whether this is justified.

Recommended dietary allowances derived from results obtained in goats will be given here whenever it is possible; when not, data obtained from other ruminants and adapted to goats will be used.

(a) *Energy Requirements*

Recommended energy allowances for maintenance and lactation (Tables VI, VII, VIII) were obtained from weekly dietary balances in 94 lactating goats recorded daily over 15–19 weeks (Morand-Fehr and Sauvant, 1978b). Energy expenditure for lactation was estimated from daily milk yield, weekly fat percentage and from the regression equation of gross energy of 1 kg goat milk on fat percentage (Sauvant, Fehr, Charlet-Lery, unpublished). This equation is very similar to the one Gaines (1928) obtained for cows, probably because of the comparable dry matter compositions of cows' and goats' milk (Fehr, 1974). Empty live weight was determined from the equation based on results obtained in sheep (Church, 1975), which seemed to apply to goats too:

$$\text{E.L.W.} = \text{L.W.} - 3{\cdot}7 \text{ kg D.M.I.} \qquad (1)$$

Where E.L.W. = empty live weight, L.W. = live weight, D.M.I. = daily dry matter intake.

Table VI. Recommended allowances for maintenance.

Live weight (kg)	Net energy (MJ/day)	Digestible crude protein (g/day)	Ca (g/day)	P (g/day)
40	4·44	32	3·0	2·0
50	5·16	40	3·5	2·5
60	5·89	48	4·0	3·0
70	6·61	56	4·5	3·5
80	7·33	64	5·0	4·0

Table VII. Recommended allowances for gestation (last 2 months)

Live weight in early lactation (kg)	Net energy (MJ/day)	Digestible crude protein (g/day)	Ca (g/day)	P (g/day)
40	7·91	85	9·0	3·5
50	8·63	103	9·5	4·0
60	9·35	120	10·0	4·5
70	10·07	138	10·5	5·0
80	10·80	155	11·0	5·5

Energy balance in the goat calculated from the amounts of feed ingested, milk yield and live weight is more closely related to the empty live weight gain than to the live weight gain.

The feedstuffs were analysed by official methods. Metabolizable energy content of the forages was calculated from digestibility trials or estimated from Demarquilly and Weiss's (1970) tables. Likewise, metabolizable energy content of the concentrates was calculated using the regression equation of Sauvant (Demarquilly *et al.*, 1978).

The efficiency of metabolizable energy (M.E.) utilization for lactation (*kl*) was calculated from the equation of Van Es (1975) mentioned by Vermorel (1978).

The ingested net energy was adjusted by multiple regression to live weight, empty weight gain and 3·5% milk production. The energy requirements for maintenance were expressed in MJoules of net energy for lactation, assuming that the ratio *km* (the efficiency of M.E. utilization for maintenance) to *kl* (the efficiency of M.E. utilization for milk production) is constant at 1·2

$$\text{N.E.I.} = 0{\cdot}272\ W^{0{\cdot}75} + 1{\cdot}131\ \text{E.W.G.} + 2{\cdot}928\ \text{M.Y.} \qquad (2)$$

where N.E.I. = net energy intake (MJ/day), $W^{0.75}$ = metabolic weight (kg), E.W.G. = empty weight gain or loss (kg/28 days ≏ kg/month), M.Y. = 3·5% milk yield (kg/day); multiple correlation coefficient (r) = 0·90, residual variation: 6·1%.

Table VIII. Recommended allowances for lactation

Milk yield (kg of 3·5% fat corrected milk)	Net energy (MJ/day)	Digestible crude protein (g/day)	Ca (g/day)	P (g/day)
(A) 40 kg goat				
1	7·38	88	7·5	4·0
2	10·32	144	11·5	5·5
3	13·27	202	15·0	7·0
4	16·21	258	18·5	8·0
5	19·15	314	22·0	9·5
6	22·09	370	25·5	11·0
(B) 50 kg goat				
1	8·10	96	8·0	4·5
2	11·04	152	12·0	6·0
3	13·99	210	15·5	7·5
4	16·93	266	19·0	8·5
5	19·88	322	22·5	10·0
6	22·82	378	26·0	11·5
(C) 60 kg goat				
1	8·83	104	8·5	5·0
2	11·77	160	12·5	6·5
3	14·71	216	16·0	8·0
4	17·66	272	19·5	9·0
5	20·60	328	23·0	10·5
6	23·54	384	26·5	12·0
(D) 70 kg goat				
1	9·55	110	9·0	5·5
2	12·49	166	13·0	7·0
3	15·44	222	16·5	8·5
4	18·38	278	20·0	9·5
5	21·32	334	23·5	11·0
6	24·26	390	27·0	12·5

(i) *Maintenance.* The recommended dietary allowance for maintenance calculated from equation (2) is 0·272 MJ net energy per kg metabolic weight ($W^{0\cdot75}$). This value is close to those of other authors particularly that from Rostock mentioned by Skjevdal (1974) and that of Brody (1945); it is close to the recommended allowances of Fehr and Disset (1969) used in France up to now. Although these recommendations exceed by 12% those used in ewes, it should not be concluded that the requirements of the goat are higher, as this study was carried out with lactating goats and a 10–15% increase of the maintenance metabolism may generally result from lactation.

Table IX shows fasting metabolism and maintenance requirement values estimated by some authors. Goat fasting metabolism is about 0·251 MJ net energy/kg $W^{0\cdot75}$, a value close to sheep values and which agrees with the maintenance requirements from equation (2) (0·272 MJ/kg $W^{0\cdot75}$).

Table VI shows the recommended allowances for goats of different weights

Table IX. Estimation of energy requirements for maintenance and fasting metabolism.

Reference	Techinque	Goats	Value of maintenance requirement (M.R.) or fasting metabolism (F.M.)	Requirement of 60 kg goat in MJ metabolizable energy
Brody (1964)		Dairy goats	126 $W^{0\cdot55}$ (pounds) kcal N.E. (F.M.)	
Armstrong and Blaxter (1964)	Calorimetry	Dairy goats	56·4 $W^{0\cdot73}$ (F.M.) (51·9 $W^{0\cdot75}$) (F.M.)	1529
Fujihara *et al.* (1973)	Calorimetry	Dairy goats	85·2 $W^{0\cdot75}$ (F.M.)	2509
Skjevdal (1974)	Experimentation	Dairy goats	$0\cdot6 \frac{(\text{Live weight})^{0\cdot75}}{50}$ (M.R.)	2580
Sauvant and Morand-Fehr (1978)	Experimentation	Dairy goats	65·3 kcal/milk/$W^{0\cdot75}$ (M.R.)	2318
Opstvedt (1967)	Experimentation	Dairy goats	119·94 Nkf/kg $W^{0\cdot56}$ (M.R.)	2492
Experiments of Rostock	Experimentation	Dairy goats	1000 kcal N.E./kg $W^{0\cdot75}$ (M.R.)	2156
Flatt *et al.* (1972)	Application of dairy cow coefficients	Dairy goats	117 kcal N.E./kg $W^{0\cdot75}$ (M.R.)	2522
Huston *et al.* (1971)	Application of Brody's results (1954)	Angora goats		3470
Devendra (1967a)	Experimentation	Meat goats	0·41 S.E./100 lb L.W. (M.R.)	2896
French (1970)	Recommendations		0·9 S.E./100 kg L.W. (M.R.)	2896

as calculated from equation (2). A 60 kg goat requires 5·89 MJ net energy (i.e. 0·81 L.F.U.).* The recommended allowance for maintenance increases by 0·723 MJ N.E. (0·10 L.F.U.) for every additional 10 kg live weight.

The recommended allowances of Table VI apply to confined goats having little physical activity. They must be increased for goats on pasture or on range. In spite of a lack of data, it can be estimated that on pasture, maintenance requirements increase by 15–25% according to activity and by 30–60% on range according to the length of walks and the topography of the ground.

(ii) *Lactation*. Milk components are synthesized in the same way in goats and cows (Bickerstaffe *et al.*, 1974). Moreover, in France the average composition of goat and cow milk dry matter is very similar (Fehr, 1974). It is therefore likely that the quantity of energy necessary to synthesize 1 kg milk of the same concentration in the two species is the same. From equation (2), this value for synthesizing 1 kg 4% milk is 3·18 MJ (0·44 L.F.U.), a value close to the requirement of the cow for 4% fat corrected milk production. Devendra and Burns (1970) and Huston *et al.* (1971) suggest rather similar values from bibliographical and experimental studies. However, Opstvedt (1969), Skjevdal (1974) and especially Kalaissakis (1959) indicate milk energy values that are 0·21–0·42 MJ/kg lower. As the average fat content of goat milk in France is 3·3–3·4%, the recommended allowances of Table VIII are given for 3·5% standard milk. It is easy to modify milk production allowances according to fat percentage. A 0·5% variation of the fat percentage corresponds to 0·234 MJ N.E. (0·03 L.F.U.) for production of 1 kg milk.

(iii) *Weight Variation*. Variations of live weight during lactation influence the energy requirements of goats. Like cows and ewes, goats frequently cannot satisfy their requirements during early lactation, and must use body reserves. They may begin to restore these reserves and to increase in weight from the fourth month of lactation onwards. These weight variations must be taken into account when dietary allowances for different stages of lactation are recommended, and the values of Table VII must be modified. From equation (2) the energy value of 1 kg empty live weight gain was estimated to be 31·57 MJ. This value is close to that recorded in the cow but slightly higher than the one obtained by Devendra with few goats of different breeds, and higher than the value used in sheep. Hence, 1 kg empty weight gain during 1 month results in an increase of the energy requirements over that for maintenance of about 20%.

* Feed unit for milk production of 1 L.F.U. is equal to 7·23 MJ net energy i.e. the net energy of 1 kg standard barley.

(iv) *Gestation.* Live weight in pregnant goats as in other mammals does not markedly increase except during the last third of gestation (Blanchard and Sauvant, 1974). It therefore may be assumed that extra energy is required only during the last 2 months of gestation. Because of the lack of information about the rate and the composition of weight gain of the gravid goat uterus, energy needs of pregnant goats were estimated in feeding trials.

In 147 Alpine goats, weighed weekly and feed recorded daily, over the last 6 weeks of gestation (Sauvant and Morand-Fehr, 1978, unpublished) weight increased by about 1·5 kg per week. The specific requirement for gestation was calculated from daily feed intakes and energy values at an average of 3·47 MJ N.E. (0·48 L.F.U.) per day. Brody (1964) observed an increase of the basal metabolism by about 2·90 MJ per day for 1 kg gain per week over the last 3 weeks of gestation. At 1·5 kg gain per week, the increase of the basal metabolism was 3·61 MJ per day (0·50 L.F.U.), very near the value of 3·47 MJ above.

The supplementary intake may vary greatly around 0·48 L.F.U. (0·30–0·60 L.F.U. on average). Goats bearing twins or triplets do not consume more than those of the same age bearing one kid (see Section A2bi) but they mobilize their reserves early (Sauvant and Morand-Fehr, 1978a). As for lactation, larger goats consume more feed, at a rate of 100 g D.M. per every additional kg L.W. The recommended dietary allowances of Table VII were obtained from data of this study.

(b) *Protein Requirements*

(i) *Maintenance.* Studies up to now show that the protein requirement for maintenance of goats is almost the same as that for other ruminants. Most authors (Table X) suggest allowances of 60–80 g digestible crude protein (D.C.P.) per 100 kg live weight for maintenance, except Majumdar (1960a, b) who obtained high values with the *N* balance method. In France, use

Table X. Estimation of protein requirements for maintenance.

Reference	Technique	Value of digestible protein requirements for maintenance (g D.C.P./100 kg L.W.)
Majumdar (1960a)	Endogenous nitrogen	65
Majumdar (1960b)	Nitrogen balance	112
Devendra and Burns (1970)	Recommendations	45–64
Devendra (1971)	Recommendations	60–80
Fehr and Disset (1969)	Recommendations	80
Delort-Laval *et al.* (unpublished)	Nitrogen balance	60

of this method seems to show that the requirement is about 0·6 g per kg L.W. (Delort-Laval *et al.*, 1974, unpublished) whereas the practical allowances are 0·8 g. The digestible crude protein allowances of Table VI were calculated by using the sheep requirement for maintenance, i.e. 2·14 g D.C.P./kg $W^{0.75}$.

(ii) *Lactation.* In the literature, protein allowances for production of 1 kg milk varying from 40 to 70 g D.C.P. have been given by different authors. Devendra and Burns (1970) specified that 70 g D.C.P. were necessary for producing 1 kg milk at 4·5% fat. So, at the same fat percentage, the allowance for milk production in goats is similar to that in cows. Consequently, the same protein efficiency for milk production (0·55) used in cattle was applied to goats. Allowances were adapted to the lower concentration of goat milk.

The mean relation between fat (F.P.) and protein (P.P.) percentages of goat's milk is (Sauvant and Morand-Fehr, 1974, unpublished):

$$\text{P.P. } (\%) = 5{\cdot}2 \text{ F.P. } (\%) + 13{\cdot}5 \; (n = 2030 \; r = 0{\cdot}96) \qquad (3)$$

This equation enables us to estimate protein requirement per kg milk when only the fat percentage is known:

$$\text{D.C.P. } (\%) = 10{\cdot}3 \text{ F.P. } (\%) + 20{\cdot}1 \qquad (4)$$

Thus, for 3·5% standard milk, the expected average protein content is 3·17% and D.C.P. allowances 56·1 g/kg milk.

(iii) *Weight Variation.* Because of the lack of information about the proportion of protein in weight gain or loss, it is not possible to take into account the weight variations in the recommended protein allowances.

(iv) *Gestation.* In order to estimate the protein requirements for gestation we considered that adult goats weighing on an average 50, 60 and 70 kg are producing about 6·5, 7·5 and 8·5 kg of kid weight, respectively, and the need for 1 kg kid weight is 9·6 g D.C.P. per day during the last 8 weeks of gestation. The protein efficiency for gestation (0·50) in ewes was used to establish the values shown in Table VII.

(c) *Mineral and Vitamin Requirements*

A goat yielding 600 kg milk per year may excrete in the milk twice the quantity of her body minerals. Even a low yielding goat on poor pasture may suffer imbalances because of soil mineral deficiencies. The risks of deficiencies are high for phosphorus, calcium, magnesium, sodium and trace-elements, as well as for vitamins A and D.

Because of lack of data, particularly on the apparent or true digestibility coefficient of calcium and phosphorus, the allowances of Tables VI, VII, VIII are calculated by the factorial method using the same true digestibility coefficient as for ewes (Table XI). These values take into account the mineral composition of goat milk recorded on 32 goats over a whole lactation (Sauvant, Gueguen and Morand-Fehr, 1975, unpublished). These allowances are similar to those of Majumdar (1962) and slightly lower than those of Mackenzie (1967) which seem to be too high.

Table XI. Utilization of minerals by goats (%).

(a) True digestibility % (values are taken from sheep and used in calculating allowances in Tables VI–VIII)

	Ca	P	Mg
Gestation	30	60	25
Early lactation	35	65	25
Late lactation	30	60	25

(b) Endogenous losses or net requirements of minerals in goats at maintenance

	Ca	P	Mg	K	Na
mg/kg L.W./day	20	30	3·5	20	8

Table XII. Recommendations for macro-element, trace-element and vitamin content of goat feeds (Quantities per kg D.M. of feed).

Sodium	2 g
Magnesium	2 g
Sulphur	1·5 g
Copper[a]	10 mg
Cobalt	0·1 mg
Zinc	75 mg
Manganese	50 mg
Iodine	0·2 mg
Selenium	0·1 mg
Vitamin A	5 000 IU
Vitamin D	1 400 IU
Vitamin E	100 IU

[a] Above 20 mg/kg of D.M., risk of toxicity.

Table XII gives the practical allowances used in France for other macro-elements, trace elements and vitamins. Only vitamins A, D and E must be added to the diet. Vitamins of group B are generally synthesized in sufficient quantities by the rumen flora.

(d) *Water Requirements*

The availability of water may be a limiting factor of the milk yield in tropical and even in temperate areas.

Total requirements for water (water of the feeds + drinking water) were about 4–4·5 kg/kg D.M. in experiments in tropical areas (Devendra, 1967; Mackenzie, 1967; Devendra and Burns, 1970). From observations of Martin *et al.* (1974) and Hervieu *et al.* (1974, unpublished), it may be concluded that non-lactating goats reared in temperate climates at temperatures between 12 and 18°C need 2 kg water/kg D.M.; lactating goats need about 3·5 kg/kg D.M. However, the true effect of the milk production level on water intake is not known because trials are mostly made on non-lactating goats.

In practice, the need for drinking water is highly variable because it depends on climatic conditions, milk production level, feed intake level and water content of the feeds. The intake of drinking water is relatively low with diets composed of green roughages or silages, but it is seldom zero even if the total water intake is higher than values mentioned here.

The intake of water is relatively constant at air temperatures of 10–20°C, and increases when the temperature rises up to about 35°C (Appleman and Delouche, 1958; Rao and Mullick, 1965). Above this temperature, the reaction of the goats is variable, particularly if they are not adapted to high temperatures. Then the intake of dry matter, and sometimes of water, may decrease. However, in comparison with other ruminants, goats can adapt to high temperatures relatively easily (Schoen, 1968) by keeping the same dry matter intake level, by reducing their water losses in the urine and faeces and by increasing their pulmonary ventilation. Moreover, they are able to increase their water intake. According to Anderson (1955), they may easily drink quantities of water representing 20% or more of their live weight without apparent disorders. Likewise, goats are able to hold out during a temporary water deficiency (Schoen, 1968). In Mexico, dry Criollo goats have been observed to remain without water for 5 days taking only plant water and dew, which may represent rather substantial quantities of water (Gall, personal communication).

As a matter of fact, goats are very sensitive to water quality (water palatibility, taste and temperature of water). Salinity and temperature may illustrate this sensitivity. While goats reared on salt grounds ingest water containing several grammes of salt per litre without difficulty, non-adapted goats mostly refuse water rich in minerals. However, the exact salt-water tolerance of goats is difficult to estimate. Generally, goats do not like water that is too cold (Lyhs *et al.*, 1962; Fehr and Disset, 1969). Consumption of water is considerably reduced when its temperature is lower than 5°C.

When possible, water should be available for the animals all day, particularly with high yielding goats. If not, it is advisable to accustom them to regular waterings during the day.

B. Effect of Diet on Goat Performance

The nature and composition of diets clearly affect milk production and composition and, in some cases, reproductive performance although less is known about the latter. Consequently, it is necessary to determine the main feeding factors necessary to obtain optimum performance and establish feeding methods well adapted to goats.

1. Reproduction

As in other ruminants, the level of energy intake may have a marked effect on goat reproductive performance. Singh (1968) observed that whatever the protein content of the diet, the prolificity of Jamnapari and Barbari goats, given 75% of sheep energy allowances according to Morrisson, was reduced and the duration between two parturitions lengthened.

A lower energy intake level than that corresponding to maintenance needs during late gestation significantly increases the number of abortions (Wentzel *et al.*, 1974). However, a too high energy intake due to a large supply of concentrates (600–700 g per day) makes parturitions more laborious with longer times of dilatation and slower deliveries (Fehr *et al.*, 1974).

2. Milk Production and Composition

(a) *Effect of Energy*

According to all our trials on goats, the energy intake level is the factor affecting milk production the most markedly and the most quickly (Morand-Fehr and Sauvant, 1978a). In mid-lactation, the correlation between energy intake and milk yield always ranges around 0·83 whatever the mode of distribution (*ad libitum* or not) and the composition of the diet (Table XIII). At this stage of lactation, a decrease in energy consumption generally has repercussions for milk production, but only about 50% of the milk quantity corresponds to this decrease in the energy intake (Fehr and Delage, 1973); the other 50% depends on the mobilization of adipose tissue, as clearly shown by the increase in the level of non-esterified fatty acids in the blood plasma. Moreover, the milk protein percentage tends to increase and the fat percentage to decrease when the energy intake level increases (Table XIII). Thus, when the amount of ingested energy increases from 85% of total requirements to 115%, milk production increases by 8% and the fat percentage decreases by 7% so that the fat production is hardly affected; the protein percentage increases by 3%. This relative constancy of the fat production masks perceptible variations in the fatty acid composition. The percentage of palmitic acid increases and the percentage of stearic and oleic acids decreases systematically and significantly. The percentage of shorter chain fatty acids (C16) tends to increase but only the percentage of medium chain fatty acids (C_{14}, C_{12} and sometimes C_{10}) is higher and the percentage of short chain acids is slightly reduced (Morand-Fehr, 1978).

Table XIII. Effect on milk production and composition of consumed energy (32 Alpine goats, 1–5 years old) (Morand-Fehr and Sauvant, 1978).

	Correlation between consumed energy and		
Stage of lactation	Milk yield	Fat percentage	Protein percentage
Early lactation (first 8 weeks)	+0·75	0	−0·19
Late lactation (9th–27th week)	+0·83	−0·14	+0·15

Although energy and dry matter intake are generally closely related, it was demonstrated by modifying the proportions of roughages and concentrates that it is the energy intake and not the dry matter intake which is the main factor affecting milk production.

Opstvedt (1969) and Kondos (1972) also observed the favourable effects of energy intake on goat milk yield in full lactation. Moreover, Sachveda *et al.* (1974), in low yielding animals, recorded the poorest results with the ticularly if these lipids are highly unsaturated. In that case, they disturb rumen

In early lactation, the effect of ingested energy on milk yield is a little less evident (Table XIII) because goats can partly intensify the mobilization of adipose tissue according to the magnitude of the energy shortage (Morand-Fehr and Sauvant, 1978a). At that stage of lactation, a high energy intake level has no effect on fat percentage and a negative effect on protein percentage.

(b) *Effect of Protein*

As in other ruminants, nitrogen supplementation is beneficial when the diet is protein deficient. In most cases, it improves the feed intake level and the activity of the rumen flora. In particular, an addition of urea to the diet improves the feeding value of poor quality forage and nitrogen retention in goats (Fujimara and Tasaki, 1975).

It seems that the responses of cows and goats to protein intake variations are identical although there is little experimental evidence. Lindhal (1955, 1956) observed that feeding lactating goats with concentrates containing either 14 or 16% crude protein together with lucerne hay containing about 15% crude protein provided similar milk yields whereas supplementing hay containing 11% crude protein with concentrates of 16% increased milk production. On the other hand, results of Sachveda *et al.* (1974), in spite of the low production of the experimental goats, show that a limited amount of protein tends to reduce the milk yield if the energy of the diet is not limitative. According to some of our observations, the effect of crude protein seems to be more marked at the beginning of lactation.

Methionine seems to be the first limiting amino acid in the lactating goat, followed by threonine, valine and isoleucine (Champredon *et al.*, 1977). However, it is difficult to understand the possible consequences of these findings for goat yield.

The protein content of the diet does not seem to influence the milk protein percentage; hence protein overfeeding does not increase milk protein percentage, but might increase the non-protein nitrogen (N.P.N.) content of the milk. This could explain the higher N.P.N. content in goats' milk than in cows' milk with diets composed of lucerne, clover and other legumes. The same results were obtained with cows when the protein intake varied and the energy level of the diet remained constant (Rook and Line, 1962; Gordon and Forbes, 1970).

(c) *Effect of Lipids*

The content and the nature of lipids in the diet have little effect on goat milk yield except when their concentration in the diet exceeds 7–9%, particularly if these lipids are highly unsaturated. In that case, they disturb rumen fermentation by reducing the acetic acid production which tends to decrease the milk production and even the fat percentage. But if the oil added to the diet is protected from the action of the rumen fauna by encapsulation in formaldehyde treated casein, this fat rather tends to increase the milk fat percentage (Astrup *et al.*, 1972).

However, dietary fat plays a particular part because it supplies the largest quantities of C18 fatty acids and a fraction of C12–C16 acids of the milk fat when the energy balance of the animal is equal to zero or positive, i.e. when adipose tissue is not mobilized. A poor lipid diet decreases the milk fat percentage (generally by 3 or 4 points) and the percentage of C18 acids, and increases the percentage of shorter chain acids (Fig. 4). In fact, it is in particular the production of C18 acids which significantly decreases whereas the production of shorter chain acids tends to slightly increase. The addition of plant or animal fat to the diet restores the milk fat percentage. It seems that 2% fat in the dietary dry matter is a minimum below which the milk fat percentage may decrease; this corresponds to about 1 g of lipids per kg live weight. Diets with poor quality roughage, beets, barley and very defatted oil meal may have these effects on milk.

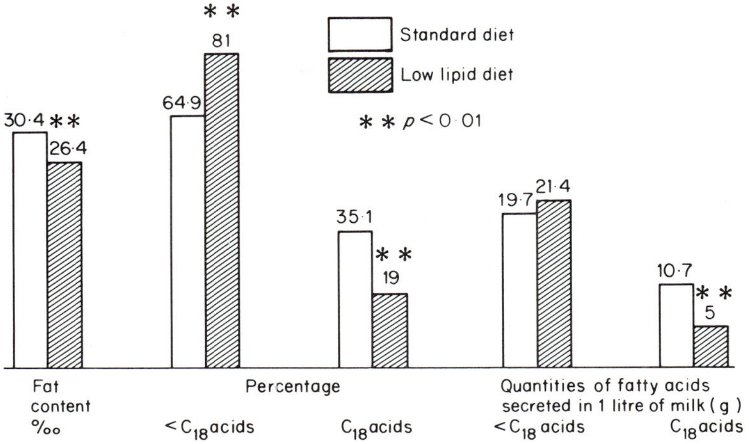

Fig. 4. Effect of a low lipid diet on the fatty acid composition of goat milk.

The kind of dietary fat affects the fatty acid composition of the milk fat, but because of fat hydrogenation in the rumen this influence is small. For example, a vegetable oil rich in linoleic acid causes a substantial increase in the milk fat of stearic and oleic acids and a decrease of shorter chain fatty acids, particularly myristic and palmitic acids (Delage and Fehr, 1967). However, if this oil is protected from rumen hydrogenation (by the treatment described above) the goat fat contains high proportions of linoleic acid (Astrup *et al.*, 1972).

(d) *Effect of the Dietary Energy Concentration*

The forage to concentrate ratio (F/C) expressed in dry matter is an important characteristic of the diet. With good quality forages, the F/C ratio allowing optimum performance is highly variable (between 0·6 and 2·3 with lucerne hay and even more if forage quality is very high; Fehr and Camara, 1972). When the forage is of poor quality, however, the optimum F/C ratio is located within a more restricted range, between about 0·8 and 1 (Sauvant and Fehr, 1974; Morand-Fehr and Sauvant, 1978a).

The energy density of the diet seems to be more suitable than the F/C ratio to characterize a diet because it includes not only the proportions of forages to concentrates, but also the quality of the forage. In different diets based on green and stored forages, the energy density of a diet allowing optimum milk production is about 9·4 MJ of metabolizable energy per kg dry matter (Morand-Fehr and Sauvant, 1978a, Fig. 5). This value is slightly higher in early lactation (second, third month) and lower in late lactation (seventh to ninth month).

At the same energy intake level, a high energy concentration of the diet tends to decrease fat percentage slightly. However, a high F/C ratio tends to increase it (Fehr and Camara, 1972).

The diet has an effect on milk production and composition, more because of its energy and protein intake than because of its energy or protein concentration or the kind of feeds used, except some feeds that can modify or disturb rumen fermentation.

C. Diets and Feeding Systems

The nutritional basis of goat feeding and dietary factors affecting the performances must be taken into account when establishing feeding schedules

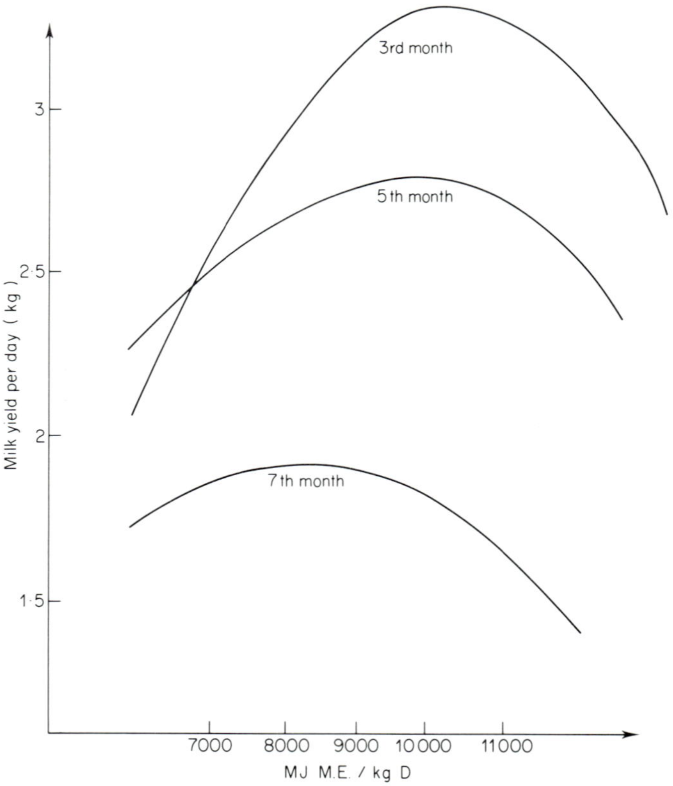

Fig. 5. Effect of energy density of the diet on milk yield in the goat.

for goats during an annual cycle. These principles will be applied to diets generally used in Western Europe with a temperate climate where production conditions range from moderate to very intensive.

As for other ruminants, the goat diet is generally composed of roughages, which constitute the basis, and possibly concentrates. The basal ration may contain one or several green or stored forages: hay, silage or artificially dried grass. Feeds like beets or cabbages etc. may also be used. Concentrates are given in well defined quantities and proportions to balance the basal ration with energy and protein and to satisfy the needs of the goats. A vitamin and mineral supply is generally necessary. The mineral feed can be given alone or with the basal ration or mixed with the concentrates.

1. Basal Rations

As goats are great dry matter consumers (see Section A2a), the aim in practical feeding is to feed as much forage as possible to satisfy the greatest part of the requirements for energy. During the past few years, many studies have been made to analyse different forage factors which may induce variations of the forage intake level (de Simiane, 1978).

(a) *Green Forages*

Goats may graze or green forage may be stall fed in a zero-grazing system. The intake level at the trough is generally much higher than at grazing. However, Masson and Decaen (1978, personal communication) recently showed that goats can consume large quantities of forages on pasture (80–97 g D.M./kg $W^{0.75}$) if pasture management is well adapted to the goats.

Green forage intake levels at stall feeding are generally very satisfactory. They vary according to plant species roughly from 60 to 100 g D.M. per kg $W^{0.75}$. Table XIV shows that the voluntary intake level of legumes,

Table XIV. Intake level of green forages used as a sole feed in the basal diet (tolerated refusals: 25–35%; concentrate intake = 0·7 kg D.M.) (de Simiane, 1978).

Forage	Stage of lactation (days)	Intake level: g/kg $W^{0.75}$	Intake level: for a 60 kg goat (kg)
Italian rye-grass	60–160	82 (60–101)	1·77 (1·3–2·18)
Orchard grass	60–160	65 (59–81)	1·40 (1·27–1·75)
Fescue	60–160	68 (62–75)	1·46 (1·34–1·62)
Lucerne	60–160	103 (69–157)	2·22 (1·49–3·38)
Red clover	60–160	93 (60–118)	2·00 (1·29–2·54)
Vetch + oats	130–150	79 (51–150)	1·70 (1·1–3·23)
Maize	160–200	81 (63–99)	1·74 (1·35–2·13)
Sorghum	160–200	85 (76–89)	1·83 (1·63–1·91)

particularly lucerne and red clover, is higher than that of grasses, and that in grasses, rye-grass is better ingested than dactyle or fescue (de Simiane, 1978).

However, within each species, the voluntary intake level varies according to the growth stage of the plant. The maximum intake is generally observed 1 or 2 weeks before the beginning of ear emergence in grasses and 1 week before the beginning of budding in legumes. The voluntary intake decreases when the crude fibre content of the forages increases and the content of protein decreases (de Simiane, 1978). On the other hand, there does not seem to be a very constant relationship between the voluntary intake and the dry matter content of the forages. However, with forages containing less than 10–12% D.M., the dry matter content appears to be the main limiting factor of the intake level.

Green forages may be fed long or chopped before feeding. While chopping reduces intake by about 15%, it is necessary with cane forages, like maize, sorghum and sunflower for facilitating handling and reducing waste.

Green forage intake level depends on the level of refusals tolerated and this is all the more marked when the animals are young. Lowering the amount of refusals from 38% to 15% may reduce the consumption by 0·5 kg D.M. per goat and thus affect milk production. An increase of tolerated refusals from 10 to 40% may save 400 g concentrates (50% of total concentrates) without reducing milk production (de Simiane, 1978). The effect of the proportion of refusals on the intake level also depends on the plant species. Thus, for the same level of refusals the amount of clover ingested exceeds that of lucerne. While farmers are becoming aware of the different levels of green forage intake of goats, the value for milk production of green forage is still less well known.

Green forages may be given alone or mixed with stored forages such as hay, silages or dehydrated and pelleted forages. The total dry matter intake level can be improved when the diet contains more than one forage, but the response of the animals is very variable and depends on the feeding value of the forage.

(b) *Stored Forages*

The storage method may substantially modify the voluntary intake level of

Table XV. Feed intake level (total D.M. kg/100 kg L.W.) of goats in mid-lactation given forages *ad lib.* + concentrate ⩽ 700 g D.M. (de Simiane, 1978).

	Green forage	Hay	Silage
Italian rye-grass	3·3 (2·1–4·7)	3·1 (2·1–3·9)	2·0 (1·7–2·6)
Lucerne	4·3 (2·3–5·7)	3·2 (2·3–4·0)	2·6 (1·9–3·2)

green forages and also their utilization by goats. Stored forage is consumed less than green forage and silage definitely less than hay (Table XV).

(i) *Hay*. The hay intake level in lactating goats varies between 50 and 100 g D.M./kg $W^{0\cdot75}$ (1·2–2·4 kg D.M. in 60 kg goats). Variation in the hay intake level essentially depends on the climate, hay-making technique and growth stage at harvesting. Exposure of lucerne hay for 15 days to alternating rain and sunshine leads to a 35% decrease in intake (Fehr and Le Jaouen, 1977). On the other hand, pelleting of poor quality hay definitely improves its voluntary intake level (Fehr, 1974). Hay-making tends to reduce the intake difference between forage species. However, hay-making conditions induce much larger variation of the voluntary intake and the milk production value of the forage than the botanical composition of the hay.

(ii) *Dehydrated and Pelleted Forages*. The intake levels of dehydrated and pelleted forages are similar to those of good quality hay. With high temperature dehydratation followed by pelleting there are no differences between the intakes according to plant species; thus, dehydrated and pelleted fescue is as well eaten as Italian rye-grass or lucerne.

Milk yields are mostly very satisfactory when goats are given dehydrated and pelleted forages. But when forages are conditioned in the same way, a balanced ration composed of grass (40%), whole plant maize (40%) and lucerne (20%) gives the highest milk yields. Diets containing grasses (fescue or rye-grass) or a maize–lucerne mixture give slightly lower results. Dehydrated lucerne alone does not give rise to difficulties but the poorer energy concentration tends to decrease milk yield. Dehydrated whole maize alone does not seem to be favourable, particularly during late lactation and during gestation because of excessive fattening of goats (de Simiane, 1978).

(iii) *Silages*. Silages in goats give rise to digestive or metabolic problems more often than other stored forages. Voluntary intake level of silages varies between 55 and 85 g/kg $W^{0\cdot75}$ and is definitely lower than that of other forages. The large variation in the silage intake levels is due to quality that depends on fermentation conditions and harvesting stage of the plant. Milk yield tends to be lower by 5–15% than with green or pelleted forage, but it is equal to or somewhat higher than with medium quality legume hay. With maize silage, fat percentage tends to be the same as, or slightly higher than, with other forages; protein percentage is the same. During late lactation and during gestation, live-weight gain tends to be larger than with green forage or hay. It is sometimes advisable to restrict maize silage intake in order to avoid fattening. However, supplying hay with maize silage, particularly during late gestation and early lactation would tend to reduce digestive or metabolic disorders and would improve milk yield. In the same way, combining hay with grass silage would improve total forage intake level and milk yield. Observations made in experimental stations or in farms indicate that digestive and metabolic or other health disorders while rare in goats fed grass silage do occur with maize silage. In particular acidosis, demineralizations, cortical necrosis, enterotoxemia and listeriosis may appear more frequently.

The use of some silages as sole forage of a basal ration may lead to satisfactory results and will correspond to the production capacity of most goat herds. Feeding exclusively silages and concentrates creates a risk of disease, but these diets can be fed during consecutive periods of 2–3 months during lactation. Feeding maize silage to high yielding goats is certainly more problematic than the use of grass silages.

(c) *Other Feeds*

Crucifers like rape (*Brassica napus*) or kale (*B. oleracea*) and beets (*Beta vulgaris*) are often added to goat diets with green or stored forages. Other plants like carrots (*Daucus carota*), swedes (*Brassica napus napo-brassica*), Jerusalem artichokes (*Helianthus tuberosus*) and turnips (*Brassica rapo rapifera*) are also used but not so often. The crucifers have the advantage of being rich in energy and relatively rich in protein. Because of anti-nutritional factors, the ingestion of large quantities of crucifers can be injurious to the thyroid gland and in lactating females may result in severe disorders. It is generally advisable to limit cabbages in lactating goat diets to 30% of the total dry matter. The voluntary intake of beets essentially depends on their physical form (cleanness, chopping). Forage-restricted dairy goats may ingest large quantities of beets, up to 1 kg D.M., without disorders. But the addition of beets to the diet must be calculated according to the characteristics of this feed (rich in carbohydrates, poor in fibre) which, apart from the dry matter content, allow them to be ranked among the concentrates.

Thus, a great number of basal diets of a single forage or a forage mixture can be used for goats, but they are seldom sufficient to obtain optimum milking performance. The forage energy intake is indeed often insufficient and imbalances between energy, protein and mineral supplies are frequent in basal diets. Consequently these diets should be supplemented with concentrates.

2. Concentrates

(a) *Type of Concentrates*

Goats may partly or completely refuse concentrates because of their physical form or defective conservation. It is better to give coarsely ground or pelleted concentrates than finely ground ones to avoid waste and risk of introducing fine particles into the lungs of the animals. Goats are apparently susceptible to concentrate quality, perhaps more than cows. They quickly reduce their intake when the concentrate is mouldy or fermented.

Cereals are generally well accepted (wheat sometimes less), likewise milling by-products (cereal shorts, screenings, brans) and oilmeals (groundnut, linseed, soybean, sunflower). Rapeseed meal may decrease acceptability of concentrates although this effect is not constant. Dehydrated lucerne flour and some animal fats may be involved in the poor acceptance of some concentrates.

(b) *Concentrate Feeding*

The amount of concentrate given to goats for supplementing the basal diet must be determined not only from a technical, but also from an economic point of view according to concentrate and goat milk prices. For analysing the economics of concentrate feeding, we assume conditions favouring highest performance, i.e. where the ratio of goat milk to concentrate prices is high.

(i) *Late Gestation.* The supply of concentrates during late gestation affects the milk production during early lactation (Table XVI). Concentrate supplementation during late gestation reduces the forage intake but, with adequate forage quality, not enough to avoid an increase in the energy intake and

Table XVI. Effect on milk yield of the concentrate supply before parturition[a] (Sauvant and Morand-Fehr, unpublished).

	Concentrate supply		
	low (150 g/day)	high (600 g/day)	Significance
Milk production (kg/day)	2·99	3·76	$p \leq 10\%$
Fat content (g/kg)	41·0	44·4	$p \leq 5\%$
Energy balance (MJ)	− 2·47	− 4·49	$p \leq 5\%$
N.E.F.A. (mg/100 ml)	90·8	110·8	$p \leq 10\%$
Ketone bodies (mg/100 ml)	27·8	3·04	$p \leq 10\%$

[a] 32 goats: lucerne hay *ad lib.*, average values of 8 weeks.

consequently in the energy reserves, i.e. storage of lipids in the adipose tissue of pregnant goats. Consequently, after parturition more adipose reserves are mobilized and milk production is enhanced (Table XVI). In Norway, concentrate feeding during the dry period increased milk production in high yielding goats but reduced it in low yielders (Table XVII). This was apparently due to displacement of forages. On average, the experimental groups consumed 20% less forage but with poor quality forage intake was reduced by as much as 50%. Thus, recommendations on feeding of concentrates during the dry period will have to consider forage quality, milk yield and physical conditions of the goats. However, concentrate intake at

Table XVII. Difference in milk yield due to additional feeding concentrates in the dry period in relation to yield in the previous year; 193 goats, 3 years and older (Skjevdal, 1979).

Yield in previous year (kg)	Quantity of concentrates fed additionally 18 kg[a]	30 kg[b]
	kg milk	kg milk
Less than 500	0	−64
500–599	−13	− 2
600–699	+15	+11
700–799	+23	+73
More than 800	+91	0

[a] Averages of 2 experiments at A.U.N.
[b] Averages of 2 experiments at Valbjør.

a forage concentrate ratio below 1 may result in digestive disorders, difficulties at parturition (see Section B1), low forage intake after parturition and low milk yield at the onset of lactation (Fehr, 1974). Thus, the forage intake of a 60 kg goat must be 1·2 kg D.M. or more during late gestation for the supply of concentrates to be efficient. In many French and Norwegian experiments, it was shown that this may not be ensured if forage quality is poor (see also Chapter 15).

(ii) *Early Lactation.* Before and after kidding goats should be fed the same feeds. Towards the end of the dry period rations are often adjusted so as to supply nutrients sufficient for 2 kg milk. After parturition the concentrate ration should be increased carefully and by not more than 0·1 kg every third day or the goats may go off feed. While feeding concentrates twice a day will be satisfactory with small amounts, more than 1·2 kg should be split into three daily portions. Goats may lose appetite a few days after parturition or later in lactation, especially if tied up in stalls. A minor disturbance causing rumen pH to fall from the average lactation value of 6·0 to 6·2 may result in reduced appetite (Skjevdal, 1974). However, in a feeding trial adding 1% urea to the concentrates as a buffer or giving hay prior to feeding, concentrates did not affect feed consumption or milk yield. Rumen pH was 6·39, 6·28 and 6·26 in the groups fed hay before the concentrates or urea and the controls respectively; there were no differences in the content of volatile fatty acids in the rumen fluid (Fauske, 1972). Appetite of goats that have gone off feed will be restored within a few days and major loss of milk prevented by reducing the concentrate ration to one half, supressing silage

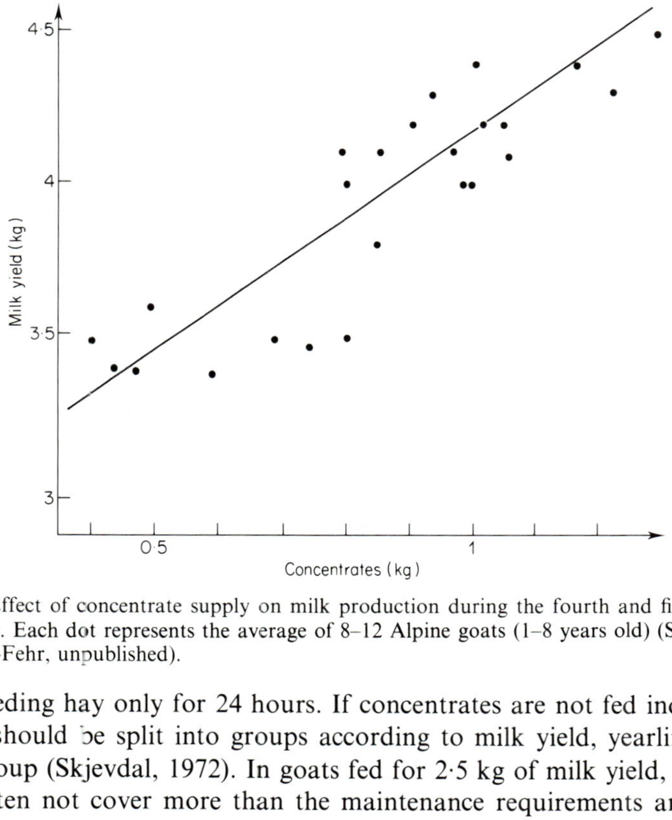

Fig. 6. Effect of concentrate supply on milk production during the fourth and fifth week of lactation. Each dot represents the average of 8–12 Alpine goats (1–8 years old) (Sauvant and Morand-Fehr, unpublished).

and feeding hay only for 24 hours. If concentrates are not fed individually herds should be split into groups according to milk yield, yearlings being one group (Skjevdal, 1972). In goats fed for 2·5 kg of milk yield, roughage will often not cover more than the maintenance requirements and 0·4 kg of concentrates will have to be fed for each kg of milk. As a consequence fat content will be reduced.

In our trials, goats fed concentrates in addition to the basal diet of lucerne hay produced 140 g additional milk per 100 g of concentrates (Fig. 6). This limited effect of concentrate feeding was due to the displacement of forage with a coefficient of 1, i.e. forage intake decreased by 100 g D.M. when 100 g of concentrate D.M. was added. The effect of concentrate feeding was even less in the first week of lactation. Comparable results were recorded by de Simiane (1978) with maize–silage diets. Nevertheless it was shown that concentrates are the most efficient in early lactation.

(iii) *Decreasing Phase of Lactation.* During this period, the response of milk production to concentrate supplementation is less obvious than during early lactation. The effect depends on production level, type of basal ration, concentrate intake level and feeding method.

In adult goats offered green grass or legume forages *ad libitum* during the third to fourth month of lactation, an increase in the amount of concentrate up to 0·7 kg D.M. does not reduce (sometimes increases) the forage intake and improves the milk yield. However above 0·7 kg D.M., the displacement of green forages by concentrates quickly equals 1 so that improvement of the milk yield is only small.

To obtain the same milk yield, the supply of concentrates must be higher with a hay diet than with a green forage diet. The response of milk yield to the supply of concentrates depends on the mode of concentrate feeding,

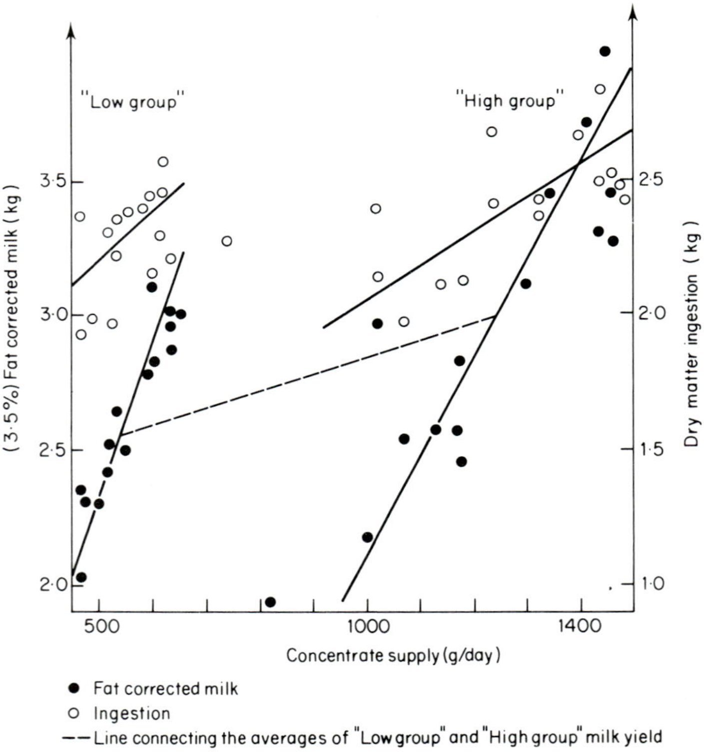

Fig. 7. Effect of concentrate supply on dry matter intake and (3·5%) fat corrected milk yield of goats (Sauvant and Morand-Fehr, 1977).

i.e. whether given according to the milk production level or not (Fig. 7). When two groups of goats received lucerne hay *ad libitum*, and 350 g ("high group") or 175 g ("low group") of concentrates per kg milk, displacement was slightly lower than 1 and only 79 g of milk were produced per 100 g concentrates ingested. Within groups, there was no displacement of hay by concentrate and milk production increased by 480 g with a concentrate supplementation of 100 g. Thus, the supply of concentrates is biologically and economically inefficient unless given according to milk yield. Consequently, it is better to give high quantities of concentrates to high yielding goats and reduce it in low producers. Unfortunately, it is difficult in practice to have different quantities of concentrates for each goat, but it is possible to form groups of goats according to their level of production.

(c) *Concentrate Feeding Plan during the Reproductive Cycle*

The supply of concentrates cannot be defined in a static way. It is advisable to determine an overall strategy of concentrate feeding for the whole reproductive cycle rather than define a succession of independent phases as in

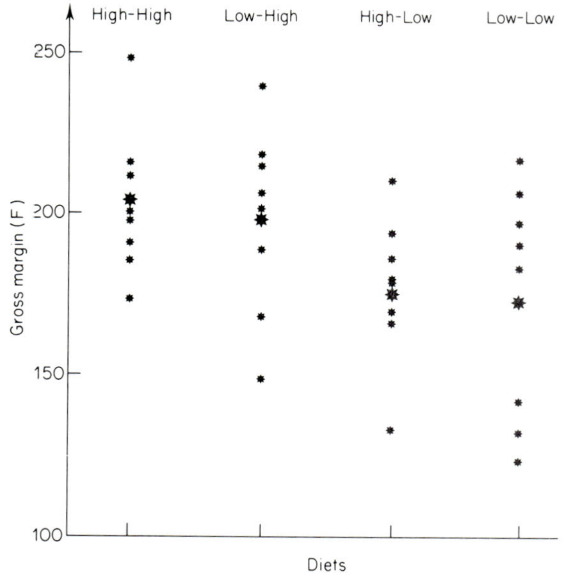

Fig. 8. Income over feed cost in dairy goats subjected to four feeding programmes over 4 months around parturition. Each star corresponds to a goat; big stars represent averages of each group. High: high concentrate diet; low: low concentrate diet. The first high or low characterizes the diet before parturition. The second high or low characterizes the diet after parturition (Sauvant and Morand-Fehr, unpublished).

the previous section. In fact, some experimental results show that the energy homeostasis is maintained over the whole reproductive cycle.

It would be interesting to define a plan of concentrate supplies before and after parturition. Figure 8 gives an economic comparison of four strategies applied to goats (Fehr *et al.*, 1974; Sauvant *et al.*, 1974). Because of current French prices of goats' milk and feeds, a high level of concentrate supply ("High group") is more favourable at the beginning of lactation than during

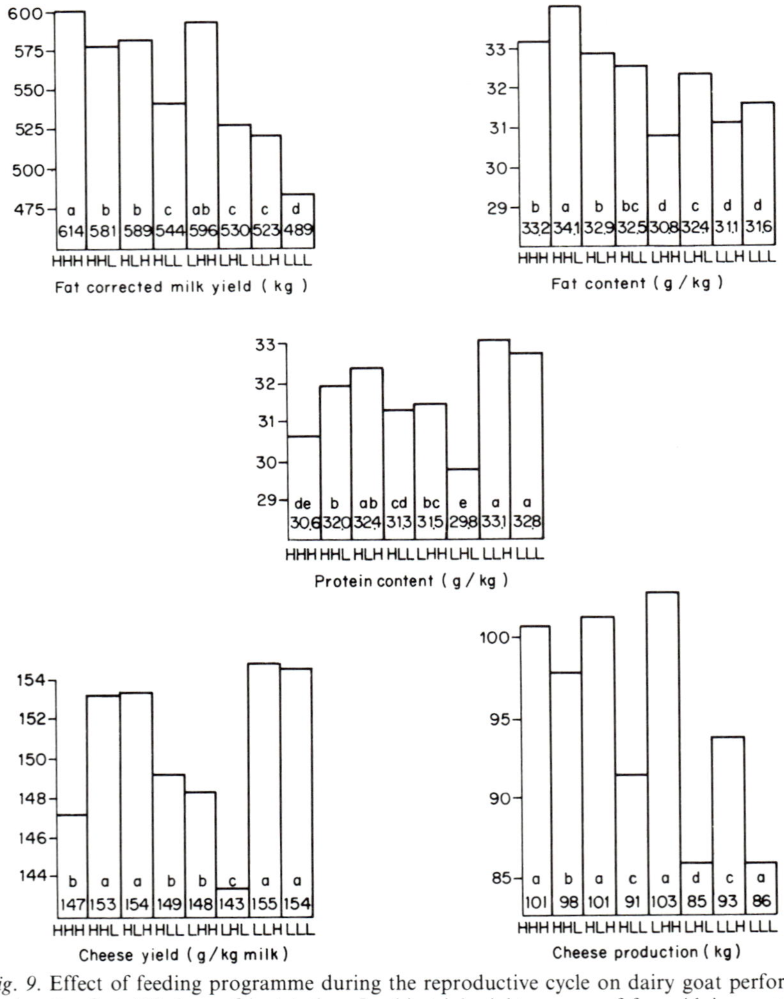

Fig. 9. Effect of feeding programme during the reproductive cycle on dairy goat performance during the first 189 days of lactatation. In this trial, eight groups of four Alpine goats each received high (H) or low (L) levels of concentrates in late gestation, in early and mid-lactation. Values with the same letters are significantly different ($p < 0{\cdot}05$) (Morand-Fehr and Sauvant, 1978a).

late gestation. Figure 8 emphasizes, however, that in each treatment the individual variations are higher than those of the diets. We can use the same approach to study the more extended periods: late gestation (8 weeks), ascendant period of lactation (8 weeks) and also the descendant period of lactation. An overall strategy of concentrate feeding during a reproductive cycle can be established from these results (Sauvant *et al.*, 1974; Morand-Fehr and Sauvant, 1978a).

In a trial with 32 goats, Morand-Fehr and Sauvant (1978a) showed that a high supply of concentrates from 8 weeks before parturition to 30 weeks of lactation is not technically and economically the most satisfactory solution (Fig. 9). A diet relatively poor in concentrates during late gestation and high during the whole lactation and a diet rich in concentrates during late gestation and in mid-lactation and poor in early lactation give the same performances; they are technically as good and economically better than the diet rich in concentrates offered over the whole cycle.

The goat dairyman may also ask whether a given quantity of concentrates during lactation should be offered at variable levels according to changes in production or at a constant level. Figure 10 shows that the second technique, easier to apply in practice, leads to a better milk performance. This difference is due to the fact that animals consuming the same quantity of concentrate during early lactation will be accustomed to a higher quantity of lucerne hay or maize silage. This high forage intake level persists during the whole lactation.

According to these experiments, feeding programmes for dairy goats must be established for the whole year and not by adjusting to their needs every day because the response of an animal to a diet depends on its nutritional past, and it is advantageous to use the capacity of the goat to store reserves

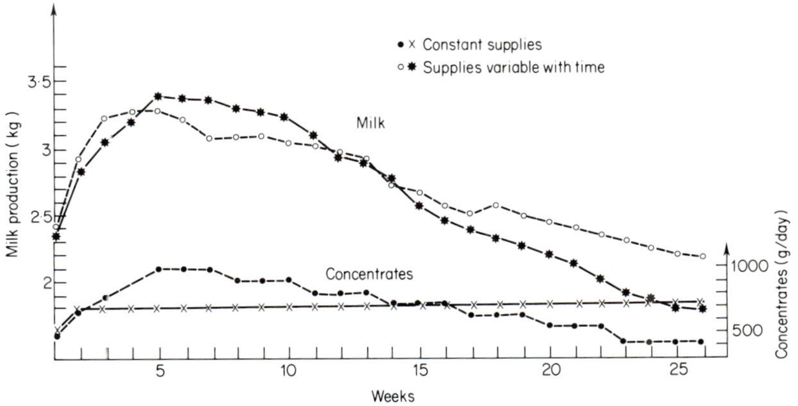

Fig. 10. Effect of the plan of concentrate supply during lactation (Morand-Fehr and Sauvant, unpublished).

during late lactation and gestation when needs are low and to mobilize them during early lactation when needs are high and appetite is low. Goats may "buffer" the rapid variations in their production needs and also the variation in feed availability during the year.

Conclusion

Although the same fundamental nutritional laws observed in other ruminants apply to goats, their particular feeding behaviour must be taken into account. With the information available on intake levels, feed efficiency and nutrient requirements, herd feeding programmes can be established for intensive production conditions (Sauvant and Morand-Fehr, 1977a). This is still difficult for pastoral conditions, for which there are few data available, particularly with respect to the use of poor forage and browse. Owing to the capacity of goats to adapt to diverse environmental conditions, goat production can develop in quite different areas where goats have advantages over other ruminants: under extensive arid conditions, where, because of their feeding behaviour, goats are the only animals able to thrive; or under intensive conditions, where they can produce most efficiently because of their high feed intake and milk potential. Goat production can be improved in various environments throughout the world if suitable rational feeding methods are applied.

References

Appleman, R. D. and Delouche, J. C. (1958). *J. Anim. Sci.* **17**, 326–335.

Anderson, B. (1955). *Acta Physiol. Scand.* **33**, 50–65.

Armstrong, D. G. and Blaxter, K. L. (1964). *In* "3rd Symposium on Energy metabolism", EAAP, Academic Press, London and New York.

Astrup, H. N., Nedkvitne, J. J., Skjevdal, T., Bakke, E., Lindstad, P. and Eckhardt, W. (1972). *Norges Landbrukshogsk.* **90**, 24–31.

Baumgardt, B. R., Byer, W. J., Jumah, H. F. and Krueger, C. R. (1964). *J. Dairy Sci.* **47**, 160–164.

Bickerstaffe, R., Annison, E. F. and Linzell, J. L. (1974). *J. Agric. Sci. Camb.* **82**, 71–85.

Blanchard, D. and Sauvant, D. (1974). *In* "Journée d'études sur l'alimentation de la Chèvre Laitière", 13 June 1974, pp. 76–85. ITOVIC, Paris.

Blanchart, G., Brun-Bellut, J. and Vignon, B. (1977). *Bull. E.N.S.A.I.A. Nancy* **19**, 1–11, 37–47.

Brody, S. (1945). *Missouri Agric. Exp. Sta. Res. Bul.* 291.

Brody, S. (1964). "Bioenergetics and Growth". Hafner, New York.

Champredon, C., Pion, R. and Prugnaud, J. (1977). *Ann. Biol. Anim. Biochim. Biophys.* **17**, 85–93.

Chenost, M. (1972). *Ann. Zootechn.* **21**, 107–111.

Church, D. C. (1975). *In* "Digestive Physiology and Nutrition of Ruminants" (D. C. Church, ed.) Oregon State University.
Devendra, C. (1967a). *Malaysian Agric. J.* **46**, 80–97.
Devendra, C. (1967b). *Malaysian Agric. J.* **46**, 191–203.
Devendra, C. (1971). *In* "2nd Int. Conf. Goat Breeding", pp. 88–115, Tours (France), July 1971, ITOVIC, Paris.
Devendra, C. and Burns, M. (1970). *In* "Goat Production in the Tropics", pp. 89–114. Commonwealth Agricultural Bureaux, Farnham Royal, Bucks, England.
Delage, J. and Fehr, P. M. (1967). *Ann. Biol. Anim. Biochim. Biophys.* **7**, 437–444.
Demarquilly, C. and Weiss, P. (1970). *In* "Etude SEI no. 40", INRA, SEI, Versailles.
Demarquilly, C , Andrieu, J. and Sauvant, D. (1978). *In* "INRA Alimentation des Ruminants", pp. 469–518. INRA, Versailles.
El Hag, G. A. (1976). *Wld. Rev. Anim. Prod.* **12**, 43–48.
Es, A. J. H. van (1975). *Livestock Prod. Sci.* **2**, 95–107.
Fauske, R. (1972). *Norges Landbrukhøgsk., Radet for Husdyrforsøk, 198–202.*
Fehr, P. M. (1971). *Ann. Zootechn.* **20**, 305–320.
Fehr, P. M. (1974). *In* "Journée d'étude sur l'alimentation de la Chèvre Laitière", June 1974, pp. 19–34. ITOVIC, Paris.
Fehr, P. M. and Camara, S. (1972). *In* "2nd World Congress of Animal Feeding", Madrid, October 1972, Vol. V. pp. 381–388.
Fehr, P. M. and Delage, J. (1973). *C.R. Acad. Sci. Paris* **276**, Série D, 3449–3452.
Fehr, P. M. and Disset, R. (1969). *Rev. Elevage*, 45th spec. no., 145–151.
Fehr, P. M. and Le Jaouen, J. C. (1976). *Rev. Laitière Fr.*, no. 338, 39–55.
Fehr, P. M., Hervieu, J. and Delage, J. (1974). *In* "Journée d'étude sur l'alimentation de la Chèvre Laitière", June 1974, pp. 88–102. ITOVIC, Paris.
French, H. H. (1944). *E. Afr. Agric. J.* **10**, 66.
Fujihara, T. and Tasaki, I. (1975). *J. Agric. Sci. Camb.* **85**, 185–187.
Gaines, W. L. (1928). *Illinois Agr. Expt. Station Bull.* 308.
Gihad, E. A. (1976). *J. Anim. Sci.* **43**, 879–883.
Gihad, E. A. and Morav, H. M. (1977). *In* "Symposium on Goat Breeding in Mediterranean Countries". Malaga, Grenada, Murcia, Spain, October 1977, pp. 161–163.
Gordon, F. J. and Forbes, T. J. (1970). *Rec. Agric. Res.* **18**, 29–37.
Geoffroy, F. (1974). *Ann. Zootechn.* **23**, 63–73.
Gray, S. V. (1947). *J. Exp. Biol.* **24**, 15–19.
Hossain, W. (1960). *Agric. Pakist.* **11**, 139–152.
Hossain, W. (1961). *Agric. Pakist.* **12**, 212–217.
Huston, J. E., Shelton, M. and Ellis, W. C. (1971). Texas Agric. Exp. Station. Document no B 1105, March 1971, 16 p.
Jang, S. and Majumdar, B. N. (1962). *Ann. Biochem. Exp. Med.* **22**, 303–308.
Jones, G. M., Larsen, R. E., Javed, A. H., Donefer, E. and Gaudreau, J. M. (1972). *J. Anim. Sci.* **34**, 830–838.
Kalaissakis, P. (1959). *Z. Tierphysiol. Tierernähr. Futtermittelk.* **14**, 204–214, 303–311.
Kondos, A. C. (1972). "Proc. Nat. Goat Breeders' Conf.", Melbourne, May 1972, pp. 7–12. Goat Breeders' Soc. of Australia.
Lindahl, I. L. (1955). Annual Report Animal and Poultry Husbandry Research Branch A.R.S., U.S.D.A.
Lindahl, I. L. (1956). Annual Report Animal and Poultry Husbandry Research Branch A.R.S., U.S.D.A.
Lyhs, L., Berger, L. and Wildenhahn, (1962). *Mh. Vet. Med.* **17**, 899.

Mackenzie, D. (1967). "Goat Husbandry", Faber and Faber, London.
Majumdar, B. N. (1960a). *J. Agric. Sci. Camb.* **54**, 329–334.
Majumdar, B. N. (1960b). *J. Agric. Sci. Camb.* **54**, 335–340.
Majumdar, B. N. (1962). *In* "Handbook of Animal Husbandry" (S. K. Sen and P. Nath, eds), p. 75. Indian Council of Agricultural Research, New Delhi.
Martin, D., Fehr, P. M., Hervieu, J. and Cucci, A. (1974). *In* "Journée d'étude sur l'alimentation de la Chèvre Laitière", June 1974, pp. 151–158. ITOVIC, Paris.
Mia, W. H., Sahai, B., Majumdar, B. N. and Kehar, N. D. (1960a). *Ind. J. Dairy Sci.* **13**, 1–8.
Mia, W. H., Sahai, B., Majumdar, B. N. and Kehar, N. D. (1960b). *Ind. J. Dairy Sci.* **13**, 9–15.
Morand-Fehr, P. (1978). 29th Annual Meeting of European Association for Animal Production, Stockholm. June 1978. Report, Study Commission on Sheep and Goat Production.
Morand-Fehr, P. and Hervieu, J. (1977). *In* "Symposium on Goat Breeding in Mediterranean Countries", Malaga, Grenada, Murcia, Spain, October 1977, pp. 156–160.
Morand-Fehr, P. and Sauvant, D. (1978a). *Livestock Prod. Sci.* **5**, 203–213.
Morand-Fehr, P. and Sauvant, D. (1978b). *In* "INRA Alimentation des Ruminants", pp. 449–467. INRA, Versailles.
Opstvedt, J. (1969). *In* "Grassland and Sheep and Goat Production" (O. R. W. Spedding ed.) pp. 89–100, EAAP Report 2 Study Commission on Animal Nutrition.
Rao, M. V. N. and Mullick, D. N. (1965). *Ind. J. Vet. Sci.* **35**, 288–296.
Ridges, A. P. and Singleton, A. G. (1962). *J. Physiol.* **161**, 1–9.
Rook, J. A. K. and Line, C. (1962). *16th Int. Dairy Congr.* 57–62.
Ruckebusch, Y. and Bost, J. (1963). *Rev. Med. Vet.* **114**, 184–195.
Sachdeva, K. K., Sengar, O. P. S., Singh, S. N. and Lindahl, J. L. (1974). *Milchwiss.* **29**, 471–475.
Sauvant, D. and Fehr, P. M. (1974). *In* "Journée d'étude sur l'alimentation de la Chèvre Laitière", June 1974, pp. 59–70. ITOVIC, Paris.
Sauvant, D. and Morand-Fehr, P. (1977a). *In* "Symposium on Goat Breeding in Mediterranean Countries", Malaga, Granada, Murcia, Spain, October 1977, pp. 174–183.
Sauvant, D. and Morand-Fehr, P. (1977b). *In* "Symposium on Goat Breeding in Mediterranean Countries", Malaga, Granada, Murcia, Spain, October 1977, pp. 184–194.
Sauvant, D. and Morand-Fehr, P. (1978a). *Anim. Prod.* (in press).
Sauvant. D., Delage, J. and Fehr, P. M. (1974). *In* "Journée d'étude sur l'alimentation de la Chèvre Laitière", June 1974, pp. 106–114. ITOVIC, Paris.
Schoen, A. (1968). *E. Afric. Agric. Forestry J.* **34**, 256–262.
Sharma, V. V. and Rajora, N. K. (1977). *J. Agric. Sci. Camb.* **88**, 75–78.
Simiane, (de) M. (1978). *In* "Advances in Goat Feeding" (P. Morand-Fehr *et al.*, eds), pp. 124–189. I.N.A., Paris.
Skjevdal, T. (1972). *Sau og Geit* **25**, 63–65.
Skjevdal, T. (1974). *Tech. Bull.* 169, Dpt. Anim. Nutr. Agric. Univ. Norway, 42 pp.
Skjevdal, T. (1979). *Meld. Norg. Landbrukshøgsk.* **58**, (24) 1–22.
Vermorel, M. (1978). *In* "INRA Alimentation des Ruminants", pp. 47–88. INRA, Versailles.
Wentzel, D., Morgenthal, J. C., Van Niekerk, C. H. and Roelofse, C. S. (1974). *Agroanimalia* **6**, 129–132.
Wilson, A. D. (1977). *Austr. J. Agric. Res.* **28**, 501–508.

7. Diet Selection, Grazing Habits, and the Place of Goats in Range Management

L. B. Merrill and C. A. Taylor

A. Diet Selection by Goats

The selection of plant species by goats is determined by several factors.

1. Vegetation Composition

Diet selection by goats is primarily determined by the variety of plant species present and the relative abundance of each species. As stated by Malechek and Leinweber (1972b), a major deficiency of many food habit studies has been the failure to include data relative to forage availability. If available vegetation is not reported, it is difficult to decide if animals used a plant species or kind of forage because it was preferred, or because there was nothing else available.

During the warm season, if available forage consists of common curlymesquite (*Hilaria belangeri*), shinoak (*Quercus mohriana*), liveoak (*Quercus virginiana*), and juniper (*Juniperus* spp.), goats will normally have a higher selection index for shinoak over liveoak. Their use of common curlymesquite and juniper will vary with rainfall and season of growth. Goats eat common curlymesquite readily when it is green, and it may make up a major portion of their diet. When the grass becomes mature or dries out from lack of rainfall, however, the diet will contain lesser amounts of curlymesquite. If only goats are on this range, the grass species will be under utilized as long as any of the shrubs or low tree species are bearing usable forage (Merrill, 1975a).

If the plant composition is changed appreciably and such species as netleaf hackberry (*Celtis reticulata*), elbowbush (*Forestiera pubescens*), and pricklyash (*Zanthozylum fagara*) are available in appreciable amounts along with greater portions of shinoak, liveoak, juniper, and common curlymesquite, then the diet will shift. Hackberry will generally be selected first, followed by elbowbush, pricklyash, and shinoak. If adequate forage is available from these species, liveoak, juniper, and common curlymesquite may receive very little use with the exception of juniper during the dormant months.

Vegetation complexes containing greater portions of hackberry, elbowbush, and pricklyash might result in little use of shinoak which was rated highest in desirability in the first vegetation complex described.

The ability of goats to adapt to their environment is truly remarkable. They have been observed doing very well and propagating their numbers under harsh conditions with very little rainfall and no source of potable water to drink. Their only requirement under these conditions is that some form of succulent plant life is available. Spanish or meat-type goats have been observed on small dry islands in the Galapagos group on which the primary edible vegetation was a giant form of pricklypear (*Opuntia* spp.) which grew to a height of over 7 m. The goats ate or pawed at the base of the plants until they fell; then they consumed the entire plant. The pricklypear appeared to contribute almost all of the moisture the goats consumed and approximately 80% of their total diet. The remainder consisted of annual grasses and forbs which furnished forage for two very brief periods of rainfall and some low growing waxy leaved trees on which no visible grazing was evident. The two periods of rainfall with approximately 7·5 cm each were about 6 months apart. The goats bred following one rainy period and their kids were started during the other (Leo B. Merrill, 1944, personal observation).

The importance of plant composition or available vegetation in determining goat diets has been pointed out by numerous investigators. Cory (1927), in a study conducted on the Texas A & M University, Texas Agricultural Experiment Station, Sonora Research Station, determined that goats were mainly browsers, spending 53% of their total feeding time utilizing woody forage species. Thirty-eight per cent of their time was spent grazing grass and forbs. The remainder was spent in supplementary feeding and miscellaneous grazing activities. Fraps and Cory (1940), in a study also on the Sonora Research Station, revealed that goats consistently used a larger number of plant species over the year than either cattle or sheep. Browse comprised more than 50% of the diets during all seasons except spring when 53% of the grazing time was spent utilizing forbs. McMahan (1964) determined that goats diets consisted of over 50% browse during all seasons of the year. On the Kerr Wildlife area near Kerrville, Texas forbs made up less then 15% of the total diet. Grasses were utilized more heavily in the spring when grass was young and succulent. The conclusion was that goats are primarily browsers, but the study failed to relate utilization with available forage. Without such a relationship, the importance of forbs was probably minimized. Askins and Turner (1972) reported that goats utilized 72% browse and 28% weeds and grass. Again, availability was not reported.

Huss *et al.* (1970) reported in studies near Montemorelos, NL, Mexico, that goats preferred browse despite the abundance of grass. Grass usage was slight even under heavy rates of grazing. No evidence was found to indicate that goats had consumed large quantities of forbs. Again in this study, availability in relation to utilization was not reported.

Carrera (1969) reported that goats ate browse almost exclusively in the arid zones of Mexico. Since goat diets change under conditions of mixed vegetation types, this again would appear to be a circumstance in which little else was available other than browse. Malechek (1970) reported that goat diets on lightly grazed pastures contained 26% shinoak while diets on heavily grazed pastures contained 18%. This difference was significant. The diets of goats on heavily grazed pastures contained 6% juniper while the diets on lightly grazed pastures contained only 3% juniper. This difference was also significant, demonstrating a difference in available forage between the two pastures. The goats were thus forced to eat less shinoak on the heavily grazed pasture because less was available. They consumed more juniper since it was more available, though less desirable.

2. Season of Use

Most studies on browsing habits of goats have demonstrated a significant correlation between diet selection and season of use. In warmer climates, cool season forbs and grasses furnish considerable forage for goats during autumn and early winter when rainfall is adequate. This reduces browse consumption. In colder climates, forbs and grasses are dormant during late autumn until early spring. Under these conditions, forbs are less important for browsing animals and browse species furnish the bulk of their diet (Dietz, 1972; Cook, 1972). Grass utilization is influenced by availability of both forbs and browse species. Where rainfall is short, preventing forb growth, grass might be heavily used during certain periods of the year.

A detailed study presented by Malechek and Leinweber (1972a) reported the available forage and botanical composition of Angora goat diets on lightly and heavily grazed ranges. From this study, conducted on the Sonora Research Station, goats could be classified as grazers rather than browsers. Grass was the major class of forage consumed on both grazing treatments throughout the period from June to mid-October, but from October to November the proportion of browse in the diet rapidly increased. Liveoak was the major browse species in diets on both treatments in January and early February, while shinoak replaced other browse species in diets from the time of leaf emergence in early April until frost kill in early November. Decrease in the availability of liveoak in late November resulted in heavy use of juniper and tasajillo (*Opuntia leptocaulis*) on the heavily grazed pasture. Forb production in this area is dependent upon fall and winter rainfall. It is important to note that rainfall was markedly below average for the period studied, and there was little forb production or consumption.

Malechek and Leinweber (1972b) reported that abrupt increases in crude protein levels of diets selected by Angora goats corresponded to shifts in forage selection from dry grass to immature leaves of shinoak and forbs on lightly stocked ranges and from pricklypear cactus and juniper to shinoak leaves on the heavily stocked range. Diet digestibility averaged 47·6% and 48·4% respectively for the lightly and heavily stocked pastures.

Palazón (1953) determined that goats in Spain ate brush in the dry season and grasses, legumes and forbs in the wet season. Fraps and Cory (1940) found that goats on the Sonora Research Station utilized more browse from July to August and from November to January than other periods of the year.

The Range Plant Handbook (1937) points out that many shrubby species are readily consumed by goats. The seasonal aspect is stressed in the case of California joint fir (*Ephedra californica*) where it was shown that this important range plant was not utilized on summer ranges but was heavily used in winter. Under those circumstances, forbs and young grasses were available in summer but not in winter, emphasizing the importance that season plays in diet composition. The same criterion can be demonstrated with winterfat (*Eurotia lanata*) which the name implies is utilized primarily in the winter.

B. Grazing Habits

1. Grazing Habits as Influenced by Kind of Goat

For many years ranchmen have utilized the Spanish or meat-type goat when there was need to keep brush in check. It was felt that the Spanish goat is a more effective browse eater and that there was less economic loss when Spanish goats were placed under grazing or browsing stress than when Angora goats were used. The meat goat, therefore, gradually acquired a reputation as a better browser than the Angora or hair goat. Since the milk-producing goat was not used extensively on range lands of this area except as a cross to increase the size of the meat goat, it was never included in range management or brush control programs.

In recent years prices increased for both mohair and meat products of goats. Prices of mature goat hair increased to as high as $7.70 per kg grease weight while kid hair went as high as $9.90 or even higher in 1977. At the same time the value of goats in biological brush control was rediscovered and suddenly a great interest was developed in the dietary habits of Angora versus meat goats as well as the economic returns from each.

Merrill (1975a) pointed out that preliminary studies on the Sonora Research Station demonstrated that Angora goats grazed Texas wintergrass (*Stipa leucotricha*), Texas cupgrass (*Eriochloa sericea*), and sideoats grama (*Bouteloua curtipendula*) to a height of 4 to 5 cm. At the same time Spanish or meat goats permitted all these grasses to grow to a height of 20 cm with no evidence of grazing use. The meat or Spanish goat is short haired; therefore, this animal can graze dense brush without danger of getting its hair tangled in thorns or twigs. A meat goat is taller, giving it an advantage in browsing over the hair goat. Angora goats utilized all forbs 100% during the study. Meat goats utilized forbs heavily but permitted a few to reach heights of 5 to 10 cm. Angora goats permitted shinoak sprouts to produce shoots 5 to 15 cm long and utilized from 15 to 30% of current growth. The Angoras utilized 80% of current growth of both hackberry and pricklyash. Meat goats on the other hand utilized all shoots of shinoak, hackberry, and pricklyash 100%.

Research into the diets of the two kinds of goats conducted on the Sonora Research Station has indicated that there was a difference in the diets of the two types of goats. This research at the Sonora Research Station will be continued.

C. The Place of Goats in Range Management

The goat has adapted itself to almost all vegetation types and into the grazing habits or characteristics of all other grazing or browsing animals. It has, in fact, been so adept that claims have been made that the goat not only consumed vegetation not eaten by other animals, but in taking such vegetation, it reduced competition to the vegetation eaten by other animals, i.e. cattle, and actually permitted an increase in the cattle numbers. Jackson (1923) stated that some ranchmen declared that more cattle could be pastured on a section ($= 1$ mile^2) with sheep and goats than without them. Mackenzie (1956) stated that goats can live where cattle and sheep would starve, and can share a pasture which cattle and sheep are over-grazing and actually improve the pasture for cattle and sheep by reducing brush competition. This belief was once very widespread and still persists in some areas of the United States. Unfortunately the concept is entirely false. When goats occupy an area for the first time, they consume forage species not heavily selected by other animals; however, this condition has been shown to be short lived. As soon as goats utilized the forage species not selected by other livestock, they began to compete with the other animals for their preferred forage (Merrill. 1975a).

1. Goats in Common Use Grazing

Common use is the grazing of more than one species of livestock on a given area at the same time. The advantage of such use have been amply demonstrated over the world, and especially on the Edwards Plateau in Texas where cattle, sheep, goats, horses, deer and other kinds of wildlife utilize a given range at the same time. It has been shown that common usage increased the number of animal units which could be grazed, where an animal unit was an equivalent of 1 cow and calf or 5 ewes and lambs or 6 goats (does) and kids. Carrying capacity where common use was practised was increased from 10 to 25% over grazing with a single kind of livestock with an equivalent increase in economic returns (Merrill and Miller, 1961; Huss and Allen, 1969).

Particular advantages of grazing goats in the common use program was pointed out by Merrill and Young (1954). They found a significant beneficial response in both sheep and cattle when either or both of these species were grazed in combination with goats, but there was no significant response in goats when grazed with cattle or sheep compared to grazing with goats only. This would demonstrate the adaptability of goats in fitting themselves to grazing or browsing any vegetation complex. With single use, both cattle and sheep grazed their preferred species, then were limited in their gains when they were forced to shift their grazing or browsing to other species. Goats, on the other hand, shifted easily and did not demonstrate a limitation of their diets when they went to other vegetation species in the complex.

Although it has been pointed out that goats are more adapted to adjusting their diets to existing conditions than other kinds of livestock and that returns from goats are about as great when grazed alone as when grazed with cattle and sheep or other animals, it does not follow that goats will be less destructive in a vegetation complex than other animals. The goat has greatly limited or destroyed many species of vegetation when the grazing was with goats only. These animals utilize seed to a greater extent than other domestic animals and when it is available will take a major portion of their diets from seed. They also utilized certain forbs and several palatable browse species to the point of near extinction at heavy stocking rates with goats only, and greatly reduced several browse species even at light rates of grazing (Malechek, 1970). Rather than being considered an animal which should be grazed alone, the adaptability of the goat admirably fits in the role of being the most efficient common use animal of all the domestic or even wild species. It has been shown to utilize vegetation niches which were very lightly or totally unused by other domestic animals. If the vegetation complex is diversified and in a vigorous condition, the goat provides less competition with other grazing or browsing animals in the vegetation complex. Such vegetation conditions are encountered in well planned and properly used grazing management systems.

It has been demonstrated that grazing two or more kinds of animals on mixed vegetation had two benefits. First, several kinds of plants were utilized. The grazing or browsing pressure was, therefore, equalized over several plant species. Thus, when cattle, sheep, and goats were grazed together, grass, forbs, and browse were utilized much more uniformly than they would have been by any one animal species including goats. Where cattle were grazed alone at moderate rates, grass was used properly; but forbs and browse were under utilized. Where sheep were grazed alone, all forbs were taken first, then grass was spot grazed and coarse material was not used. Browse was used very little and trees were permitted to grow freely. Goats grazed alone, first utilized forbs and most desirable browse, then less desirable browse species, but grass was under utilized (Merrill and Young, 1954; Merrill and Miller, 1961; Merrill 1975a).

Each kind of animal was found to benefit from the grazing habits of the other two. Cattle benefited sheep and goats by grazing coarse grasses, leaving them shorter and less coarse, thus making them more palatable to sheep and goats. Cattle generally removed coarse forage left ungrazed by sheep, thus reducing spot grazing. Sheep benefited the cattle by grazing forbs or weeds not utilized by cattle, thus reducing competition to the grasses preferred by cattle. Goats utilized more browse, reducing shading and competition with grasses and forbs, which produced more forage for cattle and sheep. Cattle and sheep utilized forage species less preferred by goats and since much of the grazing pressure was exerted by sheep and cattle, the goats in the complex experienced less competition for their preferred species due to fewer goats (Merrill, 1975b; Merrill *et al.*, 1966). Similar results were reported by Huss *et al.* (1970). They found that practices which improved goat pasturage and performance also improved the pasturage and performance for other kinds of livestock. The same criterion was reported by Merrill *et al.* (1957) for livestock and deer. As range condition improved and carrying capacity increased for cattle, sheep and goats, deer numbers likewise increased. Although deer and goat diets were found to be very similar, the deer numbers appeared to suffer no adverse affect due to the presence of goats in combination with cattle and sheep. Deer numbers did decline when pastures were

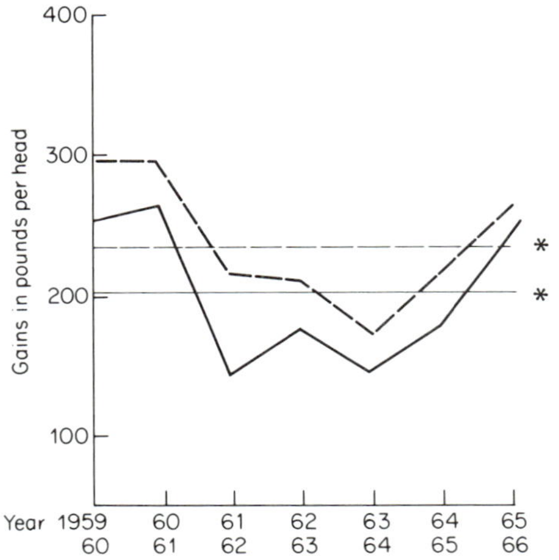

Fig. 1. Cattle gains per head when grazed alone or with sheep and goats, 1959–66 (Merrill, 1975b). —— Cattle only; ----- cattle, sheet and goats. * Horizontal lines represent the average gain per head for 7 years.

under heavy continuous grazing by goats only, but not as much as with sheep only.

Figures 1 and 2 show the consistent difference in cattle and sheep gains when all three species of livestock were grazed together as compared to grazing cattle or sheep alone.

There was no appreciable difference from grazing goats with cattle and sheep as compared to grazing goats alone.

Wilson *et al.* (1975) studied diets of sheep and feral goats in Western New South Wales using esophageal fistulated animals. They found that goat diets consisted largely of browse, with the leaves of rosewood (*Heterodendrum oleifolium*) a consistent component, although a large proportion of herbaceous material (mainly, *Bassia* spp.) was eaten on occasions. Nitrogen content of the diets ranged from 1·6 to 3·0%, with that of the goats generally being greater than that of the sheep. Sheep and goat diets, although competitive for some herbaceous plants such as *Bassia* spp., were partly complementary indicating that animal productivity from the existing present vegetation could be improved by grazing those animals together. Plant species eaten by both sheep and goats were also affected by grazing pressure. Goats tended to eat considerable amounts of tree and scrub leaves at low grazing pressures and could obtain more of those leaves at high grazing pressures because of their ability to graze up to 2 m from the ground. The

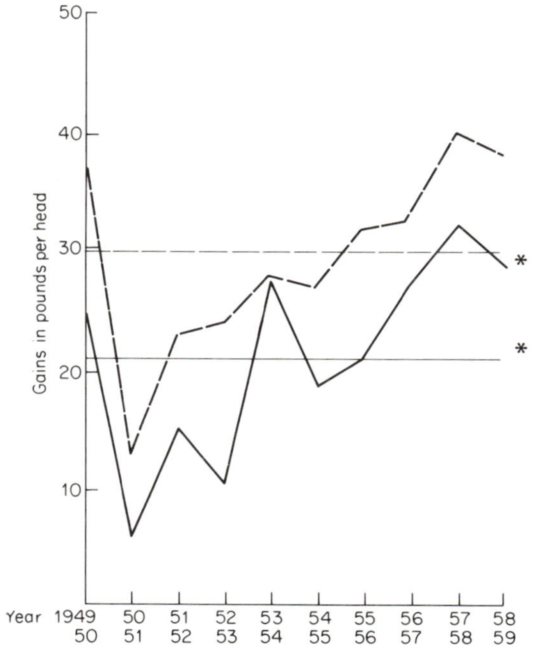

Fig. 2. Sheep gains per head when grazed alone or with cattle and goats during the 1949–59 period (Merrill, 1975b). ——— Sheep alone; ----- sheep, cattle and goats. * Horizontal lines represent average gains for 10 years.

researchers concluded that goats consume a wider range of the available vegetation than sheep and would make better use of the vegetation as a whole.

According to Strang (1973), mixed herds of cattle and goats gave promising results in maintaining open pastures, and per unit area were producing more edible protein than either cattle or goats grazed separately.

Naveh (1974) described the use of goats in management schemes to improve vegetation in Mediterranean upland ecosystems. He stated that due to poor livestock and grazing management, the level of production of goats was not only low, but their impact on the land in combination with uncontrolled grazing by cattle and sheep had disastrous results for vegetation in all Mediterranean countries.

Since these range goats (Black Mamber) are able to survive and produce in poor conditions, too harsh for either cattle and sheep, they have been blamed as the main cause of Mediterranean land ruin. However, they are the last link in the land devastation brought on by indiscriminate burning, cutting, grazing, slope denudation and cultivation. This has led to the conclusion that a complete ban of goats on Mediterranean upland was the only recourse for improvement of vegetation. Naveh (1968, 1970, 1972) discussed problems of conserving, restoring and rationally using these threatened upland ecosystems. He warned of the fallacy of regarding their complete protection into impenetrable forests as the only alternative to present misuses. Instead he proposed other approaches of integrated ecosystems planning, in which their functional diversity could be conserved and their biological productivity increased. He concluded that moderate goat grazing in conjunction with proper grazing management systems would be a part of the overall strategy to improve the ecosystem. According to Naveh (1960), feeding habits of Arab cattle and goats on brush ranges in the Western Galilee differed significantly. Out of 40 woody species, 30 were eaten regularly by goats. Cattle consumed only 10 woody plants, chiefly trees, during the dry season and after brush fire.

Au Toit (1972), working with goats in South-west Africa, concluded that in a brush grass community cattle would give a higher level of production when grazed in a combination with goats than when grazed in combination with sheep. He was able to get 98 cattle grazing days per hectare when grazing with goats compared with 65 grazing days per hectare when grazing with sheep.

2. Economic Benefit from Common Use Grazing

Various studies have shown a positive advantage in carrying capacity of ranges when more than one species of livestock is grazed in combination and especially if goats are included in the combination. Therefore, it would logically follow that economic returns would be greater where common use grazing was practised as opposed to one species use.

When goats are to be used in common grazing, any breed which fits local vegetation conditions and requirements of people of the area is capable of producing livestock products and economic returns equivalent to or better than other kinds of livestock. However, under average range conditions of the Edwards Plateau, the Angora goat has produced economic returns superior to other kinds of goats or to cattle or sheep over a 25-year period (Merrill and Taylor, 1976). As range conditions improved, the performance of all animals also improved; but the Angora goat responded most favorably because of its high nutrient requirements (Merrill and Taylor, 1976).

Merrill *et al.* (1966) found that grazing cattle alone produced the least gross returns of any kind or combination of livestock. Sheep alone produced next lowest returns followed by goats. Gross returns from goats alone, however, were not significantly lower than the gross produced by cattle and goats or cattle, sheep and goat combinations. Cattle, sheep and goats in combination had slightly higher returns than cattle and goats but the difference was not significant. Gross returns from cattle alone were about 28% less over a 16-year period than the returns from cattle and goats or cattle, sheep and goats. Returns from sheep were slightly more than those from cattle but were significantly less than those from cattle and goats or cattle, sheep and goats.

Merrill and Miller (1961), studying production during a 7-year period in which all but one year was in a period of severe drought (rainfall for a 7-year period was less than half of long time average), found that cattle, sheep and goats in combination produced \$6·52 per hectare compared with \$5·98 for cattle and goats, \$5·78 for sheep alone, and \$4.42 for cattle alone. Throughout those studies, it was evident that cattle were greatly benefited by grazing with sheep and goats, but cattle also contributed greatly to the value of common use grazing since they consumed coarse grass material not readily taken by either sheep or goats. Goats contributed heavily to the value of common use since they consumed coarse browse or brush forage materials not taken by cattle or sheep, thus producing animal products which would not have been produced if goats had been absent. The goats, however, received little advantage from common use. In a cattle, sheep, goat combination, sheep contributed to common use because they consumed some broadleaf forbs or weeds not taken by goats, but their contribution was less than either cattle or goats. Since sheep controlled plants which were weeds on a cattle range, they made a much greater contribution to common use grazing when cattle and sheep were the only kinds of animals present. It could be concluded that goats were the most valuable contribution to common use because they took less from the other users than either cattle or sheep. Since sheep were not required to take any forage species not taken by cattle or goats, they were probably the least valuable.

3. Goats as They Influence Vegetation Composition Change

The most common concept of goat influence on vegetation is that of biological brush control. Many would prefer eradication in the terminology. A more desirable concept would be the role of goats in shrub management. This approach has been very effective and has been used with less ecological upset than any other brush management measure. This concept requires a thorough knowledge of the ecological complex. With careful manipulation of goat numbers, the change in the vegetation complex can be slowed, stopped or even reversed. Thus, there is less chance of doing permanent damage to the vegetation complex than in other methods of brush control.

In order for goats to exert desirable changes in vegetation, browse species to be changed must first be placed in reach of goats. The method most used for this has been chaining. This method consists of dragging an anchor

Table I. Percent canopy cover of major browse plants before and after brush control (Merrill, 1975a).

	Liveoak		Shinoak		Mesquite		Juniper		Sacahuiste		Opuntia		Total	
	Pastures' number													
	20	27	20	27	20	27	20	27	20	27	20	27	20	27
1970	1·9	8·2	4·4	3·3	1·4	3·5	1·3	8·2	2·4	3·3	1·2	0·5	14·7	27·3
1974	0·0	1·4	0·0	0·1	0·4	0·6	0·0	0·5	1·1	0·3	0·9	0·0	3·0	3·8
% Average reduction	91·5		98·5		77·0		97·0		72·5		62·5		83·0	

chain weighing 44·6–74·5 kg/m between two crawler tractors spaced 20–30 m apart, pulling down all trees and shrubs between them. Merrill (1975a) pointed out that goats preferred browse plants to grass and would eat browse and permit grasses to grow as long as edible forage remained on the browse. There were exceptions in that grasses were preferred during short periods when they were green, but as a whole, browse was taken first. Table I shows canopy cover change on several brush species before chaining and then after chaining and browsing for 4 years using one goat per hectare. Shinoak was almost entirely eliminated. Liveoak was reduced 91·5%; mesquite, 77%; juniper, 97%; sacahuiste, 72·5%; pricklypear, 62·5%; and all brush was reduced 83%. A deliberate attempt was made to eliminate brush in this study. Economic returns from goats were so high, however, that early in the study it was concluded that goats and other animals should be grazed together on such a controlled area to manage the mixed vegetation for maximum economic returns. The cost of chaining was paid for in 4 years by increased production from grazing and browsing.

An investigation was conducted by Au Toit (1972) to determine whether goats could control brush regrowth and reinfestation after initial clearing. A dense stand of acacia karroo was chopped and the regrowth grazed either rotationally or continuously by sheep or goats (Table II).

Table II. Average regrowth of sampled coppices compared with that of the control, 1969–71 (Au Toit, 1972).

Grazing treatment	1968–69 Regrowth		1969–70 Regrowth		1970–71 Regrowth	
	g	%	g	%	g	%
Continuous, sheep	287·8	84·6	262·6	84·1	513·8	73·9
Rotational, sheep	297·5	87·3	223·0	71·4	553·4	79·6
Continuous, goats	111.3	32·6	79·2	24·4	130·2	18·7
Rotational, goats	146·2	42·9	99·3	31·8	165·0	23·7
Control: no graze	340·9	100·0	312·3	100·0	695·0	100·0

At the conclusion of each grazing season a number of dead sample trees were encountered in all pastures. These data are presented in Table III.

Results of this trial provided no evidence to support any prejudice against the goat as a grazing agent, when compared to the merino type sheep. Rather, by restricting brush competition and thus encouraging grass production, the goat would have a beneficial effect on the vegetation complex.

Davis *et al.* (1973) found that goats would effectively control oak on rangelands, but it was first necessary to make the brush available to the goats by mechanical treatment. Adams and Hughes (1977) found that goats could

Table III. The mortality as affected by grazing treatment (Au Toit, 1972).

	Number of dead trees		
Grazing treatment	1968–69	1969–70	1970–71
Continuous, sheep	1	6	9
Rotational, sheep	3	13	20
Continuous, goats	10	21	32
Rotational, goats	3	18	21
Control: no graze	1	1	5

effectively control brush on forest lands as well as produce food at the same time.

4. Dangers of Goat Grazing in Common Use or Alone

Dangers involved in goat grazing have been pointed out by many authors. There is no doubt that the dangers are very real and that they have not been eliminated. This has only been true because the goat can eat nearly all available vegetation and do quite well on it. It has been this ability to survive on what was left that has made the goat dangerous. Hall (1936) stated that over-grazing may so destroy vegetation and bare the surface that severe erosion results. Of all livestock, the goat was the worst offender. Although Hall placed little blame on other animals for starting destructive erosion, he is correct in his view that the goat will carry on the process, if forced. Huss (1972) pointed out that many countries are still blaming the goat for all problems on rangelands. In some countries efforts have even been made to prohibit goat grazing entirely. In other countries government policies encourage controlled usage of goats and through development programs try to increase their productivity. Although Huss has pointed out that the controversy is still going on, there should be none. It has been adequately demonstrated that the goat can destroy vegetation when forced, but it has also been demonstrated that the goat lends itself to conservation grazing much better than any other domestic animal when properly utilized. It is only destructive when man has imposed the conditions for destruction.

5. Management on Rangelands

The handling requirement of goats is dependent on a number of conditions. If the goat is the meat type, it will require a minimum of care and a minimum of supplemental feed compared with the Angora goat. The meat-type goat does require good fences. It readily finds and uses small openings to move from one area to another. Most commonly used fences are net wire approximately 900 mm high with two barbed wires on top—one roughly 100 mm above the net wire, the second about 150 mm above the first. Such fences will hold cattle, sheep, goats, and horses. Almost all goats in the United States are grazed under fence rather than being herded.

6. Supplemental Feeding on Rangelands

All goats have been notoriously neglected in the United States. Because of their ability to survive under poor range conditions, most ranchmen have felt that goats could take care of themselves under any conditions. They, therefore, placed them on their poorest pastures and seldom provided supplemental feed when vegetation was dry or dormant. Spanish or meat-type goats performed rather well under such conditions, since they utilized browse heavily and were able to select a diet that contained most of their nutrient requirements. For this reason a Sutton County, Texas, ranch committee recommended that meat goats required a supplement of not more than 150 g of 20% protein grain cubes daily for 90 days when ranges were in fair condition and producing adequate forage except when dormant from cold or dry weather (Sutton County Sheep and Goat Sub-committee, 1967).

The Angora doe, however, has been a different problem. Its protein requirements are the highest of any domestic ruminant animal. Huston *et al.* (1971) pointed out that although Angora goats could be maintained on rations as low as 7% protein, mohair production at this level was very poor. Mohair production increased as protein in the diet increased. The requirements of various classes of Angora goats are shown in Table IV.

Nutritional stress has been found to produce abortion in pregnant nannies. It has also been found that pregnant Angora does or those producing milk were placed under food stress at any time when vegetation was dry due to lack of rainfall or to winter dormancy. It has, therefore, been concluded that Angora does under average range conditions should receive a protein and probably a phosphorus supplement whenever there has been a prolonged period in which vegetation has not produced green growth (Sutton County Sheep and Goat Sub-committee, 1967). A small amount of high protein supplement fed to doe kids will frequently provide sufficient growth in kids on range to cause them to breed as yearlings rather than as two year olds. The increased production from early breeding or in increased mohair production will pay for the cost of the supplement (Taylor and Merrill, 1977).

7. Predators

All young kids of any breed are highly vulnerable to numerous predators.

Table IV. Required composition of diets for Angora goats (Huston *et al.* 1971).

Classes and weights (lb) of goats	TDN (%)	DE (Mcal/lb)	CP (%)	P (%)
Wethers and dry does				
50	50–60	1·0–1·2	9·0–10·4	0·21
60	50–58	1·0–1·2	9·0–10·4	0·19
80	50–54	1·0–1·1	9·0–9·7	0·17
100	50	1·0	9·0–9·5	0·16
120	50	1·0	9·0	0·16
Pregnant does (last 8 weeks)				
50	54–62	1·1–1·2	10·0–11·1	0·18
60	54–60	1·1–1·2	10·0–10·5	0·17
80	54–56	1·1	10·0	0·16
100	54	1·1	10·0	0·16
Lactating does (16 weeks)				
50	58–65	1·2–1·3	11·0–12·0	0·22
60	58–62	1·2	11·0–11·5	0·21
80	58–60	1·2	11·0–11·2	0·20
100	58	1·2	11·0	0·20
Developing billies				
80	65	1·3	11·4	0·19
100	60	1·2	10·5	0·16
120	55	1·1	9·8	0·14
Growing kids and yearlings				
20	68	1·4	13·3	0·19
40	65	1·3	11·6	0·18
60	60	1·2	10·6	0·18
80	58	1·2	10·0	0·18

TDN = total digestible nutrients.
DE = digestible energy.
CP = crude protein.
P = phosphorus.

Beasom *et al.* (1977), studying predator losses in goats in very high coyote and bobcat densities, found that reducing predator numbers by 60 to 85% increased kid crops 27 times. However, the remaining coyotes and bobcats only permitted raising a 13·5% kid crop. With the high predator density, there was virtually no kid crop. This work demonstated that the worst predators were coyotes (*Canis latrans*), bobcats (*Lynx baileyi*), and foxes (*Urocyoncinereo argenteus*), but raccoons (*Procyon lotor*), and skunks (*Mephitis mesomelas*), will also kill substantial numbers of very young kids (Merrill and Taylor, 1976). Goat kids are more vulnerable to predators than lambs under range conditions and must receive protection to produce satisfactory economic returns.

Conclusion

Based on research results, it is evident that with mixed vegetation types and comparable rates of grazing, goats spread their grazing pressure much more evenly over all kinds of vegetation than cattle or sheep. This results in lighter utilization of the grass component, which is beneficial both for vigor and seedling establishment. Also, by restricting brush growth, the competitive effect of the brush is minimized.

Since goats have a wide grazing spectrum, there is no doubt that in a mixed vegetation complex, the inclusion of the goat in the grazing system would give far superior utilization of the vegetation than is possible with only sheep or cattle. Provided a sensible stocking rate is applied, the inclusion of the goat is not detrimental to its associated grazers or ecosystems.

When animals with very high nutritional requirements such as Angora goats are grazed on rangelands, careful considerations should be given for development and maintenance of the ultimate vegetation complex required for maximum production from the animal. Vegetation composition most desirable for Angora goats will almost certainly vary for different soils and climatic conditions. Despite such variations, a theoretical ideal vegetation complex based on diet research might be of interest. Taylor (1977), studying Angora goats which were being grazed on the most highly developed vegetation complex known in the Edwards Plateau region of Texas, found that the Angora goat diet consisted of 25 to 30% forbs, 45 to 50% browse, and 25 to 35% grass. Such a diet could be obtained from vegetation with a composition similar to that shown in Table V.

With a vegetation composition as shown in Table V, Angora goats could readily obtain a diet consisting of 15–20% forbs in summer and fall and 25 to 35% forbs in spring and winter. This would result in crude protein intake generally exceeding 11%. This diet from range forage accompanied by judicious supplemental feeding would result in production of 4·5–5·0 kg of mohair and a kid crop approaching 100%. Results of this nature are considered outstanding since the average kid crop on the Edwards Plateau was less than 50% and mohair production was 3·5–4·0 kg during 1977 indicating poor nutrition from poor range.

Brush encroachment is ecologically inevitable in many regions of the world, and no satisfactory means of checking it has yet been developed. Although current control techniques are not completely ineffective, the utilization of brush regrowth with browsing animals presents an inexpensive means of control and at the same time affords a program for vast expansion of red meat production in an era of constantly increasing demand.

Table V. A vegetation complex with a composition suitable to all kinds of livestock which would result in maximum kid and mohair production in the Edwards Plateau.

Grasses 53% Composition				Forbs 21% Composition				Browse 26% Composition			
Warm season 43%		Cool season 10%		Warm season 14% Mostly perennials		Cool season 7% Mostly annuals		Warm season 15%		Cool season 11%	
Plant	%	Plant	%	Plant	%	Plant	%	Plant	%	Plant	%
Curlymesquite (*Hilaria belangerii*)	20	Texas wintergrass (*Stipa leucotricha*)	4	Orange zexmenia (*Zexmenia hispida*)	2	Redseeded plantain Tallow weed (*Plantago rhodosperma*)	2	Shinoak (*Quercus mohriana*)	8	Liveoak (*Quercus virginiana*)	8
Sideoats grama (*Bouteloua Curtipendula*)	4	Plains lovegrass (*Eragrostis intermedia*)	1	James nailwort (*Paronychia jamesii*)	2	Texas bluebonnet (*Lupinus texensis*)	1	Elbowbush (*Foresteria pubescens*)	1	Juniper (*Juniperus* spp)	2
Cane bluestem (*Bothriochloa barbinodis*)	3	Canada wildrye (*Elymus canadensis*)	1	Texas snoutbean (*Rhynchosia texana*)	1	Rosering gaillardia (*Gaillardia pulchella*)	1	Skunkbush sumac (*Rhus aromatica* var. *flabelliformis*)	2	Vine ephedra (*Ephedra antisyphlitica*)	1
Little bluestem (*Schizachyrium scoparium*)	1	Wrights threeawn (*Aristida wrightii*)	3	Velvet bundleflower (*Desmanthus velutinus*)	1	Texas croton (*Croton texensis*)	1	Netleaf hackberry (*Celtis reticulata*)	1		
Fall witchgrass (*Leptoloma cognatum*)	3	Carex (grasslike) (*Carex* spp.)	1	Mexican sagewort (*Artemesia ludoriciana* var. *mexicana*)	1	Other forbs	2	Catclaw (*Acacia* spp.)	1		
Hairy tridens (*Erioneuron pilosum*)	4			Sweet gaillardia (*Gaillardia suavis*)	1			Littleleaf sumac (*Rhus microphylla*)	1		
Hairy grama (*Bouteloua hirsuta*)	1			Bushsunflower (*Simsia calva*)	1			Purple dalea (*Dalea lasiathera*)	1		
Green sprangletop (*Leptochloa dubia*)	1			Little mallow (*Malva parviflora*)	1						
Vinemesquite (*Panicum obtusum*)	1			Pitcher gaura (*Gaura coccinea*)	1						
Slim tridens (*Tridens muticus*)	1			Hedeoma (*Hedeoma drummondia*)	1						
Texas cupgrass (*Eriochloa serecia*)	4			Other forbs	2						
Total	43	Total	10	Total	14	Total	7	Total	15	Total	11

References

Adams, T. E. and Hughes, C. L. (1977). *Rangeman's J.* **4**(1), 4.

Askins, G. D. and Turner, E. E. (1972). *J. Range Mgt* **25**, 82–87.

Au Toit, P. F. (1972). *In* "Proceeding of the Grassland Society of Southern Africa", Vol. 7, pp. 46–50.

Beasom, S. L., Gober, D. R. and Guthery, F. S. (1977). *In* "Sheep and Goat, Wool and Mohair", pp. 32–33. TAES Cons. PR 3445–3470.

Cory, V. L. (1927). *Texas Agric. Exp. Sta. Bull.* 367.

Carrera, C. (1969). *In* "Simposio internacional sobre el aumento de la producción de alimentos en Zona Aridas", Monterrey, Mexico, 1968 Proc. Texas Tech. College. Lubbock, Tex., pp. 219–224.

Cook, C. W. (1972). *In* "Wildland Shrubs—Their Biology and Utilization", pp. 303–310. Int. Symp. Utah State Univ. Logan, Utah, July 1971. USDA Forest Service, Gen. Tech. Report INT–1.

Davis, G. G., Bartel, L. E. and Cook, C. W. (1973). *J. Range Mgt* **28**, 216–218.

Dietz, D. R. (1972). *In* "Wildland Shrubs—Their Biology and Utilization", pp. 289–302. Int. Symp. Utah State Univ. Logan, Utah, July 1971. USDA Forest Service, General Tech. Report INT–1.

Fraps, G. S. and Cory, V. L. (1940). *Tex. Agr. Exp. Sta. Bull.* 586, p. 39.

Hall, A. D. (1936). "The Improvement of Native Agriculture in Relation to Population and Public Health". Oxford University Press, London.

Huss, D. L. (1972). *In* "Wildland Shrubs—Their Biology and Utilization". Int. Symp. Utah State Univ. Logan, Utah, USA, July 1971. USDA Forest Serv. Gen. Tech. Report INT–1, pp. 331–338.

Huss, D. L. and Allen, J. V. (1969). *Tex. Range Sta. Tex. Agr. Exp. Bull.* 1089, p. 14.

Huss, D. L., Zertuche, J. M., Juárez de Leøn, J. L. and Carrera, C. (1970). Informe sobre proyecto de experimentación para el aprovechamiento con ganado caprino de zonas con vegetación tipo matorral alto subperenifolio en la Región de Linares, N. L. Dep. Zootécnia. ITESM, Monterrey, Mex., pp. 48.

Huston, J. E., Shelton, M. and Ellis, W. C. (1971). Nutritional requirements of the angora goat. TAES Bul. 1105 16 pp.

Jackson, A. D. (1923). TAES. Press Release.

Mackenzie, D. (1956). "Goat Husbandry". Faber and Faber, London.

Malechek, J. C. (1970). Diss. Range Sci. Dept. TAMU, College Station., 61 pp.

Malechek, J. C. and Leinweber, C. L. (1972a). *J. Anim. Sci.* **35**, 1014–1019.

Malechek, J. C. and Leinweber, C. L. (1972b). *J. Range Mgt* **25**, 105–111.

McMahan, C. A. (1964). *J. Wildlife Mgt* **28**, 798–808.

Merrill, L. B. (1975a). *In* "Beef Cattle Sci. Handbook", Vol. 12, pp. 372–376.

Merrill, L. B. (1975b). *In* "Beef Cattle Sci. Handbook", Vol. 12, pp. 368–371.

Merrill, L. B. and Miller, J. C. (1961). Economic Analysis of Yearlong Grazing Rate Studies on Substation No. 14, Near Sonora, Tex. Agr. Exp. Sta. MP 484, p. 8.

Merrill, L. B. and Taylor, C. A. Jr. (1976). *Rangeman's J.* **3**(3), 74–76.

Merrill, L. B. and Young, V. A. (1954). Results of Grazing Single Classes of Livestock in Combination with Several Classes When Stocking Rates Are Constant. Tex. Agr. Sta. PR 1726, p. 7.

Merrill, L. B., Teer, J. G. and Wallmo, O. C. (1957). Reaction of Deer Population to Grazing Practices. *Tex. Agr. Progr.* **3**(5), p. 10–12.

Merrill, L. B., Reardon, Patrick O. and Leinweber, C. L. (1966). Cattle, Sheep, Goats ... Mix 'Em Up for Higher Gains. *Tex. Agr. Prog.* **12**(4), pp. 13–14. TAP 447.

Naveh, Z. (1960). Ph.D. Thesis. Hebrew University of Jerusalem.

Naveh, Z. (1968). *Ann. Arid Zone* **7**, 163.

Naveh, Z. (1970). *In* "Proc. XII. Int. Grassland Congress", p. 59.

Naveh, Z. (1972). "The Roles of Shrubs and Shrub Ecosystems in Present and Future Mediterranean Use", pp. 414–427. Faculty of Agr. Eng., The Technician, Haifa, Israel.

Naveh, Z. (1974). *J. Environmental Mgt* **2**, 351–371.

Palazón J. (1953). "Ganado Cabrio". Salvat S. A., Madrid.

"Range Plant Handbook" (1937). Revised June, 1940. USDA.

Strang, R. (1973). *Bio. Cons.* **5**, 96–104.

Sutton County Livestock and Production and Range Management Guideline. Prepared for Sheep and Goat Sub-Committee of Sutton County, Texas, Program Building Committee (1967).

Taylor, C. A., Jr. (1977). Mimeographed Sheet.

Taylor, C. A., Jr. and Merrill, L. B. (1977). *In* "Sheep and Goats, Wool and Mohair", pp. 22–25. TAES. Cons. PR 3445–3470.

Wilson, A. D., Leigh. J. H., Hindley, N. L. and Mulham, W. E. (1975). *Austr. J. Exp. Agr. Anim. Husb.* **15**(72), 45–53.

8. Growth

P. Morand-Fehr

Growth of animals is not limited, as is often believed, to a simple increase in body weight; it is also accompanied by a modification of the proportions of different parts of the organism and in particular of some tissues or organs, relative to the whole body, and even by involutions like that of the thymus, or by a fast development at a given growth stage like that of the gonads.

Thus, to apply satisfactory production methods to young goats, growth should not only be analysed through the body weight of the animal or the development of the skeleton, but also by taking into account the whole succession of physiological transformations in the animal. In this chapter we shall try to give an anatomical and physiological characterization of the growth of kids as compared to that of other ruminants and we shall analyse the factors affecting growth performances. Then we shall give some examples of production techniques intended for replacement animals in particular. They may also be used for meat kids but specific techniques for goat meat production are given in Chapter 9.

A. General Description of Growth in Kids

In order to characterize the growth of kids, Morand-Fehr and Duborgel (1976) analysed the growth of 170 Alpine kids weaned at 5 to 6 weeks, receiving a milk replacer from week 2 to weeks 5–6 and then hay *ad libitum* as well as variable amounts of concentrate feeds (600–100 g/day). The feed supply always satisfied the requirements for growth. During the first 30 weeks, the growth curve was rather rectilinear (Fig. 1). The mean live weight was in the range of 31 kg at the age of 30 weeks. Individual variability of live

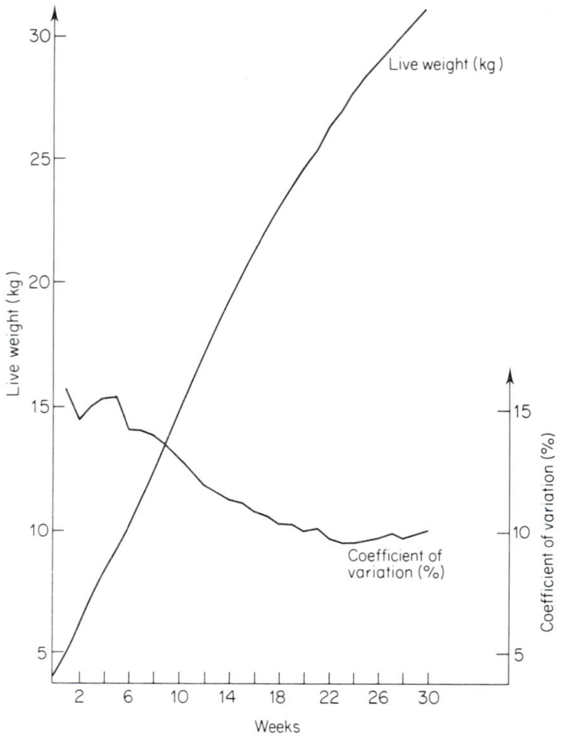

Fig. 1. Average growth curves of 170 French Alpine female kids from birth to 30 weeks old (from Morand-Fehr and Duborgel, 1976).

weight decreased during the growth period parallel to the regression of the role played by some factors favouring this variability, i.e. birth weight, litter size, weaning conditions.

During the first 12 weeks, the weight gain curve (Fig. 2) was characterized by an almost constant daily average gain of 170 g, leaving out the weaning period, and then by a daily weight gain decreasing linearly to 75 g at week 30. This constant weight gain during the first 12 weeks and the subsequent decrease are in agreement with observations of Restall (1976) on feral goats and of Singh and Singh (1974) on Jamnapari goats. Moreover, the logarithm of the fraction of the weight gain over body weight is a linear function of the time, as shown in Fig. 3 from a study on 32 Alpine female kids receiving goats' milk or a milk replacer and weaned at 3 or 5 weeks (Fehr and Disset, 1971).

From the age of 7 months onwards, growth depends on the age at which the young goat is mated. In France where kids are generally mated for the

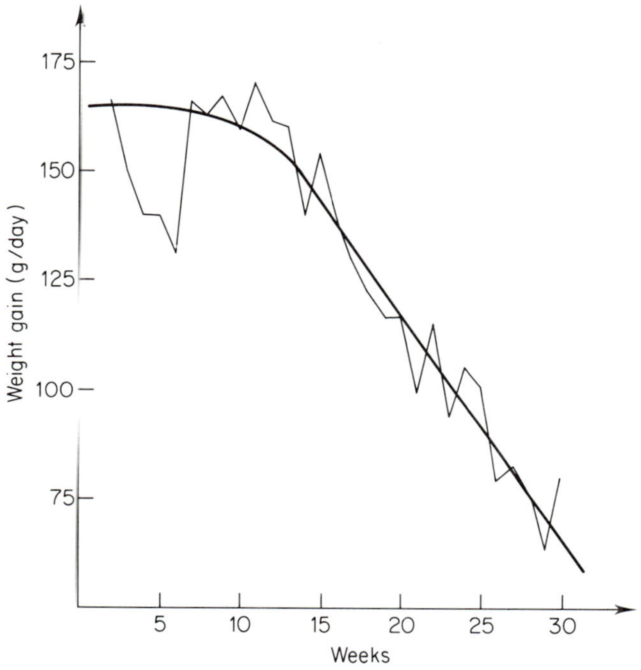

Fig. 2. Weight gain curves of 170 French Alpine female kids (from Morand-Fehr and Duborgel, 1976).

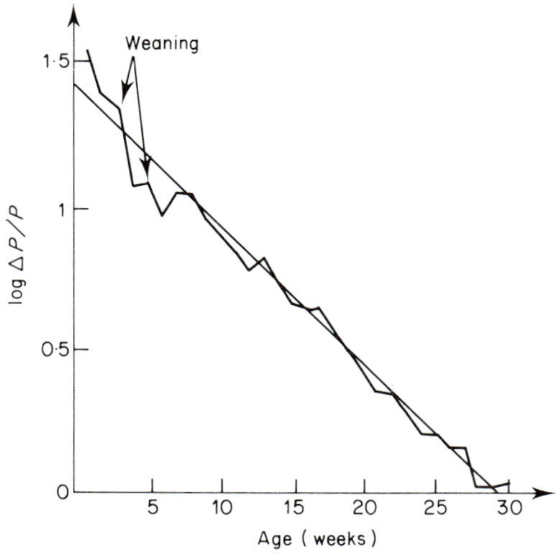

Fig. 3. Changes of weight gain with age of 32 French Alpine female goats (from Fehr and Disset, 1971). $y = -0{\cdot}049t + 1{\cdot}44$; $y = \log \Delta P/P$; t = time in weeks; P = live weight: ΔP = difference in weight between 2 weeks.

first time between 7 and 10 months of age, the total weight gain during the first gestation is 18 kg, on average (Fig. 4; Fehr *et al.*, 1971; Blanchard and Sauvant, 1974), and that of non-pregnant animals, 10 kg. Accordingly, it may be considered that during the 153 days of gestation, the true weight gain of the goat itself ranges from 8 to 12 kg. Its body weight increases from 10 to 14 kg during the first lactation. During its second reproductive cycle (gestation–lactation), the growth of the goat is still variable between animals (2–7 kg). In our experimental herd, the adult weight (59 kg, on an average) seems to be reached at an age varying between individuals from 2 to 3 years.

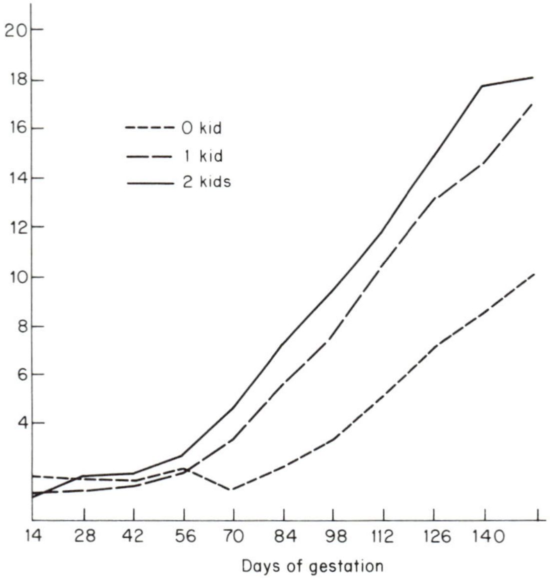

Fig. 4. Weight gain during gestation in 1-year-old French Alpine female kids (from Blanchard and Sauvant, 1974).

B. Growth Stages in Kids

1. Fetal Stage

Although there has been no systematic study on fetal development of goats, our own observations on live weight variations and on metabolism seem to show that their gestation is very similar to that of ewes. Thus, growth of the fetus seems to be very slow during the first 100 days of gestation;

more than 80% of its growth is observed during the last 8 weeks and its maximum daily gain during the last 3 weeks. As in the other ruminants, dry matter, protein, lipid and mineral contents of the fetus increase during the last stage of gestation.

2. Birth Weight

Birth weight is highly variable as shown in Table I, reporting the observations made in different breeds kept in very variable environmental conditions. This table gives additional information to the data compiled by Devendra and Burns (1970). The birth weight of a kid depends primarily on the form and size of the adults of the breed to which it belongs. On average, the birth weight represents 1/15 of the adult weight of the goats of the breed studied (see Chapter 3). However, within breeds, the birth weight also depends on the weight of the parents and especially the adult weight of the dam (Prasad *et al.*, 1971; Morand-Fehr and Hervieu, 1978, unpublished; Devendra and Burns, 1970) but not on the age of the dam (Prasad *et al.*, 1971). In fact, the adult weight of the dam and the mean weight of the kids born are positively correlated irrespective of the litter size (Morand-Fehr, and Hervieu, 1978, unpublished); this partly reflects the nutritional status of the dam.

Birth weight is also markedly influenced by sex (Devendra, 1966; Singh, 1973; Prasad *et al.*, 1971; Nikitenko *et al.*, 1972; Castillo *et al.*, 1972; Choueiri, 1973; Buvanendran and Jalatge, 1974; Montemurro, 1966; Wilson, 1958a; Moulick and Syrstad, 1970; Singh *et al.*, 1977; Morand-Fehr and Hervieu, 1978, unpublished) and litter size (Choueiri, 1973; Castillo *et al.*, 1973; Belichon and Marques, 1971; Moulick and Syrstad, 1970; Singh *et al.*, 1977; Morand-Fehr and Hervieu, 1978, unpublished). The effect of sex varies according to breed and litter size. Birth weight of males generally exceeds that of the females by 5–15%. Birth weight decreases as litter size increases (Table I). The season during which the kiddings take place, does not seem to affect the birth weight (Singh, 1973; Moulick and Syrstad, 1970).

From birth to the adult stage, the growth of kids depends on some environmental factors, e.g. climate or housing, feeding, diet, level of feed intake, and internal factors such as genetic factors on which the growth potential, health, sex and litter size depend.

3. Milk Feeding Period

The period during which the kid consumes liquid feed (goats' milk, milk from the cow or ewe, milk replacer) may last 1 to 5 months. With the aim of studying growth during the first month, a period where feeding is essentially

Table 1. Birth weight of kids (kg).

A. Irrespective of sex and litter size

Breeds	Male and female	References
French Alpine	3·5	Gill and Dev (1972)
Anglo-Nubian	2·9	Gill and Dev (1972)
West African Dwarf	1·4	Sada and Vohradsky (1973)
Malaysian	3·3	Devendra (1966b)
French Alpine	3·9	Morand-Fehr and Semptey (unpublished)

B. Irrespective of litter size

Breeds	Male	Female	References
Kambing Katjang	1·5	1·4	Devendra (1966)
Jamnapari × indigenous goat of Indonesia (half bred)	2·9	2·6	Devendra (1966)
Anglo-Nubian × indigenous goat (half bred)	2·4	2·3	Devendra (1966)
Barbari	1·8	1·6	Singh *et al.* (1977)
Indian local goat	1·4	1·4	Singh *et al.* (1977)
East African dwarf	2·3	2·0	Wilson (1958a)
Black Bengal		0·7	Ali *et al.* (1973)
Black Bengal	1·2	1·1	Moulick and Syrstad (1970)
White Angora × Goat of Don	3·5	3·0	Nikitenko *et al.* (1972)
Maltese	3·3	3·3	Montemurro (1966)
French Alpine	4·0	3·7	Morand-Fehr, Semptey and Hervieu (1978, unpublished)

C. According to sex and litter size

Breeds	Single		Twin		Triplet		References
	Male	Female	Male	Female	Male	Female	
Black Bengal	1·3	1·1	1·2	1·1	1·1	1·0	Moulic and Syrstad (1970)
Criollo × Nubian	3·6	3·1	3·2	3·0			Castillo *et al.* (1973)
Criollo	2·6	2·4	2·3	2·0			Castillo *et al.* (1973)
Nubian	3·3	2·9	2·9	2·7			Castillo *et al.* (1973)
Damascus	4·4	4·0	3·6		3·1		Choueiri (1973)
Murcian	2·9	2·8	2·8	2·5	2·5	2·5	Belichon and Marques (1971)
French Alpine	5·5	4·0	4·0	3·7	3·4	3·1	Morand-Fehr, Semptey and Hervieu (1978, unpublished)

liquid, an experiment was carried out on 32 male Alpine kids separated from their dams immediately after birth and receiving a milk replacer *ad libitum* from 8 days of age (Fehr and Sauvant, 1974). Using these experimental conditions, it was possible to eliminate the effect of the milk production level of the dam and the effect of insufficient feeding.

Irrespective of the ingestion level, growth during the first month mainly depended on the birth weight and the weight gain during the first week, the period of adaptation to the milk replacer. With the same litter size, the growth rate of kids with a reduced birth weight tended to be lower during the first month; it seemed to depend on individual, probably genetic, factors. Conversely, if the low birth weight was due to a large litter size, the weight gain of the kid was higher and it thus tended to overcome its handicap and to exhibit a higher degree of fattening. However, the effect of litter size disappeared rapidly and this is confirmed by results of Singh and Singh (1974). Furthermore, a poor early growth during the first week, due to a difficult adaptation to milk feeding, may be followed by a partially compensatory body growth which does not favour deposition of body fat, except

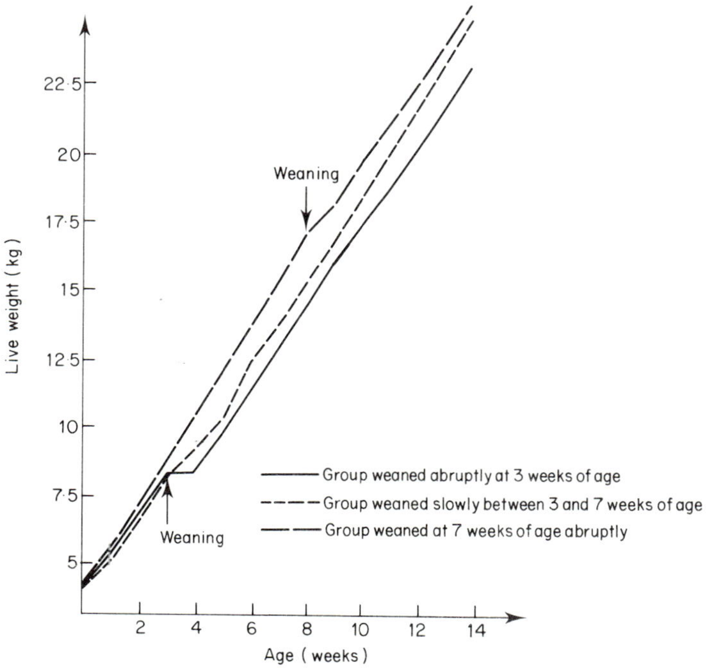

Fig. 5. Influence of the weaning type on growth of 24 French Alpine kids until 14 weeks of age (from Fehr and Sauvant, 1976).

abdominal fat and especially mesenteric fat; in addition, the different gastric parts seem to be well developed in these animals. Growth performance of kids during the milk feeding period was linearly correlated with the intake of milk dry matter and more accurately with the energy intake level. The feed conversion rate during this first month ranged between 1·1 and 1·3 and depended on the one hand on the energy concentration of the milk and more precisely its fat content, and on the other, on the mode of feeding. Restricted feeding as compared to *ad libitum* feeding substantially reduced the weight gain, but definitely improved the feed conversion ratio because of a lower storage of fat in the adipose tissues (from 0·2 to 1 point according to feeding conditions).

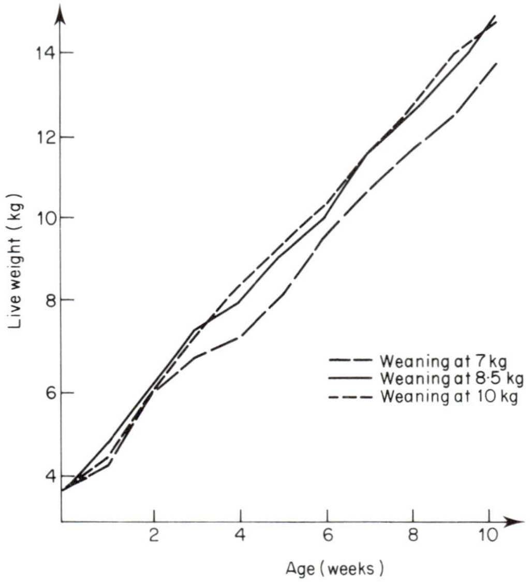

Fig. 6. Effect of weight at weaning on the growth of 24 French Alpine female kids.

4. Weaning

Replacing liquid feeding by solid feeding always causes a shock in the young, the magnitude of which depends on several factors. Most frequently the effect of this shock is to reduce and sometimes stop weight gain. The earlier the weaning, the more marked the reaction (Fig. 5). Fehr and Sauvant (1976) observed that weaning at 3 weeks causes an interruption of growth for about 1 week in Alpine kids, while weaning at 7 weeks only leads to a very slight growth retardation. In fact, the magnitude of this shock seems to be more related to the weight than to the age of the kid (Fig. 6). Fehr (1972) has shown that weaning of Alpine kids at 7 kg and 8·5 kg brings about a growth delay of 1200 g and 44 g respectively as compared with the growth of kids weaned at 10 kg which exhibit almost no shock effects at weaning. Furthermore, males seem to be more susceptible to the weaning shock than females. Skjevdal (1974) affirms that weaning does not give rise to problems if the kid is previously fed with 70–80 g concentrate feed per day. According to our own observations, 30–50 g concentrate feed per day would be sufficient for early weaning. The amounts of milk replacers and solid feeds (forage and creep feed) offered during the milk period in order to reduce the weaning

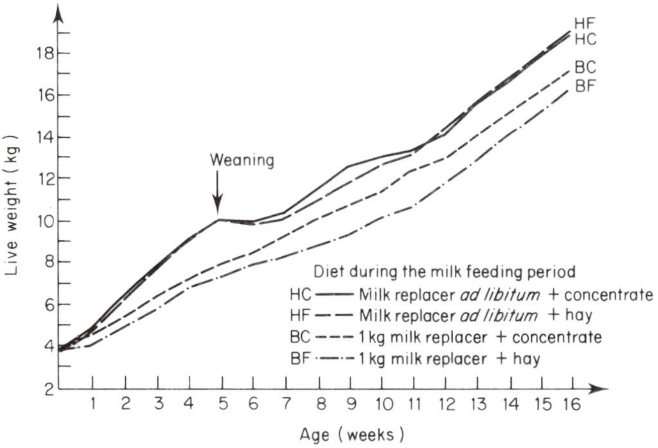

Fig. 7. Influence of the level of feed intake and the type of solid feed during the milk feeding period on the growth of 32 French Alpine female kids (from Fehr and de Simiane, 1974).

shock may affect performance of kids during and just after weaning. The results of the experiment reported in Fig. 7 indicate that *ad libitum* feeding of a milk replacer, as compared with restricted feeding, reduced the intake of solid feeds and accordingly led to a very marked weaning shock. However, the kids consuming most milk and thus the least amounts of solid feeds before weaning tended to exhibit the highest growth rate probably because of a larger development of the rumen.

According to Fehr (1975), it is possible to characterize the metabolic profile of these animals according to the type of weaning (Fig. 8). A slight shock at weaning is accompanied by a slight mobilization of fat reserves revealed by the low blood level of non-esterified fatty acids (NEFA) and by a very progressive decrease in the blood level of glucose. On the other hand, during a severe weaning shock the kid mobilizes much more fat reserves; its glycaemia rapidly falls and then tends to remain constant and even to increase slightly. Thus, the metabolism of the kid during weaning appears to be a highly valuable index accounting for the growth rate differences observed.

5. Post-weaning Growing Period (third to seventh month)

Growth is often slower after weaning (Romagosa Vila, 1975; Fehr *et al.*, 1976; Singh and Singh, 1974). As a matter of fact, weaning frequently coincides with the period where the growth rate begins to decrease, i.e. towards the end of the third month (Fig. 2). A partial compensatory growth is sometimes observed after weaning if the weaning shock has been severe. When weaning has taken place satisfactorily, the age at weaning does not affect the subsequent growth rate (Opstvedt, 1969; Fehr, 1972; Fehr and

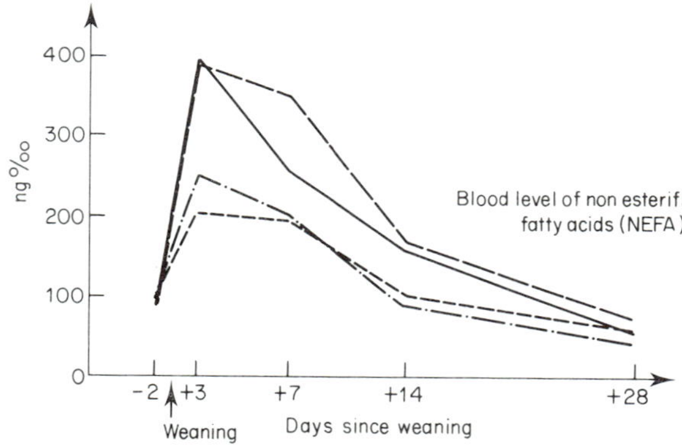

Fig. 8. Effects of weaning on blood parameters (same key as Fig. 7) (from Fehr, 1975).

Sauvant, 1976). During this period (third to seventh month) live weight and weight gain differences between male and female kids are increasing (Louca and Hancock, 1977; Singh and Singh, 1974). Litter size has no influence any more (Prasad *et al.*, 1971; Singh and Singh, 1974). But the weight of the dam affects the weight gain of the kids very much and more obviously than during the second and third month (Table II, Romagosa Vila, 1975).

There is a limited residual effect of diet during the milk feeding phase on subsequent growth. However, post-weaning growth of kids that have received liberal quantities of milk is better because of a higher weaning weight. The nature of the solid feeds offered during the milk feeding period (hay or concentrate feed) has no effect on post-weaning intake. However, the post-weaning growth performance of kids fed with concentrate feed before weaning is higher than that of kids fed with hay only. Wilson (1958a) and Lautier and Morand-Fehr (1976) showed that the energy ingested after weaning, depending mainly on the quantity of concentrates fed, strongly affects weight gain. In two experiments with 52 Alpine kids, between 13 and 30 weeks of age (Lautier and Morand-Fehr, 1976, 1977) a decrease of 300 g in the concentrate supply (from 500–200 g up to the twentieth week and from 400–100 g thereafter) accompanied by an *ad libitum* lucerne hay feeding reduced the live weight at 7 months from 1·5 to 2 kg. Crude protein content of the diet has also a marked effect on growth rate (Louca and Hancock, 1977). Male and female Damscus kids weighing on an average 16–32 kg lost about 5 kg when the total crude protein content of the diet decreased from 16·6 to 10·9%.

Table II. Average daily weight gain of female kids according to the adult weight of dams (g/day) (Romagosa Vila, 1975).

	Adult weight of dams (kg)		
Growth period (days)	30–40	40–50	55–60
30–90	128	143	148
90–200	118	138	147

6. Final Growth Stage (seventh month–adult age)

During this period, reproductive rhythm and feeding are the factors affecting growth most. Unfortunately, as information is lacking about this period, the effects cannot be described with accuracy. The growth of young male goats is very much reduced during the breeding season and they even lose weight, not because of their sexual activity, but because of the marked reduction of their feed intake. Furthermore, an early gestation seems to retard growth of young female goats, in particular in the case of 2 or 3 fetuses. During the first lactation there seems to be a sort of competition between milk production and growth. Growth of a high-yielding goat is delayed during lactation, and adult weight is therefore reached later.

The effect of nutrition on growth depends on the physiological stage of the goat. At some stages, nutrient requirements for pregnancy or lactation prevail over requirements for weight gain of the dam. A rather abundant feeding during late lactation and early gestation is more advantageous for the dam and tends to affect its growth favourably (see Chapter 6).

Furthermore, there seems to be a relation both between and within breeds in time taken to reach mature weight and mature size, large animals growing for a longer period of time.

C. Development and Body Composition During Growth

Description of changes in the conformation of the growing animal and in body proportions (carcass and offals), as well as of the variations of proportions of muscle, bone and adipose tissues, must be based on the results of carcass dissection and measurement trials because of the accuracy of this kind of information. Unfortunately, only studies on a limited number of breeds are available: the East African dwarf breed (Wilson, 1958a, b, 1960), the Alpine breed (Morand-Fehr *et al.*, 1977; Fehr and Sauvant, 1976; Sauvant and Morand-Fehr, 1977, unpublished) and to a lesser extent, feral goats (Lapido, 1974) Kambing-Katjang goats (Devendra, 1966) and Sudanese desert Goats (Gaili, 1976).

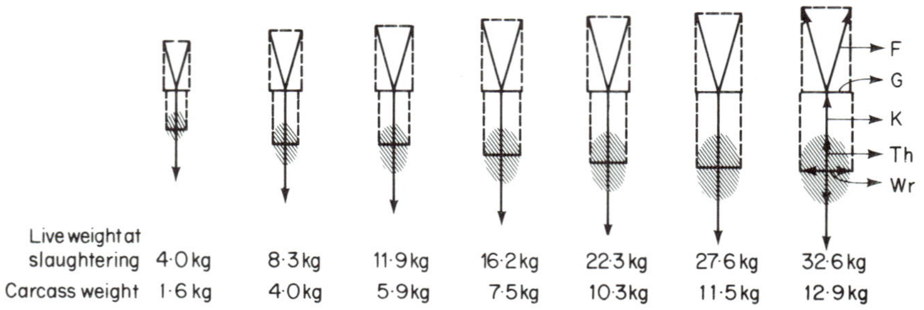

Fig. 9. Change in the carcass characteristics in real values. K: length of the carcass; Wr: width at the rib level; Th: depth of the breast; G: trochanter width; F: leg-length (between perineum and tarsometatarsal articulation) (from Fehr *et al.*, 1976a, b).

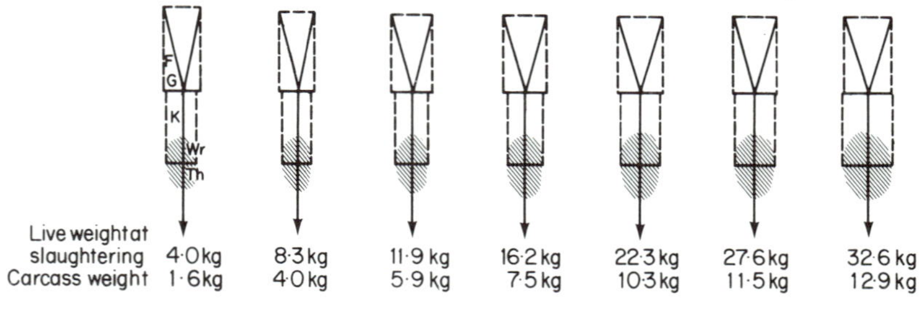

Fig. 10. Change in the carcass characteristics with constant length (K). K: length of the carcass; Wr: width at the rib level; Th: depth of the breast; G: trochanter width; F: leg-length (between perineum and tarsometatarsal articulation) (from Fehr *et al.*, 1976a, b).

1. Measurements and Body Development

Fifty-four Alpine males slaughtered between birth and a weight of 32·6 kg were used to study the main carcass measurements according to weight and at constant carcass length (Fig. 10) (Morand-Fehr *et al.*, 1977). The width and depth increased more rapidly than length. The thigh became more and more compact and the chest became a little rounder when the live weight increased from 4 kg to 32 kg. This change in body conformation with live weight can be observed in most animal species and in particular in sheep and cattle. These results confirm observations of Wilson (1958a) in this field.

At the same age, all body measurements of male kids are larger than those of female kids because of their higher weight. But at the same weight, females exhibit the largest measurements especially the length. Likewise, a low feeding level tends to increase the values of all the measurements recorded either at the same weight or at the same age (Wilson, 1958a).

Experiments dealing with body measurements (Wilson, 1958a), as well as with carcass cutting and dissection (Wilson, 1958b; Lapido, 1974; Morand-Fehr *et al.*, 1977; Sauvant and Morand-Fehr, 1977, unpublished), clearly show that, as in other ruminants, growth of the kid is centripetal. In fact, the terminal parts of the body, notably feet and head, are relatively well developed at birth. The growth "wave" moves progressively from the extremities to the thorax and more particularly to the dorso-lumbar part and the breast of the animal. The allometric coefficient of head and feet is definitely lower than 1 indicating that their growth is slower than that of the whole body while the carcass cuts with the highest allometric coefficients are the breast, saddle and loin (Morand-Fehr *et al.*, 1977, Table III). The relatively fast growth of the dorso-lumbar region is principally due to a very active but late proteinogenesis, whereas that of the breast is due to fat storage and an important proteinogenesis (Sauvant and Morand-Fehr, 1977).

2. Development of Viscera

Relative growth of the different parts of "offals" (measured by the allometric coefficients or by changes in their proportion of empty live weight in the kid) are comparable to young animals of other domestic species, especially the calf and the lamb. They may be ranked according to an order of growth ranging from the earliest to the latest (Wilson, 1958b; Sauvant and Morand-Fehr, 1977):

head—skin—red offals (kidneys, heart, liver, lungs)—gastro-intestinal organs (abomasum, intestine, omasum, rumen + reticulum).

At the same live weight, head and skin of males are relatively larger, but the proportion of all their red offals is lower than that observed in females. This difference between red offals of male and female goats seems to be mainly due to a larger fat layer on the viscera of females (Wilson, 1960). Although Wilson (1960) showed that feeding level affects weight of offals, the effect does not seem to be great, except for the liver which seems to be more sensitive to the production method and feeding. According to Fehr and Sauvant (1976), the liver of kids weaned at 3 weeks is heavier than that of kids weaned later. Weight of the liver might depend on the degree of energy deficiency during weaning, leading to a more or less pronounced mobilization of body fat reserves.

Table III. Allometric coefficients[a] of cuts and tissues of kid carcasses[b] (between 4 and 32·6 kg live weight). From Fehr *et al.* (1976).

Carcass cuts	Allometric coefficients
Poitrine (Breast)	1·175
Filet (Loin)	1·155
Selle (Saddle)	1·081
Carré couvert (Best-end neck)	1·034
Gigot (Leg and part of chump)	0·988
Collier (Scrag)	0·957
Epaule (Shoulder)	0·921
Carré de couverture (Middle neck)	0·911
Tissues	
Fat	1·823
Muscle	1·067
Bone	0·741

[a]% weight increase of cuts or tissues relative to weight increase of the whole carcass
[b]54 Alpine male kids

Growth of the gastro-intestinal organs is late but rapid due in particular to the very large increase in the volumes of the reticulo-rumen and large intestine during and after weaning while the growth of the abomasum and the small intestine, more involved in the digestion of milk, is relatively less marked (Wilson, 1960; Gihad and Morad, 1977). There is a close correlation between type of diet and the weight proportions of these organs. Milk feeding favours development of the abomasum whereas roughage feeding leads to a faster development of the reticulo-rumen and omasum. The effect of feed intake level depends on the composition of the diet (Wilson, 1960; Fehr and Sauvant, 1976).

3. Body Composition

Changes in body composition during growth are similar to these observed in other ruminants. The water content in the body of the kid decreases as live weight increases, but more in bones and adipose tissue than in muscle, skin and viscera (Wilson, 1958b). In addition, growth of bones, already well developed at birth, is slower than that of the rest of the carcass; growth rate of muscles is comparable to the whole carcass while fat, present at birth in small amounts, is deposited more and more during late growth (Wilson, 1958b; Morand-Fehr *et al.*, 1977). These observations are confirmed by the allometric coefficients of the kid (Table III). Their values are close to those recorded in lambs (Tulloh, 1963; Boccard *et al.*, 1964). In a similar study

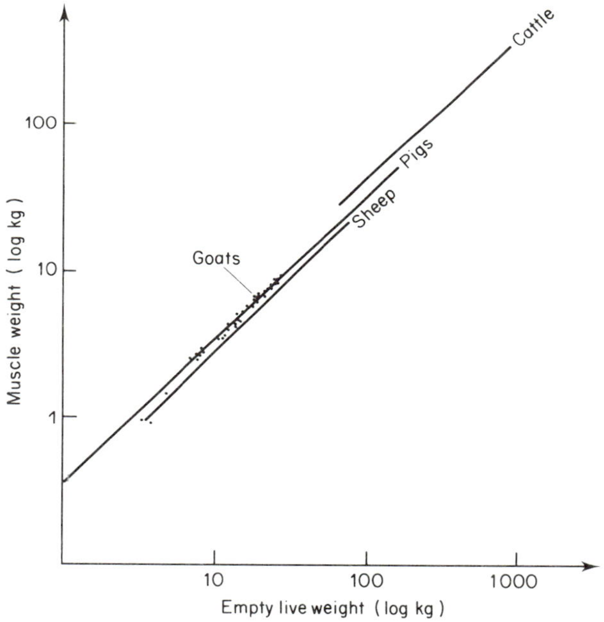

Fig. 11. Muscle development of 44 French Alpine kids as compared to Tulloh's mean results on sheep, pigs and cattle (from Fehr *et al.*, 1976b).

performed on Ile de France lambs and Alpine kids (Sauvant and Morand-Fehr, 1977), it was observed that the variations of muscular proportions in the different parts of the body are larger in the kid, especially in the anterior part of the carcass: breast, shoulder, middle neck (*carré découvert*), scrag (*collier*). The variations of carcass fat proportions are rather different in the two species. In lambs, the evolution of variations of each type of adipose tissue follows rather well that of the corresponding part of the body. This is not the case in kids since fat is deposited rather late in the *gigot* (the leg and part of chump) developing rather early, and rather early in the slow developing areas such as the *selle* (saddle).

Morand-Fehr *et al.* (1977) attempted to characterize the development of various tissues of kids from a dairy breed in relation to that of other domestic animals. They compared their results on kids with those of Tulloh (1963) obtained in cattle, sheep and pigs concerning the relationship between live weight and the weight of bones, muscles and fat during growth. The development of the muscle masses of kids is rather similar to that of the pigs and exceeds that of sheep (Fig. 11). The proportion of bone mass of goats also resembles that of pigs (Fig. 12). The relative volume of adipose tissue is

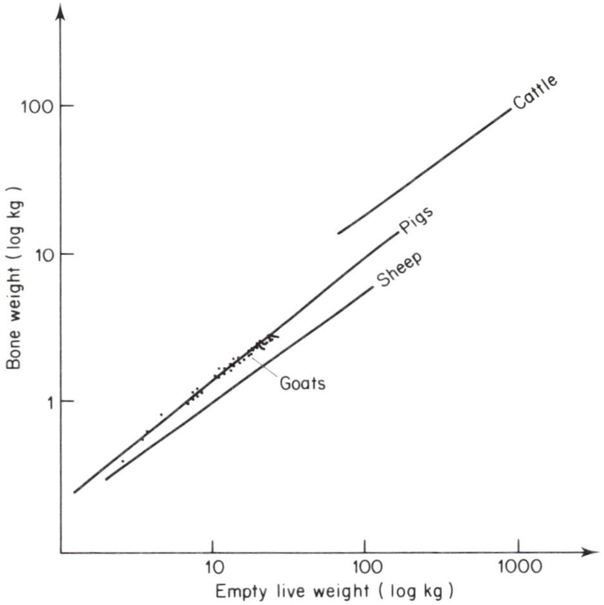

Fig. 12. Bone development of 44 French Alpine male kids as compared to Tulloh's mean results on sheep, pigs and cattle (Fehr *et al.*, 1976b).

definitely smaller in the kid than in the pig or lamb and is rather similar to that of cattle (Fig. 13). These results clearly show that the kid has particular carcass characteristics and that its growth cannot be equated with that of other species, especially ovine.

The proportion of bone to live weight or to carcass weight and its changes during growth are barely affected by dietary factors. The proportion of bone in females seems to be slightly lower than in males (Wilson, 1960). The muscle tissue of the kid is characterized by low lipid content and high water, protein and ash contents (Gaili *et al.*, 1972). During late growth the percentage of muscle in the female tends to be a little higher than that of the male, at

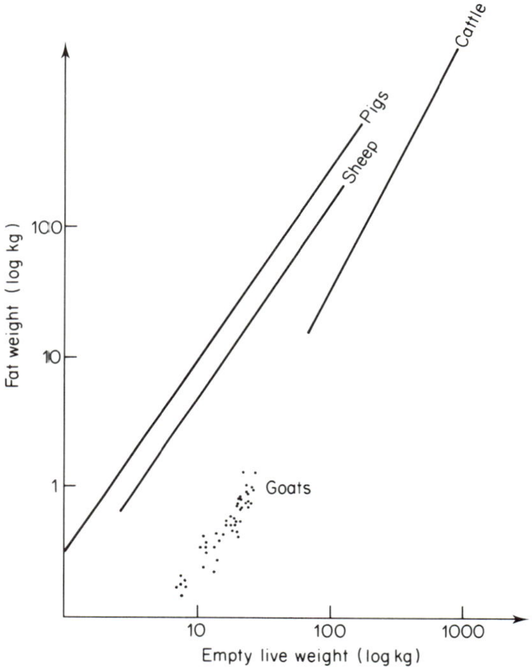

Fig. 13. Fat development of 44 French Alpine male kids as compared to Tulloh's mean results on pigs, sheep and cattle (Fehr *et al.*, 1976b).

the same weight (Wilson, 1960). Dietary factors seem to have little effect on the proportion of muscle in the carcass if the comparison is made at the same weights.

Of all the different components of a kid's body, the adipose tissue is the most sensitive to the various extrinsic and intrinsic factors. Storage of fat in the different adipose tissues may vary from one kid to another. During growth, the deposition of subcutaneous fat seems to be very late, and much later than the internal fats such as the renal or mesenteric fat. Females contain generally more fat than males (Wilson, 1960). The volume of adipose tissue mainly depends on feeding conditions. The higher the feeding level, especially that of concentrates, the more rapid the deposition of fat, and accordingly the faster the growth (Wilson, 1960; Fehr and Sauvant, 1976; Fehr *et al.*, 1976a) and the larger the protein deficiency of the diet relative to energy supply.

The fatty acid composition of the various adipose tissues changes during growth, notably at the moment of weaning and during the post-weaning period, but there is a difference between internal and external tissues

(Sauvant *et al.*, 1979). During the milk feeding period, the fatty acid composition of the adipose tissues of the kid depends on the fatty acid composition of the milk fat. In addition, increase in the feed intake level tends to reduce the saturation of the adipose tissues. After weaning, the composition of intramuscular fat does not change much. Internal fat such as perirenal and mesenteric fat becomes richer in stearic acid (C18:0) because of hydrogenation by the ruminal flora of the dietary fatty acids unsaturated in C18; subcutaneous fat accumulates acids with lower carbon numbers and minor acids (odd and branched acids) to the detriment of palmitic acid (C16:0) and oleic acid (C18:1). For that reason, the kid, like the lamb (Duncan and Garton, 1967) and unlike cattle, exhibits subcutaneous depositions which are rich in odd or branched acids with a low melting point and whose presence favours the increase in tissular water content (Molenat and Theriez, 1973).

The general physiological laws governing growth are valid for the kid as for other young animals: centripetal body growth, successive growth of bone, muscle and fat, etc. However, the kid has its own characteristics, in particular the body mass contains more water and proteins and less lipids; a weight gain of 1 kg in the kid would theoretically require less energy and more protein than 1 kg weight gain in lamb, at the same growth rate. These characteristics should be taken into account to define kid production techniques and more particularly feeding plans.

D. Kid Production Techniques

Kid production techniques have to be adapted to the growth potential of the caprine species, but principally to the breed used. Moreover, they have to take into account some other characteristics of young goats such as their reaction to the environmental conditions, their feeding behaviour and their sensitivity to pathogenic agents.

1. Environmental Conditions

The kid at birth is frail and its thermoregulation is sometimes delicate. Housing conditions must therefore be good, draughts must be avoided and a minimum temperature of 14°C maintained for the first days after birth and of 10°C after about 10 days, for European high yielding breeds. The temperature can be lower for rustic breeds. The kid should be reared with its dam or in a group as it does not like to be isolated. As with lambs, there are "suckers" among the kids sucking the umbilical cord remnants, ears, scrotum of other kids. From their very first days of life, kids like to jump and climb on things around them; for that reason there are often accidents; feet and head blocked, accidental strangulations. This must be taken into account when choosing the housing installations. Houses for kids older than 3–4 months may be open (semi open-air) or closed if well ventilated. The density must not exceed 2·5 kids per m^2 for animals weighing 30 kg at 7 months.

Although the behaviour of the kid may be different between breeds, when on the pasture or ranges it leaves the herd more easily than the lamb. Thus, the kid may get lost or be attacked by predators if the herd is not well protected. When kids are less than about 4 months old (depending on breeds), the daily distance covered by the herd on range has to be moderate and the kids have to rest at regular intervals to prevent their growth rate from being retarded by too much strenuous exercise. The social order is very rapidly established in a kid herd on the basis of fights organized by the very young kids and resembling those of the adult animals. The oldest kids, i.e. kids born early in the season, have the best chance of becoming the dominant animals of the herd. When this order has been established, the behaviour of the herd may be rather homogenous and there is no fighting for access to feed. The onset of sexual behaviour occurs very early at about 3–4 months, so it is advisable to separate males from females at that age to avoid early matings and consequent growth delay.

2. Feeding

(a) *Feeding Behaviour*

Generally kids quickly begin to suck their dams, with the exception of some which may have buccal malformations resulting in a "hanging tongue". They may, however, be rapidly trained to suck from a teat bottle or from a recipient. When concentrate feed or hay is offered from the age of 8–15 days onwards, the consumption starts between 15 and 30 days of age depending on the behaviour (or boldness) of the dominant kids of the herd. The forage should be chopped, and the concentrate feed offered in form of pellets or rolled grains and not in the form of fine flour. Kids try to draw out the forage and scatter it on the litter resulting in much wastage. The feeding system should therefore be adapted so as to prevent this. As kids grow, their behaviour resembles more and more that of the adult goats (see Chapter 6).

(b) *Feed Intake Level*

The level of feed intake depends on many factors: age and growth rate of the kids, nature of the feeds and mode of feeding. Levels of feed intake of the 170 Alpine female kids used to establish the growth curve of Fig.

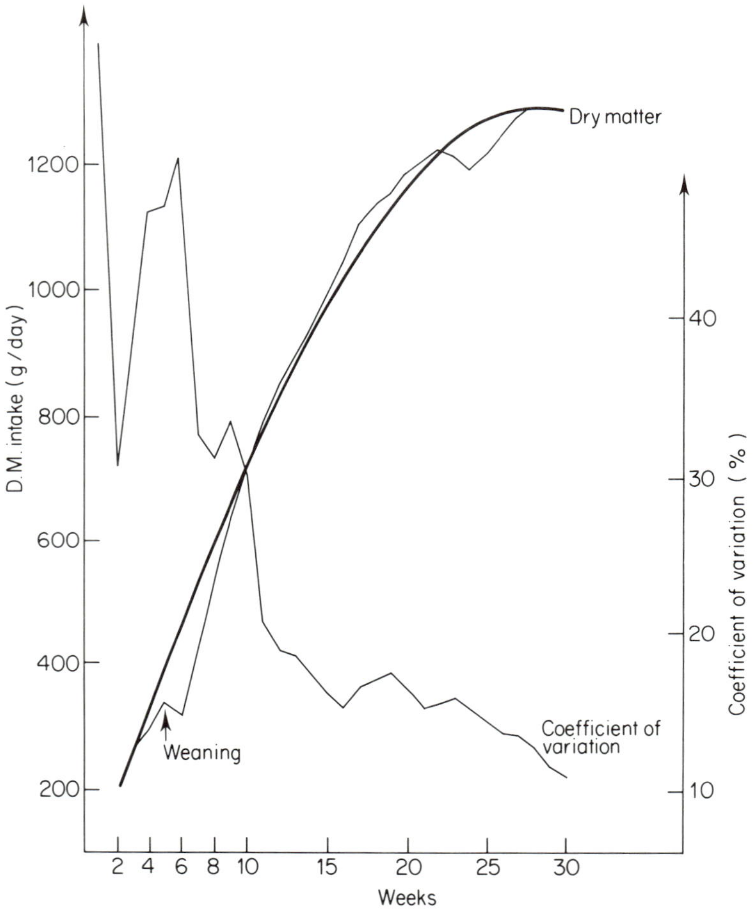

Fig. 14. Dry matter intake in 170 French Alpine female kids (Morand-Fehr and Duborgel, 1976).

1 (Morand-Fehr and Duborgel, 1976) were recorded daily. They were weaned at 5–6 weeks and their post-weaning diet was composed of hay and concentrate feed. The dry matter intake (Fig. 14) increased regularly from 100 to 1200 g/day with a slackening from week 12 which became very marked from week 25 onwards.

The variability which was moderate in the second week increased markedly during the weaning period because of the different weaning methods and then decreased very rapidly until week 30. This decrease in the variability probably accounts for that recorded on live weight (Fig. 1). The same phenomena were observed with respect to the total dry matter intake relative to the metabolic weight (Fig. 15). However, the maximum intake expressed

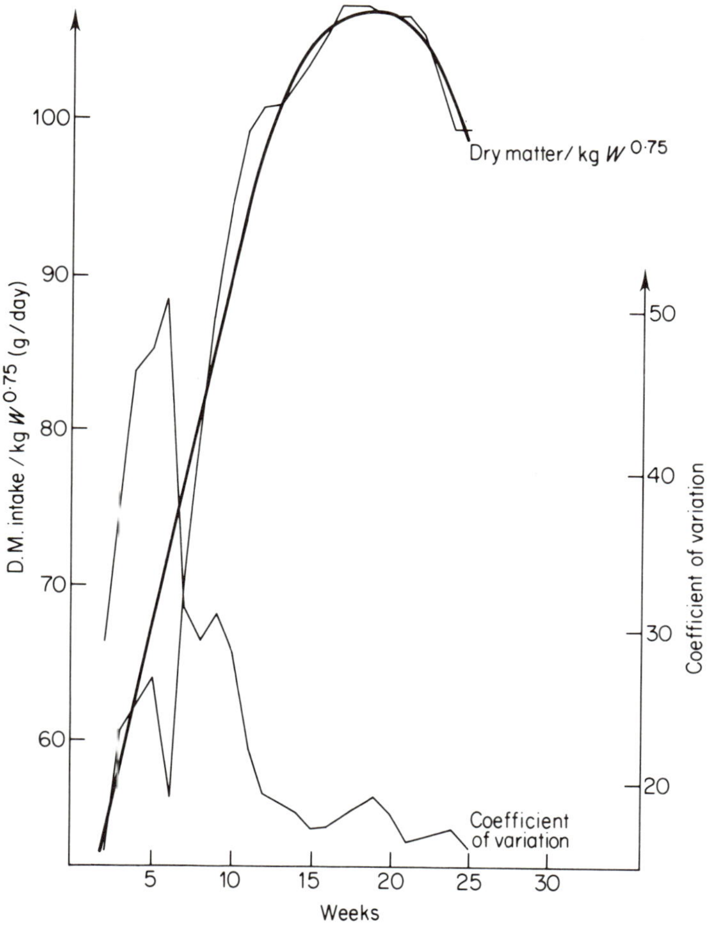

Fig. 15. Curve of dry matter intake in 170 French Alpine female kids relative to the metabolic live weight (Morand-Fehr and Duborgel, 1976).

in this way was located between 16 and 20 weeks. The values recorded in this study have to be adjusted for breed differences and palatability of the feeds offered.

(c) *Recommended Nutrient Allowances*

Only very insufficient data are available about the requirements of young goats. Most published recommendations are extrapolations of those used in lambs. Using nutritional experiments on young goats, we have established recommendable nutrient allowances for kids (Morand-Fehr and Sauvant, 1978). The recommended energy allowances (Table IV) were derived from experimental results obtained in 60 Alpine kids during their first 3 months of life and in 150 Alpine kids from 3 to 7 months whose growth was approximately similar to that reported in Fig. 1. They received a milk replacer from the age of 7 days and were weaned during the fifth week. Thereafter, they were offered dry feed including lucerne or pasture hay and variable amounts of concentrates (0 to 500 g/day). When the supplies were small or when no concentrates were given, they received maize plant wafers and a mineral-vitamin mixture. Energy value of the diets was calculated in net energy as described by Morand-Fehr and Sauvant (1978). When the digestibility of the forages had not been measured, it was estimated by means of the equations suggested by Demarquilly *et al.* (1978). Recommended energy allowances for kids were expressed in MJ net energy milk with the aim of simplifying calculations. As a matter of fact, nutrient allowances for dairy goats are expressed in the same units. Using "net energy milk", as done for lactating goats, for calculation of rations for growing kids results in a slight overestimation of energy in roughage available for growth, but practical consequences are minor.

Table IV. Recommended dietary allowances for breeding female kids (from Morand-Fehr and Sauvant, 1978).

			Recommended allowances			
Mean age (months)	Mean weight in mid-month (kg)	Growth rate (g/day)	MJ net energy milk	D.C.P.[a] (g)	Ca (g)	P (g)
1	6·5	165	3·18	80	3·1	1·3
2	11·5	165	3·62	79	3·4	1·5
3	16·3	155	4·12	77	3·5	1·6
4	20·7	140	4·63	74	3·5	1·7
5	24·5	115	4·92	68	3·2	1·7
6	27·6	90	5·06	62	2·9	1·7
7	30	70	5·14	60	2·7	1·6

[a] Digestible crude protein.

Energy supplies suggested in Table IV are similar to those of French (1944) and Huston *et al.* (1971). The latter were obtained by the factorial method on the basis of the requirements for maintenance (Brody, 1964), the amount of energy retained per kg gain and the metabolizable energy available for growth. They are slightly lower than the former French recommendations established in Leroy F.U. by Morand-Fehr and Duborgel (1976), in particular for the older animals. (Leroy F.U. = former net energy system proposed by

Leroy used for pigs and ruminants where 1 F.U. corresponds to net energy of one kg of barley.) This small difference is probably due to the differences in the diets used during these two studies. The previous allowances were defined on the basis of diets relatively rich in concentrates which might have changed the composition of the gain by increasing the amount of lipids retained.

Trodahl *et al.* (1979) have given allowances (Chapter 15) for Norwegian kids of smaller size than Alpine kids and with a different growth curve. However, they are not very different from ours if compared at the same weight gain.

The recommended supplies (in digestible crude protein (D.C.P.), Ca and P) for kids were calculated by using the metabolic utilization coefficients of nitrogen and the apparent digestibility coefficients (A.D.C.) of minerals applied to lambs, and by assuming that the percentage of proteins in the overall gain is slightly higher than that of lamb because of the smaller proportion of fat in the overall gain of kids.

(d) *Feeding Methods*

The kid may be separated from its dam immediately after birth (in that case it receives the milked colostrum), or after having suckled the colostrum (Skjevdal, 1974; Fehr and Disset, 1968; Fehr, 1975), or it may be reared by its dam. Intermediate solutions may also be used: the kid may stay with its dam which is milked once a day, or it may stay with its dam during the night and be separated from her during the day, the dam being milked once or twice a day.

During the milk feeding period kids may receive either goats' milk, cows' milk or a milk replacer. The growth performances may be the same with these three types of milk (Opstvedt, 1969; Skjevdal, 1974; Fehr, 1971; Louca *et al.*, 1975), but the intake of milk replacer dry matter must be 10–25% higher than that of goats' milk (Fehr, 1975), probably because of a lower digestibility. Milk replacer may be offered only 4 days after birth. The number of daily meals may vary from three to one. In the latter case the milk replacer must be very concentrated—up to 24%—and drinking water must be available for the kids all day; it may be offered hot (35–40°C) or cold (6–10°C) (Fehr, 1975) and at a concentration of 12–16% (the concentration must be calculated so that the fat content of goats' milk is not exceeded).

Milk replacer may be offered *ad libitum* or restricted. However, growth rate is closely related to the level of dry matter intake. Contrary to that, the efficiency of feed conversion is inversely proportional to the amount of feed allotted. *Ad libitum* feeding may cause bloating of the abomasum due to an excessive milk intake. Thus, on account of the advantages and

Table V. Examples of feeding methods in kids:

	Milk feeding period		
	Period (days from birth)	Quantity of milk (kg)	Solid feeds
Technique A			
(goats' milk–	till day 3	colostrum (1·5 kg max.)	
2 meals)	31	1·5 kg goats' milk	
	32	1 kg goats' milk	from day 7:
	33–42	0·5 kg goats' milk (1 meal)	hay
	43	weaning	concentrates } *ad libitum*
			water
Technique B			
Milk replacer	till day 3	colostrum (1·5 kg max.)	
(2 milk meals,	4–6	goats' milk (1.5 kg max.)	
hot or cold)	7	1 kg goats' milk— 0·5 kg milk replacer	from day 7: hay
	8	0·75 kg goats' milk 0·75 kg milk replacer	concentrates } *ad libitum* water
	9	0·50 kg goats' milk 1 kg milk replacer	
	10	1·5 kg milk replacer	
	16–32	1·8 kg milk replacer	
	33	1·5 kg milk replacer	
	34	1 kg milk replacer	
	35–42	0·5 kg milk replacer (1 meal)	
	43	weaning	
Concentration of 13–15% according to the composition of the milk replacer.			
Technique C			
Milk replacer	till day 3	colostrum (1 kg max.) (2 meals)	
1 hot milk meal	4–6	goats' milk (1 kg max.) (2 meals)	
	6	0·5 kg goats' milk— 0·5 kg 12% milk replacer	
	7	0·5 kg goats' milk— 0·5 kg 14% milk replacer	
	8	1 kg 15% milk replacer (1 meal)	from day 7:
	9–32	1 kg–1, 2 kg 16–20% milk replacer (1 meal)	hay concentrates } *ad libitum*
	33	0·8 kg 16–20% milk replacer (1 meal)	water
	34	0·5 kg 16–20% milk replacer (1 meal)	
	43	weaning	
N.B. When using this method, special attention should be paid to the watering of the animals. Furthermore, it is advisable to choose a milk substitute relatively rich in fat and protein.			
Technique D			
Milk replacer automatic or	till day 3	colostrum (offered in a teat-bottle)	
semi-automatic	4–41	automatic machine feeding	from day 7:
milk feeding	42	abrupt weaning	hay
			concentrates } *ad libitum*
			water

Post-weaning period
Technique E: Diet based on forages of poor quality
Post-weaning and Month 3:
Pasture hay of poor quality, concentrates including 6000 MJ milk N.E. and 160 g D.C.P./kg *ad libitum* up to a maximum of 500 g
Months 4 and 5:
Hay *ad libitum*
400 g concentrates including 6700 MJ milk N.E. and 140 g D.C.P./kg
Months 6 and 7:
Hay *ad libitum*
300 g concentrates including 6700 MJ milk N.E. and 120 g D.C.P./kg
Technique F: Diet based on forages of good quality
Post-weaning and Month 3:
Pasture hay of good quality
400 g concentrates including 6000 MJ milk N.E. and 160 g D.C.P./kg *ad libitum* until a maximum of 400 g
Months 4 and 5:
Pasture hay or green forage *ad libitum*
300 g concentrates including 6000 MJ milk N.E. and 130 g D.C.P./kg
Months 6 and 7:
Pasture hay or green forage *ad libitum*
100 g concentrates including 6000 MJ milk N.E. and 100 g D.C.P./kg

disadvantages of the different methods, restricted feeding of animals reared for reproduction and *ad libitum* feeding of animals reared for meat production is advisable.

Milk replacers of variable composition are utilized well by the kids. They may contain 16 to 24% fat and 20 to 28% protein (Lautier *et al.*, 1977) and may be composed of various proportions of reconstituted milk powder, various crude or treated starches or whey (Lautier *et al.*, 1976), but use of vegetable protein and especially soybean protein in pre-weaning feeding seems to reduce performance (Tanabe and Kameoka, 1977).

Weaning (i.e. the change from liquid to dry feeding) may be practised from the age of 35 days and even earlier if the birth weight of the animal has increased 2·5-fold (Fehr and Disset, 1969; Fehr, 1975) or if the kid has ingested 6 kg milk feed dry matter (5 kg would probably be sufficient) (Morand-Fehr and Hervieu, 1975, unpublished) but 2·65 kg as suggested by Ben Asher and Volcani (1975) seems to be insufficient.

Weaning may be prepared by offering forages during the milk feeding period or by using concentrates. However, earlier weaning is possible when using concentrates.

After weaning, kids may be reared on pasture or on range, but then their growth is generally lower because of the rather low energy intake level and in some cases because of parasitosis.

Growth performance is often higher when kids are reared in goat houses. The basal diet may be composed of green forage, hay, silage (although level of intake may be a limiting factor), artificially dried forages and straw. Growth may be improved by concentrate feeding (Wilson, 1958a; Devendra, 1966; Lautier and Morand-Fehr, 1976) through an increased energy intake or similarly a rise in crude protein content of concentrates (Louca and Hancock, 1977). Thus, the allowance of concentrates and their composition should be determined according to the growth objectives and the nature of the basal diet.

The techniques shown in Table V are applied in France to kids reaching a weight of 30–32 kg at 7 months (Fehr, de Simiane, 1974).

3. Health

Very little is known about the pathology of young goats and it is often equated with that of lambs. Our purpose here is not to describe the pathology of young goats, but to draw attention to the effect of rearing conditions on health and so determine which conditions reduce losses or diseases to a minimum during the growing period. Guillimin (1977) analysed kid losses in a herd of 250 goats, kept on zero-grazing (Table VI). Kids were fed by their dam for 24 hours and thereafter offered a milk replacer, which was hot for 5 days and then cold on a free-service system. The animals were weaned at 45 days at 12–13 kg body weight. They were vaccinated after birth against pasteurella, pyogenes and streptococci and at 3 weeks against enterotoxaemia. Accidents in relation to kidding represented a rather important cause of losses as did coccidiosis, septicaemia and paralytical myopathies. It is obvious that various problems due to the rearing techniques, in particular the difficulty in adapting animals to the milk feeding technique, lead to rather large losses.

In other herds where animals are not vaccinated, diseases such as pneumonia and diarrhoea are often observed in the young. The colostrum constitutes an essential element in the defence of the young against these diseases. When kids are fed by their dam on pasture in moist areas, parasitosis such as strongylosis and fasciolosis may bring about a large growth reduction.

Coccidiosis is very frequent among animals kept in goat houses. Ten species have been described in sheep, some of them exist in goats (Yvoré, 1977). Unfortunately, there is little relationship between the presence of oocysts in the faeces and the pathogenic character of the infestation. The contamination may occur very early and cause diarrhoea and a marked growth retardation. Only a few disinfection methods are efficient. An excessive animal density should be avoided and the kids should not be reared in rooms with a high degree of humidity. Treatment of the dams before kidding and of

Table VI. Kid mortality before birth in a herd of 250 dairy goats over 7 years (from Guillimin, 1977).

Causes	1971	1972	1973	1974	1975	1976	1977	Total
Abortus	18	6		5	7	22	2	60
Mummified fetus		4	9	7	1		2	23
Dam mortality before kidding (gestation toxaemia, listeriosis and various accidents)		3		12	18	5	5	43
Total	18	13	9	24	26	27	9	126
Kid mortality between birth and weaning								
dystocia	2	3	1	4	6	5	10	31
embryotomy	1		2					3
very slow parturition	8	4		2	4	7	1	26
insufficient licking by dam	6	3			5	5		19
breathing difficulties		3	1		1	1		6
suckling inadaptation	6		2	3	1		4	16
pleuropneumonia	3							3
gastroenterotoxaemia				1				1
paralytic myopathy	1						3	4
coccidiosis	1			10				11
accidents	8			2	4	2	1	17
colisepticaemia	6		1	1		2	3	13
undetermined causes	8	2		1	4	6	5	26
necrosis of cortex							1	1
Total	50	15	7	24	25	28	28	177
rate of losses (%)	10·3	3·2	2·6	7·2	6·1	6·5	6·9	6·4
Kid mortality between weaning and first mating								
pleuropneumonia	4		1					5
coccidiosis				19	8	4		31
paralytic myopathy	15		2					17
accident	1							1
gastroenterotoxaemia	1		3	2				6
colisepticaemia			1					1
urinary calculi					1			1
Total	21		7	21	9	4		62

the young before weaning and thereafter at regular intervals if necessary, gives satisfactory results.

Like lambs, kids may suffer from ecthyma. Vaccination of the dams seems to be very effective since antibodies are found in the colostrum.

Like other young ruminants kids are sensitive to nutritional mineral and vitamin deficiencies. When early weaning fails, i.e. when the animal barely eats for more than 8 days, several symptoms occur, the effects of which persist for a very long time. Kids, like lambs, are also very sensitive to high copper supplies and to excessive supplies of phosphorus which often lead to urinary calculi.

In most cases, neonatal diseases are the main causes of economic losses. They are due to low viability of the kids and often depend on inadequate feeding of the dams during late gestation (malnutrition or dietary imbalances) or on bad rearing conditions (places where the kids do not succeed in regulating their internal temperature, lack of control at the moment of kidding).

In France, we recommend the following practices for rearing kids:

(a) Adequate feeding of pregnant animals. Energy and protein malnutrition during late gestation as well as gestation toxaemia should be avoided. Administration of a polyvalent vaccine just before the last month of gestation.

(b) Supply of colostrum as soon as possible after parturition. To prevent the kids from receiving from their dam a colostrum which is too weak in antibodies (colostrum of primiparous animals), the colostrum of all the multiparous goats should be mixed together, kept cold and offered to all the kids after birth.

(c) Mixed vitamin treatment during the first 48 hours and in some cases administration of polyvalent vaccine (particularly if the dams have not been vaccinated).

(d) Excessive kid population densities should be avoided and the ventilation of the rooms must be efficient.

(e) Vitamin and cocciostatic treatments before weaning.

(f) Verification of growth and solid feed intake conditions in the animals which are going to be weaned.

(g) After weaning, treatment against parasitoses if necessary. Vaccination against ecthyma is given in critical situations.

Conclusions

Growth conditions not only affect the performance of meat kids, but may also have repercussions on the performance of replacement kids. Poor development, pulmonary or hepatic diseases, an insufficiently developed rumen might, for instance, lead to low performance or probably a reduced longevity. The production techniques used during the growing period of goats should, therefore, be established according to well determined objectives taking into account the type of animal to be produced and the environmental conditions. The kid is extremely adaptable; it is able to adapt itself to the most varied rearing conditions, but some basal rules have to be applied, principally with respect to the environmental, feeding and sanitary conditions for obtaining satisfactory results. This adaptability of the goat partly explains the inhabitation of goats in very different environments all over the world.

References

Ali, S. Z., Hoque, M. M. and Hasnath, M. A. (1973). *Indian Vet. J.* **50**, 438–440.

Belichon, P. and Marques, F. (1971). *In* "2nd Int. Conf. Goat Breeding", Tours, France, July 1971, pp. 339–343. ITOVIC, Paris.

Ben Asher, A. and Volcani, R. (1975). Annual Meeting of European Association of Animal Production—Sheep and Goat Commission, Report 6, Warsaw, June 1975.

Blanchart, D. and Sauvant, D. (1974). *In* "Journée d'études sur l'alimentation de la Chèvre Laitière" June 1974, pp. 76–87. ITOVIC, Paris.

Boccard, R., Dumont, B. L. and Peyron, C. (1964). *Ann. Zootech.* **13**, 367–378.

Buvanendran, V. and Jalatge, E. F. A. (1974). *Ceylan Vet. J.* **22**, 14–15.

Castillo, J., Garcia, O. and Osal, N. (1972). *Agronomia Trop.* **22**, 251–259.

Choueiri, E. (1973). *In* "III World Conf. Animal Production" Pre Conf. Vol. I. Short contributions, Melbourne (Australia), **1** (B) 1.

Demarquilly, C., Andrieu, J. and Sauvant, D. (1978). *In* "INRA—Alimentation des ruminants", pp. 469–518. INRA, Versailles.

Devendra, C. (1966). *Malaysian Agric. J.* **45**, 345–369.

Devendra, C. and Burns, M. (1970). "Goat Production in the Tropics". Commonwealth Agricultural Bureau, Farnham Royal Bucks, England.

Duncan, W. R. H. and Garton, G. A., (1967). *J. Sci. Food Agric.* **18**, 99.

Fehr, P. M. (1971). "Xème Congr. Int. Zootechnie", Versailles, July 1971.

Fehr, P. M. (1972). "IIème Congrès Mond. Alim. Anim.", Madrid, October, 1972, Vol. V, pp. 353–373.

Fehr, P. M., (1975). *In* "L'allaitement Artificiel des Agneaux et des Chevreaux", pp. 83–105. SEI, Versailles.

Fehr, P. M. and Disset, R. (1968). *Rev. Elev.* **12**, 109–114.

Fehr, P. M. and Disset, R. (1969). "Journée Ass. Fr. Zootechnie", Vol. II, Report 39. AFZ.INA, 16, Paris.

Fehr, P. M. and Disset, R. (1971). *In* "2nd Int. Conf. Goat Breeding", Tours, France, pp. 181–193. ITOVIC, Paris.

Fehr, P. M. and Sauvant, D. (1974). *Ann. Zootechn.* **23**, 503–518.

Fehr, P. M. and Sauvant, D. (1976). *Ann. Zootechn.* **25**, (2), 243–257.

Fehr, P. M. and de Simiane, M. (1974). *Bull. Tech. Inf.* **190**, 425–437.

Fehr, P. M., Guessous, F., Tissier, M. and Sauvant, D. (1971). *In* "2nd Int. Conf. Goat Breeding", pp. 211–222. ITOVIC, Paris.

Fehr, P. M., Sauvant, D., Delage, J., Dumont, B. L. and Roy, G. (1976a). *Livestock Prod. Sci.* **3**, 183–194.

Fehr, P. M., Sauvant, D. and Dumont, B. L. (1976b). *In* "2ème Journées de la Recherche Ovine et Caprine" December 1976, Paris, pp. 166–189. ITOVIC, Paris.
Gaili, E. S. E., Ghanem, Y. S. and Mukhtar, A. M. S. (1972). *Anim. Prod.* **14**, 351–357.
Gaili, E. S. E. (1976). *World Rev. Anim. Prod.* **12** (4), 83–87.
Gihad, E. A. and Morad, H. M. (1977). "Symposium on Goat Breeding in Mediterranean Countries", Malaga, Granada, Murcia, Spain, October 1977, pp. 161–163.
Gill, G. S. and Dev, D. S. (1972). *Indian J. Anim. Prod.* **3**, 173–178.
Guillimin, P. (1977). *In* "3ème Journées de la Recherche Ovine et Caprine", November/December, 1977, pp. 279–286. ITOVIC, Paris.
Huston, J. E., Shelton, M. and Ellis, W. C. (1971). Texas Agric. Exp. Sta. Bull. 1105, p. 6.
Lapido, J. K. (1974). *Diss. Abstr. Int. B*, **34**, (12) 5755.
Lautier, C., Morand-Fehr, P. and Hervieu, J. (1976). Document ITOVIC, Paris.
Lautier, C., Morand-Fehr, P. and Hervieu, J. (1977). Document ITOVIC, Paris.
Louca, A. and Hancock, J. (1977). *J. Anim. Sci.* **44**, 927–931.
Louca, A., Maurogenis, A. and Lawlor, M. J. (1975). *Jap. J. Zootech. Sci.* **48**, 361–370.
Molenat, G. and Theriez, M. (1973). *Ann. Zootechn.* **22**, 279–293.
Montemurro, O. (1966). *Ann. Fac. Agr. Univ. Bari* **20**, 1–17.
Morand-Fehr, P. and Duborgel, E. (1976). ITOVIC, Paris.
Morand-Fehr, P. and Sauvant, D. (1978). *In* "INRA Alimentation des Ruminants", pp. 449–467. INRA, Versailles.
Morand-Fehr, P., Sauvant, D. and Dumont, B. L. (1977). *In* "Symposium on Goat Breeding in Mediterranean Countries", Malaga, Granada, Murcia, Spain, October 1977, pp. 338–348.
Moulick, S. K. and Syrstad, O. (1970). *J. Agric. Sci.* **74**, 409–414.
Nikitenko, V. A., Zaporo, Z. H. and Sev, E. B. (1972). *Ovtsevodstvo* **18**, 12–14.
Opstvedt, J. (1969). *In* "Grassland in Sheep and Goat Production", pp. 89–100, (O. R. W. Spedding, ed.). EAAP Report N°. 2, Study Commission on Animal Nutrition.
Prasad, S. P., Roy, A. and Pandey, M. D. (1971). *Agra Univ. Res.* **20**, 45–54.
Restall, B. J. (1976). *In* "2ème Journées de la Recherche Ovine et Caprine", Paris, France, December 1976, pp. 155–165. ITOVIC, Paris.
Romagosa Vilá, J. A. (1975). *Av. aliment. mejora anim.* **16**, (2). 73–75.
Sada, I. and Vohradsky, F. (1973). Institut Tropického a Subtropickèho Zemèdèlstvı, **3**, 173–185.
Sauvant, D., Bas, P. and Morand-Fehr, P. (1979). *Ann. Zootechn.* **28**, 73–92.
Singh, C. S. P., Mishra, H. R., Singh, S. K., Mukmerji, D. J. and Prasad, B. (1977). *Indian J. Sci.* **47**, 592–594.
Singh, B. B. (1973). *Indian Vet. J.* **50**, 1103–1106.
Singh, B. B. and Singh, B. P. (1974). *Indian Vet. J.* **51**, 322–326.
Skjevdal, T. (1974). *Meldinger Norges Landbrukshogsk.* **53**, (39), 19 pp.
Tanabe, S. and Kameoka, K. (1977). *Anim. Prod.* **20**, 213–218.
Trodahl, S., Skjevdal, T. and Steine, T. A. (1981). *In* "Goat Production", p. 511. Academic Press, London and New York.
Tulloh, N. M. (1963). *In* "Carcase Composition and Appraisal of Meat Animals", (D. E. Tribe, ed.), No. 5, pp. 1–30. CSIRO, East Melbourne Victoria, Australia.
Wilson, P. N. (1958a). *J. Agric. Sci.* **50**, 198–210.

Wilson, P. N. (1958b). *J. Agric. Sci.* **51**, 4–21.
Wilson, P. N. (1960). *J. Agric. Sci.* **54**, 105–130.
Yvore, P. (1977). *In* "3ème Journées de la Recherche Ovine et Caprine", November/December, 1977, pp. 3–15. ITOVIC, Paris.

9. Meat Production

R. T. Naudé and H. S. Hofmeyr

A. Introduction

There is a dearth of information on meat production from goats. Interpretation of the relatively few published reports on this topic is complicated by the fact that there are so many goat breeds kept under such widely different conditions that comparisons of results are not always meaningful. It was therefore considered necessary to do some classification. It seems possible to classify effectively meat produced from goats according to the main purpose for which goats are kept. The following categories are suggested:

Angora Goats: These animals are kept primarily for mohair production. Old animals and culls are slaughtered for their meat.

Dairy Goats: Male kids form the main meat outlet from the dairy goat enterprise, traditionally slaughtered at a very young age (Fehr *et al.*, 1976). In such enterprises up to 50% of the meat production is from culled older animals. Unfortunately meat from old animals is rather unpopular and thus not in high demand.

Meat Goats: This category comprises all goat breeds kept and bred mainly for meat production. The South African Boer goat is probably the breed that exhibits the most pronounced meat production qualities (Devendra and Burns, 1970). At the same time goats owned by people in developing communities should also be included in this category. These animals are usually not purposefully selected in any specific direction, and vary tremendously in size, shape and colour (Devendra and Burns, 1970), but they all supply meat, skins and sometimes fibre to their owners. A large proportion of these animals is part of a subsistence type of farming and therefore never reaches a market, but is consumed by their owners. As a matter of fact butchers in general are not very enthusiastic about goat meat from animals beyond one year of age. This could be ascribed to unfair generalization on goat meat quality, following previous experience with low quality meat from underfed, unfinished slaughter stock.

Feral Goats: In Australia and New Zealand there is a small but significant meat export trade, based on cropping feral goats, descendants of imported goats, mainly of the dairy goat type (Kirton, 1970; Restall, 1976; Moorhouse and Batten, 1978; Batten and Moorhouse, 1979).

The various aspects of meat production from goats will be considered with this classification in mind where available information allows such an approach. A study of meat production in any particular species, breed or type should comprise an account of reproduction, growth rate, feed conversion, dressing percentage and carcass and meat quality. Some of these aspects have been dealt with elsewhere in this book, and will therefore only be referred to in the context of meat production.

B. Reproduction

One of the most favourable attributes of goats as meat-producing animals is their high rate of reproduction. A study of world literature shows that most goat breeds have average litter sizes of approximately 1·5 and higher as a result of high twinning rates; triplets and quadruplets are frequent (Devendra and Burns, 1970; Wilson, 1976). The goat is the most prolific of all domesticated ruminants under tropical and subtropical conditions. Does reach puberty as early as 6 months of age when well fed and young bucks can successfully mate at an age of about 168 days (Skinner, 1972). There are goat breeds able to breed throughout the year (Hofmeyr *et al.*, 1965; Devendra and Burns, 1970) but Angora goats and dairy goats originating from temperate regions have limited breeding seasons (Hofmeyr, 1962; Hofmeyr *et al.*, 1965, Devendra and Burns, 1970). Where meat production is to a greater or lesser extent a by-product as is the case with the Angora and dairy goat, limited breeding seasons might be tolerated. However, where goats are kept primarily for meat production, restricted breeding seasons are a definite disadvantage. Fortunately the occurrence of extended breeding seasons is widely spread amongst goat breeds in subtropical and tropical areas (Devendra and Burns, 1970). The Boer goat of South Africa also breeds

Table I. Oestrus periods recorded in Boer goat does, relative to possible periods per month (Hofmeyr *et al.*, 1965).

Month	%	Month	%
Jan.	33	July	56
Feb.	30	Aug.	71
March	52	Sept.	40
April	52	Oct.	48
May	63	Nov.	16
June	59	Dec.	44

throughout the year, although its sexual activity becomes rather low during the late spring and summer months, as indicated in Table I (Hofmeyr *et al.*, 1965). But this cycle apparently is not entirely governed by the light regime since it does not coincide with decreasing and increasing length of daylight.

The main advantage of extended breeding seasons is that the number of kids per doe produced in one year could be dramatically increased by mating twice a year or three times in two years. This was actually achieved with Boer goats, where the number of kids per doe per year was increased from an average of 1·89 per year (Hofmeyr, 1962) to 3·60 by mating twice a year (Skinner and Hofmeyr, 1969). This was accomplished by keeping the does on a very high nutritional regime and by weaning the kids at 6 weeks of age. The fact that the mere introduction of the male goat brings does into oestrus (Shelton 1960; Lamont, 1964), usually within 8 days after the introduction (Skinner and Hofmeyr, 1969), is of great advantage when breeding twice a year, although it still has to be confirmed that this phenomenon takes place out of the breeding season. Special care has to be taken, however, that bucks are fit to serve does when they are bred in the late spring or early summer, since the male apparently has a peak sexual activity during the autumn and this declines in summer (Skinner, 1972). Skinner and Hofmeyr (1969) successfully overcame this by treating Boer goat rams with 500 IU PMSG (Pregnant mare serum gonadotrophin) prior to their introduction to the does in early summer. These PMSG-treated bucks had sufficient libido to serve successfully the large number of does that simultaneously came into oestrus due to the introduction of bucks.

It is a common experience that multiple births in goats are associated with a high mortality rate (Devendra and Burns, 1970). There seems to be, however, no biological reason why mortality should be so high, provided that management and nutrition is adequate. Without discussing this aspect in detail it should be pointed out that the full meat production potential of goats can only be utilized by exploiting their prolificacy. To this end intensive management and high nutritional levels might be economically worthwhile. In an experiment with 30 Boer goat does, kept under intensive conditions in a barn, we successfully raised twins and triplets. It was, however, necessary to pay special attention to triplets during the first few days after birth. Our experience was that under such conditions multiple suckling freely took place, with triplets tending to suckle mothers of singles (Hofmeyr and Grey, unpublished).

C. Growth Rate

The advantage for meat production of high prolificacy in goats is unfortunately not matched by high growth rates. Growth figures quoted by Devendra and Burns (1970) are disappointingly low.

Where goats are kept primarily for purposes such as milk and mohair production, high growth rates are not necessarily very important. This is also true where goats are kept under poor management and nutrition conditions and where their adaptability is of more importance than growth rate. Wilson (1958) measured growth rate in the East African goat in Uganda. On a high plane of nutrition kids gained 91 g/day to reach 15 kg at 20 weeks of age. This is in close agreement with the growth rates of Sudanese goat kids of 87 g/day to 3 months and 67 g/day to 6 months of age (Wilson, 1976). Devendra (1966) reported body weights at 30 weeks of age of 24·1 and 22·4 kg respectively for Anglo-Nubian and Jamnapari crossbreds. Assuming birth weights of 2·5 kg, this would imply growth rates of about 102 and 94 g/day. These growth rates are considerably lower than the normal growth rates in lambs of many sheep breeds.

Recently more favourable results were obtained with dairy goats and Boer goats which might serve as an indication of the growth potential of goats for meat production. Norwegian kids fed until 5 to 10 weeks of age gained 150 to 210 g daily (Skjevdal, 1974; Robstad, 1976). Fehr *et al.* (1976) studied the effects of feeding methods and age at slaughter on the growth performance of entire male Alpine (dairy goat) kids. The kids received milk replacer *ad libitum* from day 8 after birth. They were slaughtered at body masses of 16, 22, 28 and 34 kg and received either milk replacer, milk replacer plus concentrates (20% protein) or concentrates plus maize wafers. Growth rates before weaning in excess of 200 g/day were easily obtained, but after weaning at 42 days of age, gains were significantly reduced. Average daily gains for groups slaughtered at 28 and 34 kg body mass thus dropped to 172 and 176 g/day.

We have done an experiment with 30 Boer goat does under intensive conditions. The does were kept in a barn where they had free access to a complete ration with a digestible D.M. content of 60% and crude protein of 14%. The kids were left with their dams till 38 days of age, but had free access to a creep feed with 70% D.M. digestibility and 14% crude protein. During a 12 week post-weaning period they were fed on *ad libitum* basis a ration with 65% D.M. digestibility and 14% crude protein. The average growth rate of the 54 kids from birth till weaning was 227 g/day and over the 12-week period after weaning it dropped to 200 g/day.

The best recorded growth figures for Boer goats were recently obtained under the Boer Goat Improvement Scheme run by the Department of Agricultural Technical Services of South Africa. Highly selected (show quality) Boer goat kids were left with their dams until weaning at 100 days and then had free access to a concentrate ration while kept on good mixed shrub savannah grazing. The growth figures are given in Table II, and should be considered as rather exceptional for goats.

Table II. Body mass and growth rate recorded in highly selected South African goat kids (Campbell, 1977).

Age (days)	Body mass (kg)				Average growth rate from birth (g/day)	
	N	Males	*N*	Females	Males	Females
100	5	30·65	5	29·00	291	272
150	5	42·70	5	37·70	272	240
210	5	53·60	5	45·00	245	204
270	5	69·50	5	51·80	250	186

Thus the majority of published reports indicate that a major limiting factor in meat production from goats is their inherently low growth rate. The few favourable reports published, however, seem to indicate that progress in this regard is possible. Butten and Moorhouse (1979) mention studies with a group of Saanen-cross feral kids where the best animals grew 1·8 times faster up to 135 days of age than the poorest.

D. Feed Conversion

Feed conversion (kg feed dry matter per kg mass gain) in meat production is basically a function of feed composition, level of feed intake relative to maintenance needs and body composition. The influence of feed composition on feed conversion is via the amount and type of end products of digestion available from a given amount of feed. Since the goat has the same digestive system as have sheep and cattle, it can be accepted that the role of feed composition in feed conversion will be similar for these three species. Furthermore, since very little has been published on this aspect in goats, it will not be dealt with here any further.

The influence of level of feed intake on feed conversion has been dealt with in detail in the literature (*vide* Blaxter, 1969). It is well known that, with all other factors kept constant, the higher the feed intake, the better the feed conversion, simply due to the fact that a larger proportion of the total feed is diverted from maintenance to growth. This will obviously go hand in hand with higher growth rates.

As far as body composition is concerned, the general trend is that if the fat content of body gain in animals is low then feed conversion is better than when the fat content of growth is high (Meissner *et al.*, 1975). The role of body composition is, however, not as important as level of feed intake in the actual determination of feed conversion (Meissner *et al.*, 1975).

Very few published reports are available on feed conversion in goats. Ueckermann (1969) fed Boer goat kids three diets: a 60% concentrate diet, a 60% roughage diet and a total roughage diet. The animals were slaughtered on reaching a body mass of 31·8 or 45·4 kg. The feed conversion ratios on the three diets were 8·6; 8·1 and 10·3 for the 31·8 kg slaughter group and 9·5; 9·5 and 13·2 for the 45·4 kg group. These figures are by no standards satisfactory relative to that of other domestic animals. However, they fit in with the low growth rates reported by Ueckermann (1969). Even lower efficiencies are recorded in the literature (Wilson, 1958) but then again associated with low growth rates.

The question now arises whether the reported high feed conversion ratios in goats are due to high maintenance needs, low feed intake or to high fat content in body mass gains. As far as maintenance needs are concerned, Hofmeyr and Lategan (1964) could find no significant differences between 30-month-old Boer and Dorper sheep castrates. There seems to be no obvious reason why maintenance needs of goats should differ from that of sheep.

Contrary to popular opinion fat content of body mass gain in goats is quite high and might even be higher than in sheep (this is discussed in full in Section E). This could be attributed to the rather high feed conversion ratios in some goats.

Our own results with Boer goat kids and those of Fehr *et al.* (1976) with Alpine kids put the matter of feed conversion perhaps into perspective. Boer goat kids, growing at a rate of 200 g/day over a period of 12 weeks consumed an average of 790 g/day of a growth ration which was about 3·9% of their average body mass during that period. The average feed conversion ratio over the same period was 3·90 g feed/g weight gain which compares very well with what could be expected of lambs growing at the same rate. These figures seem to indicate that for a given growth rate or feed intake kids are as efficient as lambs. Average daily weight gain and feed expenditure in Norwegian goats fed until 5 to 10 weeks of age were 160 to 200 g and 2 to 2·5 F.U. respectively (Skjevdal, 1974; Robstad, 1976). Fehr *et al.* (1976) found feed conversion rates in Alpine kids of between 2·41 and 3·83 F.U./kg gain when they grew at ratios of between 209 and 172 g/day. These figures again compare favourably with the position in lambs growing at the same rates.

It can thus be concluded that feed conversion in the goat is often unfavourable due to low growth rates. These could be due to poor nutritional conditions as well as to a limited genetical potential for growth. When, however, the goat grows at a faster rate, available information shows that feed conversion improves, to compare with the values found in sheep. It thus indicates the need to breed and select for higher growth rates. The relative lack of subcutaneous fat in the goat wrongly gives the impression that body mass gain is low in energy value and thus favours low feed conversion ratios. In fact, available evidence with breeds like the Boer goat points to the contrary.

E. Carcass Characteristics

The economic value of a carcass depends upon its yield of saleable meat as well as the cutting and processing quality of the meat. In industry the former is evaluated by subjectively assessing the conformation of the carcass and by the amount and distribution of visible carcass fat. Evaluation of the cutting and processing quality of the meat is strongly influenced by local customs and preferences. For instance in South Africa the meat from young Boer goat kids is sold as an alternative to lamb, and is thus compared on a quality basis with lamb. Meat from mature goats is specifically sought after by the local Indian community who prefer it to beef and lamb. In France and other parts of Europe as well as in Latin America young milk fed kids are considered a delicacy. In Texas, USA, the meat of Angora goats is used in sausages. In large parts of the world, developing communities are to a greater or lesser extent dependent upon the goat as meat supplier, and make use of every part of the goat, regardless of its quality (Devendra and Burns, 1970). It is thus obvious that any study of carcass characteristics must be done in the context of a particular goat meat production system. Goat and kid carcasses have certain characteristics which are, to a greater or lesser extent, typical of all types of the species. These are a thin, shallow carcass which is not compact as in the case of meat animals, but which becomes thicker and more compact as carcass weight increases (Fehr *et al.*, 1976). Similarly the proportions of muscle and fat vary greatly according to weight and age as evidenced by differences between the results of Kirton (1970), Gaili *et al.* (1972), and Naudé and Venter (1977). Composition of goat carcasses at birth and at three stages of growth is given in Table III.

In order to produce larger and heavier carcasses, it would be preferable to castrate males. Intact males during the mating season stop growing and may even lose weight. This phenomenon occurs at 9 months of age in the

Damascus breed in Cyprus (Louca *et al.*, 1977) and at 5 to 6 months of age in the Alpine breed in the west of Europe (Hervieu and Morand-Fehr, unpublished). Castration as early as 7 to 15 days of age yields satisfactory results (Louca *et al.*, 1977; Castellanos *et al.*, 1979) and is preferable to castration at 7 months of age. The carcasses of castrates are fatter and the lean meat has less flavour than that of intact males (Louca *et al.*, 1977; Castellanos *et al.*, 1979; Nitter, 1975).

Table III. Composition of goat carcasses at varying ages (Robstad, 1976, Skjevdal, 1974).

	Age			
	0 weeks	7 weeks	22 weeks	1 year
Carcass weight (kg)	1·3	5·1	8·4	18·0
Dressing percentage	41	46	44	45
Percent bone-free meat in carcass	53	70	77	82
Content of bone-free carcass				
Dry matter (%)	24	30	35	40
Protein (in % of dry matter)	78	68	59	46
Fat (in % of dry matter)	8	29	36	53
Ash (in % of dry matter)	1·2		1·1	

The published work on carcass characteristics in goats is fairly limited. It is considered worthwhile to report fully on the studies by Naudé and Venter (1977) on Boer goats and Fehr *et al.* (1976) on dairy goat kids in France.

1. South African Boer Goat

As stated previously, in South Africa meat of young Boer goats competes with lamb, which is very popular. The relative importance of kid and goat meat production in South Africa is well illustrated by the most recent statistics quoted in the annual report of the Meat Control Board of South Africa (1975–1976). In the nine major urban areas of the country, where the largest proportion of meat producing livestock is being slaughtered and marketed in carcass form, 22 850 kid carcasses and 88 200 goat carcasses were sold by auction during this period. The average carcass weights were 15·6 and 18·7 kg for kids and goats respectively and 14·5 and 17·9 kg for lamb and sheep. Auction prices of the various types of meat carcasses were as follows: (US $/kg): veal 1·04; beef 1·00; lamb 1·48; mutton 1·30; kid 1·17; goat 1·06; pork 1·04. From these figures it is evident that the South African consumer considers kid and goat meat quite acceptable.

At the Animal and Dairy Science Research Institute, Irene, a comparative trial was recently conducted on the growth and meat production of Boer goat kids and lambs of three sheep breeds. Male kids and lambs were separated from their dams 48 hours *post partum* and artificially reared on cows' milk fed from a bottle with a nipple until 6 weeks of age. Castration was performed at 3 weeks of age but tails were not amputated. A complete ration containing 23% finely chopped hay and with a crude protein content of 15·9% was fed *ad libitum* with drinking water available in individual feeding pens from birth onwards.

South African mutton merino (S.A.M.M.), Dorper, Merino and Boer goat wether lambs and kids were slaughtered at approximately 10, 23, 32 and 41 kg live weight. Several carcass measurements were recorded and right sides jointed and dissected into bone, subcutaneous fat and "meat", which contained muscle and intermuscular fat. Chemical analysis of the fat plus "meat" was performed in order to calculate "muscle" and "fat" content of the carcass (Carroll, 1967).

The results of the experiment are presented in Figs 1 to 15 illustrating the mean values of the various characteristics studied. Statistically significant ($p < 0{\cdot}01$) age trends were observed in all parameters. These trends within a particular parameter, were different for the four groups compared. The major object of the discussion of the results of this experiment will, however, be to compare the characteristics of the two species.

The mean birth weights were 4·17; 4·07; 3·65 and 4·15 kg for the S.A.M.M., Merino, Dorper and Boer Goat. Epstein and Herz (1964) reported on the birth weights of eight goat breeds of subtropical countries. Twin male kids weighed between 2·20 and 3·50 kg. Twin and triplet Boer goat male kids weighed 4·22 kg indicating a greater growth potential than that of the goats mentioned by Epstein and Herz (1964).

The growth rates between birth and 10 kg live weight were remarkably similar for the three sheep breeds and were considerably superior to that of the Boer goat. Joubert and Hofmeyr (1967) also found Dorper lambs to gain more rapidly than Boer goat kids. In the present study kids were very slow in adapting to the artificial rearing conditions, especially with regard to the consumption of the concentrate ration. Once they started feeding however, they gained almost as rapidly as their sheep contemporaries (Fig. 1) resulting in a final slaughter weight of 42 kg at 248 days as compared to 235 days for the Merino. The growth trends of the carcasses (Fig. 2) of the S.A.M.M. and the Dorper were very similar and those of the Boer goat and the Merino were of the same order, even though the live weight gains of the latter two groups were different. Differences in the dressing percentage between the two groups were obviously the explanation.

The Boer goat had a remarkably high dressing percentage (Fig. 3) (48·3%) compared to that of the S.A.M.M. (46·6%) and the Merino (41·0%) but was almost similar to that of the Dorper (48·5%).

Fig. 1

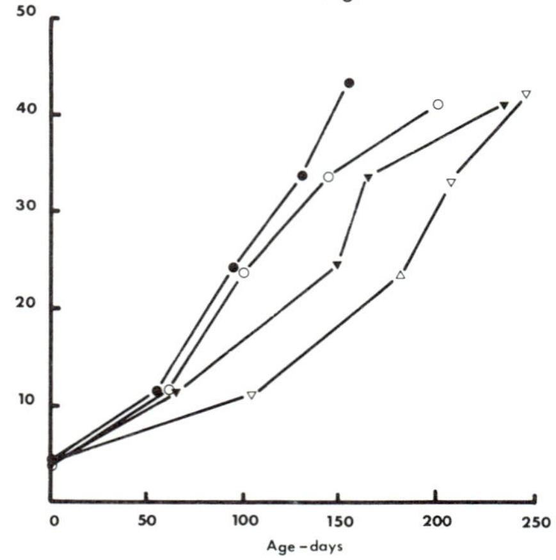

Fig. 2

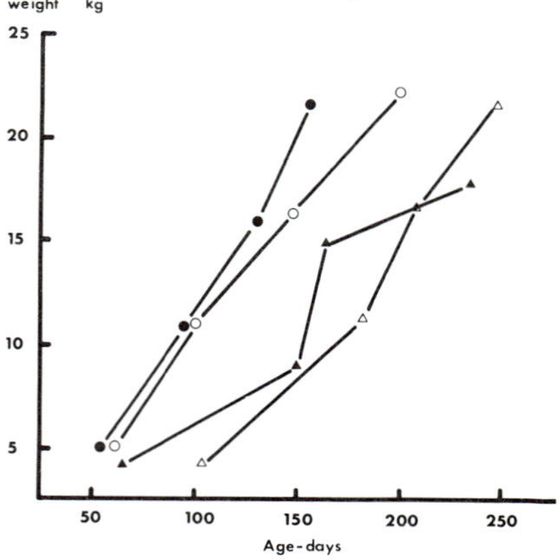

Fig. 3

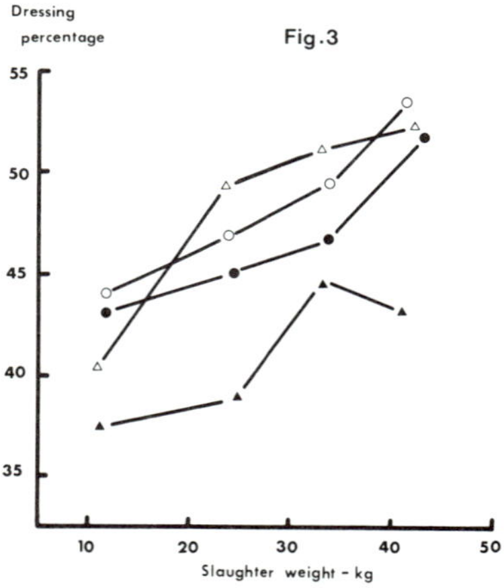

Fig. 4

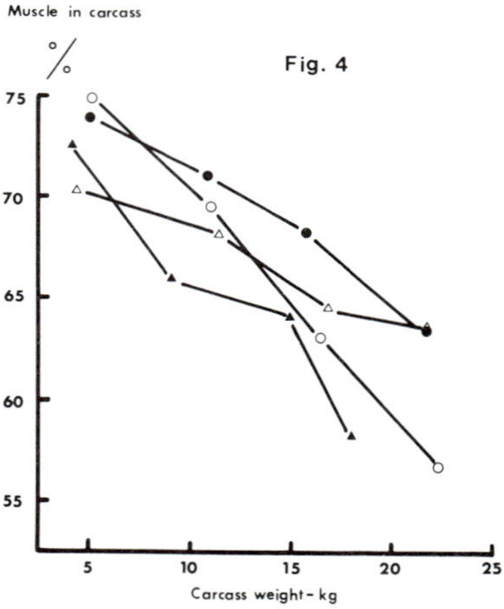

Fig. 5

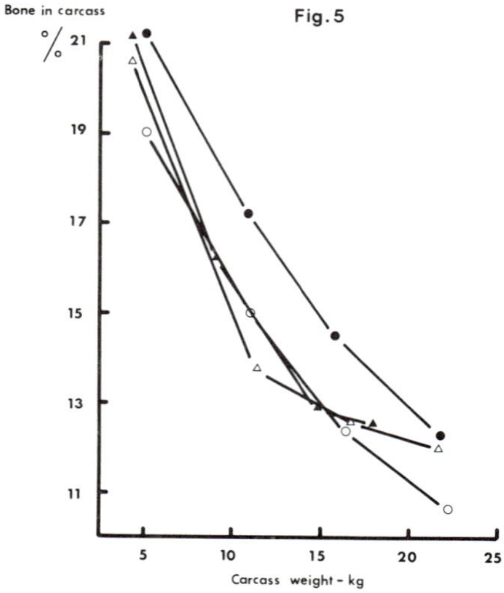

Fig. 6

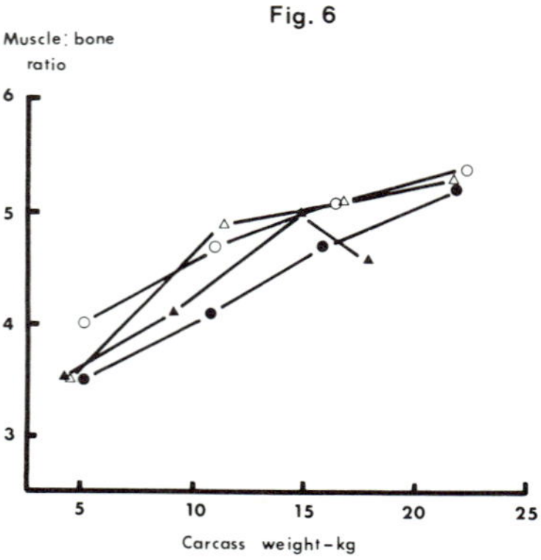

Figs 1–15. Growth and carcass characteristics of South African Mutton Merino (S.A.M.M. ●——●), Dorper (○——○), Merino (▲——▲), lambs and Boar goat (B goat △——△), kids slaughtered at live weights of 10, 23, 32 and 41 kg.

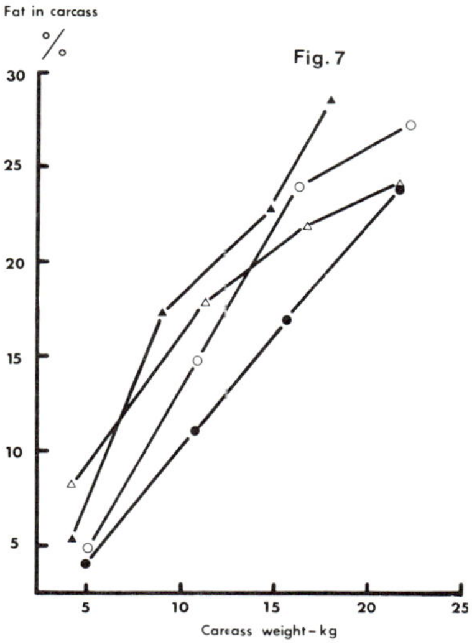

Fig. 7

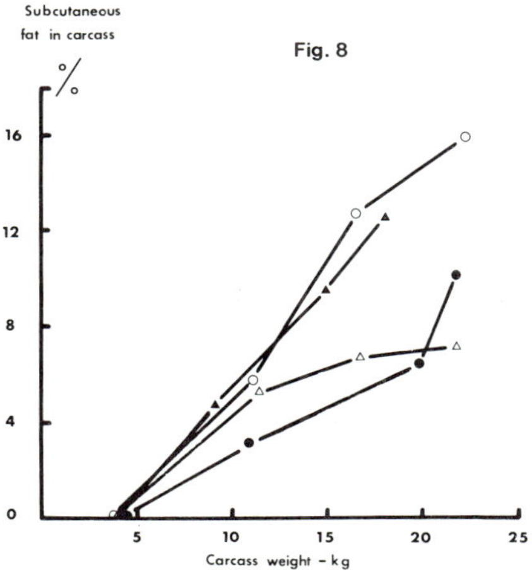

Fig. 8

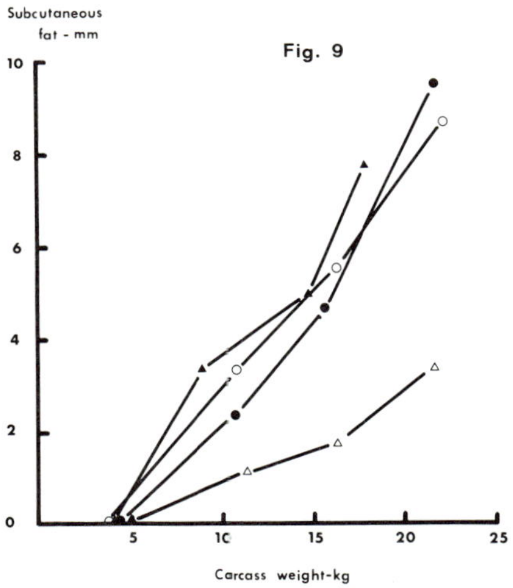

Fig. 9

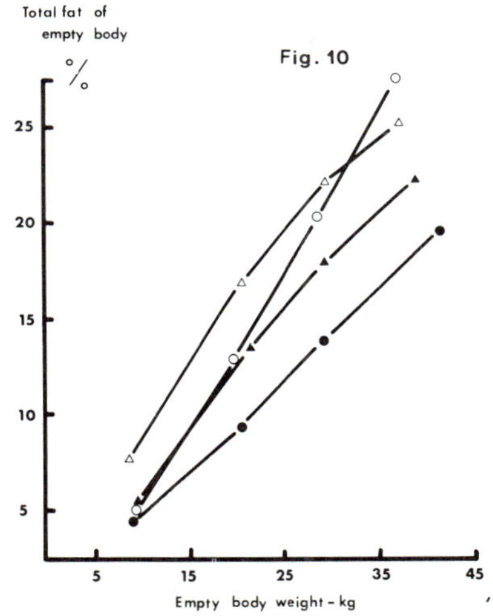

Fig. 10

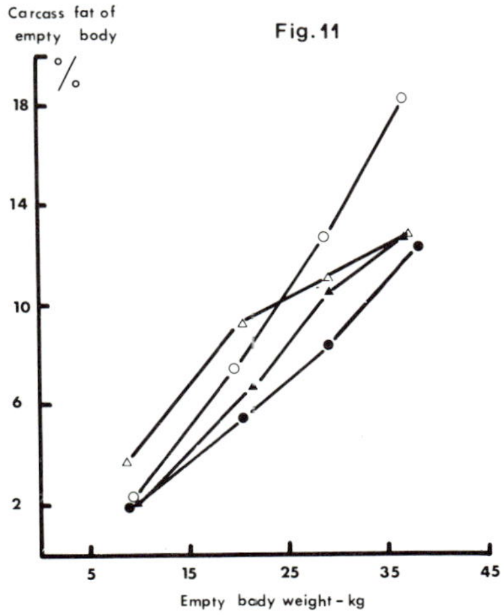

Fig. 11

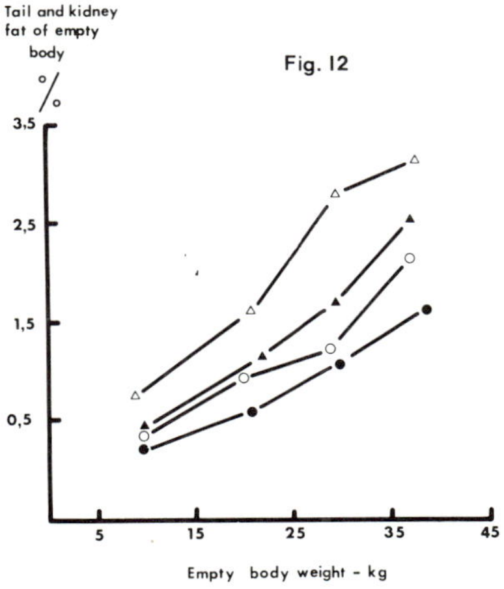

Fig. 12

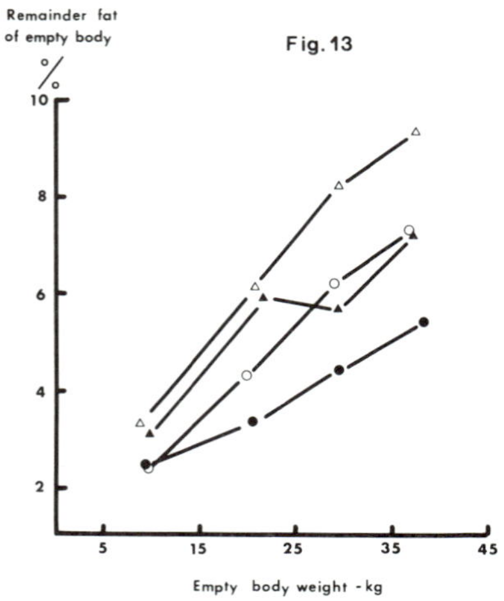

Fig. 13

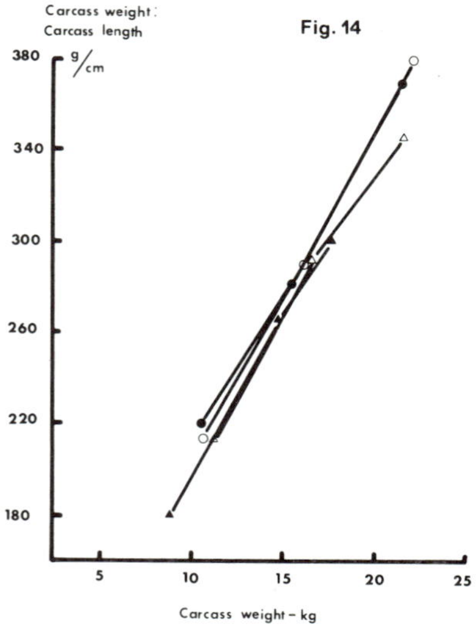

Fig. 14

Fig. 15

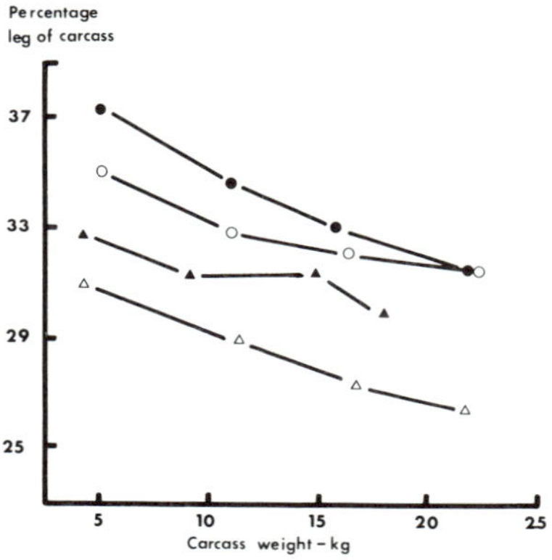

It is rather difficult to compare figures of dressing percentage from different countries because different methods of slaughter are applied (kidneys and kidney fat and also the head are not removed from the carcass in all cases). From a number of reports it is evident, however, that values vary between 44 and 55% (Richetti, 1975; Gaili, 1972, 1976; Ganekar *et al.*, 1973; Srivastava *et al.*, 1968; Devendra, 1966; Owen, 1975; Fehr *et al.*, 1976; Fehr and Sauvant, 1976; Naudé and Venter, 1977).

Some of the factors influencing dressing percentage are the following:

—weaners have a higher dressing percentage than sucklers (Fehr and Sauvant, 1974);

—dressing percentage increases as the energy concentration of the ration increases (Fehr and Sauvant, 1974);
—dressing percentage increases with age, weight of animal and fat deposition in the carcass (Gaili, 1972, 1976; Ghanekor *et al.*, 1976; Naudé and Venter, 1977);
—dressing percentages of kids are usually comparable to those of lambs, but Naudé and Venter (1977) have found dressing percentage of woolled Merino lambs to be significantly ($p < 0{\cdot}01$) lower than that of Boer goat kids, mainly due to the heavier fleece of the lambs.

One of the main results of this breed comparison was that Boer goat carcasses had a high muscle and low bone content (Figs 4 and 5), resulting in a high mean muscle to bone ratio of 4·71:1. Fehr *et al.* (1976), however, reported that lambs had a higher muscle to bone ratio than Alpine kids. Owen (1975) reported lean to bone ratios in the indigenous Malawi goat of between 3·14 and 4·18. The Boer goat therefore seems to be a more muscular type than the Alpine breed of France. The figure for S.A.M.M. carcasses was 4·39:1, for the Merino 4·32:1 and for the Dorper 4·76:1 (Fig. 6).

The most interesting results were those of the different measures of body fat of the Boer goat relative to sheep. The goat species is generally regarded as a lean meat type. This assumption is, however, only valid in terms of a particular fat depot of the body of the Boer goat. The mean fat content in the carcass of animals of all age groups was found to be 18·0% for the Boer goat, 18·9% for the Dorper, 18·5% for the Merino and 13·9% for the S.A.M.M. (Fig. 7). The impression that the Boer goat is a lean breed originates from the fact that it has a very low content of subcutaneous fat (Fig. 8); 6·4% as compared to 6·5% for the lean S.A.M.M., 8·9% for the Merino and 11·5% for the Dorper. When the thickness of the fat layer on the carcass was measured at the traditional position (C) of the *M. longissimus thoracis* in the region of the tenth and eleventh rib, the superficial leanness of the Boer goat was illustrated dramatically (Fig. 9). The mean value for the three slaughter groups (23, 32 and 41 kg) of Boer goats was 2·3 mm as compared to values of between 5·4 and 5·9 mm for the three sheep breeds. These findings are in agreement with those of Fehr *et al.* (1976) who also found a very low proportion of subcutaneous fat in dairy goat carcasses. The fat content of the empty body (entire body less the content of the digestive system) was the highest (18·0%) for the Boer goat (Fig. 10), and the lowest for the S.A.M.M. (11·8%). The total body fat was distributed between the carcass, tail and kidney depots as well as the "remainder" of the empty body, in such a manner (Figs 11, 12 and 13) that the Boer goat had the highest proportion kidney and non-carcass fat of the four types compared.

The greater carcass and leg length of the Boer goat in comparison to sheep resulted in a less compact carcass (Fig. 14), which corresponds with the results of Fehr *et al.* (1976) who reported on the comparatively poor carcass conformation of kids of Alpine dairy goats. The relatively high muscle to bone ratio of the Boer goat carcass did therefore not correspond with a less compact carcass which is in the market place traditionally but erroneously associated with poor muscling. The content of the high priced leg joint in the Boer goat carcass was however found to be significantly inferior to that of the carcasses of the various types of lambs (Fig. 15). The stage of muscle development in the body of a growing animal is closely related to its level of fat deposition which signifies physiological development of the body tissues (Naudé, 1977). The distal muscles of the forelimb and the hind limb, and the proximal muscles of the forelimb in the ox are moderately early developing (Butterfield, 1963), resulting in these muscles comprising a smaller proportion of the total muscle weight as the animals become physiologically more mature in terms of total body fat content. Berg and Butterfield (1976) provided evidence that these relations are largely the same in other closely related domestic ruminants.

The meat production potential of the Boer goat therefore compares reasonably well with that of sheep breeds with regard to post weaning growth performance, dressing percentage and muscling of the carcass. The low content of subcutaneous fat in the Boer goat creates the impression of a lean meat producer which was, however, proved to be an erroneous assumption. By comparison the breed actually has a high content of non carcass as well as intermuscular fat. The thin fat cover and the lanky appearance of the carcass however results in a poorer commercial value in comparison to lamb carcasses at similar weights. One characteristic which is undoubtedly inferior in terms of carcass value, is that of the lower yield of the high priced leg joint at comparable carcass weights of lambs.

2. Dairy Goats

Fehr *et al.* (1976) conducted an experiment with 36 Alpine male kids. They were slaughtered at 16, 22, 28 and 34 kg after receiving highly concentrated diets. Dressing percentage varied between 51·0 and 53·9. This can be considered typical for goats on such high concentrated diets.

The carcasses of these kids exhibited poor conformation, being long, narrowly and poorly muscled. They were also very lean. Subcutaneous fat cover was lacking to such an extent that it influenced storage qualities (severe dessication during storage) and also commercial acceptability. This finding corresponds with that of Owen (1975) who also reported excessive dessication during storage due to lack of fat cover in the indigenous Malawi goat. Due to the low fat content the muscle and bone proportions were high and the muscle to bone ratio were lower than those found in lamb. This is in contrast to our own findings with Boer goats. The Boer goat, a meat producing type, seems to be a more muscular goat than the milk producing Alpine breed. The percentage muscle in the latter carcasses varied between 67·2 and 68·6; bone percentage was between 21·9 and 24·4 and fat between 5·1 and 7·1.

Fehr *et al.* (1976) also gave the allometric coefficients of the various commercial cuts in the carcass relative to carcass mass. The breast ("poitrine") had the highest coefficient of 1·230, indicating that it grew about 23% faster than the carcass itself. Then followed saddle ("selle") with a coefficient of 1·210, shoulder with 0·903 and middle neck with 0·855.

3. Other Breeds

Gaili *et al.* (1972) and Gaili (1978) published studies on the type of goat found in the desert and semi-desert areas of North Africa and the Middle East. These studies give a good indication of what can be expected from fattened and unfattened goats, found in large parts of the world, and not bred for any specific purpose.

Table IV. Some carcass characteristics in Sudan desert sheep and goats (Gaili *et al.*, 1972).

	Goats (unfattened)			Sheep (unfattened)		
	I	II	III[a]	I	II	III
Dressing Percentage	40·36	41·33	43·01	40·71	40·92	43·23
Bone in carcass (%)	43·60	41·10	36·35	40·73	33·75	32·79
Eye muscle area (cm^2)	2·11	4·03	6·16	4·15	6·82	8·67
	Goats (fattened)			Sheep (fattened)		
	I	II	III	I	II	III
Dressing Percentage	49·01	51·11	54·08	48·28	50·01	52·96
Bone in carcass (%)	27·33	21·77	21·15	24·98	19·33	18·98
Eye muscle area (cm^2)	4·75	6·53	9·03	9·22	10·83	13·39

[a]I, II, III refers to young, yearling and mature.

These authors obtained 66 male Sudan desert goats at a local market, 22 of each of three age groups, young, yearling and old. For comparative purposes the same number of desert sheep were obtained. About half of the animals were slaughtered shortly after being obtained, while the rest were fattened for sixty days on a concentrate ration and then slaughtered. Table IV provides a summary of their main observations.

From Table IV it is obvious to what a large extent dressing percentage in the goat is influenced by nutritional status and diet. It increased from between

40·36 and 43·01% in unfattened goats to 54·08% in fattened goats. However, the data on desert sheep indicate that in this respect goats are not different from sheep. The data also agree with the results of Fehr *et al.* (1976) with dairy goats that most goat breeds are poorly muscled, as indicated by the high percentage of bone in the carcass and the small eye muscle area relative to that of desert sheep. Gaili *et al.* further state that fat development was relatively poor in the goats.

This same picture emerges from the results of Kirton (1970) on feral goats in New Zealand. He commented on the almost complete absence of subcutaneous fat cover over the eye muscle of feral goat carcasses. Owen (1974, 1975) reported on the meat producing characteristics of the indigenous Malawi goat, which can be considered as fairly typical of the "small East African goat" described by Mason and Maule (1960) which is widespread throughout East and parts of Central Africa. Contrary to the results of Gaili *et al.* (1972), Fehr *et al.* (1976) and Kirton (1970), the Malawi goat carcass is well muscled with percentage lean content in the carcass of between 64 and 66% and lean to bone ratios of between 3·14 and 4·18. Subcutaneous fat cover is, as with other goats, seriously lacking, especially in carcasses of younger animals where severe hanging losses due to excessive moisture loss were recorded. Preliminary indications are that the meat producing characteristics of this type of goat can be improved by crossing with Boer goats (De Haas, 1978).

Eggen *et al.* (1973) obtained data from 238 Angora goats slaughtered commercially. Aged does and younger does culled from experimental flocks were studied. In these animals, with mean live weights of 30·7 kg, dressing percentage was only 37·3%. Forty-eight of the carcasses were boned. Total boneless yield was 57% of chilled carcass mass. On a live weight basis, 19% of the goat is fat free, easily removeable muscle. This is 5 to 15% lower than that for cows of Cutter and Canner grades. They also found sheep to be higher yielding than goats.

McDowell and Bove (1977) refer to results of a thesis by Ladipo (1973) on carcass composition of goats in comparison with lambs. The proportion of edible tissue was about the same in both species but the proportions of bone and of lean meat were higher in goats. Goat meat was less acceptable than lamb, mainly because it was tough.

4. Prediction of Carcass Composition

The part–whole relationships of a carcass are useful aids in growth and carcass studies in which the muscle, fat and bone contents are measures of treatment differences. Gall *et al.* (1972) found in German Fawn-coloured goats high correlation coefficients for the regression of total body fat on physically separable and on abdominal fat. The multiple regression of total muscle weight on forearm muscle weight, body weight and forearm length had R^2 values of 0·80. When the predictor part is obtained without any considerable damage to the carcass, and its tissues have a high correlation with the tissues of the entire body or carcass, application of this technique becomes particularly useful. The leg and especially the shoulder joints of a kid carcass are two very suitable joints which are being used in this regard. The joint tissue content is highly correlated with the carcass tissue content and has a low residual variance as well as a low coefficient of variation. The

Table V. Part–whole relationships in goats (Morand-Fehr *et al.*, 1977).

Predictor parts	Carcass components		
	Muscle	Bone	Fat
	Correlation coefficient	Correlation coefficient	Correlation coefficient
Shoulder			
Muscle	0·995	—	—
Bone	—	0·989	—
Fat	—	—	0·953
Leg (gigot)			
Muscle	0·995	—	—
Bone	—	0·979	—
Fat	—	—	0·921
Four trotters	0·980	0·985	—
Caul fat	—	—	0·954

results of Morand-Fehr *et al.* (1977) in this regard are summarized in Table V. The shoulder, which is easy and quick to remove accurately, is very useful and reliable for predictions of this nature. In Table VI Naudé and Venter (1977) report similar results for the Boer goat in South Africa. Certain parts of the "fifth quarter" of the carcass are also useful and reliable predictors of carcass composition; the four trotters give a very good indication of muscle and bone content and the caul fat of the fat content of a carcass. The trotters–caul fat combination is a very useful one because of the ease in obtaining it and the low cost of these parts of the animal.

Table VI. Part–whole relationships in carcasses of Boer goat slaughtered at 10, 23, 32 and 41 kg live weight ($N = 20$).

Carcass joint	Muscle	Fat	Bone
Leg	$r = 0{\cdot}988$ $y = -10{\cdot}30 + 3{\cdot}28x$	$r = 0{\cdot}987$ $y = -25{\cdot}16 + 6{\cdot}57x$	$r = 0{\cdot}989$ $y = 28{\cdot}81 + 3{\cdot}25x$
Dorsal trunk	$r = 0{\cdot}997$ $y = 107{\cdot}48 + 4{\cdot}98x$	$r = 0{\cdot}986$ $y = 13{\cdot}94 + 6{\cdot}03x$	$r = 0{\cdot}995$ $y = 15{\cdot}80 + 4{\cdot}33x$
Ventral trunk	$r = 0{\cdot}988$ $y = 312{\cdot}50 + 3{\cdot}95x$	$r = 0{\cdot}994$ $y = 49{\cdot}42 + 1{\cdot}92x$	$r = 0{\cdot}915$ $y = 237{\cdot}70 + 6{\cdot}37x$
Shoulder	$r = 0{\cdot}995$ $y = 146{\cdot}13 + 6{\cdot}01x$	$r = 0{\cdot}991$ $y = 28{\cdot}17 + 10{\cdot}05x$	$r = 0{\cdot}991$ $y = 1{\cdot}23 + 4{\cdot}11x$
Neck	$r = 0{\cdot}981$ $y = 36{\cdot}23 + 10{\cdot}67x$	$r = 0{\cdot}975$ $y = 48{\cdot}45 + 12{\cdot}33x$	$r = 0{\cdot}990$ $y = 3{\cdot}25 + 9{\cdot}85x$

Table VII. Amino acid composition of meat samples (g/16gN) (after Srinivasan and Moorjani, 1974).

Amino acid	Sheep meat	Goat meat	Beef
Arginine	6·6	7·4	6·8
Histidine	2·8	2·1	3·0
Lysine	7·9	7·5	8·1
Tryptophan	1·4	1·5	1·4
Phenylalanine	3·3	3·5	3·4
Methionine	3·1	2·7	2·9
Threonine	4·6	4·8	4·5
Leucine	7·6	8·4	7·5
Isoleucine	4·6	5·1	4·5
Valine	5·5	5·4	4·9
Tyrosine	3·0	3·1	3·4
Cystine	1·3	1·2	1·1

F. Acceptability of Goat Meat

While some countries and communities discriminate against goat meat, other communities even prefer goat meat to meat of cattle and sheep (Devendra and Burns, 1970). Acceptability of meat is so much influenced by local custom and preference that it is almost impossible to apply a universally acceptable standard for acceptability of goat meat. However, the question might be put if goat meat has some intrinsic peculiarities which make it unacceptable. As far as essential amino acid composition is concerned it closely resembles that of beef and mutton, although it is higher in arginine, leucine and isoleucine, as indicated in Table VII (Srinivasan and Moorjani, 1973). It has already been shown that goat carcasses are poorly covered with fat compared to those of sheep and cattle. This is probably attributable to its lack of popularity in countries where carcass standards have been drawn up with reference to sheep and cattle.

A few studies have been carried out to test the acceptability of goat meat (Kirton, 1970; Gaili *et al.*, 1972; Pike *et al.*, 1972; Fehr *et al.*, 1976). In all these studies use was made of taste panels. Pike *et al.* (1973) carried out a series of sensory panel tests to compare meat from goats with beef, pork and lamb. Since this study was carried out in the USA it might be assumed that the members of the taste panels were Americans, not very used to goat meat. They found goat less desirable than beef and pork in flavour and tenderness and less desirable than lamb in flavour, juiciness and tenderness. Their data suggested that chops or steaks from goats are not interchangeable with those from lamb of similar age. The group studied the possibility of replacing lean beef with goat meat in Frankfurters. They found little effect on cooking yield, stability, flavour desirability or overall satisfaction when goat meat was included (Eggen *et al.*, 1973; Marshall *et al.*, 1977).

Kirton (1970) used taste panels to compare meat from feral goats with that from sheep. This comparison was done in a country (New Zealand) where sheep meat is the most important meat on the market. Taste panel results were in favour of meat from domestic sheep and the discrimination against goat meat was due to lack of tenderness. Kirton explicitly indicated that there was no indication of any strong flavour or odour associated with the meat from goats, even from male animals.

Table VIII. Acceptability of goat meat ($N = 33$) (5 = Very favourable impression; 1 = Very unfavourable impression) (Morand-Fehr *et al.* (1977)).

	Loin chops	Shoulder joint
Flavour	2·16	2·26
Tenderness	2·13	2·06
Aroma	2·20	2·26
Juiciness	2·19	2·35
General Impression	3·55	3·71

Gaili *et al.* (1972), with their study of desert sheep and goats in the Sudan, also found that the taste panels found a tendency for mutton to be more tender than goats' meat, while no flavour differences could be detected. They also commented on the absence of strong flavoured meat in adult goats, which is contradictory to the current concept of goats meat. Recently Wong *et al.* (1975) did chemical analyses of volatile acids from fat hydrolysate of minced mutton samples and one goat meat sample, and correlated it with organoleptic evaluation of the meat. A "goaty" odour note common to goat meat and mutton was correlated with the presence of 4-methylocanoic acid. Levels of this acid were much higher in the sample of goat meat than in the mutton samples.

Morand-Fehr *et al.* (1977) conducted a taste panel on goat meat (carcass weight 7 to 13 kg) in order to assess the consumer's general opinion of the meat as well as his view regarding the organoleptic acceptability of the meat. For 6% of the panelists chops were unacceptable and for the shoulder joint the figure was 9%. The taste panel results are given in Table VIII. These values are in accordance with those found in other species at the same laboratory. In goat meat a frequent complaint from taste panelists is a lack of flavour often associated with a lack of tenderness which leads to a general unfavourable impression.

According to Morand-Fehr *et al.* (1977) and Smith *et al.* (1974) the only satisfactory way of assessing the tenderness of meat is by comparing the shear force of various uncooked muscles from the carcasses of well reared young animals. Fairly high values between 35 and 62 N/cm^2 were obtained by Morand-Fehr *et al.* (1977). Sauvant *et al.* (1976) investigated the influence of certain muscle characteristics on the organoleptic impression that consumers have of cooked meat. Flavour, tenderness, aroma and juiciness in decreasing order influenced meat acceptability. The pH and colour of meat, which generally does not vary very much between animals, had very little organoleptic influence.

The meat of older goat kids was regarded as being more juicy, with a more favourable aroma than that of younger kids, which was however more tender and to a lesser degree somewhat less palatable. The results of Smith *et al.* (1978) underline the age dependence of tenderness and palatability. Yearling Angora goats produced roasts that were juicier and more tender than those from younger goats, and in Spanish goats quality was better in meat from 6-month-old animals than from younger or older animals. Meat from very young kids (3 to 5 months of age) was tough and that from aged goats was decidedly inferior in palatability. Goat meat, like mutton and beef, can be tenderized by means of electrical stimulation of the carcass (Savell *et al.*, 1977) or by using reciprocating blade tenderizing machines (Bowling *et al.*, 1976).

It can thus be concluded that goat meat can hardly be distinguished from mutton or lamb as far as flavour is concerned provided that it is obtained from relatively young animals. The long-held idea that goat meat is strongly flavoured is not well founded. However, goat meat seems to be definitely less tender than lamb. Muscle cold shortening* of the poorly covered kid and

* *Pre-rigor* muscles rapidly chilled to temperatures below 10°C, within the first 10 hours post slaughter tend to contract extensively, causing severe muscle toughening.

goat carcasses in comparison to lamb and sheep carcasses seems to have had a toughening effect on goat meat. Preliminary results from a study by Naudé and Smit (1977) do not indicate that muscles from kid carcasses have a markedly lower solubility of collagen than those of lamb carcasses when compared at similar ages. Intrinsic differences in muscle tenderness between sheep and goats therefore seem improbable but the higher chilling rate which is possible for the poorly covered pre-rigor kid and goat carcasses could be the cause of muscle cold shortening and toughening.

G. Conclusion

The great advantage, from the viewpoint of meat production, of the high prolificacy of goats is not always matched by fast growth rates. It has however been shown that goats can grow up to 250 g/day provided that they are adequately fed. The low growth rates go hand in hand with poor feed conversion ratios.

Kid and goat carcasses often are less acceptable to the butcher than those of lamb and sheep carcasses, mainly because of their poorer carcass conformation and leaner appearance, hence lower carcass grading in many countries, resulting in poorer commercial value on the meat market.

The carcasses of most goat breeds tend to be leaner than those of sheep breeds and to have poorer muscling. There is a lack of subcutaneous fat cover on carcasses of all goat breeds especially over the loin area. Compared with sheep, body fat is concentrated in the non-carcass parts. The South African Boer goat seems to yield a superior carcass than Angora, dairy and other meat goats.

This conclusion seems justified because of the fact that certain carcass characteristics such as the fat content and muscling of Boer goat carcasses compared favourably with those of specialized mutton-producing breeds such as the South African mutton Merino and the Dorper.

Kid and goat meat is quite acceptable for human consumption and may in some communities replace mutton, lamb and beef to a certain extent, although taste panels find it less tender than lamb and mutton, which may however be due to post mortem carcass chilling conditions.

References

Batten, G. J. and Moorhouse, S. R. (1979). *N.Z. J. Agric.* **138**, 16–18.

Berg, R. T. and Butterfield, R. M. (1976). "New Concepts of Cattle Growth", pp. 176–200. Sydney University Press.

Blaxter, K. L. (1969). *In* "Energy Metabolism of Farm Animals" (K. L. Blaxter, J. Kielanowski and Greta Thorbek, eds), pp. 21–28. Eur. Ass. Anim. Prod. Publ., 12.
Bowling, R. A., Smith, G. C., Carpenter, Z. L., Marshall, W. H. and Shelton, M. (1976). *J. Anim. Sci.* **43**, 122–130.
Butterfield, R. M. (1963). "Proc. Carcase Composition and Appraisal of Meat Animals, East Melbourne, Australia", pp. 7–1 to 7–20.
Campbell, Q. P. (1977). Unpublished data.
Carroll, M. A. (1967). Beef carcase evaluation: Fat, lean and bone. Proc. 14th Easter School, (Univ. Nottingham, 389).
Castellanos, A., Morand-Fehr, P. and Hervieu, J. (1979). *Ann. Zootechn.*, **28**.
De Haas, H. J. (1978). *Der Tropenlandwirt.* **79**, 7–12.
Devendra, C. (1966). *Malaya Agric. J.* **45**, 345–369.
Devendra, C. and Burns, M. (1970). "Goat Production in the Tropics". Commonwealth Agricultural Bureaux, Farnham Royal, Bucks, England.
Eggen, N. R., Carpenter, Z. L., Smith, G. C. and Shelton, M. (1973). *J. Anim. Sci.* **37**, 260. Abstr. 126.
Eggen, N. R., Smith, G. C., Carpenter, Z. C., Berry, B. W. and Shelton, M. (1973). *J. Anim. Sci.* **37**, 260, Abstr. 127.
Epstein, H. and Herz, A. (1964). *J. Agric. Sci.* **62**, 237–244.
Fehr, P. M. and Sauvant, D. (1974). *Ann. Zootechn.* **23**, 503–518.
Fehr, P. M. and Sauvant, D. (1976). *Ann. Zootechn.* **25**, 243–257.
Fehr, P. M., Sauvant, D., Delage, J., Dumont, B. L. and Roy, G. (1976). *Livestock Prod. Sci.* **3**, 183–194.
Gaili, E. S. E. (1976). *Wld Rev. Anim. Prod.* **12**, 83–87.
Gaili, E. S. E. (1978). *Trop. Anim. Hlth Prod.* **10**, 103–108.
Gaili, E. S. E., Ghanem, Y. S. and Makhter, A. M. S. (1972). *Anim. Prod.* **14**, 351–357.
Gall, C., Frahm, K., Graf, K. and Osterkorn, K. (1972). *Z. Tierzücht. ZüchtBiol.* **89**, 123–128.
Chanekar, V. M., Bhatawadekar, S. D. and Soman, B. V. (1973). *Indian J. Anim. Sci.* **43**, 133–135.
Hofmeyr, H. S. (1962). M.Sc. (Agric.) Thesis, University of Pretoria.
Hofmeyr, H. S. and Lategan, P. J. (1964). *Proc. S. Afr. Soc. Anim. Prod.* **3**, 117–119.
Hofmeyr, H. S., Joubert, D. M., Badenhorst, F. J. G. and Steyn, G. J. van D. (1965). *Proc. S. Afr. Soc. Anim. Prod.* **4**, 191–195.
Joubert, D. M. and Hofmeyr, H. S. (1967). *Tydskr. Natuurwetenskappe* **7**, 190.
Kirton, A. H. (1970). *N.Z. J. Agric. Res.* **13**, 167–181.
Lamont, J. L. (1964). *S. Afr. J. Agric. Sci.* **7**, 305–310.
Louca, A., Economides, S. and Hancock, J. (1977). *Anim. Prod.* **24**, 387–391.
Marshall, W. H., Smith, G. C., Dutson, T. R. and Carpenter, Z. L. (1977). *J. Food Sci.* **42**, 193–196.
McDowell, R. E. and Bove, L. (1977). The goat as a producer of meat. Cornnell Int. Agric. Mimeo. Nr. 56 pp. 23–27.
Meissner, H. H., Roux, C. Z. and Hofmeyr, H. S. (1975). *Agroanimalia* **7**, 105–114.
Moorhouse, S. R. and Batten, G. J. (1978). *N.Z. J. Agric.* **137**, 22–25.
Morand-Fehr, P., Sauvant, D. and Dumont, B. L. (1977). "Symposium on Goat Breeding in Mediterranean Countries", Malaga-Grenade-Murcia (Spain), October 1977, pp. 338–348.
Naudé, R. T. (1977). Unpublished data.
Naudé, R. T. and Smit, M. C. (1977). Unpublished data.

Naudé, R. T. and Venter, H. J. (1977). Unpublished data.
Nitter, G. (1975) *Kleinvichzüchter* **23**, 316–320.
Owen, J. E. (1974). *Trop. Sci.* **16**, 75–83.
Owen, J. E. (1975). *Trop. Sci.* **17**, 123–138.
Pike, M. I., Smith, G. C., Carpenter, Z. L. and Shelton, M. (1973). *J. Anim. Sci.* **37**, 269, Abstr. 158.
Restall, B. J. (1976). *In* "2èmes Journées de la Recherche Ovine et Caprine", pp. 155–165. ITOVIC, Paris.
Richetti, F. (1975). *Acta Med. Vet.* **21**, 25–39.
Robstad, A. M. (1976). *Sau og Geit* **29**, 162–164.
Savell, J. W., Smith, G. C., Dutson, T. R., Carpenter, Z. L. and Suter, D. A. (1977). *J. Food. Sci.* **42**, 702–706.
Shelton, M. (1960). *J. Anim. Sci.* **19**, 368–375.
Skinner, J. D. (1972). *Trop. Anim. Hlth Prod.* **4**, 120–128.
Skinner, J. D. and Hofmeyr, H. S. (1969). *Proc. S. Afr. Soc. Anim. Prod.* **8**, 155–156.
Skjevdal, T. (1974). *Sau og Geit* **27**, 102–104.
Smith, G. C., Pike, M. I. and Carpenter, Z. L. (1974). *J. Food. Sci.* **39**, 1145–1147.
Smith, G. C., Carpenter, Z. L. and Shelton, M. (1978). *J. Anim. Sci.* **46**, 1229–1235.
Srinivasan, K. S. and Moorjani, M. N. (1974). *J. Food Sci. Tech.* **11**, 123–124.
Srivastava, V. K., Raizada, B. C. and Kulkarni, V. A. (1968). *Indian Vet. J.* **45**, 219–225.
Ueckermann, L. (1969). "Produksiestudies met Boerbokke". (M.Sc.) (Agric.) thesis. Univ. Pretoria.
Wilson, P. N. (1958). *J. Agric. Sci.* **50**, 198–210.
Wilson, R. T. (1976). *Trop. Anim. Hlth Prod.* **8**, 221–232.
Wong, E., Johnson, C. B. and Nixon, L. N. (1975). *N.Z. J. Agric. Res.* **18**, 261–266.

10. Milk Production

C. Gall

A. Lactation

1. General

In temperate countries, goats are considered as dairy animals and milk is their main product. Most of the more productive dairy goat breeds of the world have been derived from European breeds. By contrast, in many tropical countries the main product of goats is meat. However, many of the tropical goat breeds are also milked eventually, albeit for short periods or during favourable years only.

Dairy characteristics are very marked in dairy goats. They are readily milked and milk production is not affected by time changes in milking routine, by switching from suckling to milking, or from hand to machine milking. Goats are known to recover lactation after even extended periods during which yield has gone down to a few hundred ml due to restricted feeding. This capacity is important under range conditions with fluctuating feed supply and in feeding experiments (Chalmers, 1972). They may continue lactating at appreciable levels for extended periods of much over 2 years. Chalmers reported average daily yields of 5·8 kg in a goat producing a total of 3975 kg during 687 days of her first lactation. Linzell (1973a) has kept goats for 3 years without new pregnancies. They continued lactating at between 1 and 3 litres per day. The importance of hormones for lactation is well established. While hypophysectomy in lactating goats completely inhibited lactation, it was restored to pre-treatment levels by joint administration of growth hormones, insulin, prolactin, thyroxin and glucocorticoids (Cowie *et al.*, 1964). It has, however, not been possible to prove that differences in milk yield are related to differences in hormonal activities. Milk yield falls after the onset of oestrus (Peaker and Linzell, 1974) but there are individual variations which seem to be related to mammary blood flow (Burvenich, 1980).

Spontaneous lactation of male and of juvenile goats is occasionally observed. Fautz *et al.* (1970) summarized cases reported in the literature. They described a kid in which milk secretion was observed immediately after birth. The kid was first hand-milked at 7 days of age at intervals of 3 to 4 days. After day 42 milk was withdrawn with a milking machine. During a 148-day lactation the kid was milked 37 times altogether, producing a total of 13·67 kg milk. The percentages of fat, protein, lactose, dry matter and energy content were 3·11, 3·32, 4·48, 12·50 and 634 kcal/l respectively which was not significantly different from average first lactation values in the same herd. The animal had her first parturition at 10 months of age weighing 34·1 kg, which was about the herd average, with a normal kid weighing 3·25 kg. Her first lactation with 1·48 kg per day during 333 days was above the herd average of 1·13 kg.

Recently Larson (1978) has reviewed similarities and dissimilarities of lactation in goats and cows. Goats are widely used as experimental animals in studies on lactation or other studies involving lactation, and considerable basic knowledge about the mechanisms of lactation has been derived from these (Peaker, 1977). The whole of the literature reporting on such studies still needs to be systematically scanned for compiling all the results which may add to our knowledge of the goat. Much of this information is unidentified because paper titles do not mention the fact that experiments have have been carried out with goats. Such studies will not be reviewed here, however, unless they have some bearing on the breeding and management of the goat.

2. Milk Ejection and Withdrawal

The neuro-hormonal complex regulating milk ejection and its importance for milk withdrawal has been reviewed by Denamur (1965) and Martinet and Richard (1973). As in the cow, ejection of milk occurs in the goat after stimulation of a neuro-endocrine reflex involving the release of oxytocin; fright, agitation and unusual conditions at milking inhibit both the liberation of oxytocin in the hypothalamus and the action of oxytocin on the myoepithelial cells of the mammary gland. Intravenous injection of 50 mU oxytocin resulted in milk ejection within 0·33 min and 1000 mU resulted in ejection of about 20% of the milk present, within 0·21 min (Svoboda and Bilek, 1976).

Two minutes after stimulation by machine milking oxytocin was down to quite low levels (Dyusembin, 1974) but during suckling and hand milking oxytocin release occurred at any time (McNeilly, 1972). The time of oxytocin release during milk withdrawal and the blood levels reached varied between goats and from one milking to another in the same goat. Dyusembin (1974) reported a small oxytocin release after thermal and mechanical stimulation of the udder. Stimulation was much more effective by milking or milking combined with massage, resulting in oxytocin levels 10 and 20 times higher than after massage alone. Oxytocin was released by conditioning the goats before suckling in only about 6% of all goats investigated but in no case even after very intensive and prolonged conditioning before hand milking. In some animals, oxytocin was not released until the final stripping and no oxytocin at all was released during suckling or milking in about a third of all goats studied. Oxytocin was released in 62% of the milkings in weeks 1 to 6 of lactation (McNeilly, 1972) but only in 32% in weeks 5 to 104 (Cleverley and Folley, 1970). This declining response to milking stimuli of oxytocin release with the course of lactation may explain some of the differences in persistence and lactation length. Thus, in the goat the release of oxytocin during milk withdrawal does not appear to be a regular feature and the necessity of oxytocin for milk withdrawal has been questioned, since, in denervated udders, normal milk yields were possible by careful hand milking (Linzell, 1963) and in goats milked under cyclopropane anaesthesia milk yield was about the same as in the control periods (Yokoyama and Ota, 1965). In suckling and hand milking there was no relation between milk yield and oxytocin release (McNeilly, 1972).

Fractionated ejection from the alveoli of milk which is stored between milk withdrawals in the cistern may be a normal feature in the goat. Svoboda and Bilek (1976) observed milk ejection of 2 to 3% of the alveolar milk at feeding. In other experiments, however, oxytocin releases have not been observed before the milk withdrawal routine began.

In addition to milk ejection, oxytocin may have a direct stimulating effect on milk secretion. Although earlier experiments which showed this effect used doses of oxytocin that exceeded the quantity physiologically released at milking (Denamur, 1953), the effect has been observed with small doses too (Linzell and Peaker, 1971a). Milk secretion increased by about 20% in goats which were milked every hour after being given small doses of oxytocin. The effect was reduced if milk was not removed every hour, and it did not persist. It was explained by a local direct effect of milk evacuation in which a lipid seems to be involved, and a systematic stimulation of pituitary prolactin release (Linzell and Peaker, 1971a). These results show that in the goat the neuro-hormonal milk ejection reflex may not be essential for normal milk withdrawal by both suckling and milking. This may be related to the particular structure of the mammary gland of the goat, in which the volume of the cistern in relation to the gland volume is greater than in other mammals. Furthermore, it seems possible that milk ejection is assisted by direct neural stimulation of the contractile elements of the udder, as postulated by Russian workers (Zaks, 1962). However, udder stimulation seems to be important for sustained lactation yield.

In the lactating goat prolactin level increases after milking. But it seems that high levels of prolactin are not necessary to maintain milk yield in mid-lactation (Hart, 1974) and correlations between prolactin levels and milk yield do not exist (Hart, 1975a). Prolactin concentration after milking declines in autumn and this is governed by day length but it is not the major factor causing the decline of milk yield during late lactation (Hart, 1975b).

Udder growth and lactation may be induced in juvenile or non-pregnant adult goats by administration of oestrogen and progesterone (Hart, 1976). Levels of production in induced lactation and in subsequent lactations after pregnancy are about half normal lactations. Fertility seems to be unaffected (Delouis, 1975). Thus, the method deserves some interest, mainly for using goats who have failed to conceive.

The suckling kid extracts milk from the teat sinus in two phases: a suction where pressure falls to -100 to -300 mmHg and a compression of 80 to 100 mmHg within the teat sinus (Ardran *et al.*, 1957, 1958). Since present-day milking machines normally do not imitate this change to a positive pressure, normal blood circulation is not maintained during machine milking, whereby pain and tissue alterations may be produced.

During experimental frequent milkings (1–4 times per hour) cell counts rose to as much as 5×10^6 cells/ml without any clinical signs of inflammation. Linzell (1967a) assumed that contraction of the alveoli in response to oxytocin and massage increases the loss of cells into milk. By contrast, in experiments by Caruolo (1974) with hourly milkings, cell concentration decreased with time in a curvilinear manner, reaching a minimum after about 31 hours. This would suggest depletion of stored cells, and afterwards a steady cell movement in the order of 3·2 million cells/ml. Milk volume and cell concentration were inversely related ($r = -0{\cdot}52$). Epithelial cells and cell debris are more frequent in all goat milk samples and have to be given special consideration when interpreting cell counts for assessing udder health (Roguinsky, 1974).

B. Factors affecting Milk Yield

1. Body Size and Weight

Mature dairy does may weigh between 30 and 80 kg. Variation of size within breeds has received major attention. Apparently, larger does will have to produce more milk than smaller animals in order to pay for the higher maintenance costs, but Brody *et al.* (1938) and Ormiston and Gaines (1944) found gross energetic efficiency of dairy goats to be independent of body weight. However, with high levels of milk production at the beginning of lactation, goats may not be able to consume sufficient energy and may have to draw on their body fat stores (Morand–Fehr and de Simiane, 1977). Feed intake capacity is thus an important breeding goal, along with lactation milk yield.

Since the feed intake capacity, which is determined by physical and behavioural factors, is difficult to assess directly, body size is commonly used as an indicator. Paunch girth and externally measured abdomen volume are closely related to rumen volume (Table I). Size will determine the ability

Table I. Correlation coefficients between external body measurements and rumen volume (n) (Gall, 1969).

		Rumen contents	
	Rumen volume	Fresh matter	Dry matter
paunch girth	—	0·87 (64)	0·521 (64)
abdomen volume	0·981 (40)	0·6560 (20)	—

Table II. Relations between body weight and milk production in goats.

Record	Time of measurement	b	r^2	n	Reference
Yearly production (kg milk)	after kidding	4·76	0·231	104	Nyland (1936)
Yearly production (kg milk)	in winter	9·585	—	150	Metzger (1964)
Yearly production (kg milk)	1 May	7·68	0·145	549	Rønningen (1964)
Yearly production (kg milk)	1 May	—	0·16	5795	Rønningen (1967a)
—	June	1·25[a]	0·353	22038	Steine (1975a)
Lactation yield (kg milk)	7 months of age	—	0·122	—	Ricordeau and Bouillon (1975)
Lactation yield (kg milk)	average of monthly weighings	—	0·49	—	Mittal (1979)
Yearly production (kg fat)	1 May	0·48	0·272	549	Rønningen (1964)
8 Month lactation (kg F.C.M.)	31 days p.p.	0·91	0·325	318	Ormiston and Gaines (1944)
200 Day lactation (kg F.C.M.)	after kidding	5·889	0·335	120	Lampeter (1970)
200 Day lactation (kg F.C.M.)	5 weeks p.p.	9·946	0·288	120	Lampeter (1970)
35 Days (kg F.C.M.)	5 weeks p.p.	1·74	0·295	108	Gall (1963)
70 Days (kg F.C.M.)	10 weeks p.p.	1·764	0·127	139	Gall (1963)
140 Days (kg F.C.M.)	20 weeks p.p.	3·099	0·173	141	Gall (1963)

[a] Age and month of kidding held constant.

to consume coarse feeds since time needed for clearing rumen contents by rumination is dependent on body weight, with r^2 of a third degree ploynomial of 0·78 (Welch *et al.*, 1975). Some results of investigations on phenotypic relations between milk production and body weight are compiled in Table II. There is a general positive regression of milk yield on body weight, but weight variation accounts for only about 10% of the variation of milk yield. Relationships may differ mainly because animals were weighed at different stages of lactation and therefore different body condition, and they may be underestimated if, without individual feeding, low yielding goats tend to grow fat (Steine, 1975a). There is a significant but small influence of body weight on fat content. In the course of the year and related to nutrient intake, lactation and gestation, body weight changes considerably due to accumulation and depletion of body fat. During the dry period fat is stored in the body cavities. Body fat in week 10 of lactation in yearlings was 4% of live weight (Gall *et al.*, 1972a) with a coefficient of variation of 51·7%. Skjevdal (1974) found 22% of the body weight of yearlings to be fat of which 64% was stored in the body cavities; after starving, body fat was reduced to 7·4%. About 10% of the body weight may be lost at the beginning of lactation (Gall, 1963; Skjevdal, 1972a, 1974c) but is recovered about 4 months post partum. Storage of body fat during the dry period seems to influence milk production positively at the beginning of lactation. The capacity of the animal to mobilize body fat is greater in animals which have been fed liberally during the dry period and, therefore, have stored more fat. There seems to be a genetic disposition towards mobilization of fat at the beginning of lactation (Sauvant *et al.*, 1979).

Mobilization of fatty tissue seems to begin during the last third of pregnancy (Chilliard *et al.*, 1978) and to be related to the level of milk production (Chilliard *et al.*, 1979). Fehr and Sauvant (1975) found weight losses of adult goats, at the beginning of lactation, to be related with milk yield of the first week only. Body weight varies with season which, however, in investigations by Rønningen (1976a), accounted for only 2% of the total weight variation. If goats are weighed at fixed dates rather than at fixed intervals from kidding, season will be confounded with stage of lactation. Body weight taken in May was related with season of kidding, least square constants decreasing from December to May from 0·7 to −0·5 (Steine, 1975a).

2. Age

Age is another source of variation of milk yield which is, however, closely related to body weight. In 22 038 Norwegian goats, body weight taken in June increased up to 6 years of age and decreased in years 7 and 8. Age accounted for 45% of body weight variation. Least square constants for the effect of age on body weight are given in Table II on page 493.

As early as 1936 Cunningham and Addington showed that milk yield in lactations begun at 24 months of age was higher in does which had kidded and started a first lactation at about 12 months than in does which first kidded at 24 months. From the data of Iloeje *et al.* (1980) (Table III) it can be seen that first lactation yield increases with age at kidding but goats freshening at 13 to 16 (those kidding between April and June even up to 19) months do not produce more than the 12-month-old does. This may be due to unfavourable conditions for the development of kids born between October and February. Peak milk yield is attained between 4 and 8 years of age

Table III. Effect of age and season on milk yield in Alpine goats (compiled from Iloeje *et al.*, 1980).

Age at kidding (months)	Season of kidding: January to March			April to July		
	n	Milk (kg)	Month of birth	*n*	Milk (kg)	Month of birth
10	25	630	March/May	25	551	June/Sept.
11	146	669	Feb./April	88	557	May/Aug.
12	197	677	Jan./March	261	595	April/July
13	79	621	Dec./Feb.	336	574	March/June
14	13	615	Nov./Jan.	210	544	Feb./May
15	1	422	Oct./Dec.	127	473	Jan./April
16	9	600	Sept./Nov.	57	445	Dec./March
17	5	687	Aug./Oct.	19	514	Nov./Feb.
18	9	690	July/Sept.	20	438	Oct./Jan.
19	42	727	June/Aug.	17	510	Sept./Dec.

(Alderson and Pollak, 1980; Dickinson and King, 1977; Gall, 1969; Neseni and Freimuth, 1952; Parkash *et al.*, 1972; Ricordeau, 1963; Riggert, 1978; Rønningen and Gjedrem, 1966; Steine, 1975a). Age was the most important source of variation in milk yield while it ranked second (after season) in fat yield (Alderson and Pollak, 1980). Steine (1975a) found that milk from younger goats tended to have a higher fat content than that from older ones, but the effect of the kidding season was not significant.

Weight and age are confounded and different results have been obtained as to which of the two is the primary factor influencing milk yield. Rønningen (1967a) and Lampeter (1970), using multiple regression analysis, compared the relative influence of body weight and age on milk yield. Rønningen concluded from his results that variation of milk yield was mainly due to age if weight was held constant, but Lampeter, conversely, showed that weight was the main source of variation if age was held constant. These seemingly contradictory results may be explained by the different stages of lactation at which the animals were weighed. Lampeter weighed immediately after kidding and 5 weeks post partum, while Rønningen's goats had been weighed during the third to fourth month of lactation.

Since body weight is not regularly available in population analyses, the effect of size on milk yield is usually corrected for on the basis of age. Steine (1975a) calculated least square constants for the effect of age on milk yield independent of season of kidding (Table IV); R^2 was 0·35. The relationship may be biased by selection (Rønningen and Gjedrem, 1966). There were interactions of age with herd level of production and region, which may necessitate the use of different age correction factors by level of production and region (Rønningen and Gjedrem, 1966; Steine, 1975a). Age adjustment factors were calculated by Alderson and Pollak (1980) and Iloeje *et al.* (1980) for five dairy breeds in the USA. An important interaction between age and season of freshening required consideration of seasons, while there was no significant interaction between breed of doe and age in Alderson's data.

Table IV. Effect of age on milk yield (kg). Season of kidding kept constant (Steine, 1975a).

Age (years)	Least square constants for age groups	
	Milk (kg)	Standard error
Mean	576	± 0·96
1	– 157	± 2·3
2	– 28	± 2·0
3	31	± 2–1
4	50	± 2·3
5	52	± 2·5
6	43	± 2·9
7	19	± 3·5
8	7	± 4·5
9	– 16	± 6·6

R^2 0·353

3. Udder Size and Shape

Udder attachment is of particular importance in grazing goats. In some breeds the prevalence of weak udder attachment is considered a major defect. It may be a limiting factor to the use of such breeds on ranges with thorny brush and cacti.

The size of the teat and gland cistern in relation to the volume of the secretory tissue seems to be greater in the goat than in the cow although this characteristic is not well established. As a consequence, more milk would be available for milking before the onset of oxytocin action for expulsion of milk from the alveoli. On the other hand, this particularity seems to favour extension of the cistern walls which causes broken teats and bottle-shaped udders.

Udders of Czech white polled goats were mainly round or egg-shaped, with cylindrical or funnel-shaped teats. Goats with higher milk yield tended to have pendulous udders; the correlation between yield and ground clearance of the udder was − 0·97 (Horák, 1971). Udder volume was closely related with milk yield. Correlations between linear udder measurements and milk yield were in the order of up to 0·8 in Barbari and Bengal goats (Das and Sidhu, 1975) and German fawn-coloured goats (Junge, 1963). Udder volume may be measured directly (Junge, 1963; Linzell, 1966) with high accuracy in the live goat. Contained milk volume has been measured with dilution techniques whereby polyethylene glykol (PEG) as non-absorbable marker was infused into the udder (Knutson and Sperber, 1961). Snipes and Lengemenn (1970) used this technique with radioactive-labelled PEG.

Udder volume increases during pregnancy. Although part of the more spectacular growth towards the end of pregnancy is due to increased lymph flow, accumulation of extracellular fluid and oedema, there is still a net increase in secretory tissue at each pregnancy (Linzell, 1966). The decline of milk yield in later lactation is due to both loss of secretory tissue and a fall in the rate of secretion per unit tissue.

From measurements of the udder volume it may be concluded that between milkings milk secretion continues linearly up to 18 hours (Linzell, 1966) or 19 hours (Junge, 1963). However, lactation milk yield was reduced to about 50% by milking once a day only (Mocquot and Auran, 1974). This may be explained by a reduction in mammary blood flow with increasing intramammary pressure (Pearl *et al.*, 1973) while there is no evidence for the hypothesis that intramammary pressure directly affects the secretory activity of the cell. Furthermore, any such effect would have to be divided into the inhibition of milk synthesis and of milk secretion which, in the light of newer research done to a great extent with goats, are likely to be two different functions, controlled through different mechanisms and thus possibly genetically independent (Patton, 1978).

Correlations of udder volume with milk yield are given in Tables V and VI. Due to the correlation between the volume of the milk-filled udder and the quantity of milk present in the udder, correlations between milk yield and udder volume are, in part at least, organic. However, Gall *et al.* (1972b), found a correlation of 0·73 between empty udder weight and 10 week milk energy yield, in 64 goats killed in week 10 of their first lactation.

Table V. Correlations between measurements of the isolated udder and milk production in 18 German fawn-coloured goats killed during week 8 of lactation and 19 hours after the last milking (Junge, 1963).

Measurement	Quantity of milk present in the udder	Daily milk yield	Part lactation milk yield up to day of slaughter
Volume of milk filled udder	0·723	0·700	0·722
Volume of empty udder	0·378	0·498	0·554
Volume difference	0·809	0·514	0·490

Table VI. Correlation between milk yield and udder volume[a] in 28 German fawn-coloured goats in lactation 1 to 3 (Junge, 1963).

	Weeks post partum		
Measurement	5	10	20
Volume	0·77	0·78	0·86
Index of linear measurements	0·71	0·64	0·84

[a] Measured on the live animal.

Similarly, Linzell (1966) measured the udder volume of 21 Welsh* and cross-bred goats and 34 Saanen goats at the height of lactation and calculated correlations with maximum milk yield of 0·796 and 0·878 respectively. Maximum yield (g/day) per g udder tissue was 1·66 and 1·99 in both groups, which suggests very similar secretory activity in both breeds. Linzell (1972) confirmed this between species as well, and concluded that differences in milk yield, both between and within species, are a function of the size of the mammary gland.

This is corroborated by the finding that milk yield is proportional to the mammary alveolar surface area (Naito *et al.*, 1955; Richardson, 1953). Traditionally, the productive capacity of dairy goats—like that of other dairy animals—is judged to a good extent by the physical appearance and size of the udder. Since research has shown a close relationship between the volume of the udder and the quantity of milk secreted, this practice seems to be quite well based, especially if consideration is given to age, stage of lactation, interval since last milking and expected lactation length. We have calculated

*This is a non-dairy breed imported from Kashmir in 1823 and kept since in Zoological Gardens (Fleet *et al.*, 1975).

the rank correlation coefficient for 200 days fat corrected milk (F.C.M.) yield within one herd, on the basis of one single recording of daily milk yield in week 8 of lactation, or on maximum daily yield with 0·8875 and 0·9077 respectively (Gall *et al.*, 1972).

4. Multiple Relations

Since many body measurements tend to be interrelated, multiple regressions of milk yield on various body measurements should yield more information about type-production relationships than simple correlations (Gall, 1980). Individual body traits that are positively correlated with milk yield, like muscle weight, may have a negative influence on yield in a multiple relationship, i.e. when other traits are held constant. In goats of the same body size, abdominal volume and udder volume, higher body weight should be mainly due to more body fat and more muscle. Since these tissues expend additional energy, it seems plausible that they are negatively related with milk yield. On the other hand, body traits which do not involve such secondary effects in their variation, will continue to be in positive correlation with milk yield. This is the case with udder volume; others, like abdominal volume, due to interrelationships with other traits, may not continue to be a significant source of variation in combination with other traits. Relationships between body conformation and milk yield may in part reflect energy relationships. Body size and abdominal volume determine the quantity of energy available for milk yield and other body functions. On the other hand, udder size and volume of musculature determine the quantity of energy needed for different body functions. Further research is necessary to identify and make available for genetic improvement the factors which determine whether energy is used to grow a voluminous musculature and store body fat or grow a large udder and produce an abundance of milk.

5. Growth

In ruminants there seems to be a general relationship between the type of growth and mature size: rapidly growing animals tend to become large at maturity; early maturing animals which deposit fat at an early age and at the same time reduce their growth rate, tend to be smaller at maturity. In dairy animals, there seems to be a tendency, both within and between breeds, for the more productive animals to be late maturing. Body weight between 2 and 7 months of age is in positive correlation ($r\mathfrak{T} \sim 0{\cdot}5$) with milk yield during the first 100 days of lactation (Bouillon and Ricordeau, 1975). Weight gains during pregnancy tended to positively influence lactation yield of French dairy goats in an experimental herd (Fehr and Sauvant, 1975).

We have analysed the growth of 64 German fawn-coloured goats by the method of Taylor and Fitzhugh (1971) where certain degrees of maturity are defined as fractions of mature weight and the average time taken to reach these stages (ln t_u) describes the growth rhythm; high values indicate a flat growth curve in late maturing animals. Multiple regression equations of milk energy yield on body measurements, taken on goats killed after a 10-week lactation period, were calculated (Gall, 1980). In combination with udder volume, $\ln_u$ did not significantly contribute to a reduction of the deviations from regression, but if udder volume was omitted, the contribution of $\ln_u$ was significant. The positive regression coefficient of ln t_u indicates that late maturing animals with a flat growth curve tend to produce more milk.

Table VII. Effect of number of kids born on milk and milk fat yield (kg). Age and season of kidding held constant. Least square constants ± standard error (Steine, 1975a).

Number of kids born	Milk yield[a]	Milk fat yield[a]
Mean	576± 0·96	20·6 ± 0·03
1	−17± 2·6	−0·57 ± 0·092
2	5± 2·5	0·11 ± 0·089
3	12± 4·7	0·46 ± 0·169

[a] Differences are statistically significant ($p < 0{\cdot}01$).

6. Litter Size

There are indications that mammary growth during pregnancy is regulated by the number of kids born and that plasma lactogenic activity of placental origin plays a role in this regulation (Hayden *et al.*, 1979). This would be a plausible effect since milk yield would be adapted to the needs of the young to be suckled. A relationship between the number of kids born and subsequent milk yield was found by Montaldo *et al.* (1978), Ricordeau *et al.* (1979) and in a more detailed analysis by Steine (1975a) (Table VII). There was a significant regression of number of kids born on body weight; dams with twins and triplets were 0·5 and 2·0 kg heavier respectively, than dams with singles. In a multiple regression of milk yield on number of kids and body weight, the variation of milk yield due to body weight was negligible (Steine, 1975b). Thus, the influence of number of kids on milk yield appears to be independent of age, body weight and season.

7. Season of Kidding

Lactation milk yield is to some extent influenced by season of kidding. In

Europe for example, lactations starting early in the year will benefit from better feeding conditions, endocrine control and, given a fixed mating season, a longer lactation period. Irrespective of the season of kidding, seasonal fluctuations occurred under experimental conditions with recurrent peaks of milk yield in summer (Linzell, 1973a). These cycles were independent of pregnancies, feeding and temperature; apparently light regime was the controlling factor.

The influence of season on lactation performance has been statistically analysed in Norwegian goats. It varied by age and region (Rønningen and Gjedrem, 1966; Steine, 1975a). Least square constants for lactations starting in December to April were 41, 30, 12, −23 and −60 respectively (Steine, 1975a). There were significant interactions between season and age, but the effect was small and no great error may be incurred by neglecting it (Rønningen and Gjedrem, 1966; Steine, 1976). In France about 75% of lactations start in January, February and March. Milk yields of lactations starting in October, November and December are about 200 kg higher than those started at the beginning of the year (Sigwald and Lequenne, 1976, cit. LeJaouen, 1978).

8. Nutrition

The main limiting factor of milk secretion may be the availability of glucose, because in the udder it forms lactose, which largely controls the movement of water into milk (Linzell and Peaker, 1971b). The mammary gland takes up about 70 g glucose per kg milk formed. Of the glucose entering circulation, 60 to 85% is used by the mammary tissue of the goat (Annison and Linzell, 1964). Since there is little glucose stored in the body, a reduction in feed intake will quickly affect milk yield, through lowered blood glucose levels. When lactating goats were fasted, milk yield was reduced to about 90% at 8 hours and 56% at 24 hours; it remained at about 48% for another 12 hours if the goats were milked every hour (Linzell, 1967a). Lowering the input of glucose to isolated perfused mammary glands stopped milk secretion almost entirely (Linzell, 1972) and infusing glucose into the blood stream of high producing goats stimulated milk yield by as much as 62% (Linzell, 1967b).

Most of the amino acids entering circulation are used by the mammary tissue (Linzell, 1973b) which apparently has the capacity to increase the extraction of amino acids if their concentration in the blood is low. Although needed in high quantities for milk secretion, amino acids cannot easily become rate limiting (Linzell, 1973b); in normally fed goats the additional supply to the blood vessels of amino acids did not directly affect milk yield (Champredon and Pion, 1979).

The energy expenditure for milk secretion has been calculated on the basis of the respiration of mammary tissue (Linzell, 1960; Reynolds, 1967). From Reynolds' calculations, it follows that 83 kcal would be the energy expenditure of the mammary gland to produce 1 kg of milk. Milk composition was not given in these papers but, assuming average values, about 10% of the energy secreted with milk would be required by the cells doing that secretion. This would confirm results obtained much earlier by the group working with Peterson and Turner at the University of Missouri, Columbia (Graham *et al.*, 1938). On the other hand, gross efficiency of milk production was found quite early on to be about 35% and net efficiency 61%, values very similar to those in cattle (Brody *et al.*, 1938).

9. Temperature

When lactating goats are exposed to cold, milk secretion is reduced. This is not a consequence of reduced udder blood flow only (Thompson and Thompson, 1977). At −0·5°C mammary glucose uptake, lactose secretion and milk yield were only 30% of the values at a thermoneutral temperature (20°C); these changes seem to be major factors responsible for reduced milk secretion (Faulkner *et al.*, 1980).

C. Milk Composition

1. Colostrum

In a goat milked throughout pregnancy, milk composition changed very little and immunoglobulin concentration remained very low (Linzell and Peaker, 1974). In the dry goat, udder volume and intramammary pressure start increasing about 10 and 12 weeks before parturition. The pre-colostral fluid accumulating in the teats has a high content of lactose and immunoglobulins. The rate of secretion up to a few days before parturition is only a small

Table VIII. Characteristics of mammary secretion in goats (Fleet *et al.*, 1975).

	Dry period	*Pre-colostrum*	*Colostrum*	*Milk*
IgG	Low	High	High	Very low
Citrate	Low	Low	High	High
Lactose	Low	---------Rising---------→		High
K^+	Low	---------Rising---------→		High
Na^+	High	---------Falling---------→		Low
Cl^-	High	---------Falling---------→		Low
HCO_3^-	High	Falling	Low	Low
pH	High	Falling	Low	Low

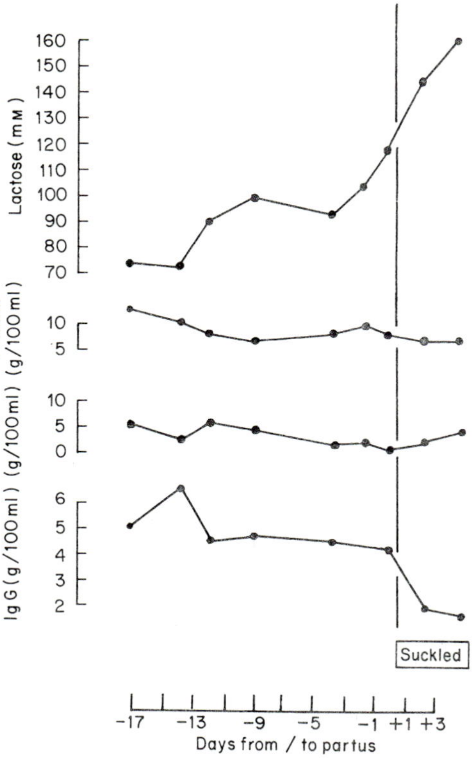

Fig. 1. Changes in the composition of colostrum in late pregnancy (after Linzell and Peaker, 1974).

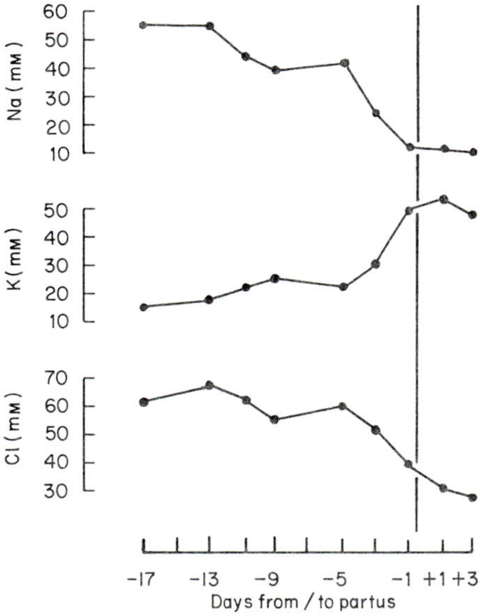

Fig. 2. Changes in mineral content of colostrum (after Linzell and Peaker, 1974).

fraction of that immediately after parturition. Two to three days before parturition the concentration of citrate rises sharply, 3–11-fold, to about 160 mg/100 ml; this level is maintained for 2–3 days post partum and, together with the high level of immunoglobulins, may be considered the true characteristic of colostrum in the goat (Fleet *et al.*, 1975) (Table VIII). When goats were milked about 2 weeks pre-partum, the concentrations of sodium, chloride, protein and immunoglobulins decreased, while those of potassium and lactose increased (Figs 1 and 2). Thus, composition changed

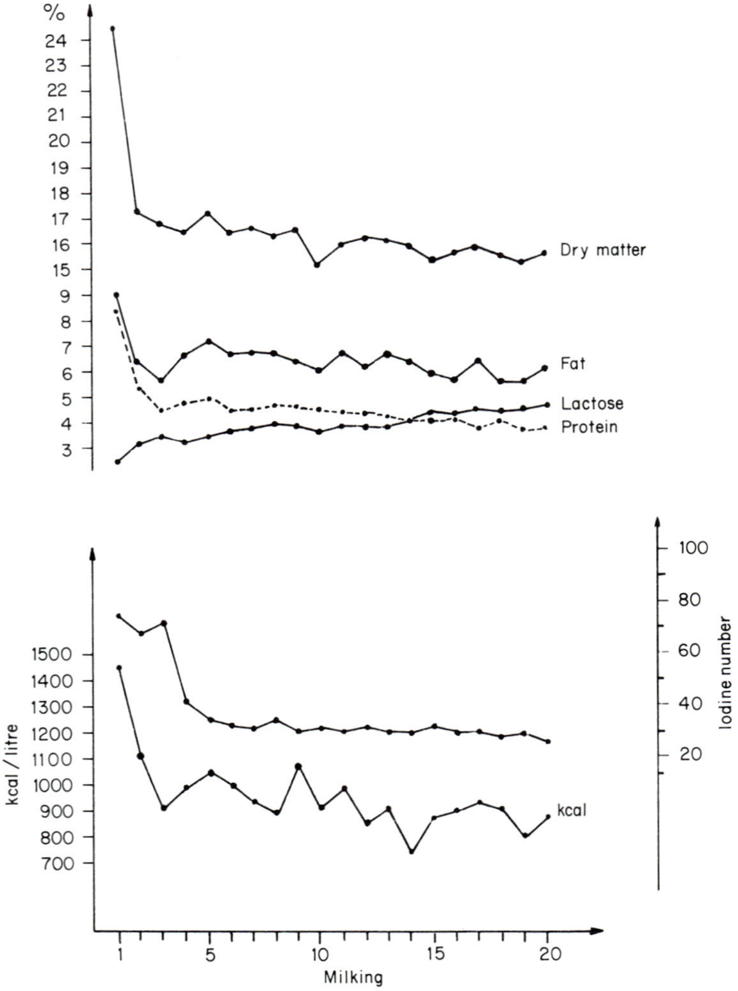

Fig. 3. Composition of colostrum in four German coloured goats (Graf *et al.*, 1970b).

to that of normal milk and the quantity secreted progressively increased. Mean yield on average reached only 48% of that in the second week of lactation. These changes with milking were apparently independent of hormonal actions since they occurred in the same manner in denervated udder halves (Linzell and Peaker, 1974).

After parturition the composition of the colostrum changes very quickly (Graf *et al.*, 1970b) (Fig. 3). The milk quantity of these German fawn-coloured goats rose from an average of about 660 g at the first milking to

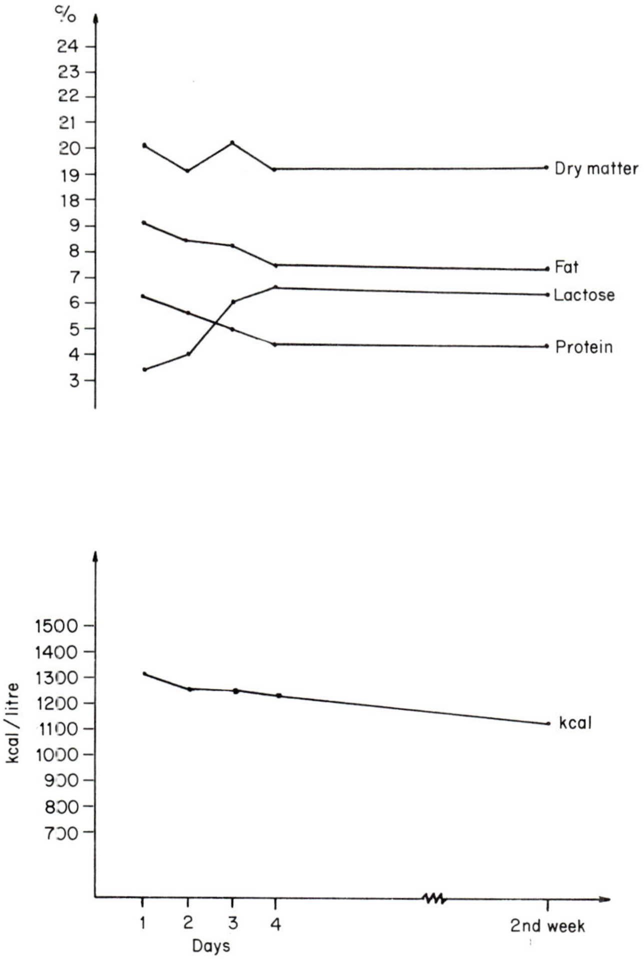

Fig. 4. Composition of colostrum; average of two lactations of six West African dwarf does (after Akinsoyinu *et al.*, 1977).

about 1325 on day 10 post partum. Calorific value in the same period fell from 1450 to 850 kcal/kg milk.

Iodine numbers indicate a higher proportion of unsaturated fatty acids in colostrum. The proportion of oleic acid decreased from 25·7% in the first week to 18·0% in the tenth week of lactation (Sauvant and Fehr, 1978). This is not confirmed by Klobasa and Senft (1970) who found less oleic, linol and linoleic acid in colostrum than in milk. In the colostrum of West African dwarf goats (Akinsoyinu *et al.*, 1977) (Fig. 4) total solids, protein and fat content declined after parturition and reached a level at day 4 which was kept for about 12 weeks. Lactose rose steadily until week 6. It is interesting to note that the composition of the first milkings was almost the same as in

Table IX. Mineral content in milk of West African dwarf goats (Akinsoyinu and Akinyele, 1979, Akinsoyinu *et al.*, 1979).

Period after parturition (days)	Ca	P	Na	K	Cl	Fe	Cu	Mn	Zn	I
	g/kg					mg/litre				μg/litre
1	0·81	0·40	1·20	3·22	5·21	2·68	3·25	0·20	13·32	320·22
2	0·81	0·52	1·51	3·50	5·00	2·60	3·26	0·21	14·11	340·11
3	0·52	0·30	1·55	3·71	4·82	2·52	3·04	0·25	14·32	300·25
4	0·60	0·30	1·60	3·24	4·51	2·50	3·15	0·18	14·40	320·06
5	0·51	0·30	1·35	3·25	4·60	2·00	3·00	0·18	14·38	317·81
Mean[a]	0·65	0·36	1·44	3·38	4·83	2·46	3·14	0·20	14·11	319·69
2nd week	0·51	1·00	0·50	2·01	1·01	0·67	0·57	0·20	3·10	149·52
Mean[b]	2·01	1·18	0·65	1·57	2·46	0·43	0·28	0·05	4·01	108·52

12 goats (trace elements in 10 goats)
[a] 1st week.
[b] 3rd to 18th week.

French goats (Sauvant and Fehr, 1978) and in German fawn-coloured goats (Graf *et al.*, 1970b) although the average total solids during lactation was 18·7% as compared with 12·5% in the dairy goats. Changes of minerals were observed in West African dwarf goats (Akinsoyinu and Akinyele, 1979; Akinsoyinu *et al.*, 1979). Concentrations of K, Na, Cl, Fe, Cu, Zn, J declined markedly until the second week. Concentrations of Ca and P declined but thereafter rose steadily and their average concentration was higher in milk than in colostrum (Table IX). Unfortunately, the authors did not indicate the quantity and gross chemical composition of the milk.

2. Milk Energy Content

The energy content of milk has received similar attention as in dairy cattle, since the energy expenditure for milk production is closely related with its energy content. Assuming that the energy content of the main milk con-

Table X. Composition and energy value of goats' milk average values of samples analysed.

	(1)	(2)	(3)	(4)			(5)	
				(a)	(b)	(c)	(a)	(b)
Number of samples	295	18	408	30	48	24	—	—
Milk quantity (g/day)	1322	—	2110	—	(6)	—	1550	1640
Fat content (%)	3·92 (2·38–7·30)	4·80	6·90	3·41	7·10	4·86	3·57	3·45
Protein content (%)	2·90	3·57	3·91	3·07	4·71	4·38	2·68	2·70
Lactose content (%)	4·01	5·06	6·30	4·47	5·59	4·72	4·32	4·45
Total solids (%)	12·50	15·12	18·68	12·15	17·78	15·28	11·39	11·38
Energy (kcal/kg)	681	833	1084	603	1020	780	—	—
Energy, calculated	685	840	—	—	—	—	651	638

(1) Graf *et al.* (1970a): 97 fawn-coloured goats, 10 weeks of first lactation, 1 sample every three weeks.
(2) Peterson and Turner (1939): 4 goats, different stages of lactation. Calorific value is kcal/1000 ml.
(3) Akinsoyinu *et al.* (1977): 6 West African dwarf goats, 2 lactations of 18 weeks, 2 samples per week, first week excluded.
(4) Calculated from Mba *et al.* (1975) (a) 3 Saanen, (b) 4 West African dwarf, (c) 2 Maradi goats.
(5) Opstvedt (1969) (a) feed level 0·38 FU/kg F.C.M.; (b) 0·48 FU/kg F.C.M.
(6) Between 500 and 1000 g.

stituents is quite constant, milk energy may be calculated from fat, protein and lactose contents. Gaines (1928) has shown that there is a close correlation between the energy content of milk calculated in this way and energy determined directly by combustion calorimetry. Perrin (1958) and Kalaissakis (1959) reported on direct determination of milk energy in small numbers of goats. Results of analyses of milk composition and calorific value of appreciable numbers of samples are given in Table X. In addition to calorimetric determination, energy values of milk were calculated from milk composition and standard energy values for the constituents by Peterson and Turner (1939), Opstvedt (1969) and Graf *et al.* (1970a) (Table X). The calculated values coincide very closely with those determined directly. In the study by Graf *et al.* (1970a) the calculated energy was only 0·58% higher than the true energy value. From the regression equation of milk energy on milk quantity and milk fat a formula for calculating F.C.M., using Gaines' energy value of milk with 4% fat, was derived: F.C.M. = 0·34 M × 16·42 F. Opstvedt (1969) and Skjevdal (1974c) found Gaines' formula well suited to goat milk. Carrying Gaines' considerations further, Osterkorn and Graf (1970b) derived a formula for calculation of milk energy value on the basis of total milk solids. These comprise all the milk energy, and their determination is as simple as that of milk fat but more accurate. In analogy to F.C.M. they propose the term T.S.C.M. for which the formula calculated from direct determination of energy is:

$$\text{T.S.C.M.} = -0{\cdot}04\ \text{M} + 8{\cdot}33\ \text{T.S.}$$

where T.S.C.M. = total solids corrected milk with 12·5% T.S.; M = kg milk; T.S. = kg total solids.

12·5% was the average total solids in the milk of the goats analysed. It contained 687·5 kcal. In contrast to the F.C.M. formula, milk quantity contributes little to the T.S.C.M. value, since all the energy is accounted for by the total solids. Thus,

$$\text{T.S.C.M.} = 8 \times \text{total solids kg}$$

gives a rough approximation of the energy content.

3. Relations between Milk Constituents

The results of different investigations on relationships between milk yield and yield of milk constituents, between the yield of various constituents and between the concentration of various constituents are quite consistent. Lactation milk yield is positively correlated with all milk constituents ($r = 0{\cdot}6$–$0{\cdot}9$) but negatively with the concentration of fat, protein and fat free dry matter ($r = 0{\cdot}2$ to $-0{\cdot}6$) while the correlation of yield and lactose percentage is zero. Both the concentration and quantity of milk constituents are in positive correlation with each other ($r = 0{\cdot}6$–$0{\cdot}9$) with the exception of lactose percentage and protein percentage which are not correlated. The flavour score is in negative correlation with the concentration of milk constituents. These phenotypic correlations have been shown to be mainly of genetic origin; they are discussed and documented in Chapter 4.

D. Length of Lactation and Lactation Curve

Length of lactation as reported in different breeds and under different environmental conditions varies between 200 and 300 days. It seems to depend much more on environmental conditions, feeding in particular, than on breed. Lactation reaches its peak at about weeks 2 to 10 (Bonnekamp, 1939; Fehse, 1970; Rako, 1950; Rønningen, 1967b; Sachdeva *et al.*, 1974; Sauvant *et al.*, 1976) but two peaks may be observed as well (Enz, 1968). There seems to be a tendency for the peak to be reached earlier in later lactations. After the peak, production declines almost linearly (Enz, 1968) at about 10% monthly (Sauvant *et al.*, 1976). The season of kidding may influence the rate of decline; the persistence index was 86, 95, 86, and 83% in goats who had kidded in February, March, April and May respectively (Horák and Pinďák, 1969). High yielding goats reached higher peak production than low yielding goats but the lactation curve levelled off at about the same slope (Sauvant *et al.*, 1976). Persistence was greatest in the first lactation (Ricordeau, 1963). The rate of decline, measured by the coefficient of linear regression, increased from about 15 g per week in the first lactation to about 40 g in lactation seven. Regression coefficients did not differ significantly between three groups of half-sibs (Riggert, 1978). The coefficient of variation of milk yield was about 25% during the first 4 months and then increased about linearly to nearly 40% at the end of lactation. Third degree regression equations accounted for about 40% of the variations in milk yield (Sauvant *et al.*, 1976). The curves of fat and protein tests are inverse to those of milk yield; they declined during the first month, and then remained stable until the fifth month when they slowly started to rise again (Rønningen, 1976b; Riggert, 1978). Tests are less variable than yield, coefficients of variation being 12 and 8% for fat and protein content during the second to sixth month of lactation but up to 20% at the beginning and the end of lactation (Sauvant *et al.*, 1976).

E. Level and Efficiency of Milk Production

There is a large variation in milk yield both between and within breeds. Sands and McDowell (1978) have compiled production data of breeds in temperate and tropical areas. This compilation will not be repeated or amended here. There are distinct dairy breeds and breeds which have not been selected for milk yield. Marked differences in production exist between these two groups but the latter are still capable of considerable milk production over and above the needs of their kids, if fed sufficiently. For instance, Boer goats raising twins during the first 17 weeks of lactation produced an average of 1·8 kg per day (Ueckermann *et al.*, 1974). Under experimental conditions in Europe, West African dwarf goats produced on average 240 kg milk (138–369) in a 210 day lactation. Under the same conditions dairy goats (Czech White) produced 595 kg (583–893); live weights of the two breeds were 19 and 56 kg (Cumlivski, 1976).

The average production of dairy breeds in European countries is between 400 and 500 kg, it is estimated at 650 kg in Israel; it is higher in goats under

Table XI. Milk production of six goat breeds in the USA (Dickinson and King, 1977).

	n	Milk (kg)	Fat (kg)	Fat (%)
Nubian	7059	752	34	4·5
Alpine	6980	916	32	3·5
Toggenburg	4133	878	29	3·3
Saanen	3373	921	33	3·6
La Mancha	1846	800	31	4·0
Experimental	524	778	29	3·8

Official dairy Goat recording, 1966–1976.
Breed averages across first five lactations of 276 to 305 days.

Table XII. Milk production in German coloured improved goats, not selected for production (Gall, 1979).

	Experimental groups[a]							
	I				II			
Lactation	Milk	Days	kg/day	*n*	Milk	Days	kg/day	*n*
1	230	267	0·86	86	326	286	1·14	117
2	285	302	0·94	70	412	287	1·44	86
3	299	275	1·09	52	443	282	1·57	67
4	307	267	1·15	42	422	279	1·51	50
5	291	267	1·09	25	434	277	1·57	39
6	269	261	1·03	17	393	266	1·48	26
7	238	264	0·90	11	365	266	1·37	21
8	204	262	0·78	7	333	249	1·34	10
9	213	263	0·81	3	195	220	0·89	30
Total	260	270	0·96	313	396	268	1·38	419

[a] I: hay only; II: supplemented with 1 kg of concentrates (oats with 5% soya).

milk control and in registered goats. Data of official milk recording of dairy breeds in the USA are listed in Table XI. An experimental herd of German coloured improved goats was kept at the University of Munich between 1956 and 1970 (Gall, 1964). For experimental reasons animals were not selected for production during the whole experiment. One group was fed on hay only, a second group received in addition 1 kg of concentrates per day. Average production of these goats is given in Table XII. Thus, under commercial conditions dairy goats may be expected to produce between 400 and 500 kg per lactation if managed in larger herds. Animals kept individually and given special care may attain 750 to 920 kg. Record productions of more than 1000 kg are not unique and the world record lactation yield is 3400 kg produced by an Australian Saanen goat. However, such yields should not be taken as yardsticks when establishing production or breeding goals.

Production of indigenous breeds in tropical countries is referred to in other chapters of this book. While many experiments have shown that the productivity of these breeds may be enhanced by crossbreeding with dairy breeds of European origin, not all such attempts at crossbreeding have been successful. Often the productivity of goats was limited by environmental conditions, nutrition, climate or managerial skill, and genetic potential was only of secondary importance.

Commercial milk production from goats is burdened with high labour costs. If dairy goat operations are to compete with dairy cattle, prices for goat products must be considerably higher. In small scale family operations, where labour cost is not important, the biological efficiency of goat milk production in comparison with cattle and possibly sheep or buffaloes is of interest. Devendra (1975) has compiled available data from tropical conditions. The relation of milk yield and digestible organic matter (D.O.M.) intake was between 67 and 145 in goats, 66 and 101 in cows. Spedding (1968) reports milk yield/D.O.M. in goats and cows as 185 and 162 respectively.

The efficiency of dairy goats managed under intensive conditions has been investigated in a large scale crossbreeding experiment at the Indian National Dairy Research Institute (Sundaresan, 1978) and compared with dairy cattle. The forage:concentrate ratio was 50:50, 80:50, and 60:40 in Friesian, Karan Swiss Cattle and Goats respectively. The efficiency of feed utilization of high- and low-yielding cattle and goats is given in Table XIII. In this comparison goats do not appear to be more efficient producers of milk than cows. It thus seems that care must be taken in comparing data from goats and other species which result from different experiments. There is a tendency for goats to be more efficient converters of feed; this probably applies mainly to coarse feeds of low nutrient contents.

Table XIII. Efficiency of milk production with cattle and goats (Sundaresan, 1978).

	Cattle		Goats	
	Holstein Friesian	Karan Swiss	Alpine Beetal	Beetal
Average body weight (kg)	600	375	49	47
Lactation yield (kg)	6 000	3 300	490	290
D.M. intake (kg/lactation)	6 600	3 650	565	440
D.M. intake (kg/litre milk)	1·10	1·09	1·16	1·53
T.D.N. intake (kg/lactation)	3 870	2 380	330	270
T.D.N. intake (kg/litre milk)	0·64	0·72	0·67	0·93
T.D.N. output/input (%)	23	25	30	21
Protein intake (kg/lactation)	450	470	95	75
Protein intake (kg/litre milk)	0·075	0·142	0·194	0·259
Protein output/input (%)	40	24	20	16
D.M. intake for 25×10^6 tons	28	28	29	38

F. Milk Recording

1. Methods and Extent of Recording

The traditional procedures in official milk-recording schemes with dairy goats are essentially the same as with dairy cattle, i.e. monthly recordings by an official recorder of the quantity of milk produced during a 24-hour interval, fat or protein test in a composite proportional sample drawn at each milking, calculation of lactation (305 day) yield or of the yield during the operational year. For the determination of fat and protein content of goats' milk, the Milko-Tester, Milko-Scan, infra-red analyser and the amido black method are as accurate as the standard Gerber and Kjeldahl methods. For all the methods, specific calibrations are necessary for cows' and goats' milk. Since the non-protein nitrogen content is high in goats' milk, direct measurement of protein is preferable to total nitrogen measurement (Grappin and Jeunet, 1979).

The cost of milk recording in goats with the traditional methods is even higher than in cattle. Since herds are small and dispersed, transport cost is high and since expenses per record are the same, the burden on the quantity of milk produced is more important. This is one reason why milk recording in goats so far has been very limited in all countries. Furthermore there is little benefit from milk recording for most goat-keepers because of prevailing small herd sizes. With only a few animals in a herd the owner has little difficulty in following the productivity of each individual goat and has no need for any formal procedure.

If composition is the base for milk payment, this is tested by the dairy. Unless sales of breeding stock benefited through it, there is little interest in milk recording.

In 1972 the percentage of goats recorded in Germany was 4·8, in Israel 0·05, in the Netherlands 5 to 10, in Norway 12·8, in Switzerland 6·7, and in the UK 1·0 (Fehse, 1972). In the USA, from the period of 1966 to 1976, 45 188 records were available (Dickinson and King, 1977), i.e. about 4000 records per year out of an estimated dairy goat population of 300 000 or 1·3%. This very limited participation in milk control is in conflict with the need for extended recording in the interests of breed improvement programmes. Ways and means are therefore discussed for extending Milk recording by simplified recording methods.

2. Simplifying Milk Recording

Essentially milk recording can be simplified by reducing the number of samples, shortening the part of lactations recorded, reducing tests for milk constituents to a part of the lactation or omitting it entirely and replacing official sampling by owner samples. As early as 1939 Bonnekamp suggested that recording in goats may be limited to four controls at about 80 days interval. According to his observations with 44 daily records within one herd and 300 lactations of a milk recording association, there were only small differences between the simplified records and those calculated from monthly recordings.

Data of complete milk recording have been used to simulate simplified recording procedures and study the accuracy of the records obtained. Gall *et al.* (1972b) calculated 300 records of a herd of about 100 German coloured improved goats in various ways (Table XIV). None of the simplified records had errors of more than 1% and all the r^2 with the standard record were 0·99. If fat was tested only once or twice during lactation, error and r^2 were 4·4, 0·92 and 2·6, 0·97 respectively. If goats were ranked either on the basis of the standard record or on the simplified records, rank correlation was greater than 0·99. If goats were ranked on the basis of their highest daily yield or on yield in week 8 of lactation, rank correlation was still 0·91 and 0·89 respectively.

Table XIV. Accuracy of 200 day lactation records calculated by simulation of simplified procedures. 300 records of German coloured improved goats in one herd (Gall *et al.*, 1972b).

	r^2	$d\%$
1	0·993	0·84
2	0·996	0·57
3	0·988	0·92
4	0·990	0·25
5	0·968	2·59
6	0·924	4·39

1. One control every 8 weeks.
2. Alternating control of evening and morning milk yield every 3 weeks.
3. Alternating control of evening and morning milk yield every 6 weeks.
4. Every fourth of the weekly controls, three controls plus one additional control 8 weeks later.
5. Standard record with only two fat tests in week 8 and 16.
6. Standard record with only one fat test in week 8.

Standard record is calculated on the basis of weekly milk yield and fat test.
r^2: correlation between standard and simplified record.
$d\%$: average deviation of simplified record from standard record.

In French dairy goats correlations between milk yield, fat percentage and protein percentage during 7 months of lactation and one single control in the fourth month were 0·94, 0·70 and 0·82 respectively (Sauvant *et al.*, 1976).

Ammann (1972) used data of 4500 Saanen goats and calculated records in various simplified ways, and compared them with records calculated on the basis of monthly recordings of morning and evening yield. Records calculated with (1) alternating data from morning and evening milkings (2) recordings with 8 week intervals plus length of lactation and (3) only one single recording around day 140 of lactation plus length of lactation, had coefficients of determination (r^2) with the standard record of (1) 0·994, (2) 0·911 to 0·983 and (3) 0·811 to 0·825 respectively. Month of kidding accounted for the range of accuracy in the latter two methods.

Ricordeau and Mocquot (1967) studied the relationship between cheese yield and different constituents in goats' milk by a multiple regression analysis. They found that about 75% of the variation in cheese yield is explained by the variation in protein content of the milk, and that fat content gives no significant additional information. They suggest that fat tests be replaced by protein tests. Since protein content varies less than fat content, sampling procedures could be simplified.

Carta *et al.* (1975) simulated different methods of sampling for fat and protein with data from fortnightly controls of 76 and 87 goats of French fawn-coloured and Saanen goats within one herd. They concluded that one protein test every 2 months was sufficiently accurate. Sampling could even be limited to one equal quantity (i.e. not necessarily proportional) at the morning milking. A method for estimating milk composition on the basis of a simple dry matter determination as experienced in Sweden (Sjödin, 1976) is referred to in Chapter 15, page 496.

Simplified recording procedures are already practised in Israel, where the owner samples once to twice a month and 3 to 5 official samples are made as a check; fat is tested in only about 15% of the records (Laor, 1971). In Norway only 6 recordings are required and only 3 fat tests, the latter in the first and second lactation only (see page 494). In California a simple group testing programme has been introduced (Holtz *et al.*, 1973).

Recently the International Committee for Recording the Productivity of Milk Animals (ICRPMA) made recommendations for minimum requirements for official milk recording in goats based on the following principles (Fai, 1972; Disset and Gall, 1972):

—at least four recordings during the first 5 months of lactation;
—records may be on milk quantity only, tests for constituents are optional;
—recordings by the owner may be accepted, provided that there is one official check during the first 5 months of lactation.

While it is expected that most schemes will employ more extensive recording procedures, the possibility of such simple recordings being officially acknowledged should in the future increase the number of goats recorded.

3. Use of Part Lactation Data

Since goats' lactation periods are more or less synchronous, the question may be raised for what part of the lactation period recording has to be extended in order to give reliable estimates and avoid those extra tests that do not yield an increase in accuracy sufficient to justify the effort. Furthermore, there is a particular interest in using part lactations in progeny testing of goat sires. Since in northern countries the breeding season is quite limited, with the majority of matings taking place in August/September, and parturitions starting in February, sire evaluations based on 5 month first lactation records will be available before the onset of the mating season in that same year. The sire will then be 3½ or with artificial insemination 2½ years of age.

Ricordeau and Bouillon (1967) established correlation coefficients between 260 day records and part records of 100 and of 60 days. Correlation coefficients between total, 100- and 60-day yield of milk quantity, fat percentage and protein percentage were 0·85, 0·77, 0·90, 0·81, 0·89 and 0·83 respectively. Correlation coefficients between 200 day first lactation and 35, 70 and 100 day part lactation F.C.M. yield in German fawn-coloured goats were about 0·85, 0·87 and 0·93 respectively (Enz, 1968). Rønningen (1967b) calculated within herd correlations between cumulative test-day production and yearly production. Coefficients of correlation were 0·74 and 0·80 for 2 and 3 cumulative months respectively. Steine (1975b) found similar correlations increasing from 0·66 for the relationship between total lactation yield and a part lactation record based on a single daily record to 0·76 for 2-monthly records and 0·82 for 3-monthly records. By extending records based on 3-monthly records and using a multiplicative extension factor for the last record, estimated total yield is related to actual lactation yield with $r = 0{\cdot}8$.

Following the findings of Steine (1975b), Wiggans *et al.* (1979) calculated projection factors on the basis of 3707 lactation records of Alpines, Saanen, Toggenburg, LaMancha, Nubian and crossbreds in the USA. The same factors could be applied to all breeds, except Nubians. Different factors are given for two herd production levels, two seasons of kidding and four different ages at kidding. These projection factors are considered by the authors as provisional; the greatest source of error is length of lactation, which varied in the material between 30 and 305 days.

G. Milk Production and Raising of Kids

Milk feeding of kids in goats kept primarily for milk production is important for raising kids, both for replacement and for fattening. Combining highest possible quantities of saleable milk with adequate growth rate of kids, therefore, is an important goal of dairy goat management. Immediate separation of kids after birth, as is common practice in dairy cattle, is feasible in goats only where prices for milk are high and milk replacers are available at reasonable cost. This is not normally the case in tropical countries where, therefore, there is much interest in combining milking with nursing of kids. Lactation in Barbari and Jamnapari does whose kids were separated immediately after birth was compared with does which nursed their kids until 4 months of age and were milked from weaning onwards (Prabhu and Amble, 1946). Lactation lasted on the average 56 days longer in the does milked throughout, but yield adjusted for lactation length did not differ significantly. Growth rate of kids suckling all the milk of their dams until 4 months of age did not differ from those raised by hand-feeding but was quite low in both groups (less than 60 g per day). Damascus goats suckled daily for 8 hours only and then separated from their kids for the remaining 16 hours, produced 38% more milk (350 *vs* 253 kg) for sale during 10 weeks of lactation than does suckled continuously. While kids suckling continuously were heavier at 10 weeks than the group suckling partially, differences had disappeared at the age of 150 days (Hadjipanayiotou and Louca, 1976). Weaning at 5 weeks increased the quantity of saleable milk over weaning at 10 weeks but adversely affected growth rate of the kids (Louca *et al.*, 1975).

On page 260 Morand-Fehr shows that weaning of Alpine kids at 5 to 6 weeks of age and 8·5 to 10 kg body weight does not impair growth provided concentrate intake is about 50 g daily. In specialized dairy operations kids are traditionally slaughtered at a very early age (see page 285) although there is some interest in higher slaughter weight. While sales of kids contribute considerably to total revenue of milk production, no effort is made in dairy breeds to improve this meat output through breeding. In the light of the findings discussed in Section B4 any attempt to increase the slaughter value of cull does would tend to negatively affect milk production due to the negative relationships of muscle and body fat with milk yield. Similarly, selection for improved fattening ability of kids because of the relationship between growth rhythm and milk yield should tend to reduce milk production.

In tropical areas, some goats are used for milking in years with good fodder supply only, and in less favourable years will be just allowed to raise a kid. In such areas a more fleshy breed of goat will better resist years of scarce feed supply and produce a valuable carcass.

H. Economics of Milk Production

Many goats are kept in family holdings for subsistence or as a side activity without cash input and opportunity costs. In these, economic evaluation is hardly feasible (Guss, 1977) and considerations have to be limited to the contribution to family nutrition and a possible net income. But for larger operations economic analysis is important. There are two main aspects which have to be considered: (1) comparison with other possible uses of the means of production especially land and labour and (2) the returns from investment in feeding, management and breeding. Some data on economics mainly in tropical countries are discussed on page 586ff. Here we shall discuss some data available on more intensive types of operations.

For a herd of 100 milking does in the USA, Guss (1977) has given factors for calculating costs and benefits. He confirms the old rule that "the selling price for goats' milk must be approximately twice that of cows' milk for profitable production".

1. Switzerland

Economic accounts of seven herds in Switzerland are reported by Fehse *et al.* (1976). Average herd size was 22 goats; 1·4 kids were raised per doe (Table XV).

In this analysis only cash expenses for feed are considered. It would, therefore, apply only to those conditions where available feed resources have no alternative use and therefore no opportunity cost. With this limitation, income seems quite appreciable.

For production conditions in Switzerland, Ammann (1979) reports

Table XV

	Swiss francs per goat
expenses	
feed	72·10
veterinary	2·20
depreciation	18·35
various	21·60
total cost	114·25
income	
milk	238·85
kids and replacements	175·20
various	15·10
total	432·85
net income	318·60

calculations comparing rentability of dairy cattle, goats and sheep. Gross returns at medium productivity of pastures (3000 starch equivalent per ha) are about equal with cattle and goats but lower with sheep. For a parity family income on 15 ha land, about 120 goats would be necessary. This means that 5 to 6 goats would replace one cow. Because of higher labour requirements, income per hour would be lower with goats. In order to equal income from cows producing 4200 kg of milk per year, goats would have to produce 765 kg.

2. France

Comparative economic analyses of two different systems of milk production in France have been made by Sigwald and Lequenne (1971). Type 1 is a family operation with 60 does and cheese production on the farm. Type 2 has 140 does; milk is sold to a dairy for cheese production and in addition to the herd owner one full-time labourer is employed. Both herds produce 600 kg of milk and raise 1·54 kids per doe. Replacement rate is 15 and 17%. In type 2 returns on milk per unit (doe) are only half of type 1 and with labour costs total expenses exceed the income.

In a more detailed analysis Sigwald and Sauvant (1976) examined data of some hundred operations during 2 consecutive years in one area of France. Some characteristics of these operations are shown in Table XVII.

The income from the goat operations was mainly determined by the level of production which accounted for two-thirds of income variation. By comparison, cost of production was much less important. There is, therefore, much interest in improving production of forage and milk and improving

Table XVI

	Type of operation 1	2
	French Francs	
expenses		
feeding	14707	34365
veterinary care	930	2175
various	620	5800
replacements	5040	10134
cheese manufacture and marketing cost	10260	—
labour	—	66844
total	31557	119318
income		
milk (cheese)	54720	63840
kids	4480	9128
cull animals	600	1200
total	59800	74168
net income	28234	−45150

Table XVII

	1973	*1974*	*1976*
Total agricultural surface	27	26	28·9
Surface used for goat forage production	9	12	12·2
Number of goats	51	57	59
Does in first lactation	24	23	24·5
Labour units	1·97	1·86	1·96
Milk yield (kg)	570	557	625
Grains and concentrated fed per goat (kg)	270	259	
Goats/ha	6·9	6·5	
Milk per area used for goats (kg/ha)	3832		

the organization of the operation. Overall, milk yield per ha was modest and few of the farms attained margins of Francs 6000/ha.

Since most of the farms have more area available than they use for goats and since labour demand by the goats is unequal during the year, there is interest in evaluating other complementary activities. Possibilities tested for complementing goat milk production were: beef fattening, cereal production and tobacco cultivation. Beef was competitive only if milk was sold instead of being processed at the farm and production was low. Cereals were competitive for relatively large farms, where about 140 goats would have to be kept and where labour becomes the limiting factor. Labour-intensive tobacco cultivation may be competitive on very small farms and if family labour in excess of 1·5 units is available. Gross margins increased up to lactation yields of 650 kg. At further increases to 800 kg, diminishing returns were noticeable. The higher inputs in breeding, management and know-how necessary for this level of production, which at present is attained by very few farms, are hardly justified by extra returns. The smallest returns per goat were achieved with intensive forms of production (zero-grazing) and sales of fresh, unprocessed milk. Higher stocking rates were possible and therefore higher gross margins per ha were attained with zero-grazing than with grazing, namely F 9900 vs. 6500.

Labour and animal productivity tends to be higher in specialized goat operations (Sigwald and Sauvant, 1977). In southern France where previously cereals were cultivated with very low yields (1000 kg/ha) and where Mediterranean bush vegetation (garrigue) had developed, Dreyer (1975) kept about four goats per ha with an average annual production of 580 kg milk. Net returns per ha were 425 Francs which was considerably higher than previous agricultural production. With adequate fertilization the stocking rate could have been doubled.

References

Akinsoyinu, A. O. (1977). *J. Dairy Res.* **44**, 57–62.

Akinsoyinu, A. O. and Akinyele, I. O. (1979). *J. Dairy Res.* **46**, 427–431.

Akinsoyinu, A. O., Tewe, O. O. and Mba, A. U. (1979). *J. Dairy Sci.* **62**, 921–924.

Alderson, A. and Pollak, E. J. (1980). *J. Dairy Sci.* **63**, 148–151.

Ammann, P. (1972). *Schweiz. Landwirt. Monatsh.* **50**, 239–247.

Ammann, P. (1979). *Schweiz. Landwirt. Monatsh.* **57**, 161–167.

Annison, E. F. and Linzell, J. L. (1964). *J. Physiol.* **175**, 372–385.

Ardran, G. M., Cowie, A. T. and Kemp, F. H. (1957). *Vet. Rec.* **69**, 100–101.

Ardran, G. M., Cowie, A. T. and Kemp, F. H. (1958). *Vet. Rec.* **70**, 808–809.

Bonnekamp, H. (1939). *Z. Tierzücht. ZüchtBiol.* **42**, 29–462.

Bouillon, J. and Ricordeau, G. (1975). "lères Journées de la Recherche Ovine et Caprine", December, 1975, Vol. 1: "Espèce Caprine" (Reproduction Selection), pp. 149–155. INRA, ITOVIC, Paris.

Brody, S., Mrs. Sandburg, C. and Asdell, S. A. (1938). *Res. Bull. Mo. Agric. Exp. Stn.* No. 291, 64 pp.

Burvenich, C. (1980). *Z. Tierphysiol. Tierernähr. Futt.* **43**, 18–27.

Carta, R., Casu, S., Flamant, J.-C. and Ricordeau, G. (1975). *Ann. Génét. Sélec. Anim.* **7**, 293.

Caruolo, E. V. (1974). *Br. Vet. J.* **130**, 380–387.

Chalmers, M. I. (1972). Symposium on Milk Recording Practice for Sheep and Goats. Israel, March.

Champredon, C. and Pion, R. (1979). *Ann. Rech. Vet.* **10**, 379–381.

Chilliard, Y., Durand, G., Sauvant, D. and Morand-Fehr (1978). *C. R. Acad. Sci. Paris* **287**, Série D, 1131.

Chilliard, Y., Morand-Fehr, P., Durand, G. and Sauvant, D. (1979). *Bull. Acad. Vét. France* **52**, 417–422.
Cleverley, J. D. and Folley, S. J. (1970). *J. Endocr.* **46**, 347–361.
Cowie, A., Knaggs, G. and Tindal, J. (1964). *J. Endocr.* **28**, 267.
Cumlivski, B. (1976). 27th Annual Meeting, EAAP, Zürich, August 1976, Sheep and Goat Commission.
Cunningham, O. C. and Addington, L. H. (1936). *J. Dairy Sci.* **19**, 405–409.
Das, D. and Sidhu, N. S. (1975). *Indian J. Hered.* **7**, 1–9.
Delouis, C. (1975). "lères Journées de la Recherche Ovine et Caprine", December, 1975, Vol. 1: "Espèce Caprine", pp 64–72. INRA, ITOVIC, Paris.
Denamur, R. (1953). *C. R. Séance. Soc. Biol.* **147**, 78–82.
Denamur, R. (1965). *Dairy Sci. Abstr.* **27**, 193–224, 263–280.
Devendra, C. (1975). *Wld Rev. Anim. Prod.* (FAD) **11**, 46–53.
Dickinson, F. N. and King, G. J. (1977). *J. Dairy Sci.* **60**, Suppl. I, 104.
Disset, R. and Gall, C. (1972). 23rd Annual Meeting, EAAP, Verona, Italy, Oct. 1972, Sheep and Goat Commission.
Dreyer, P. (1975). *La Chèvre* No. 86, 15–21.
Dyusembin, (1974). "Symposium sur la Traite Mécanique des Petits Ruminants", Millau, France, May 1973. *Ann. Zootech.*, numéro hors série, 1974, 77–81.
Enz, H. (1968). *Vet. Diss.*, München.
Fai, I. (1972). International Committee for Recording the Productivity of Milk Animals, 18th Session, London.
Faulkner, A., Thomson, E. M., Bassett, J. M. and Thomson, G. E. (1980). *Brit. J. Nutr.* **43**, 163–170.
Fautz, J., Erfle, V., Frahm, K., Gall, C., Graf, F. and Osterkorn, K. (1970). *Arb. Inst. Tierzucht, Vererbungs- u. Konstitutionsfor. Heft* **8**, 75–82.
Fehr, P. M. and Sauvant, D. (1975). "lères Journées de la Recherche Ovine et Caprine", December, 1975, Vol. 1: "Espèce Caprine", pp. 73–84. INRA, ITOVIC, Paris.
Fehse, R. (1970). 21st Annual Meeting, EAAP, Gödollö, Hungary, Sheep and Goat Commission.
Fehse, R. (1972). *Schweiz. Landwirt. Monatsh.* **50**, 230–238.
Fehse, R., Ammann, P. and Item, C. (1976). 27th Annual Meeting EAAP, Zürich, Sheep and Goat Commission.
Fleet, I. R., Goode, J. A., Hamon, M. H., Laurie, M. S., Linzell, J. L. and Peaker, M. (1975). *J. Physiol.* **251**, 763–773.
Gaines, W. L. (1928). *Illinois Agr. Exp. Sta. Bull.* **308**.
Gall, C. (1963). *Arb. Inst. Tierzucht, Vererbungs- u. Konstitutionsfor., Univ. München, Heft* **5**, 1–72.
Gall, C. (1964). *Tierärztl. Umsch.* **19**, 516–520.
Gall, C. (1979) *In* "Producción Caprina en Medios Difficiles". Seminario, VII Reunión ALPA, Panama, September, 1979.
Gall, C. (1980). *J. Dairy Sci.* **63**, 1768–1781.
Gall, C., Fautz, J., Frahm, K., Osterkorn, K. and Graf, F. (1972a). *Z. Tierzücht. ZüchtBiol.* **89**, 181–190.
Gall, C., Osterkorn, K. and Graf, F. (1972b). Unpublished data, University of Munich.
Graf, F., Osterkorn, K., Fautz, J., Frahm, K. and Gall, C. (1970a). *Arb. Inst. Tierzücht, Vererbungs- u. Konstitutionsfor., Univ. München, Heft* **8**, 45–52.
Graf, F., Fautz, J. and Frahm, K. (1970b). *Arb. Inst. f. Tierzucht, Vererbungs- u. Konstitutionsfor., Univ. München, Heft* **8**, 45–52.

Graf, F., Fautz, J. and Frahm, K. (1970b). *Arb. Inst. f. Tierzucht, Vererbungs- u. Konstitutionsforsch., Univ. München, Heft* **8**, 65–74.
Graham jr., W. R., Houchin, O. B., Peterson, V. E. and Turner, C. W. (1938). *Am. J. Physiol.* **122**, 150–153.
Grappin, R. and Jeunet, R. (1979). *Le Lait* **59**, 345–360.
Guss, S. B. (1977). "Management and Diseases of Dairy Goats", 222 pp. Dairy Goat Publishing Corporation, Scottsdale, Arizona, USA.
Hadjipanayiotou, M. and Louca, A. (1976). *J. Agric. Sci.* **87**, 15–20.
Hart, I. C. (1974). *J. Reprod. Fert.* **39**, 485–499.
Hart, I. C. (1975a). *J. Endocr.* **64**, 313–322.
Hart, I. C. (1975b). *J. Endocr.* **64**, 305–312.
Hart, I. C. (1976). *J. Endocr.* **71**, 41–42.
Hayden, T. J., Thomas, C. R. and Forsyth, I. A. (1979). *J. Dairy Sci.* **62**, 53–57.
Holtz, E. W., Murrill, F. D., Weisheit, H. A. and Oliver, J. C. (1973). *J. Dairy Sci.* **56**, 655.
Horák, F. (1971). *Chovatel* **10**, 162–164.
Horák, F. Pindák, A. (1969). *Acta Univ. Agric. Fac. Agron., Brno* **17**, 299–306.
Iloeje, M. U., Rounsaville, T. R., McDowell, R. E., Wiggans, G. R. and VanVleck, L. D. (1980). *J. Dairy Sci.* **63**, 1309–1316.
Junge, F. (1963). *Vet. Diss.* Univ. München.
Kalaissakis, P. (1959). *Z. Tierphysiol. Tierern. Futtermittelkde.* **14**, 303–311.
Klobasa, F. and Senft, B. (1970). *Milchwiss.* **25**, 453–456.
Knutson, P. G. and Sperber, I. (1961). *Kungl. Lantbrukshögsk. Ann.* **27**, 33–40.
Lampeter, W. (1970). *Vet Diss.* Univ. München.
Laor, M. (1971). "2nd Int. Conf. Goat Breeding", Tours, France, July 1971, pp. 295–301. ITOVIC, Paris.
Larson, B. L. (1978). *J. Dairy Sci.* **61**, 1023–1029.
Linzell, J. L. (1960). *J. Physiol. Lond.* **153**, 492–509.
Linzell, J. L. (1963). *Q. J. Exp. Physiol.* **48**, 34–60.
Linzell, J. L. (1966). *J. Dairy Sci.* **49**, 307–311.
Linzell, J. L. (1967a). *J. Physiol.* **190**, 333–346.
Linzell, J. L. (1967b). *J. Physiol.* **190**, 347–357.
Linzell, J. L. (1972). *Dairy Sci. Abstr.* **34**, 351–360.
Linzell, J. L. (1973a). *J. Physiol.* **230**, 225–233.
Linzell, J. L. (1973b). *In* "Production Disease in Farm Animals" (J. M. Payne, K. G. Hibbit, and B. F. Samson, eds), pp. 89–106. Ballière and Tindall, London.
Linzell, J. L. and Peaker, M. (1971a). *J. Physiol.* **216**, 717–734.
Linzell, J. L. and Peaker, M. (1971b). *Physiol. Rev.* **51**, 564.
Linzell, J. L. and Peaker, M. (1974). *J. Physiol. Lond.* **243**, 129–151.
Linzell, J. L., Fleet, I. R., Mepham, T. B. and Peaker, M. (1972). *Q. J. Exp. Physiol.* **57**, 139.
Louca, A., Mavrogenis, A. and Lawlor, M. J. (1975). *Anim. Prod.* **20**, 213–218.
Martinet, J. and Richard, Ph. (1973). "Symposium sur la Traite Mécanique des Petits Ruminants", Millau, France, May. *Ann. Zootech.*, numéro hors série, 1974, 29–49.
Mba, A. U., Boyo, B. S. and Oyenuga, V. A. (1975). *J. Dairy Sci.* **42**, 217–226.
McNeilly, A. S. (1972). *J. Endocr.* **52**, 177–188.
Metzger, C. (1964). *Schweiz. Arch. Tierheilkd.* **106**, 414–429.
Mittal, J. P. (1979). *Indian J. Anim. Sci.* **49**, 757–759.
Mocquot, J. C. and Auran, T. (1974). *Ann. Génét. Sél. Anim.* **6**, 463–476.
Montaldo, H., Juárez, A., Forat, M., Berruecos, J. M. and Villareal, M. (1978). Abstr. 70th Ann. Meet. ASAS, Mich. State Univ.

Morand-Fehr, P., de Simiane, M. (1977). "Symposium on Goat Breeding in Mediterranean Countries", Malaga-Granade-Murcia, Spain. pp. 101–145. EAAP and the Spanish National Committee for Animal Production, Madrid.
Naito, M., Shoda, Y. and Nagai, J. (1955). *Endocr. Jap.* **2**, 205–220.
Neseni, R. and Freimuth, U. (1952). *Milchwiss.* **7**, 113–120.
Nyland, J. (1936). *Meld. Norg. Landbr. Høgsk.* **16**, 1–115.
Opstvedt, J. (1969). *In* "Grassland in Sheep and Goat Production" (C. R. W. Spedding, ed.), pp. 89–100. European Association for Animal Production, Study Commission on Animal Feeding, Report No. 2.
Ormiston, E. E. and Gaines, W. L. (1944). *J. Dairy Sci.* **27**, 243–247.
Osterkorn, K. and Graf, F. (1970). *Arb. Inst. Tierzucht, Vererbungs- u. Konstitutionsforsch., Univ. München, Heft* **8**, 52–58.
Parkash, C., Khanna, R. S. (1972). *Indian J. Anim. Prod.* **2**, 10–13.
Patton, Stuart (1978). *J. Dairy Sci.* **61**, 643–650.
Peaker, M. (1977). *Symp. Zool. Soc. Lond.* **41**, 374 pp.
Peaker, M. and Linzell, J. L. (1974). *J. Endocr.* **61** 231–240.
Pearl, S. L., Downey, H. F. and Lepper, T. L. (1973). *J. Dairy Sci.* **56**, 1319–1323.
Perrin, D. R. (1958). *J. Dairy Res.* **25**, 215–220.
Peterson, V. E., Turner, C. W. (with Ralston, N. P.) (1939). *J. Nutr.* **17**, 293–301.
Prabhu, S. S. and Amble, V. N. (1946). *Indian J. Vet. Sci.* **16**, 212–222.
Rako, A. (1950). *Schweiz. Arch. Tierheilk.* **92**, 305–320.
Reynolds, M. (1967). *Am. J. Physiol.* **212**, 707–710.
Richardson, K. C. (1953). *J. Endocr.* **9**, 170–184.
Ricordeau, G. (1963). *Bull. Tech. Inf. Ing. Serv. Agric.* No. 179, 1–16.
Ricordeau, G., Bouillon, J. (1967). *Bull. Tech. Inf. Ing. Serv. Agric.* No. 218, 303–310.
Ricordeau, G. and Mocquot, G. (1967). *Ann. Zootech.* **16**, 165–181.
Ricordeau, G., Bouillon, J., Sanchez, F., Mocquot, J. C. and Jajous, A. (1979). *In* "V Jounées de Recherche Ovine et Caprine", pp. 403–426. INRA, ITOVIC, Paris.
Riggert, E.-H. (1978). *Vet. Diss.* Hannover.
Rønningen, K. (1964). *Meld. Norg. Landbr. Høgsk.* **43**, (18), 20 pp.
Rønningen, K. (1967a). *Meld. Norg. Landbr. Høgsk.* **46** (4), 23 pp.
Rønningen, K. (1967b). *Meld. Norg. Landbr. Høgsk.* **46** (11), 15 pp.
Rønningen, K., Gjedrem, T. (1966) *Meld. Norg. Landbr. Høgsk.* **45** (21), 17 pp.
Roguinsky, M. (1974). "Symposium sur la Traite Mécanique des Petits Ruminants", Millau, France. May. *Ann. Zootech.*, numéros hors série, pp. 117–118.
Sachdeva, K. K., Sengar, O. P. S., Singh, S. H. and Lindahl, I. L. (1974). *Milchwiss.* **29**, 471–475.
Sands, M., McDowell, R. E. (1978). Cornell Int. Agric. Mimeograph 60, Ithaca, New York, 53 pp.
Sauvant, D. and Morand-Fehr, P. (1978). *XX Int. Dairy Congr.* Vol. E Paris, 48–49.
Sauvant, D., Fehr, P. M., Hervieu, J. and Blanchard, D. (1976). 27th Annual Meeting, EAAP, Zürich, August, Sheep and Goat Commission.
Sauvant, D., Chilliard, Y., Bas, P. Morand-Fehr, P. (1979). *Ann. Rech. Vet.* **10**, 404.
Sigwald, J. P. and Lequenne, D. (1971). "2nd Int. Conf. Goat Breeding", Tours, France, July 1971, 75–83. ITOVIC, Paris.
Sigwald, J. P. and Lequenne, D. (1976). cit. Le Jaouen, J. C., XX Int. Dairy Congr. Paris 8EC.
Sigwald, J. P. and Sauvant, D. (1976). 27th Annual Meeting, EAAP, Zürich, August, Sheep and Goat Commission.
Sigwald, J. P. and Sauvant, D. (1977). "Symposium on Goat Breeding in Mediterranean Countries", Malaga-Granada-Murcia, Spain. pp. 372–379. EAAP

and the Spanish National Committee for Animal Production, Madrid.
Sjødin, E. (1976). 27th Annual Meeting, EAAP, Zürich, August, Sheep and Goat Commission.
Skjevdal, T. (1974a). *Tech. Bull.* **169**, Dept. Anim. Nutr. Agric. Univ. Norway, 44 pp.
Skjevdal, T. (1974b). *Sau og Geit* **27**, 82.
Skjevdal, T. (1974c). Mim. paper. **35**, Dept. Anim. Nutr. Univ. Norway, 30 pp.
Snipes, M. B. and Lengeman, F. W. (1970). *J. Dairy Sci.* **53**, 1596–1602.
Spedding, C. R. W. (1968). *Agric. Progress* **44**, 7–23.
Steine, T. A. (1975a). *Meld. Norg. Landbr. Høgsk.* **54**, 1–29.
Steine, T. A. (1975b). *Meld. Norg. Landbr. Høgsk.* **54**, 1–25.
Steine, T. A. (1976). *Meld. Norg. Landbr. Høgsk.* **55**, 1–19.
Sundaresan, D. (1978). XX International Dairy Congress, Paris, 1978, 6 EC.
Svoboda, T. and Bilek, J. (1976). *Vet. Med.* **21**, 399–406.
Taylor, St. C. S. and Fitzhugh jr., H. A. (1971). *J. Anim. Sci.* **33**, 726–731.
Thompson, G. E. and Thompson, E. M. (1977). *J. Physiol. Lond.* **272**, 187.
Ueckermann, L., Joubert, D. M. and Seyn, G. J. van D. (1974). *Wld Rev. Anim. Prod.* **10**, 73–83.
Welch, J. G., Clark, D. B. and Rutledge, J. H. (1975). *J. Anim. Sci.* **41**, 432.
Wiggans, G. R., Vleck, L. D. van and Dickinson, F. N. (1979). *J. Dairy Sci.* **62**, 797–801.
Yokoyama, A. and Ota, K. (1965). *J. Endocr.* **33**, 341–351.
Zaks, M. G. (1962). "The Motor Apparatus of the Mammary Gland." Oliver and Boyd, Edinburgh and London.

11. Milking and the Technology of Milk and Milk Products

J. C. Le Jaouen

A. Milking

Milking is one of the main activities of the whole dairy operation. Collecting the milk is the end point of all the efforts of the breeder in managing his dairy herd. Milking results will depend on udder conformation, the milk ejection reflex, milking methods, health status, quality of milking equipment etc. Milking goats is a traditional farm activity and it is very susceptible to the consequences of rural depopulation. With increasing manpower problems, mechanical milking of goats tend to become a necessity and all the related technical, economical, labour and social problems are gaining importance.

1. Anatomy of Mammary Gland

During pregnancy the mammary gland is prepared under humoral and neural influences for its natural function, nurturing the kids. Very early during the intra-uterine period, the para-axial mammary crests are organized but the mammary gland does not develop until puberty, when oestrogens act on the duct system. For the development of the alveoli the joint action of oestrogen and progesterone is necessary and, therefore, the final mammary gland is not formed before parturition. While some milk may be formed during the latter part of pregnancy, secretion of colostrum begins at parturition. In the course of a few days, colostrum changes to milk of normal composition. Milk secretion is initiated and maintained by prolactin, the specific lactogenic hormone of the pituitary, as well as by the complex of its other hormones (GH, ACTH, TSH etc.).

The udder of the goat is composed of two separate complexes, which are physiologically and pathologically independent. Milk yield normally differs between the two complexes but too great differences are considered as a defect, since milking time will differ too much. In a study on 100 goats of the Saanen, Toggenburg, Nubian and Alpine Breeds, Rusev *et al.* (1973) found an average index of symmetry of 40%. In 23% of the goats the two complexes produced identical quantities of milk.

The udder is an exocrine gland with the following tissues, from outside to inside:

—The common skin covering the whole udder; in dairy animals a fine and soft skin is preferred.
—Elastic-muscular tissue attaching the udder to the abdomen and separating the two complexes.
—Connective tissue, sometimes filled with fat. A good udder should not have too much of this tissue and present a "fleshy" appearance. This quality can be judged by the volume of the udder which after milking decreases markedly in good dairy animals. The empty udder should feel limp and soft.
—The glandular tissue with cells specialized to synthesize the milk constituents which line the walls of the alveoli and the milk ducts, forming a dense network leading to the cistern which holds the milk. A particularity of some goat udders is the large volume of this reservoir formed by the cistern and the teat which are anatomically less differentiated than in the cow. This explains why at the end of milking the glandular part of the udder remains firm, while the cistern–teat part collapses and becomes limp.
—Arteries and veins which lead the blood to the udder. Intensive vascularization of the udder is essential since 425 litres of blood have to pass through the udder for the secretion of 1 litre of milk (Linzell, 1966).

The different forms of goat udders may be grouped into three classes:

—the lengthy pear shaped udder with teats little distinct from the glandular part (Fig. 1a). This type occurs in good dairy animals but is not preferred by breeders because the teat cups are difficult to attach to the pendulous teats (although they facilitate hand milking); on pasture this type of udder is inconvenient for the goat and is susceptible to injury; furthermore it tends to break down early;
—the oval-shaped udder (Alpine type) well attached to the abdomen with voluminous teats well separated from the glandular part and slightly inclined forward (Fig. 1b);
—the globular udder (Saanen type) attached high at the abdomen, sometimes as wide as high with teats smaller than the other types (Fig. 1c).

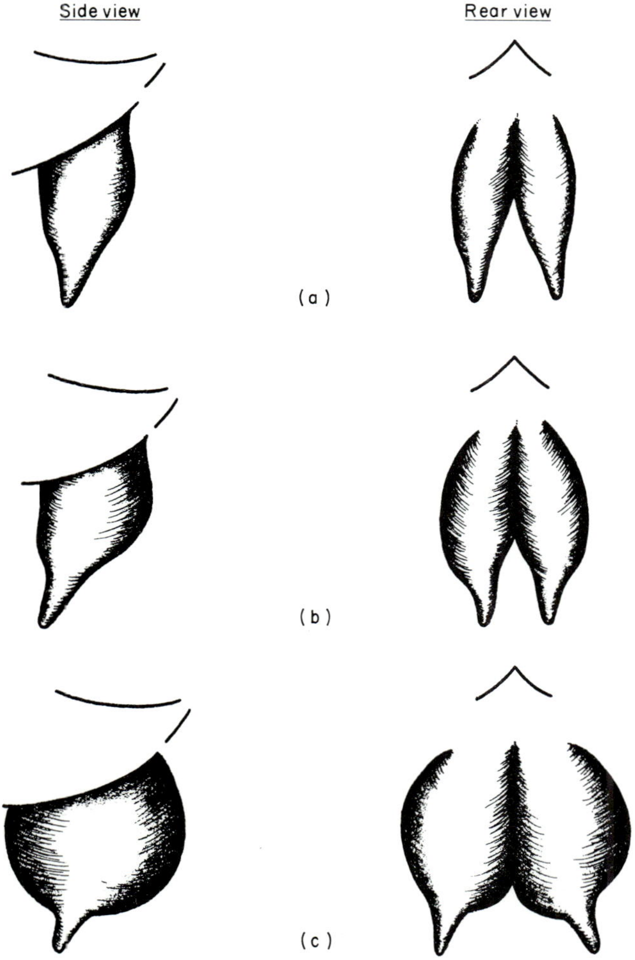

Fig. 1. Forms of goat udders, seen from the side and from behind. (a) Pear shape: typical goat udder; teats are not delineated but narrowing downwards. The combined gland and teat cistern contains about 80% of the milkable milk. (b) Oval shape: well balanced goat udder; teats are voluminous but well separated from the glandular part. (c) Globular shape: this type is very firm and similar to ewes' udders. It is less attached to the abdomen. The volume of cistern volume is smaller than in (a) and (b).

Irrespective of milk yield, the last two types have the best aptitude for machine milking.

2. Physiology of Milk Withdrawal

There are two phenomena interacting with the extraction of milk, contained in the mammary gland, for nursing the young or for use by man and which occur simultaneously or with a slight time delay: (i) the flow of cisternal milk when the teat sphincter opens under pressure of the hand of the milker, suckling, or vacuum of the mechanical milker; and (ii) the stimulation of the milk ejection reflex for collecting the milk from the glandular tissue and the milk ducts.

The milk ejection phenomenon is fairly well known in the goat since it has been used as an experimental animal (Martinet and Richard, 1973). The normal neuro-humoral reflex is initiated with the perception by the animal of a stimulus and the conduction of this information to the higher nervous centres. The stimulus originates mainly from the different manipulations of the udder preceding hand milking or attachment of the teat cups but the most effective stimulus is the suckling of the kid. However, acoustic, visual and olfactory stimuli are possible, such as the arrival of the milker or the noise of the milking machine. Furthermore, the milk ejection reflex is related with the rumination reflex; hence, goats on the milking platform are frequently seen quietly ruminating.

The stimuli received by the mammary nerves are transmitted by ascending nervous paths and the hypothalamo-posthypophysial tract to the posterior pituitary gland, liberating oxytocin which is transported with the bloodstream to the mammary gland. Under normal conditions and in the absence of any disturbance of the animal, oxytocin reaches the mammary gland within on average 24·5 seconds. (Martinet and Denamur, 1960). Oxytocin causes the myoepithelial cells surrounding the alveoli to contract and force the milk into the small and large milk ducts. The movement of milk into the cistern is facilitated by a widening of the diameter of the milk ducts.

Milk ejection under the action of oxytocin can be inhibited by interference of the sympathetic nerve system which leads to adrenaline discharge. Adrenaline discharge into the bloodstream during aggression, fright or a sudden disturbance of the environment of the dairy animal, results in a poor milk ejection reflex, or in cases of total suppression, in incomplete milking which in the long term will reduce milk yield (Labussière, 1966).

The volume of cisternal milk, particularly of dairy breeds, represents a greater proportion of the total milk than in cows. The effectiveness of the milk ejection reflex will determine which part of the alveolar milk can be extracted at milking; it depends on conditions at milking which the milker may determine. In addition to the milker and the milking machine, the goat is actively involved in the process of milking since the milk ejection reflex is necessary for a correct withdrawal of milk. While the goat is milked more easily than the cow because the proportion of alveolar milk is smaller and therefore the stimulation of the milk ejection reflex by washing and massage of the udder is less important, the reflex may be easily disturbed and the animals need suitable environmental conditions (calm, regularity, habits). Finally the ejection reflex has to be strong in order to discharge as much oxytocin as possible, since, in addition to ejecting the alveolar milk, it stimulates the mammary gland and enhances milk yield.

3. Aptitude for Machine Milking

The goat is milked more easily than the cow and the ewe because more of the milk contained in the udder is held in the cistern and therefore available immediately. Since most dairy goat herds in the world are hand milked, with labour readily available at low cost, few studies have been concerned with milking aptitude and milking methods of goats. Yet, milkability is as important for the hand milker as for the milker using a machine. The common term "milkability" actually embraces quite different characteristics such as: time needed for milking the goat (speed of milking), average milk flow rate, peak milk flow, equilibrium between halves, size, form and position of the teats, elasticity of the teat orifice and volume of strippings.

(a) *Milking Speed*

The time recordings in Table II show great variability in average time (150 to 468 centiminutes) needed to milk one goat, as measured between attaching and detaching the teat cups which occasionally includes time of overmilking. These time recordings, taken under practical farm conditions, were confirmed by recordings in an experimental herd where the average milking time per goat varied between 35 and 300 centiminutes (Bouillon and Ricordeau, 1970). In 230 Norwegian goats of 2 years and older, the average time of teat cups attached was 2 minutes. It increased by 150 centiminutes with 1 kg increase in milk yield (Skjevdal 1975).

Bouillon and Ricordeau (1970) found a correlation between milking time and milk yield of 0·62. Hence, there would be little possibility of selecting a "standard" goat with equal milking speed and adjusted to a given milking system.

(b) *Milk Flow Rate*

The precision of measuring milk flow rate depends on that of measuring milking time. With machine milking, Bouillon and Ricordeau (1970) distinguish between the true machine milk flow (milk quantity obtained with the machine/machine milking time) and approximative machine milk flow (AMF, total milk quantity/machine milking time); they found a correlation of 0·84 between these two rates. AMF varied between 0·9 and 1·3 litres per minute, increasing with milk yield (Fig. 2). Therefore, it takes more time to milk 1 litre of milk in low producers than in high producers. Apparently,

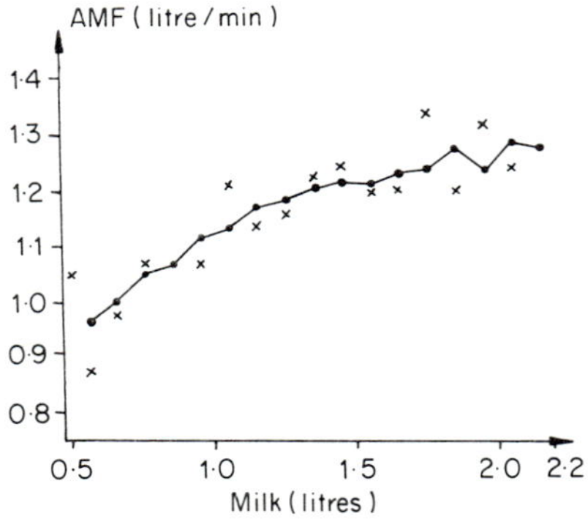

Fig. 2. Relation between milk yield (litre) and approximate machine milk flow (AMF) in litre/min (Bouillon and Ricordeau, 1970). Three measurements in different stages of lactation; 150 Saanen goats, daughters of 10 sires.

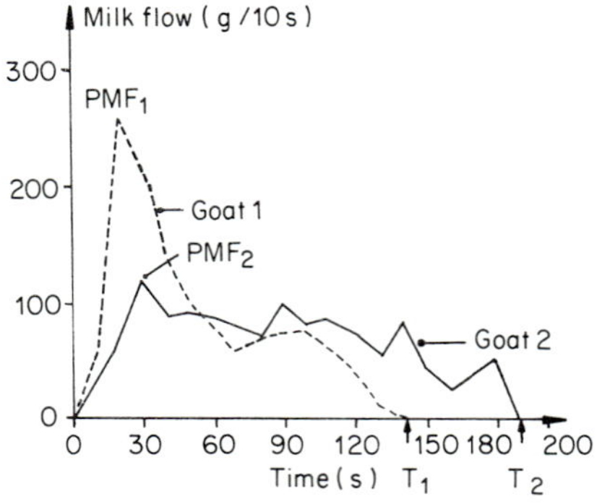

Fig. 3. Relations between milking time and milk flow rate. Goat no. 1 has a high peak flow (PMF) and short milking time (T), goat no. 2 has a low peak and long milking time (Bouillon and Ricordeau, 1970).

milk flow increases proportionally with yield until a ceiling where the opening of the teat sphincter limits milk flow.

Figure 3 shows that a goat with a high peak milk flow rate (PMF) (goat No. 1) was milked faster than a goat with a lower PMF (goat No. 2). Rusev *et al.* (1973) in four breeds (Saanen, Toggenburg, Anglo Nubian, Alpine) observed a relatively slow milk flow (0·315 kg/min) although due to low yield (81% of the animals with less than 300 g milk per milking), milking time was short (Table I).

Table I. Machine milking performance of goats in Cuba, 100 goats, 150 days of lactation (Rusev *et al.*, 1973).

	n	x ± s	S	CV	Range
Milking time (min)	100	2′33″ ± 0·051	0·507	21·79	0′25″–3′00
Milk flow (kg/min)	100	0·315 ± 0·018	0·182	57·77	0·100–0·600
Peak milk flow (kg/min)	100	0·430 ± 0·020	0·201	46·74	0·100–0·700
Udder index (%)	100	40 ± 1·120	11·20	28·00	15·00–50·00
Proportion of milk milked during the first minute (%)	100	66·90 ± 2·941	29·40	43·94	22·22–100·00
Proportion of milk milked during the second minute (%)	100	97·90 ± 1·464	14·66	14·88	77·27–100·00
Milk yield on day of measurement (kg)	100	1·34 ± 0·060	0·648	48·35	0·400–4·600

x ± s: standard deviation
S: variance
CV: coefficient of variation (%)

4. Machine Milking

Since one milker can hand milk only 30 to 80 goats, depending on their milk yield, increasing herd sizes have forced breeders to switch to mechanical milking. The necessity of providing adequate labour conditions and of reducing labour time have added to this. Machine milking of goats has been established on the basis of knowledge acquired with cows. Very few studies have been dedicated to the anatomical and physiological particularities of goats which would have facilitated the construction, from the beginning, of equipment correctly adapted to the requirements of goats.

Furthermore, the manufacturers of milking machines were not motivated to invest in research since the market was very limited. Instead they have tried to gradually adjust bovine equipment and their functional parameters. French breeders assisted by specialist technicians have over 15 years been able to try out the practical applications of improvements introduced by manufacturers on about 4500 machines. A similar development took place in other countries where goats are being machine-milked, in particular Norway and the USA, although the number of installations there is smaller. Today, with experience in practical field-work, we consider goat milking machines to be serviceable. If correctly used, they give satisfactory results but there certainly is an appreciable margin for improving efficiency. However, there are limits to technical perfection; as a matter of fact it is better to milk correctly with an imperfect machine than milk incorrectly with a perfect machine. Little time may be saved by machine milking instead of hand milking in small or medium size herds of 30 to 60 goats. But investment in mechanization may be justified because of manpower shortage and cost, as well as the increasing requirements for acceptable working conditions and social standing of milkers.

Objectives of good machine milking are:

—to milk completely, avoiding stress to the animal, where complete does not mean extraction of the last drop of milk in the udder, but maximum milk compatible with the performance of available equipment and labour costs;
—to limit the incidence of mastitis by avoiding overmilking and excessive vacuum levels;
—to enable production of clean milk meeting the bacteriological requirements. There are no great differences between cows and goats except that it is not vital to wash the udder before milking unless it is really dirty;
—to facilitate the work of the milker by using solid and reliable equipment, which is easy to handle and needs a minimum of human interference.

Purchase prices of milking machines, running and maintenance cost should be compatible with the financial capacity of goat operations. This aspect, which is essential for the breeders' decisions, is too often neglected in technical and scientific discussions.

(a) *The Milking Machine*

The concept and installation of goat milking machines follow the same principles as those for cows. Most of the components of systems produced for milking cows can be used:

motor and vacuum pump;
vacuum reserve tank;
vacuum gauge;
vacuum (pulsator) line and plugs;
milk (sanitary) pumps;
milk bucket;

for pipeline installations:
sanitary trap;
milk receiver;
releaser;
milk line;
delivery pipeline.

By contrast, some elements have to be specifically designed for goat machines:

milk bucket lid;
claw (collector) assembly;
teat cups and liner;
vacuum regulator;
pulsator.

The capacity of vacuum pumps has in the past been increased in order to reduce milking time and to avoid falling off of clusters. Most recent recommentations for air flow are:

bucket milker: 50 litres of air/minute plus 80 litres per bucket;
pipeline plants: 150 litres plus 80 litres per cluster, with a maximum of 12 clusters operated.

These high vacuum reserves are necessary since, with goats, clusters are attached and detached more frequently and air is easily admitted at the end of milking, due to collapsing teats.

The vacuum level of 38 to 44 kPa (280 to 330 mmHg) used with goats is lower than that applied to cows (50 kPa or 380 mmHg). This level is sufficient for extracting the milk because of the high proportion of cisternal milk, but there is less risk of tissue damage. However, at this low level problems occur with pneumatic pulsators. These may be overcome by changing to electric pulsators or establishing an additional vacuum supply to the pulsators at 50 kPa (380 mmHg). Pulsation ratio, i.e. the ratio of time when vacuum and atmospheric pressure are admitted to the chamber between teat cup liner and shell, varies between 50:50 and 70:30. Pulsation rate, i.e. the frequency of pulsation cycles per minute, varies between 70 and 90 and is much faster than with cows. The pulsation may be applied to both teats, alternately or simultaneously. Pulsators are pneumatic or electric.

Collectors (claws) should have sufficient capacity (interior volume) to ensure that milk flows rapidly, so that milking is fast and reflux, which may spread microorganisms, is avoided.

Teat cups are smaller and liners have a narrower bore than those for cows. While different forms of liners exist (cylindrical, conical, different lengths), results of recent investigations suggest that these differences do not have an influence on quantity of machine-milked milk, quantity of strippings and milking time.

Adequate maintenance of the milking plant and training of the milker are prerequisites for efficient machine milking. With regular checks, defects in the equipment may be detected, defective parts replaced, vacuum level and pulsation rate adjusted, advice given for proper maintenance, and thus milking performance may be improved. Observations on 230 installations in France revealed (Darracq and Le Mens, 1978) that:

—one out of four vacuum pumps was insufficient;
—more than one out of three vacuum regulators had irregular leaks;
—more than one out of two vacuum gauges was incorrect;
—more than one out of ten pulsators worked at the wrong ratio;
—in one out of two plants pulsators differed in pulsation rate and ratio;
—in more than one out of three milking units there were serious leaks.

Breeders observed after the check and regulation of their plants:

—improved performance of the plant (73%);
—better knowledge of their equipment (68%);
—better use of their equipment (74%);
—improved milking conditions (54%);
—less irritations and mastitis (24%).

5. Milking Systems

The milking system is determined by the type of milking plant (bucket milker, pipeline) and the factors which influence labour organization: kind and number of milking platforms, stalls, and units; number of animals to be milked in a group, type of animal control, means of driving and handling the animals, organization of all work ancillary to milking. Many milking systems have been devised, some inspired directly by concepts of cow milking, others resulting from ideas and practical observations of goat breeders. Following a period (1965–1975) during which numerous systems were tested, the most efficient can be singled out. They are adapted to the structure of goat operations and the milking characteristics of goats, in particular their behaviour. Three groups of milking systems may be distinguished (Le Mens *et al.*, 1978).

(a) *Milking in the Barn*

In general, an elevated floor is established along the stanchion in the barn where the animals can be secured, clear of the bedding. Pipeline or bucket-type plants may be used and the milker works his way behind the goats.

No feed is offered during milking and the goats do not have to be moved (Fig. 4a).

(b) *Milking Parlours*

(i) Simple or double platform—milking from behind: goats enter the parlour by one door and leave by another and there is a continuous flow of animal movement. Goats climb the platform, usually made of wood, where they are secured in a stanchion facing a trough for feeding concentrates. In order to facilitate the goats' entrance and departure, the platform may be enlarged by various means (hinged board, sliding board, mobile stanchion). With this arrangement goats can choose their place; milk units are attached from behind (Fig. 4b).

(ii) Simple or double platform—milking at the side: the system is similar to (i) with the following differences. The platform is narrow and goats are forced to take an oblique position. The exit door is arranged in the same oblique way, facilitating correct positioning of the animals. The goats are subject to a stricter order, because the animal entering the parlour first is obliged to take the stand near the exit door. The milk unit is attached sideways (Fig. 4c).

(iii) Milking platform—milking in front: platform and stanchion are the same as before, the difference is that the animals face the milker. By this arrangement, concentrates can be distributed directly from the corridor or the pit of the parlour, obviating the milker's need to go around the platform. The milker is positioned in front of the goats, attaching the milking units between the front feet of the animals.

(iv) Milking platform—milking at the side (tunnel) (Fig. 5): the platform has no stanchion, trough or individual stalls. The goats are not tied but stand one behind the other in a narrow tunnel with side openings. The number of goats is limited to two platforms with four to five goats, because with more animals the platform would become too long. The two platforms are used alternately; the units are attached at the side of the goats (Fig. 4d).

(c) *Rotary Milker*

The platform is a turntable with stalls in which the goats are not tied. They may or may not be offered concentrates, and the platform may be turned by hand or mechanically. The milker may be placed inside a hollow ring—or outside a solid platform; in both cases the units are attached at the side (Fig. 4e).

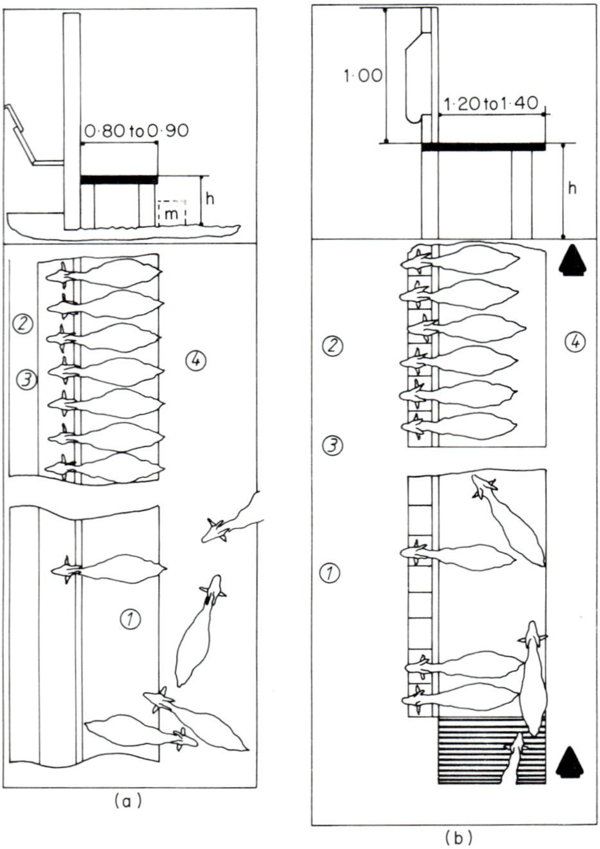

Fig. 4. Milking systems. (a) Milking in the barn; (b) platform, milking from behind; (c) platform, milking at the side; (d) tunnel, milking at the side; (e) rotary milker, milking outside. (1) Goats entering; (2) goats being milked; (3) feeding concentrates; (4) position of the milker.

6. Machine Milking *vs* Hand Milking

Time registrations in Table II reveal that with hand milking 161 to 267 centiminutes are needed per goat, excluding time for moving animals and feeding concentrates. Milking goats fixed in the stanchion within the barn takes 53 to 128 centiminutes per goat. This method has a low labour cost and may serve as a baseline for evaluating machine milking operations. Poor design or incorrect use of milking equipment can considerably reduce the advantages of machine milking over hand milking and in extreme cases these may be lost entirely. The observations indicate the minimum herd size that would justify machine milking. Furthermore they indicate that between 22

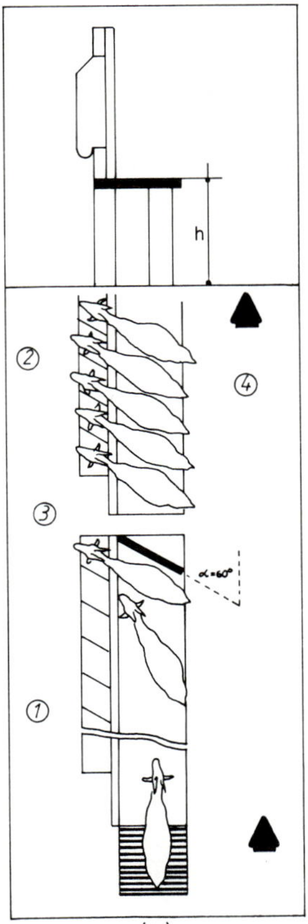

(c)

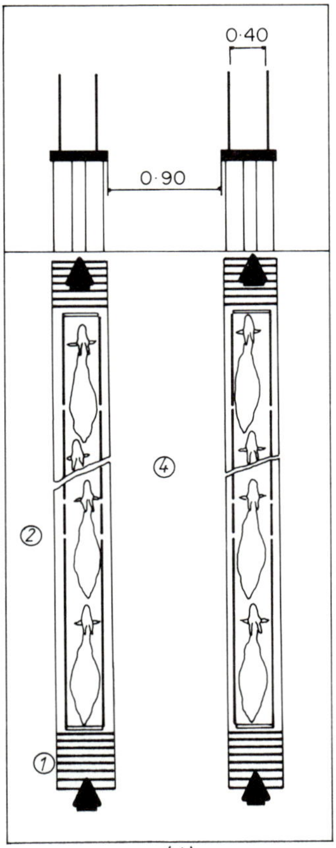

(d)

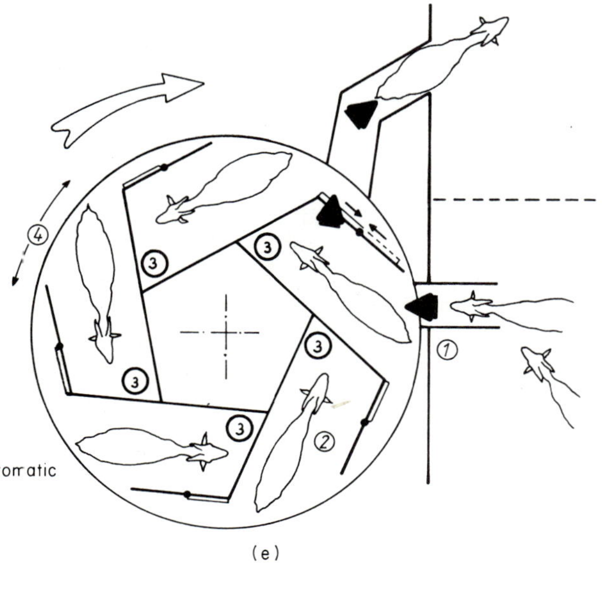

③ Automatic

(e)

and 37 goats, with average milk yields of 1 to 1·6 litres in one milking, are milked by one milker per hour, i.e. a maximum of 60 to 80 goats by two milkers. It follows that the highest labour efficiency is obtained in the barn and in double tunnels and that performance with the rotary milker differs little from that on a platform. However, using the data in the table, a direct comparison of the different systems is not possible because their efficiency depends on many quite distinct factors: number of units per milker; time needed for milking each goat; kind, duration and organization of the milking routine as carried out by the operator. It appears that labour cost of goat milking and labour organization can still be much improved and that further efforts will have to be made to fulfil the social aspirations of the people who work in this field.

7. Suppression of Stripping

Time taken for stripping in both hand and machine milking is 22% of the total milking time. It is, therefore, important to see whether the extra labour involved in the different manipulations of stripping (massage, pulling down the teat cups, stripping into the pail or the teat cups) is justified. In experiments with suppressing stripping during 4 months in Alpine and Saanen

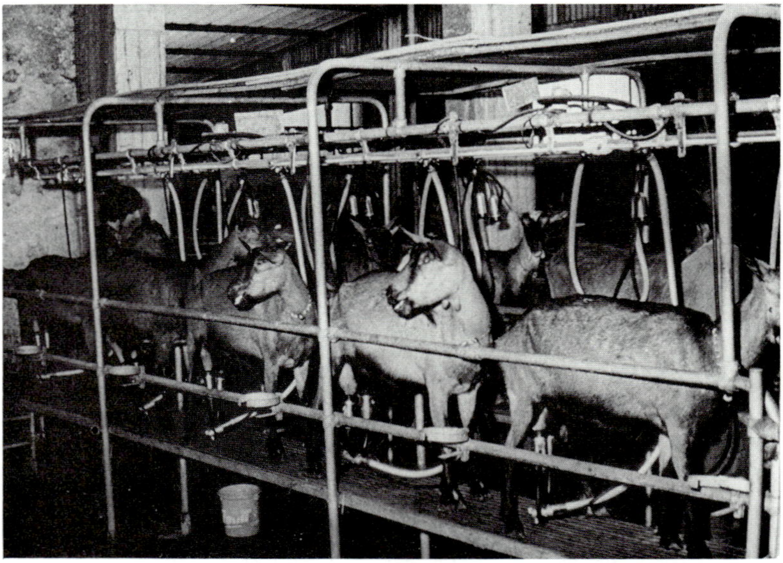

Fig. 5. Milking "tunnel", 2 × 5 goats. The animals stand free and are milked from the side; the milker works between the two platforms.

Table II. Labour efficiency in different milking systems.

	Stalls	Units	Units/ milker	Labour time/ goat (cmin)	Average milking time/goat (cmin)[a]
Hand milking	—	—	—	267	224
	—	—	—	262	220
	—	—	—	194	133
	—	—	—	161	126
Barn	—	2	2	102	165
	—	2	2	107	171
	—	2	3	171	246
	—	4	4	59	150
	—	6	6	75	254
Simple platform	8	4	4	136	330
	8	4	4	97	226
	15	4	2	256	364
	16	4	3·2	162	283
	24	6	2	152	313
Double platform	20	4	2	263	307
	28	14	7	150	253
	28	14	7	83	193
	50	6	2·5	176	202
Simple bail	10	6	6	125	324
Herringbone (1 × 12)	12	12	4·3	265	254
Tunnel, double	8	8	8	83	229
	8	8	8	69	245
Rotary milker "internal"	8	8	4	104	309
	8	8	8	77	404
	18	9	4·5	135	488
Rotary milker "external"	5	5	2·5	113	—
	5	5	2·5	118	—

[a] In machine milking, time with teat cup assembly attached; it may include time of overmilking.

goats, milk yield was reduced compared to controls by 3·4% without the appearance of mastitis (Le Mens, 1978). While strippings have a higher fat content, the total utilizable dry matter (protein and fat) did not differ between goats stripped and not stripped.

Strippings have about the same volume at morning and evening milking, but represent 8–10 and 18% respectively of total milk quantity. Volume of strippings decreased during the first 3 months of lactation and remained constant thereafter (Ricordeau and Labussière, 1970). Individual variation was high and repeatability low. Volume of stripping depended on the pulsation ratio, increasing by 24 to 30% when pulsation ratio changed from 1:1 to 3:1.

8. Suppression of one Milking a Day

Breeders are interested, mainly for social reasons, in omitting one milking, notably that of Sunday afternoon, or milking only once a day over part or the whole of lactation. The goat reacts against such a milking regime more like the cow than the dairy ewe (Mocquot, 1978). By milking goats over the whole lactation in the morning only, the quantity of milk, fat and protein, and lactation length were reduced by 35%, 37%, 33% and 16 days, respectively. Suppressing milking on Sunday evening only reduced milk yield by 5% but milk composition changed and did not return to normal values until Thursday morning. These results, obtained in an experimental herd, were confirmed under farm conditions with an average lactation yield of 553 kg (Le Mens, 1978). Milking only once on Sundays over the whole lactation reduced milk yield by 4·47%, but only by 1·22% if this regime was started after the fifth month of lactation.

9. Labour Requirements

The time needed for milking is related not only to the organization of the milking system but also to milking speed. Consequently all efforts to reduce milking time have to take into consideration the physiological mechanisms involved in the milking process, as well as practical and economic factors, comparing for example savings of labour time with losses of milk. With hand milking, a reduction in the average time needed to milk one goat will reduce the time needed for milking the herd, or else reduce the labour necessary to milk a given number of goats per hour. Apparently, milking speed differs with milk yield and the milker's skill. In France, with herds of 10 to 48 Alpine and Saanen goats, time needed per goat for hand milking and the number of goats milked per man hour increased from 133 to 249 centiminutes and from 22 to 37 respectively, when milk yields increased from 1·25 to 1·5 litres per goat (Toussaint, 1979).

The speed of milking by machine essentially depends on the characteristics of the system, the milking method and labour organization. It depends very little on the time needed for milking the individual goat (milking speed) because, with adequate equipment, one milker can simultaneously use several (2 to 10) units. In France, a milker uses 2 to 8 units, depending on the type of installation, and with an average daily milk yield of 2·2 to 3·1 kg the number of goats milked per man hour varies between 35 and 102 (average about 50). Le Mens *et al.* (1978) studied milking time in 5 hand-milking and 26 machine-milking operations in France, by measuring the time elements of the operation. In addition to overall time data, this method allows an analysis of the elements responsible for differences in milking time and simulation of operations with different milking speed of does, different work routine of the milker and different ways of organizing the work following milking as such.

In a study in Cuba, a milking plant designed for milking ewes was adapted to goats. It had 16 stalls and a device for the milker to move behind the goats on a stool sliding on a rail (Rusev and Diaz, 1973). With three different techniques, using four to eight units, the milker could milk 30 to 56 goats per hour, depending on the assistance with moving the animals. In Norway, in a herd of 200 does, milking in a parlour by two operators with four units each (Skjevdal, 1976) was not faster than the common milking in pens, where each milker operates only two units (Hilling, 1974).

B. Technology of Milk and Milk Products

1. The Place of Goats' Milk and Milk Products in the World

Since ancient times the goat has always been exploited for its milk, irrespective of the systems of goat production. There are many historical references to the place goats held in tribal life, and according to Greek mythology Zeus himself was nourished with milk of the goat Amalthaea.

World goat milk production of 7236 tons contributes only 1·6% to world milk production and ranks fourth after cows (419464 tons or 91·2%), buffaloes and ewes with 5·6 and 1·6% respectively. There are various reasons for this modest contribution:

—in most countries, goat production is oriented essentially towards meat, and milk is used only for rearing kids;

—since the level of milk production is closely related to the level of feeding and since goats are mainly found in poor zones with limited forage resources, milk yields in these unfavourable zones are low (50–150 kg/year). Furthermore in arid zones, lactation length is considerably reduced by the onset of the dry season;

—finally, the neglected status of the goat in many countries has deprived goat breeders of the benefits of research and development, in contrast to breeders of dairy cattle. Thus, goat breeders are left to manage their herds with archaic and minimal productive methods and, as a consequence, at a low level of milk production, i.e. under-exploiting the real dairy potential.

Table III indicates that Asia and Africa are the two main goat continents

Table III. Changes of goat numbers and goats' milk production between 1961–65 and 1978 (FAO, 1978).

	Goat numbers (1000 head)				Milk production (1000 tons)			
	1961–65	1978	%	Change	1961–65	1978	%	Change
Africa	109 933	143 250	33	+30%	1 101	1 413	20	+28%
Asia	206 887	244 511	56	+18%	2 982	3 217	46	+ 8%
Europe	14 437	11 466	3	−20%	1 847	1 499	22	−19%
North and Central America	15 127	11 979	3	−20%	192	308	4	+60%
South America	26 554	18 386	4	−30%	167	132	2	−21%
Oceania	255	184	—	−27%	—	—	—	—
USSR	6 422	5 586	1	−14%	788	400	6	−49%
World	379 616	435 362	100	+14%	7 077	6 969	100	+1·5%

with 56 and 33% of the world goat population. During the last 15 years their populations have increased considerably (18 and 30%), while they decreased in the other continents.

In Africa milk production increases at nearly the same rate as goat numbers in Asia. Europe contributes 22% of world goat milk production but has only 3% of the world's goats; this reflects the dairy specialization of the European goat herds. While in Northern Europe, with the exception of Norway, goat populations tend to disappear, they are stable or even increasing in Mediterranean countries (Greece, Spain, France, Italy; Charlet and Le Jaouen, 1977). Likewise, in these Mediterranean countries milk production increases both through increased goat numbers and increased productivity. In France, between 1960 and 1977, average milk yield per goat increased from 264 to 506 kg.

In many countries goats are kept as dual purpose animals for supplying meat and milk to the family. Very little is known about the contribution to the subsistence economy of goat milk and milk products. With the exception of a few countries, statistics either do not exist or are unreliable. This is explained by the general disinterest in goat production and the fact that many of the goats are kept in countries which lack the means for appropriate investigations. There are, however, some countries with specialized goat dairy operations where meat is only a by-product.

Two types of countries can be distinguished:

(i) countries with difficult climatic and agricultural conditions where goat-keeping is a tradition and goats' milk is a vital component of nutrition in the subsistence system. Very little milk—fresh or processed—is marketed. Marketing is not organized.

(ii) countries which have achieved a certain level of development and where goat milk diversifies dairy products. Choice, flavour and specific qualities of goat products predominate over the strictly nutritional aspects.

These two basic situations may be illustrated by the following examples.

In the ASEAN region (Indonesia, Malaysia, Philippines, Thailand and Singapore) goats are kept mainly for meat (Devendra, 1979). Although milk production is not recorded by official statistics, significant amounts are produced by small farmers. Its contribution of animal protein to the diet of the generally poor peasant is important, particularly since cows' milk is not always available or only at high prices.

In Iran, milk is consumed liquid by the nomads and semi-nomads but if they produce milk near a settlement they may sell some of it either fresh or processed into cheese, food casein, butter or sour milk products (Sattari, 1973).

In Greece, 409 000 tons of milk are produced annually by 4·5 million goats, of which 76% is produced in small family herds with 1 to 4 head, and 24 % in big herds in the mountains with 20 to 100 head. About one-quarter of the milk is used fresh and three-quarters are processed to soft cheese (32 000 tons), hard cheese (6 500 tons), fresh cheese (2 500 tons) and butter (1 000 tons) (Boyazoglu and Zervas, 1977).

France belongs to the second class of countries with a specialized dairy goat and cheese industry. On average 85% of the income from goats stems from milk and 15% from meat sales. Only 5% of the goats' milk is used for home consumption and 15% for feeding kids while 80% is used for processing cheese. The production of cheese from goats' milk with 35 000 tons ranks below that of cows' (1 million tons) and above ewes' milk (18 000 tons). There are 145 dairies/cheese factories of which the larger ones process 10 to 20 million litres per year. Industrial milk collection increased to a present 170 million litres per year, and cheese production to 23 000 tons. Besides, 12 000 tons of cheese are still processed by the producers themselves who produce about 70 different varieties of soft cheese.

The development and specialization of the goat industry was possible because prices for milk and cheese are higher than those for cow products, and because there is a traditional market. Following the industrialization of cheese-making, sales were expanded by a wider product distribution through the great market channels.

From the observations of these different situations we may conclude that in the few countries where a goat dairy industry exists, its development is based on consumers' preference for goat milk products which creates a specific market. Preconditions for developing a specialized intensive goat dairy industry, which can compete with bovine products are:

—a specific market, where industrial quality products fetch a price higher than bovine products;
—a volume of production sufficient to feed processing plants at conditions which ensure sufficient rentability;
—a sufficiently regular supply of the dairies throughout the year, despite the seasonality of milk production. Otherwise, high investments in installations for costly techniques, such as conservation of frozen curd, would put a heavy burden on the manufacturing price of the final product.

2. Milk Composition

Chemical composition and physical characteristics of milk determine its nutritive value and its suitability for the processing of different dairy products. Milk composition depends largely on breed, milk yield, stage of lactation, level and quality of feeding. Many publications deal with the composition of goats' milk but results are often not comparable, because they are based on small numbers of animals, sometimes of different breeds; levels of production and the methods of sampling and analysis are often not mentioned.

Like milk from other species, goat milk is a chemically and physically complex compound. Milk is an emulsion of fat in a watery solution, containing various elements, some of which are dissolved (lactose, salts), some in colloidal form (nitrogen compounds). In this chapter we shall not compile all the details of the knowledge acquired about goats' milk but mention only the principal characteristics and the important differences from milk of other species.

(a) *Physical Characteristics*

Fresh normal goats' milk, hygienically collected and conserved, does not smell any stronger than cows' milk. However, if produced by goats with a low yield, which are fed odoriferous feeds or housed in strong smelling barns, it may have a characteristic goat taste. This taste is stronger when subsequent conservation methods are poor.

Natural acidity of goats' milk is slightly lower than that of cows' milk, with 14° Dornic *vs* 16° or pH 6·4 *vs* 6·7. This difference is probably attributable to the acidity of proteins, casein in particular. Specific density is closely related to milk composition. Values vary between 1·026 and 1·042, and are comparable to those of cows' milk. Most authors, who have examined the freezing point, report values around −0·58°C, slightly lower than cows' milk.

Finally, regarding other specific characteristics such as viscosity, surface tension, electric conductivity or refraction index, it seems that goats' milk differs from cows', buffaloes' and ewes' milk but values reported in the literature are not sufficient for generalizations (Parkash and Jenness, 1968).

(b) *Chemical Composition*

Parkash and Jenness (1968) have compiled data on the composition of goats' milk. The average composition of 1 litre of goats' milk as analysed in bulk samples of French goats in full lactation with an average production of 600 kg per year is:

			g
Total weight			1030
water			914
total dry matter 116 g	fat		34
	fat free dry matter 82 g	lactose	45
		nitrogen compounds	30
		minerals	7

Table IV. Composition of milk of different species (Le Jaouen, 1974).

	Yield	Dry matter	Lactose	Nitrogen	Fat	Salt
	kg	g/kg				
Cow	3 500–5 000	115–130	45–50	30–35	35–40	7–9
Goat	500–1 000	115–130	40–50	28–35	30–38	7–9
Ewe	100–150	160–254	52–55	45–75	55–110	8–14
Mare	–	95–100	60–65	15–20	9–15	3–4

Other milk constituents are biocatalysts (pigments, enzymes, vitamins) and dissolved gases (carbon dioxide, oxygen, nitrogen). Table IV shows how the composition of goats' milk compares with that of other species. The four main components of the dry matter are lactose, fat, nitrogen compounds and minerals; vitamins have a particular dietetic importance.

(i) Lactose is the main constituent of the dry matter. It is a reducing carbohydrate; it is present in its liquid phase and belongs to the group of diholosids. It is formed by coupling of one molecule α- or β-glucose to one molecule of β-galactose. Lactose of goats' milk does not differ from that of cows' milk and its content varies very little during the course of lactation.

(ii) Fat is composed essentially of glycerides and steroids (99%). The fat forms globules, which are suspended in milk as an emulsion. Electron microscopic examination reveals that fat globules have diameters between

1 and 10 mm, as in cows' milk. However, goats' milk has a higher percentage of small fat globules. The percentage of globules of less than 1·5 mm in goats' and cows' milk is 28 and 10 respectively. This characteristic seems to be partly responsible for better digestibility of goats' milk.

Fat has largely the same composition of fatty acids as in cows' milk. However, the proportion of fatty acids is different. Goats' milk is rich in short chain fatty acids (C4 to C10) which represent 15% of all fatty acids as compared with 9% in cows' milk (Table V). At the beginning of lactation, when the doe draws on her body fat, the fatty acid composition of milk fat is different: the proportion of $C_8:0$ to $C_{14}:0$ acids is lower, that of $C_{18}:0$ and $C_{18}:1$ acids is higher. The three volatile fatty acids, caproic, caprilic and capric acid, seem to give goats' milk products their specific flavour.

Table V. Pattern of fatty acids in milk fat of different species (gas-chromatographic determination, percent of total methyl esters) (Allais, 1974).

Species	C_4	C_6	C_8	C_{10}	C_{12}
Cow	1·4	2·2	1·8	3·6	4·0
Ewe	1·1	2·7	3·3	7·6	5·5
Goat	0·7	2·4	3·2	8·7	4·7
Female	0·1	0·2	0·3	2·1	7·2
	C_{14}	C_{16}	C_{18}	$C_{18:1}$	$C_{18:2}$
Cow	13·0	30·2	13·7	27·1	3·0
Ewe	14·1	28·1	11·8	22·7	3·1
Goat	10·7	28·5	13·0	25·2	2·9
Female	10·9	29·6	7·3	35·4	6·7

A characteristic of goats' milk is that it does not contain any carotene. The absence of this chlorophyll-like pigment, which gives cows' milk its yellow colour, is responsible for the whitish aspect of goats' milk and its products.

Fat is the most variable component of the milk dry matter. Fat content is influenced by genetic factors, level of production, stage of lactation, udder health, climatic factors, management (milking technique) and nutrition.

(iii) Nitrogen compounds form the most complex part of milk, because of their diversity and their complex physico-chemical structure. Two main groups of compounds can be distinguished, namely proteins and non-proteins. Proteins are organic substances formed by amino acids with peptid bounds. They fall into three groups:

—casein which is the coagulable part of the proteins. In decreasing order it contains casein α_s, casein β and casein κ. Casein κ is susceptible to the enzyme action of rennet;

—soluble, non-coagulable proteins: β-lactalalbumin, α-lactalbumin, globulins;

—proteoses-peptones, occurring in very low concentration.

Nitrogen compounds, with the exception of the non-protein fraction, are present in the form of casein micelles. The nitrogen compounds are the raw material for cheese processing and their content determines the cheese yield of milk. Studies on the influence of seasonal variation of milk composition on cheese yield showed that 79% of the variance of cheese yield is due to variance of milk composition (Portmann *et al.*, 1968). Variance of protein content alone may explain as much as 76% of the variation in cheese yield (Ricordeau and Mocquot, 1967).

Nitrogen compounds of goats' milk seem to differ from those of cows' milk by a different distribution of the nitrogen fractions:

—Pierre (1977) indicates that the average casein α_s content is 21·2% in goats' milk *vs* 40% in cows' milk, that of casein β 67·4% *vs* 43·3%. Since the migration rates in electrophoresis of casein β_{s1} of goats' and cows' milk are different, the presence of cows' milk in goats' milk or milk products can be detected by this method.

—The percentage of coagulable nitrogen compounds in milk protein (total N-matter) is 69% in goats and 75% in cows (Ricordeau and Mocquot, 1967).

—The size of casein micelles varies considerably within and between species. They seem to be smaller in goats' milk than in cows' milk. Peak frequency of micelle diameters of cows' and goats' milk, measured with the electron microscope, was 75 and 50 nm in cows' and goats' milk respectively (Barbaro and Calapaj, 1958). Frequency distribution of micelle diameter apparently varies with stage of lactation. Non-protein nitrogen content of milk is 9% in goats and 5% in cows. This difference was found in bulk and individual samples (Mahieu and Le Jaouen, 1976).

—Protein content of milk varies much less than fat content and it seems to depend much more on genetic than on environmental factors (nutrition). However, it is clearly increased at the end of lactation.

(iv) Minerals represent only a minor fraction of goats' milk, the same as in milk from other species. Saline matter (8–9 g/litre) has to be distinguished from minerals (7–7·5 g/litre). During incineration, some of the salts are modified or destroyed, and in the ash only the minerals remain. It is, therefore, difficult to determine accurately the composition of salts in milk and, on the other hand, a list of salts existing in milk has little meaning. Not all the minerals are present in form of soluble salts but an important part exists in the colloidal phase. Thus, two-thirds of the calcium and more than half of the phosphorus are part of the calcium-phosphocaseinate complex. Table VI gives the average mineral composition of the milk of different species.

Table VI. Minerals in milk of different species (Allais, 1974).

	Goat	Ewe	Cow
		(g/litre)	
Potassium	1·6	1·5	1·6
(K_2O)	(2·0)	(1·8)	(1·8)
Sodium	0·4	0·4	0·5
(Na_2O)	(0·54)	(0·54)	(0·7)
Calcium	1·3	2·3	1·3
(CaO)	(1·8)	(3·2)	(1·8)
Magnesium	0·15	—	0·14
(MgO)	(0·25)		(0·2)
Phosphorus	1·0	1·6	1·0
(P_2O_5)	(2·3)	(3·7)	(2·3)
Chlorine	1·5	0·7	1·1
(NaCl)	(2·5)	(1·15)	(1·8)
Sulphur	0·2		0·35
Citric acid	1·5		1·8
Ash	8·0	11·0	8·0

Goats' and cows' milk have a similar mineral composition. Potassium, calcium and phosphorus are the main elements. The proportion of dissolved and colloidal elements is largely the same. By contrast, human milk has a lower mineral content. In general, the mineral composition of milk is variable. Salter (1977) gives some figures on trace elements in goats' milk (Table VII).

Table VII. Trace element content of different species (Salter, 1977).

	Goat	Cow	Human
		(ppm)	
Iron	0·25–0·68(?)	0·64	1·25
Copper	0·04–0·57(?)	0·08	0·40
Iodine	n.d.	0·04	0·07
Manganese	0·17	0·09	7(?)
Zinc	n.d.	3·2	5·3
Cobalt		0·0006	n.d.
Molybdenum	0·012	0·073	n.d.

(v) Vitamin content of goats' milk varies with season, feeding and other factors. Table VIII gives data on vitamin composition of goats', cows' and human milk. With the exception of a lower content of folic acid, Vitamin C and Inositol, goats' milk has similar vitamin contents as human milk. Goat milk contains less vitamin B6 and B12 than cows' milk but this does not present nutritional problems.

Table VIII. Vitamin contents of milk of different species (Salter, 1974).

	Goat	Cow	Human
Vitamin A (IU/100 ml)	191	159	190
Vitamin B-1 (Thiamin)	0·04	0·04	0·02
Vitamin B-2 (Riboflavin)	0·18	0·17	0·04
Niacin	0·19	0·09	0·15
Vitamin B-6	0·01	0·06	0·01
Pantothenic acid	0·34	0·34	0·18
Biotin	0·004	0·003	0·001
Folacin	0·0008	0·0059	0·0038
Vitamin B-12	0·00007	0·00042	0·00003
Vitamin C	1·5	2·1	4·3
Vitamin D (IU/100 ml)	2·4	2·2	2·2
Vitamin E	?	0·10	0·65
Other factors			
Choline	15·0	13·7	9·0
Inositol	21·0	11·0	33·0

(mg/100 ml unless otherwise indicated)

3. Characteristics and Technology of Products

(a) *Liquid Milk*

In all countries liquid goats' milk is consumed essentially by the producers themselves because only in a few countries do industrial and commercial organizations exist which process and sell milk in the same way as cows' milk. Goats' milk sold in the immediate surrounding of producers without the interference of middlemen, does not need to be processed for conservation. In many countries, goats' milk is combined and sold together with cows' milk. Sometimes there are legal limitations to the proportion of goats' milk permitted in bulk milk. Goats infected with *Brucella melitensis* transmit Malta fever (undulating fever) to man. Goats may be carriers of other *Brucellae* (*Abortus bovis* and *A. suis*) and infect cattle, sheep or pigs. Problems with the consumption of infected milk are the same as with cattle, and pasteurization is warranted if disease eradication is not possible.

The nutritive quality of milk is determined by its energy content and, more important, its content of essential nutrients including essential amino acids, fatty acids, minerals and vitamins as well as their digestibility and absorption potential. The presence of noxious substances and toxines has to be considered. Goats' milk has the reputation of possessing specific nutritional and medicinal qualities. We do not attempt to review this subject here since the available scientific references still do not allow any conclusions to be drawn about these contentions. Parkash and jenness (1968), in their review of the literature, showed that the nutritive value of goats' milk does not notably differ from cows' milk and that a better digestibility has not been proved. Yet goats' milk is used for infant feeding in the USA and in South Africa. In these two countries, goats' milk powder is produced. Furthermore in the USA, two manufacturers process 2750 tons of goats' milk into "evaporated sterile milk" and about 15 tons of this milk are produced by some smaller processors. It appears that the use of goats' milk in pediatrics could be an interesting outlet, notably for children who are allergic to cows' milk and show symptoms of intolerance (Mare, 1978).

(b) *Fermented Milk*

Since ancient times, milk from goats, ewes, buffaloes and mares has been fermented by certain oriental people. Yoghurt is the result of fermentation by *Lactobacillus bulgaricus* and *Streptococcus thermophilus* which develop at 45 to 50°C and acidify the milk. Apart from the traditional production of yoghurt from goats' milk in Mediterranean and Middle East countries, small-scale commercial production has developed in the USA. Although this yoghurt is prepared by a technique similar to that used with cows' milk, it differs in its specific goat flavour which is appreciated by consumers (Aggrawal, 1974). In the Middle East, *Doughe* is prepared from yoghurt (*L.acidophilus*) by adding water and salt. It is almost a national drink in Iran (Sattari, 1973).

Leben and *Ajran* are quite popular in the Near East, although at present they are more often produced from cows' milk than goats' milk. They result from acidification with a lactic starter.

(c) *Butter and Ghee*

Butter and ghee (rendered butter) are means of preserving the milk fat. It is not very common to produce butter from goats' milk except in some developing countries with both nomadic and sedentary populations, e.g. in Iran where it is much appreciated (Sattari, 1973). Butter is used in some typical local dishes. Since butterfat is colourless, butter from goats' milk is white. Sometimes it is artificially coloured in order to make it look more like butter from cows' milk. Goat butter is reputed to be an effective remedy against diseases, such as rheumatism and arthritis. Small scale home-production of butter for this purpose has developed in Switzerland, in the Appenzell Canton in particular. However, it is very unlikely that a significant industrial and commercial butter production can be expected, because the characteristics of goat butter are unfavourable compared with cow butter: difficulty in separating cream (because of the size of the fat globules), soft texture and susceptibility to hydrolytic rancidity.

(d) *Various Milk Products*

In Latin America goats' milk is processed to sweets. In Mexico the *"Cajeta"*, a thick liquid of caramelized milk with sugar added, is sold as such or dried as small tarts. In other Latin American countries, other sweets—*"dulces"*—are produced in a similar way. Their specific taste of goats' milk is much appreciated.

A similar product in Norway is called *geitost*, brown sugar cheese. It is a sweet caramel-coloured product with a texture in which lactose crystals may sometimes be noted. The processing is similar to the *Cajeta*, except that the casein is removed and the original lactose used instead of sugar. For processing, milk is separated and curdled rapidly. The serum is collected and mixed back with the cream. This mixture is then heated under vacuum until the desired brown colour and the typical flavour are obtained. In order to maintain this flavour, specific breeding efforts are being made which are discussed in Chapter 15. A similar dried by-product of buttermaking, the *Kurt*, is known in Central Asia.

In the Middle East, casein is produced from skimmed milk. In Iran it is called *Kashk* or dried butter. It is used as food ingredient or in the form of meal as animal feed. Casein yield depends on the degree of drying, between 50 and 100 g of casein are obtained from 1 litre of skimmed milk (Sattari, 1973).

(e) *Cheese*

In many countries, particularly in the Mediterranean basin, goats' milk is used to produce various cheeses which are much appreciated. Depending on local tradition and consumer habits this cheese is produced with more or less elaborated techniques. Furthermore, some particular types of cheese have to be produced because of local climatic conditions and the necessity of preserving cheese in different forms, or to diversify the line of flavours offered in the trade (Fig. 6). The technology of cheese-making from goats' milk is less well known than that from cows' milk, since goat cheese pro-

Fig. 6. Types of goat cheese cherished by gourmets in France (ITOVIC).

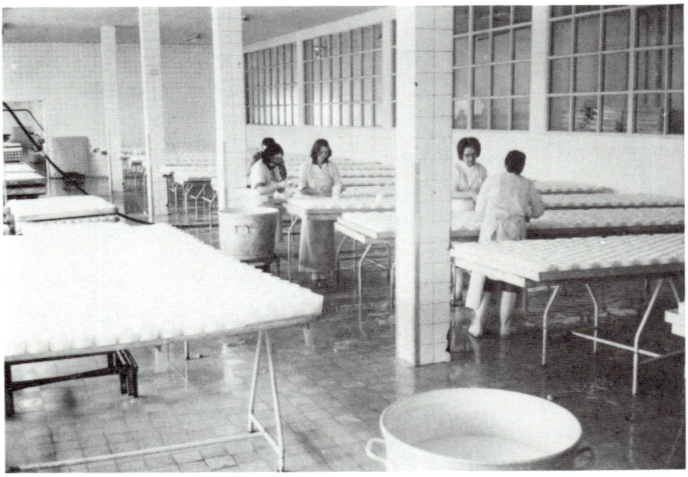

Fig. 7. Cheese factory in France where curd is put into moulds and drained (Picodon cheese in south-eastern France).

duction is less industrialized and therefore less standardized. Furthermore, many home-made varieties exist of which an inventory has never been made. It is, however, possible to classify the main cheese types:

(i) *Fresh cheese*: cheese prepared by curdling for 15–24 hours with lactic acid and a small dose of rennet added. In warm countries, curdling may be more rapid (1–2 hours) in order to avoid excessive acidity. With simple technology the curd is set to drain moderately and then put in mould (Fig. 7). The cheese has a high water content of 60 to 80% and is consumed the day following preparation. The goat taste is not strong which may be advantageous when attracting new customers. The cheese has a fine texture which is characteristic mainly for the products of lactic acid. This fresh cheese is produced in many countries. In Latin America, a particular white cheese, the *"Queso Blanco"*, can be processed in only a few hours. Its characteristic is produced by acid curdling through addition of icy acetic acid.

(ii) *Soft cheese*: the processing of soft cheese, as for fresh cheese, consists of a slow acid (or alternatively a rapid) curdling phase followed by a draining phase, for separating from the curd part of the serum (Fig. 8). In addition a third phase of enzymatic ripening follows. It is caused by the original

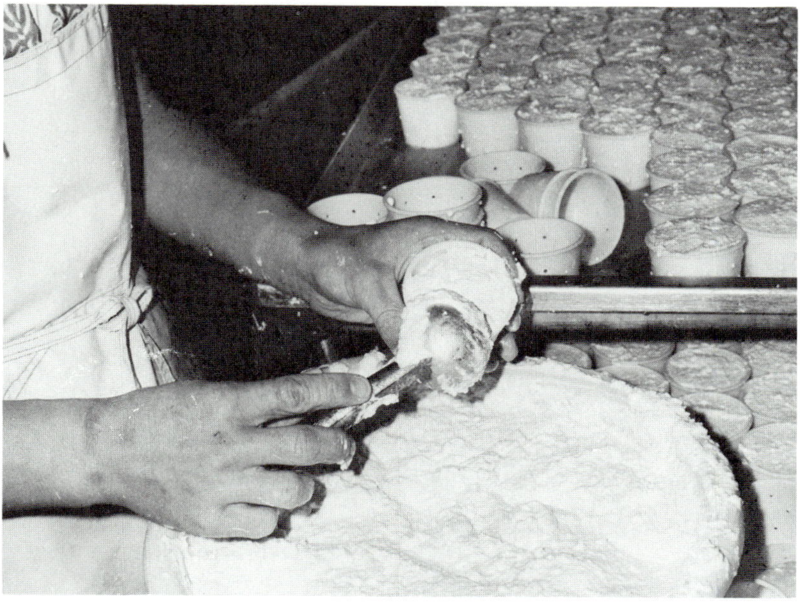

Fig. 8. Pre-drained curd is put into moulds; ladle moulding of the "Crottin de Chavignol".

Fig. 9. Saint Maure. The cut surface shows the fine texture of this cheese.

Table IX. Production of a soft goat cheese.

Day	Operation	Duration	Temperature (°C)	Moisture
1	Filtering milk renneting curdling	18 to 24 h	 20 20	
2	Moulding straining turning	24 h	 20	
3	Taking out of mould salting drying	12 to 48 h	 12–18 max.	 80–85%
4	Ripening	5 to 30 days	8–12 max.	90–95%

enzymes of the milk and by rennet but mainly by enzymes which are secreted by the microbial internal and external flora that develops in the course of ripening. Ripening of cheese is a very complex phenomenon and each cheese is characterized by its own ripening mechanism. The type of cheese and the flavour desired will determine which surface flora is used for ripening. Often these are *Penicillium* (flowery crust) as in most of the French goat cheeses (*Sainte-Maure, Valencay, Crottin* etc.). Since proteolysis and lipolysis, the main reactions of ripening, start at the surface of the cheese, ripening takes only 10 to 20 days. The resulting cheese is soft, with a moisture content of 55 to 60% (Fig. 9). In other cheese varieties, generally of small size, drying is extended and moisture content reduced to 50 to 55%. Thereby, the activity of the internal and external flora is restricted and the duration of preservation is extended. In Table IX the recipe for processing the traditional *Saint-Maure* is given.

Greek *Feta* is a soft cheese often prepared with a mixture of goats' and ewes' milk. After rapid curdling, the curd is cut, stirred, packed into moulds and then cut in big lumps, salted and put into wooden kegs. After washing and draining, the cheese pieces are put into other kegs submerged in brine. The kegs are closed and the cheese ripens for about one month, at 12°C.

(iii) *Blue mould cheese*: the curd of this cheese is prepared with lactic acid and rennet. The cheese has inside a greenish or bluish marbled appearance which originates from the mycelial filaments of the mould *Penicillium glaucum*. After 1 to 2 hours of curdling, the curd is cut in cubes, put in moulds and inoculated with *Penicillium*. After salting and piercing, the cheese is put to ripen in caves, where a temperature of 9 to 10°C and 90 to 95% relative humidity ensure the development of the moulds within the cheese. Ripening takes 1 to 5 months, depending on the variety and type of inoculation. There are only a few mould goat cheeses, like the French *Mould Savoy*, because the development of *Penicillium* in combination with the goat taste gives the cheese a strong flavour unfamiliar to most people.

(iv) *Hard cheese*: hard goat cheese is chiefly produced in warm countries or in mountainous areas. Examples are the *"Chevrotin"* produced in the French Alps and the *"Queso de Cabra de Malaga"*. This typical product of the province of Malaga in southern Spain, is produced in pieces of 2 to 2·5 kg. After rapid curdling, the curd is stirred until grains begin to form about the size of a rice corn which then are put into moulds and strained under mild pressure. The cheese is salted and ripened for five days in a fresh and well ventilated place. This cheese is also preserved for 2 to 3 months, submersed in olive oil.

(v) *Serum cheese*: the ancient cheese dairy learnt early on to use the nutritive value of serum proteins obtained during cheese-making from cows', goats' and ewes' milk. In many countries, various products are prepared from the sweet (not acidified) serum which is heated to 80–90°C with a diluted organic acid added (sour serum, vinegar, sour fruit juice). When the coagulated proteins rise to the surface the liquid is lightly boiled. Then the protein is collected, ladled into moulds, drained or lightly pressed. Serum cheeses have different names, such as *Athotyros* in Greece, *Mysost* in Sweden, *Ricotta* in Italy, *Brousse* and *Broccio* in France. While the principal raw material is serum, sometimes whole milk is added in order to obtain a product of less granulous texture.

References

Aggrawal, M. L. (1974). *Cult. Dairy Prod. J.* **8**, 131–132.

Allais, C. (1974). SEP, Paris.

Barbaro, A. d'A. and Calapaj, G. G. (1958). *Acta Med. vet. Napoli* **4**, 9.

Bouillon, J. and Ricordeau, G. (1970). *Bull. Tech. Inf.* No. 251, 417–424.

Boyazoglu, J. G. and Zervas, N. (1977). *Elev.-Ed. Bovin-Ovin-Caprin* No. 67, 66–75.

Charlet, P. and Le Jaouen, J. C. (1977). "Symposium on Goat Breeding in Mediterranean Countries", Malaga-Grenade-Murcia, Spain, October 1977, pp. 365–371. EAAP and Spanish National Committee Animal Production, Madrid.

Darracq, J. and Le Mens, P. (1978). *In* "Symposium International sur la traite mécanique des petits ruminants", Alghero, Italy, pp. 324–345.

Devendra, C. (1979). *Wld Anim. Rev.* (FAO) **32**, 33–41.

Hilling, V. J. (1974). Mimeograph Paper 122, 80 pp., Dept. Agr. Struct. Agr. Univ. Norway.

Labussière, J. (1966). "17 Int. Dairy Congress", Munich, Section A1, pp. 43–51.

Le Jaouen, J. C. (1974). "La Fabrication du Fromage de Chèvre Fermier", pp. 213, ITOVIC, SPEOC, Paris.

Le Mens, P. (1978). *In* "2ème Symposium International sur la traite mécanique des petits ruminants", Alghero, Italy, pp. 277–288.

Le Mens, P., Lequenne, D. and Toussaint, G. (1978). *In* "2éme Symposium International sur la traite mécanique des petits ruminants", Alghero, Italy, 253–262.

Le Mens, P., Toussaint, G. and Le Jaouen, J. C. (1978). *In* "2ème Symposium International sur la traite mécanique des petits ruminants", Alghero, Italy, 289–305.

Linzell, (1966). *Circulation Res.* **18**, 745–754.

Mahieu, H. (1976). Mimeograph ITOVIC, Paris, 34 pp.

Marree, H. P. (1978). *Dairy Goat J.* **56** (5), 69–72.

Martinet, J. and Denamur, R. (1960). *Arch. Sci. Physiol.* **14**, 35–96.

Martinet, J. and Richard, Ph. (1973). *In* "Symposium sur la traite mécanique des petits ruminants", Millau, France. *Ann. Zootechn.* Num. hors série, 29–49.

Mocquot, J. C. (1978). *In* "2ème Symposium International sur la traite mécanique des petits ruminants", Alghero, Italy, pp. 175–201.

Parkash, S. and Jenness, R. (1968). *Dairy Sci. Abstr.* **30**, 67–87.

Pierre, A. (1977). *Fals. Exp. chim.* No. 752, 213–222.

Portmann, A., Pierre, A. and Vedrem, P. (1968). *Rev. Lait. Franc.* No. 251, 97–101.

Ricordeau, G. and Labussière, J. (1970). *Ann. Zootechn.* **17**, 245–256.
Ricordeau, G. and Mocquot, G. (1967). *Ann. Zootechn.* **16**, 165–181.
Rusev, G. and Diaz, D. (1973). *In* "Symposium sur la traite mécanique des petits ruminants", Millau, France, *Ann. Zootechn.*, Num. hors série, 277–280.
Rusev, G., Ivanova, E. *et al.*, Diaz, D. (1973). *In* "Symposium sur la traite mécanique des petits ruminants". Millau, France. *Ann. Zootechn.*, Num. hors sèrie, 71–76.
Salter, D. L. (1977). *Dairy Goat J.* **55** (12), 12–16.
Sattari, M. (1973). *In* "2nd International Conference in Goat Breeding", Tours, France, 1971, pp. 105–113. ITOVIC, Paris.
Skjevdal, T. (1975). *Sau og Geit* **28**, 12.
Skjevdal, T. (1976). Mimeographed paper, 5pp.
Toussaint, G. (1979). Document ITOVIC, Paris, France.

Note added in proof

The review on milk composition and characteristics has been supplemented for the period 1968–1979 (Jenness, 1980). Furthermore, a review and bibliography have been prepared on goat milk products (Loewenstein *et al.*, 1980) and experiences have been reported from a pilot project on marketing dairy products in the USA (Harris, 1980).

References

Harris, E. (1980). *J. Dairy Sci.* **63**, 1649–1654.
Jenness, R. (1980). *J. Dairy Sci.* **63**, 1605–1630.
Loewenstein, M., Speck, St. J., Barnhart, H. M., and Frank, J. F. (1980). *J. Dairy Sci.* **63**, 1631–1648.

12. Fiber Production

M. Shelton

A. Introduction

At the time of domestication it seems likely that fiber production from the goat was more important than at present. The ancestry of the domestic goat tended to be restricted to colder climates or higher elevations of Eurasia and would no doubt have possessed a well developed hair coat. Even today, certain species of wild Ibex have been identified as producing some of the finest fibers (Burns *et al.*, 1962). Early man used goat hair extensively for a variety of domestic purposes such as clothing, ropes, rugs, blankets and containers. The black hair of the goat is preferred for making the tents used by desert nomads. As the goat population migrated towards the equatorial regions the relative importance of fiber production decreased and milk and meat became the primary products. The result is that the commercial value of fiber produced by the goat is small relative to the value of meat or milk. However, statistical data do not account for the fiber utilized by flock owners which does not enter commercial trade. For the most part, this type of hair is coarse, variable in quality and color and is shorn from the goat only if and when a need or a market for the fiber exists.

Two specialized types of fibers, i.e., cashmere or pashmina and mohair, are produced by highly selected breeds or genotypes. Although limited in distribution, the industries based on these fibers are the most developed of any based on the goat.

Essentially all land mammals are fiber producers to some degree. Fibers apparently serve a function of protection from extremes in temperature, sunburn and also to some degree serve as a sense organ. Animal fibers are poor conductors of heat, thus serving a unique function in protection from temperature changes. Those animals, such as sheep, goats, camels, llamas, alpacas and vicunae, from which man harvests fiber are those producing fibers with special qualities or in greater amount. The pattern of shedding is also important. One of the major results of selection for fiber production probably is the alteration of the pattern of shedding in order that man can harvest the fiber.

B. Biology of Fiber Growth

The biology of hair growth has been extensively studied and no effort will be made to treat this subject in detail. Much of the basic work utilized laboratory animals such as rats as the experimental subjects. Sheep have also been extensively studied (Duerden and Ritchie, 1924; Carter, 1939a, b; Carter and Hardy, 1947). Goats have received much less attention, but it is unlikely that they differ markedly from other species with respect to the biology of fiber production. The principles of fiber growth have been reviewed by a number of authors in the publication edited by Montagna and Ellis (1958).

Fibers are produced in a structure known as a follicle or "hair follicle". These follicles are initiated and largely developed prenatally, but some maturation of follicles continues into early postnatal life (Dreyer and Marinconitz, 1967; Gallagher, 1972). It has been shown with sheep that unfavorable conditions in prenatal and early postnatal life can adversely influence the number of secondary follicles which develop and thus potential fiber production later in life (Short, 1955; Shelton, 1964). However, the extent to which follicular development can be altered is minor. Follicles are initiated in the embryo near the start of the second trimester of pregnancy by an invagination of the epidermis and a protruding downward through the corium and into the dermis layer (Pinkus, 1958). As the follicle develops, it grows downward at an angle resulting in a slanted fiber. As the follicle grows, other structures known as suint or sudoriferous glands, sebaceous glands and the arrector pili muscle develop (Fig. 1). The glands secrete sweat, various salts, fatty acids and lanolin which are thought to provide a protective coating for the fiber to prevent the weathering or deteriorating effects of the sun rays. Other suggested benefits from these secretions are prevention of matting or felting of the fibers and provision of a water repellancy in the coats of the animals.

The fiber itself is produced by a group of rapidly dividing cells in the bulb at the base of the follicles. The rate of fiber production suggests that these are among the most rapidly dividing body cells. Once formed, the fiber is devoid of nerve cells and is not considered to be live tissue. It consists of keratinized protein similar to other tissues such as horns and nails. The

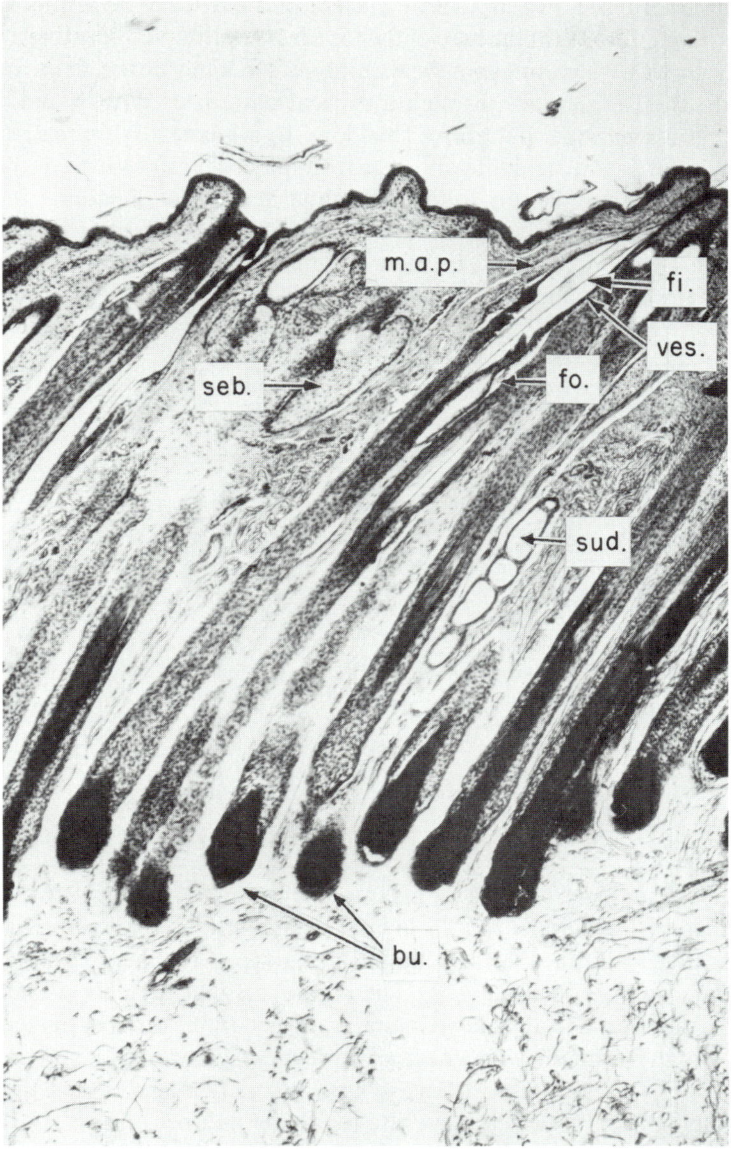

Fig. 1. Vertical section through skin of Angora goat showing the follicle bulbs and the position of the sudoriferous gland distal to the bulb. m.a.p. = arrector pili muscles; seb. = sebaceous gland; fi. = fiber; ves. = vestibule; fo. = follicle; sud. = sudoriferous gland; bu. = follicle bulb. Photo courtesy of J. H. Dreyer, ADSRI, Dept. of Physiology, Private Bag X2, 1675, Irene, South Africa.

amino acid composition of animal fibers, in this case wool, as given by Spector (1956) is shown in Table I. Except in the inactive or resting follicle, growth from the bulb at the base of the follicle is continuous. Any interruption of this growth will result in a weakening or shedding of the fiber. In most animals, the follicle goes through a cyclical pattern of growth and resting resulting in a periodic pattern of shedding. In a laboratory animal, such as the rat, a pattern of shedding and regrowth may occur in as little as 30 days. Other animals may largely follow an annual pattern of shedding. In sheep or goats which have been selected for a high level of fiber production, shedding is much reduced. In most animals, some follicles become non-functional with advancing age resulting in a reduced level of fiber production.

Table I. The amino acid composition of animal fibers.

Amino acid	% air dry basis
Glycine	6·8
Alanine	4·0
Valine	5·4
Leucine	8·6
Isoleucine	4·3
Phenylalanine	4·0
Proline	8·0
Serine	8·0
Threonine	6·5
Tyrosine	5·5
Aspartic Acid	7·4
Glutamic acid	14·0
Arginine	10·6
Lysine	3·3
Histidine	1·1
Trypotophan	1·5
Cystine	13·6
Methionine	0·7

In most mammals, follicles are arranged in groups or bundles. Fiber density is affected by the number of follicle bundles and also by the number of fibers in a bundle. Within a follicle bundle, two types of fibers (primary and secondary) are produced (Fig. 2). The normal arrangement consists of three primary fibers (a central and two laterals) and a variable number of secondary fibers. In the original wild goat, as with most other wild mammals, the primary follicles give rise to long, coarse guard hairs known as the outer coat. The secondary follicles give rise to a large number of fine, short fibers known as the undercoat or down. In the highly selected sheep or Angora goats the secondary fiber population has been developed to the extent that it masks the primary fibers. In other animals such as the tropically adapted goat, the fleece or coat may consist largely of the primary fibers with little evidence of the undercoat from secondary fibers. A dense coat of secondary fibers contributes greatly to cold resistance, but is an impediment in hot climates. The trio group of primaries is considered to be characteristic of the skin of sheep and related species (Carter, 1943; Carter and Hardy, 1947; Ryder, 1956). A similar grouping was confirmed for the Angora goat by Margolena (1966), for various African breeds by Burns (1965) and for the dairy types by Ryder (1966). However, the triad arrangement is not exclusive with either the sheep (Carter, 1943) or the goat (Dreyer and Marincowitz,

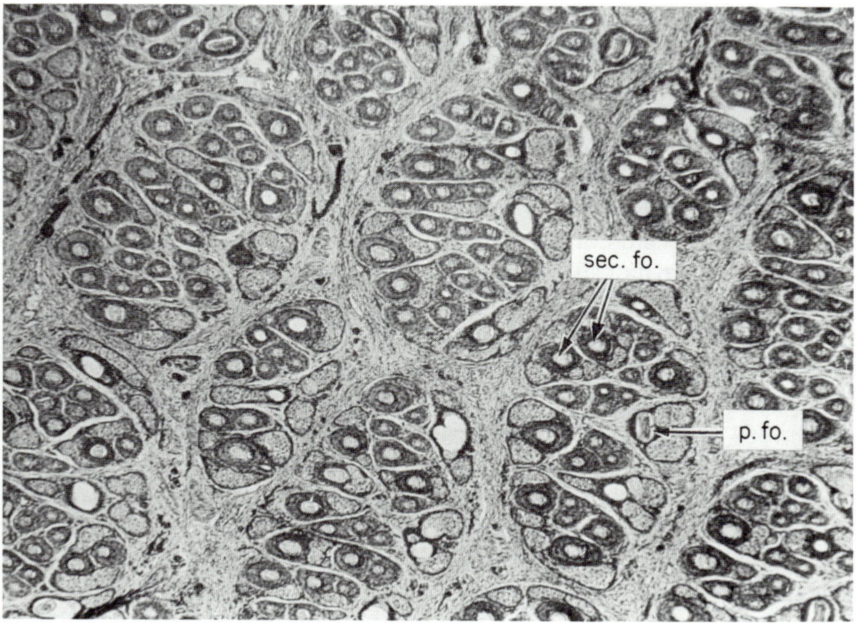

Fig. 2. Mature follicle grouping of Angora goat female approximately 15 months old. p.fo. = primary follicle; sec.fo. = secondary follicle. Photo courtesy of J. H. Dreyer, ADSRI, Dept. of Physiology, Private Bag X2, 1675, Irene, South Africa.

1967; Wentzel and Dreyer, 1967) as follicle groups may be found which contain from one to six primary follicles. In one study of the Angora, 27% of the follicle groups contained more or less than three primary follicles (Dreyer and Marincowitz, 1967). The sudoriferous and sebaceous glands and arrector pili muscle are found associated with the primary follicles, but may be absent from the secondary. Margolena (1966) states that secondary follicles "frequently have sebaceous glands, but never sudoriferous glands". The latter is associated exclusively with primary follicles.

The fiber produced by the goat is similar to that of sheep as described by Von Bergen and Mauersberger (1947), but differs in that it lacks crimp. It also has a reduced number of scales which adheres more closely to the cortex, practically without overlapping. According to Margolena (1974) "mohair has about 5 scales per 100 μ length as compared to 10 to 11 for wool". This arrangement provides a smooth surface and a glossy appearance.

In the case of the ordinary goat, it is the coarse fibers arising from the primary follicles which are used for home spinning into such articles as ropes, tents, carpets, etc. The cashmere goat is valued exclusively for the very fine fiber which arises from the secondary follicles. The Angora is a special case in which an animal has been bred for a high level of production from the secondary fibers, but their fibers are not of the real fine quality of the cashmere.

C. Cashmere

The term cashmere, as used in the trade, refers to the fine undercoat of goat hair produced by the secondary follicles. The term apparently derives from the geographic region of Kashmir in Asia. The term pashmina is also used somewhat synonymously with cashmere, especially in India. Little printed information or statistical data can be obtained relating to cashmere. The paper by Burns *et al.* (1962) recounts most of the information available on this fiber up to that time. As implied in the term, most cashmere is produced in the mountainous regions of Asia and adjoining regions. A fine fiber originating with the secondary follicles is actually produced by many goats throughout the world (Smith *et al.*, 1973; Gallagher and Shelton, 1973) but it is not generally harvested outside of the Asiatic highlands. This is apparently explained by the fact that goats in this region produce a finer fiber in larger amounts, but also the large amount of labor involved in harvesting is no doubt a significant factor in the localization of cashmere production (Fig. 3).

The cashmere goat has been transplanted to many parts of the world, but has not become established outside the areas mentioned. The heavy coat of down would be a serious problem in tropical or semi-tropical settings. Also, the different environmental conditions, in most areas to which they have been transported, England for instance, adversely influenced both the health of the goat and the type of fiber produced.

According to Forte (1977), China and the Mongolia People's Republic produce almost all of the fine cashmere which enters world trade. The amount is estimated at slightly less that 3 million kg. The major portion of the cashmere of China is produced in the north-western provinces of Ningsia, Suiyan and Kansr. The Mongolian cashmere is marketed through Russian agencies. Iran and Afghanistan produce a coarser grade in perhaps one half the amount mentioned above. The Soviet Union produces a smaller amount of a fine hair from the less than one million head of Don or other down-

Fig. 3. Spring combing time for cashmere goats in the remote mountains of Mongolia. Photo courtesy of Pringle of Scotland.

producing goats in that country. The major importers and processors of cashmere are England, the United States and Japan. Cashmere is the finest of the animal fibers used in commercial trade, being approached only by the superfine Merino wools from Australia, especially Tasmania. Clothing made from cashmere is exceptionally light, soft and warm and as such is in the luxury class. On a comparable weight basis, cashmere is reported to have three times the insulating value of fine wool. The famous "ring shawls" are made from the finest cashmere and are so named from the fact that they

Table II. Production of down from goats and selected other species.

Species, type or breed	Wt of down per fleece (g)	% down	Mean fiber diameter of down (μm)	Source
Afgan. Ibex	—	—	17·17	(1)
Chinese Cashmere	—	75·3[a]	15·16	(1)
Mongolian Cashmere	—	73·5[a]	14·34	(1)
Iranian Cashmere	—	74·7[a]	18·55	(1)
Common goat	—	0–58	15·39	(1)
Musk Ox	—	84·2	16·68	(1)
Creton Ibex	—	60·1	14·13	(1)
Bighorn sheep	—	—	11·30	(1)
Don goat	450–1150	—	18·96	(1)
Orenburg goat	220– 230	—	16·4–18·0	(1)
Orenburg × Don	250– 410	—	—	(1)
Orenburg	375	—	15·7	(2)
Kirgiz	—	39·68	14·8–18·8	(3)
Australian feral goats	—	—	15·8	(4)

(1) Burns *et al.* (1962)
(2) Shatunov (1973)
(3) Al'Meev (1973)
(4) Smith *et al.* (1973)
[a] Sorted cashmere

can be drawn through a finger ring. According to Forte (1977) "the finer Chinese/Mongolian cashmere is used primarily in the knitwear trade throughout the world, while Iranian/Afghanistan cashmere is used primarily by the weaving trade in the United States". The more common end uses are for sweaters, ladies dress goods, shawls and coatings. Fineness is one of the most important characteristics of cashmere. The superior grades range from 13 to 16 μm in diameter. The fiber is shorter than mohair, seldom more than 6 cm.

It is also generally weaker than wool or mohair. These factors make the fabric less durable as the fibers tend to pull out. Raw cashmere may be white or carry considerable color. Naturally, the white colors are preferred because they have much greater latitude in respect to the color of fabrics which can be made from them. Observation as well as analogy with other species suggest that color in goats would be strongly inheritable. This has been confirmed by Al 'Meev (1975). Thus, where controlled matings are possible, the color could be readily changed to that of white in a few generations if so desired. However, natural colors are at times used in the finished product.

Production data as well as cultural methods are difficult to obtain for cashmere or related down producing goats. Some data on levels of production are show in Table II which has been adapted from Burns *et al.* (1962) with some additions. These data confirm that other species produce down equal to or finer than cashmere, but apparently not in harvestable amounts. The Don goat of Russia appears to produce the greatest amount of down, but the quality does not approach that of Asiatic cashmere. The amount of fiber can be markedly increased by crossing with Don or Angora goats, but the fine quality of the down is lost in these crosses (Nikitenko and Zaporozhtsev, 1972).

As expected, both quality and amount of down is at least moderately inherited (Mamadaliev, 1974; Yusupov, 1974) and subject to selection; however since fiber diameter is a component of weight an inverse relationship between the two traits might be expected. Thus a heavy fleece weight and fine fiber require greater density or length of the fiber or a larger animal.

The method of harvesting the down from goats appears to be highly variable. The down fibers are normally shed in the spring, but tend to remain trapped in the fleece by the coarser fibers from the primary follicles. Some cashmere is gathered from twigs or browse where it may have been pulled out of the fleece. Apparently the traditional method in much of Asia (Burns *et al.* 1962) is to comb out the secondary fibers as they are shed from the skin in the spring. In more recent times the fibers are sheared from the heavier producing types especially in Iran and Afghanistan. The product obtained by shearing, and to a lesser extent other harvesting methods, is a mixture of fibers which must be sorted before utilization. In commercial trade channels grades are based on color, fineness and percentage of down.

Goats which are used to produce down or cashmere are seldom developed, bred or used for this purpose alone as in most areas they contribute to both the meat and milk supply. Milk production tends to be antagonistic or negatively related to fiber production.

D. Mohair Production

1. The Angora

The term mohair refers to the fiber produced by the Angora goat. The Angora is a breed of goats which has been highly developed for fiber production (Fig. 4). This animal is thought to have descended from the wild type known as the *Capra aegagrus* at some point in central Asia. However, it takes its name from the province of Angora (Ancyra) on the Anatolia plain of Turkey where they were concentrated and developed since prehistoric times. Goats with extensive production of quality hair have long been found throughout the region from Egypt to central Asia. It seems likely that one of the unique features of the Angora of Turkey was their white color. In earlier times, the inhabitants of this region apparently preferred the black

Fig. 4. Well developed Angora breeding males approximately 18 months of age.

goats or black fibers. However, the black color has serious limitations in industrial trade as it cannot be dyed the brilliant colors for which mohair is famous. There is good evidence that white fleeced goats whose hair could be used in the manufacture of altar covers and curtains in the Tabernacle existed at the time of Moses. However, it seems clear that they were preserved into historical times in the Anatolia plains of Turkey where they were discovered by modern man. Early writers believed that it was the unique environment of this part of Turkey which had created the Angora, but this is not supported by experiences in modern times as this animal has been successfully transplanted to other parts of the world.

Major centers of production outside Turkey are the Edwards Plateau region of Texas (USA) and the Bushveld of the Karoo district of South Africa. Each of these areas, especially the latter two, have some things in common and perhaps this tells something of the adaptability or utility of the Angora. Firstly, these areas are relatively dry receiving generally less than 600 mm of precipitation per year. Most areas also have some elevation of at least 500 m. From these and other observations, it is clear that the Angora is not well adapted to the wet or humid tropics. The insulating properties of the fleece is possibly a contributing factor, but the primary reason is interaction between a high protein requirement, a susceptibility to parasitism and the lower protein content of the forage of the more tropical regions. Also, the modern Angora goat is relatively poorly adapted to extreme cold climates. Its small size compounded by the practice of bi-annual shearing makes them susceptible to cold stress which can result in death losses or an increased requirement for feed, shelter and management. The high level of fiber production contributes to a high nutrient requirement, especially for protein, resulting in an animal which is poorly adapted for a multiple purpose animal. Few Angoras are utilized for milk production. The result is an animal that is relatively poorly suited for the small land holder or subsistence agriculture. Thus, they tend to be located in areas where they can be exploited on an extensive commercial scale.

Angora, along with most goats, tend to be concentrated in areas where brush or browse is to be utilized. It should not be construed from this that browse is a necessary ingredient in their diet, as goats do well on more conventional forage. The more plausible explanation is that goats have an advantage in the utilization of browse species. There is some thread of evidence suggesting advantages which may accrue to the Angora goat as a result of browse utilization. Many browse species have a high protein content which matches well the high protein requirement of the Angora (Huston *et al.* 1971). Many browse species contain tannic acid which can act as a denaturant for protein. At least one experiment (Huston and Shelton, 1967) suggests, to a degree, that this can act as a natural protectant for protein allowing it to pass the rumen intact and stimulate fiber production. However, as with artificial denaturization of protein there is a fine line between delaying availablity into the lower part of the digestive tract and making the protein completely indigestible. Thus proteins in plants having a high tannic acid level are generally low in digestibility. Angora goats in brush or browse country usually have somewhat heavier grease fleece weights apparently due in part to less leaching or weathering or loss of natural grease of the fleece where shading is available. The primary association between the goat and browse species is their advantage, both in diet preferences and foraging capabilities, in utilizing this type of forage. The primary association between goats, especially Angora, and the equatorial regions can apparently be explained by (1) the poor adaptation of the goat to cold, (2) the income from goats often does not repay for the long winter feed periods which would be necessary in more temperate or polar regions, (3) browse species, especially those which can be grazed year round, tend to be restricted to arid tropics whereas the temperate or polar regions are usually timber or farm country.

Mohair prices and thus the industry based on it, tend to be erratic and fluctuate wildly. At the time of writing, Angora numbers in Texas (USA) are at an all time low since the industry became established. However, at the present time prices are at an all time high and so long as this lasts numbers will rebuild as fast as the reproductive rate permits.

2. The Mohair Fiber

The biology of mohair production is much the same as other types of animal fiber but is a special case in which the production from the secondary follicles has been developed to a high degree. In the process, fleece weights have been increased greatly and the individual fibers have become much coarser, longer and more dense than that of the down produced by many animal populations and of which the cashmere is an example. The skin histology of fiber production has been studied by Dreyer and Marincowitz (1967) and Margolena (1974). In a well bred Angora the contrast between the fiber produced by the primary and secondary follicles is evident only at birth. The residual fibers produced by the primary follicles are referred to as kemp. These are large medullated fibers which are brittle in nature and chalky white in appearance. The kemp fiber is undesirable in good quality mohair and represents one problem to the industry especially in those areas where the animals are being graded up from non-Angora stocks. Stapleton (1976) found that partially medullated or heterotype fibers which he referred to as "gare" also arise at times from some of the secondary follicles. However, this type of fiber occurred only in certain individuals or flocks. The amount of kemp is under hereditary control. In one attempt to estimate the hereditability of kemp score (Shelton and Bassett, 1970) a moderate to a high value was obtained. It seems unlikely that complete elimination of kemp is a realistic goal, as a kemp-like fiber is the normal fiber to arise from the central primary follicle. A reduction in the amount of kemp is apparently explained by: (1) increased number or density of the secondary follicles reduces on a proportionate basis the number of fibers which arise from the primary follicle; (2) increased density of secondary follicles tends to compact the primaries to the extent that the fibers arising from them are less kemp like; and (3) shedding of the fibers from the primary follicles to the extent that these fibers are not present in the clip. Apparently all these mechanisms play a part. Kemp is most easily observed by parting the mohair along the topline. Kemp is more commonly observed in the very young animals and in the older animal. This is apparently explained by variation in the secondary to primary ratio (S/P) follicle or density of fleece. In the very young kid the secondary follicle population has not completely developed (Gallagher, 1972; Dreyer and Marincowitz, 1967) whereas in the older animal some of the follicles become non-functional (Margolena, 1974). Thus, age of the animal has an important bearing on both amount and quality of fiber produced by the Angora (Fig. 5). The number of follicles per bundle as expressed by the S/P ratio varies from 2–3 at birth to 8–10 at 1 to 2 years of age.

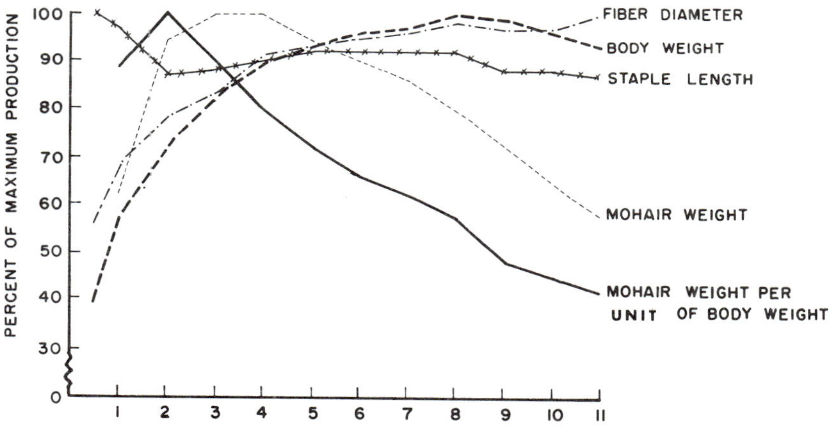

Fig. 5. Influence of age on certain production parameters for Angora does.

The mohair fiber as described by Von Bergen and Mauersbergen (1947) is similar in structure and origin to other types of animal fibers. Mohair differs from wool with respect to the absence of crimp and the epidermal scales lie close to the stem giving the fiber a very smooth appearance. The number of scales are in the order of one-half that of fine wools. The peculiar scale formation gives the fiber its smooth handle and high luster. The mohair fiber takes dye readily and since essentially all mohair is white, it can be dyed brilliant colors. In contrast to cashmere, mohair is a very strong and durable fiber.

The end uses of mohair range from upholstery, drapery materials, mens and ladies suits and coats to rugs or carpets. At times, it has also been used for wigs. In the manufacture of clothing, it can be blended with other natural as well as artificial fibers. The low volume of mohair produced dictates that it be a speciality or luxury fiber as the supply does not adequately meet the needs of any sizable industry. The result is an erratic market with widely fluctuating prices depending on fashion trends and general economic conditions. As of the present, England is the major processor of mohair.

3. Centers of Production

From their origin in Turkey, the Angora has been distributed to many parts of the world. However, significant industries have developed only in Turkey, South Africa and USA (Texas). Other countries with active interests in the Angora include Lesotho, the Soviet Union and Australia. The latter is making a serious effort to increase Angora goat numbers. Statistics on Angora goats and mohair production are difficult to obtain, because they are not often reported and fluctuate greatly over a period of years. Table III shows trends in levels of mohair production by the major producing countries. The number of Angoras in two of the major producing countries (USA and South Africa) is highly variable. For instance the number of Angora goats in USA approached five million head in 1965, but was down to only slightly more than

Table III. Mohair production in specified countries, greasy basis (metric tons), annual 1970–77.[a]

Year	United States	Turkey	Republic of South Africa	Lesotho	Total
1970	8472	6790	4173	998	20 433
1971	6960	6280	4264	816	18 320
1972	4765	5690	3674	765	14 894
1973	4536	5710	3130	569	13 945
1974	3901	5570	3697	679	13 847
1975	4037	5590	3846	750	13 473
1976	3855	5580	4136	418	13 989
1977	3900	5570	4200	397	14 067

[a] Data obtained courtesy US Department of Agriculture—Foreign Agricultural Service.

one million head in 1975. The population in South Africa has shown similar fluctuations. As of this writing, numbers are increasing in both areas. Much of this is explained by fluctuating prices.

4. Expected Levels of Production

The level of production and certain attributes of the Texas (USA) Angora goat under field conditions is revealed in the data shown in Table IV as adapted from Bassett (1966). These data were collected in the autumn of 1964 and the spring of 1965 on a random sample of animals drawn from five different ranch properties. These data show the traits considered to be important for goats and the expected level of production under average to

Table IV. Some typical production parameters for Angora goats (Texas, USA).

		Fleece wts				Fiber diameter		
Age/season	Grower	Grease wt (kg)	Clean wt (kg)	Yield (%)	Staple (cm)	(μm)	Spinning count	Body wt (kg)
6 months Autumn	A	1·3	1·0	76·1	13·2	24·8	40s	—
	B	1·5	1·1	72·9	14·0	23·6	40s	17·2
	C	1·2	0·9	79·8	14·2	24·6	40s	13·8
	D	1·4	1·1	76·9	14·7	23·5	40s	18·2
	E	0·8	0·6	83·2	14·0	21·6	40s	13·9
Average		1·2	0·9	77·8	14·0	23·6		15·8
1 year Spring	A	2·2	1·6	71·1	12·2	28·6	32s	19·5
	B	3·3	2·3	69·6	11·7	29·6	28s	21·3
	C	2·0	1·6	77·9	12·2	28·1	32s	14·1
	D	2·5	1·9	76·9	11·9	28·7	32s	19·6
	E	1·8	1·4	78·5	15·2	28·2	32s	16·0
Average		2·3	1·7	78·1	12·7	28·6		18·1
1½ years Autumn	A	3·7	2·9	78·9	14·0	34·5	24s	—
	B	3·1	2·5	78·5	14·7	30·9	28s	23·6
	C	2·8	2·1	76·3	13·5	31·0	28s	21·7
	D	2·4	1·9	79·2	14·2	29·0	32s	22·1
	E	2·2	1·8	81·9	14·0	29·6	28s	21·5
Average		2·8	2·2	78·9	14·2	31·0		22·2
2 years Spring	A	2·6	2·0	75·5	9·9	34·3	24s	32·1
	B	3·5	2·7	77·0	10·7	34·9	24s	27·8
	C	2·7	2·1	78·1	9·9	33·1	24s	24·1
	D	2·4	1·9	79·2	10·9	32·3	24s	24·1
	E	1·7	1·4	81·4	11·7	31·2	28s	22·1
Average		2·6	2·0	77·2	10·7	33·2		26·1
2½ years Autumn	A	3·7	2·8	74·3	12·2	37·1	20s	—
	B	2·7	2·2	79·2	8·9	32·2	24s	24·4
	C	2·8	2·1	75·7	13·5	32·1	24s	25·6
	D	2·9	2·2	75·7	13·0	32·6	24s	30·6
	E	2·0	1·6	82·2	12·7	29·2	28s	23·1
Average		2·8	2·2	77·5	13·2	32·6		25·9
3 years Spring	A	3·1	2·2	71·5	9·9	36·2	20s	41·7
	B	2·8	2·1	76·6	10·2	34·6	24s	25·2
	C	3·1	2·4	77·4	10·7	34·7	24s	27·0
	D	2·6	2·1	80·8	10·4	35·7	20s	33·2
	E	1·8	1·5	80·8	11·2	33·2	24s	22·6
Average		2·7	2·1	76·3	10·4	34·9		30·0

Table IV. (cont.)

Age/season	Grower	Fleece wts Grease wt (kg)	Clean wt (kg)	Yield (%)	Staple (cm)	Fiber diameter (μm)	Spinning count	Body wt (kg)
3½ years	A	3·3	2·5	75·7	11·7	38·0	20s	—
Autumn	B	2·5	2·0	79·8	14·2	31·5	28s	28·6
	C	2·7	2·0	74·3	14·2	36·7	20s	27·8
	D	3·2	2·6	78·3	13·7	36·2	20s	32·1
	E	2·3	1·9	82·7	13·0	32·8	24s	26·5
Average		2·8	2·2	78·2	13·2	35·0		28·8
4 years	A	3·0	2·2	73·8	10·2	35·8	20s	37·8
Spring	B	3·1	2·4	76·2	13·0	34·3	24s	30·1
	C	3·1	1·9	61·2	10·7	37·7	20s	30·6
	D	2·6	2·1	80·5	10·7	36·0	20s	34·6
	E	2·1	1·6	78·8	10·7	36·5	20s	26·3
Average		2·8	2·1	73·8	10·9	36·1		31·9

good range feed conditions. The fleece weights obtained in this study were in excess of 5 kg annually. This value is well above that reported as a national average and should be considered indicative only of those managed under better conditions. The weight of the first fleece produced by the weaning kid is one-half or less that of the adult animal fleece. In this study, the fiber diameter of the first fleece (weaning kid) averaged 23·6 μm whereas mean fiber diameter for the adult animals ranged from 29 to 39 μm. Heavily fed animals may run much coarser than this. Yields of clean fleece generally ranged from 75 to 80%. The reader should note the low body weights of some flocks and the high variability between flocks. Fleece weights, body weights and fiber diameters are greatly affected by the level of nutrition whereas yields and staple length are affected less by environmental factors.

The high level of fiber production of the Angora results in a relatively high nutrient requirement (Huston *et al.*, 1971), especially for protein. The high protein requirement results in an animal which is frequently or generally on a suboptimum diet. Under these conditions the animal suffers greatly from internal parasites, as it has been shown that these two factors are related. The goat is often anemic to some degree and a specific swelling condition is often observed as a result of fluid accumulation along the underline (Oedema). South African workers have attributed this to the brown stomach worm (*Ostertagia* spp.). In fact, it may result from any of a number of conditions which cause the animal to be anemic (Fig. 6).

Fig. 6. Anemia as indicated by fluid accumulation is commonly observed in young Angora goats. Internal parasites and protein deficiencies are the primary explanations.

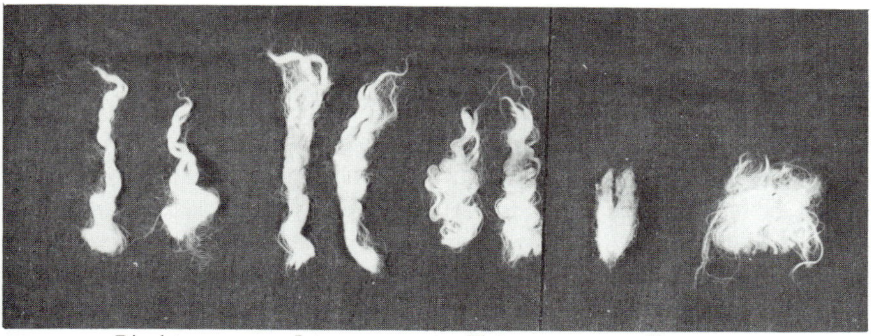

Ringlet Inter-mediate Flat Sheepy

Fig. 7. Lock types of mohair.

5. Lock Type

Lock types are generally classed as ringlet or flat locks (Margolena, 1966) or various intermediate types sometimes referred to as web locks (Fig. 7). Ringlet fleeces are indicative of finer quality hair. The same animal may be a ringlet as a kid and revert to another type at later age. In the absence of a distinct lock type, some type of mohair is largely straight. An undesirable type of hair sometimes referred to as "sheepy" is short in staple with excessive amount of curl (Fig. 7). This type of mohair, as the name implies, is more like wool than mohair. The length is inadequate and the fleece weight is much below that of normal hair. This defective type lock is highly inherited, but the exact mode of inheritance is not known. At a specific point on the animal it is an all or none condition, but normal and sheepy hair may occur at different places or in different seasons on the same animal. The condition appears to be inherited in a Mendelian manner with incomplete penetrance.

In South Africa much more grading is practised at the shearing Since the abnormal types are not normally used for breeding it is obvious that it is correlated with another trait for which breeders are selecting. It can be eliminated or reduced to a very low level by removing the parents of those showing this defect.

6. Mohair Grades

The well bred and well fed Angora produces fiber at a rate of approximately 2·5 cm per month. This is lock length and actual fiber growth may exceed this. The Angora is generally recognized as the world's best fiber producer when expressed in rate of fiber growth, amount produced per unit of body weight maintained or efficiency of production (Gallagher and Shelton, 1972).

In most production areas, apart from Turkey, goats are shorn twice per year and at times are shorn three times per year.Extremely long fibers present problems in manufacture. Well bred and well fed Angoras produce fibers of adequate length for the textile trade when shorn twice per year. However, under average or poor production conditions inadequate length can be a problem. Thus, the longer staple lengths are desirable (Shelton *et al.*, 1965) but lock length of mohair does not carry the high correlation with fleece weight as does staple length of fine wools. The major trait determining value in mohair is fineness. Others include the amount of stain, amount of vegetable matter defect and the amount of kemp. Age of the goat has an important bearing on fineness of the fiber with kid hair having a greater value. Shearing is done by age groups with the kids, yearlings and adults being shorn separately. Apart from this age distinction, most Texas (USA) hair is sold as original bags or as it comes from the goat. At times grading, based on fineness, and removal of stains is practised at the central warehouse. In general, this is done only if a substantial premium is paid for the finer grades of adult hair. When grading is practised, the grades shown in Table V are recognized.

In South Africa much more grading is practised at the shearing shed. A complete stratification based on age of the goat, length, fineness, style, character, amount of kemp, stains, colored fiber, etc. can lead to an extremely large number of grades (the South African mohair board allows

Table V. Mohair grades used in USA.

Trade terminology	Grease mohair grades		Mohair top grades	
	Spinning count	Range in fiber diam. (μm)	Bradford[a] spinning count	Range in fiber diam. (μm)
Super kid	Finer than 40s	Under 23·01	Finer than 40s	Under 23·55
No. 1 kid	40s	23·01–25·00	40s	23·55–25·54
	36s	25·01–27·00	36s	25·55–27·54
No. 2 kid	32s	27·01–29·00	32s	27·55–29·54
	30s	29·01–31·00	30s	29·55–31·54
No. 1 grown and yearling	28s	31·01–33·00	28s	31·55–33·54
	26s	33·01–35·00	26s	33·55–35·54
No. 2 grown	24s	35·01–37·00	24s	35·55–37·54
	22s	37·01–39·00	22s	37·55–39·54
No. 3 grown	20s	39·01–41·00	20s	39·55–41·54
	18s	41·01–43·00	18s	41·55–43·54
No. 4 grown	Coarser than 18s	43·01 and over	Coarser than 18s	Over 43·54

Table VI. Mohair grades used in South Africa.

Fineness symbol	Age class	Bradford[a] spinning count	Maximum fiber diam. (μm)
FH	Kid	56s–60s	30
K	Kid	50s–56s	33
YG	Young Goat	46s–50s	37
FH	Adult	40s–46s	40
H	Adult	36s–40s	43
R	Adult	Below 36s	over 43

Length codes:
A—15·2–17·8 cm
B—12·7–15·2 cm
C—10·2–12·7 cm
D—less than 10·2 cm
Quality (*character and style*)
S—Super
No symbol—average
M—poor
STN—vegetable or urine stained
LOX—cotted, urine or dung stained or otherwise defiled
SDY—seedy
X—crossbred
VEL—from skin of slaughtered Angora goats
PLK—from skin of dead Angora goats
From a combination of lettering symbols the nature of the fiber can be identified.

[a] It will be noted that Bradford spinning counts and corresponding fiber diameters used in the US and South Africa do not agree. In this connection it is important to remember that spinning counts are largely theoretical and have not been standardized.

for 67). South African mohair grades can be partially identified by symbols used in labeling according to the schedule shown in Table VI (Van Rensburg and Wessels, 1968).

Many producers set great store by the character, style and lock type, etc. of the fleece. However, in processing these attributes tend to disappear and may not contribute materially to value in the finished product. However, these traits are considered characteristic of the Angora goat and should not be totally ignored until such time as tests have been conducted on the manufacturing qualities of various types of hair.

7. Genetic Studies in the Angora

Since the primary product from this animal is fiber the important traits are those associated with the mohair fleece such as weight, length and fiber diameter. Size or growth rate is also important in that it is a measure of

Fig. 8. Hair covering on the face seriously interferes with the ability of the animal to selectively graze and reduces growth rate and reproduction.

adaptation and is also related to reproduction and resistance to certain forms of stress. Size also contributes to income since most animals are eventually marketed for meat production.

In addition to the quantitative traits breeders may be concerned with lock type, character, amount of kemp and covering of the belly, neck and face. Hair covering on the neck and belly are important if visual selection for weight is to be practised. If actual weights are to be used, then these traits can be ignored. By contrast, face covering is important in a negative manner (Shelton, 1960). Hair covering on the face significantly interferes with vision and a goat with impaired vision (Fig. 8) is at a considerable disadvantage in selective grazing. Presumably the adverse effects of hair covering on the face can be overcome by frequent removal of the hair from the face, but this has apparently not been tested.

Observation alone tells us that the fiber traits of the goat are moderate to highly inherited, otherwise man would not have been able to breed them to their present high level of performance. Few reports are available relating to genetic parameters of Angora goats. Those estimates which have been reported are summarized in Table VII. These data are preliminary in nature

Table VII. Some heritability estimates for the Angora goat.

Trait	Method	Heritability	Source
Lock length	Half sib	0·59	(1)
Fiber length	Half sib	0·22	(1)
Weaning weight	Half sib	0·33	(2)
Weaning weight	Intra sire regression	0·55	(2)
Body weight	Half sib	0·50	(3)
Grease fleece weight	Half sib	0·40	(3)
Clean fleece weight	Half sib	0·20	(3)
Yield	Half sib	0·48	(3)
Fiber diameter	Half sib	0·12	(3)
Staple length	Half sib	0·79	(3)
Face cover	Half sib	0·31	(3)
Neck cover	Half sib	0·49	(3)
Kemp score	Half sib	0·43	(3)

(1) Shelton *et al* (1965)
(2) Davis and Shelton (1965)
(3) Shelton and Bassett (1970)

but appear to indicate that the important traits are at least moderately high in heritability and good response to selection for individual traits could be expected. However, the limited data available suggest a serious negative genetic relationship between size or rate of growth and fiber production. Also, fiber diameter and fleece weight are positively related indicating difficulty in simultaneous selection for fine hair and fleece weight. Observations also suggest a negative relationship between reproductive performance and fiber production. These genetic relationships are at the root of a number of problems which are encountered in highly selected Angora goats. These problems have apparently been exaggerated by the widespread practice of selecting males from flocks which are maintained under a higher level of nutrition than that of commercial practice.

8. Some Problems in Angora Goat and Mohair Production

There are a number of problems with this industry some of which are common to all goats and some of which are unique to Angora goat and mohair production. As with most small stock, the goat is susceptible to predation and this problem may be more serious with the Angora because of the extensive conditions under which they are often produced. Labor supply may present a problem in goat production due to the necessity to confine or herd them, but mohair production adds to the labor burden because of the necessity to shear twice annually. Some of the problems which are unique to the Angora are (1) a high nutritive requirement, (2) susceptibility to death losses due to cold stress, (3) poor reproductive efficiency with abortion as a major contributing factor, (4) susceptibility to internal parasitism and (5) the problem of defects in the hair.

It is not unusual for highly selected Angora goats to approach an annual fleece production equivalent to 25% of their body weight. The yield on this hair usually ranges from 70–80% clean fiber which is almost pure protein. The nutritional efficiency of nutrient use for fiber production is not known, but a recent study reported by Shelton *et al.* (1976) suggests that the nutritional efficiency of energy use for fiber production is much higher than previously thought. However, it is clear that a high producing Angora goat has a high requirement for protein. Several studies (Shelton and Huston, 1966; Huston and Shelton, 1967) have shown that young growing Angora goats respond to ration protein levels up to 20%. Higher values have apparently not been tested. This much protein can generally be obtained only by the animal with access to high quality forage such as leguminous browse or cool season grasses. This explains in part the limited distribution of the Angora. A failure to obtain ration protein at this level predisposes the animal to many of the problems discussed later.

Low temperature or "off shear" losses with Angoras can be catastrophic if safeguards are not provided. In full fleece, goats can apparently withstand low temperatures, at least that of most environments into which they have been inserted. However, losses due to exposure have been observed with full fleeced animals exposed to prolonged rain in the temperature range of 0–10°C, but the major problem occurs with the freshly shorn animal (Stampa, 1975).

Low temperatures (such as 10°C) alone can cause losses, but these generally occur only in the impoverished or unthrifty animal or from trampling or "piling on" where several animals congregate in one place. Most generally, losses occur with drastic changes in the chill factor as a result of a combination of low temperature, rainfall and wind. In prolonged exposure to wind and rain, extreme low temperatures are not required as losses have been observed by the author in midsummer when temperatures did not drop below 15°C. It must be recognized that the Angora is extremely sensitive to exposure and this perhaps provides a further explanation for their restriction to arid

Fig. 9. Examples of intense management during kidding (kidding on the stake). Also note the cape of hair on the back of the doe to reduce cold stress.

environments. The apparent explanations for their susceptibility to cold stress are: (1) small size or small body mass as related to surface area, (2) almost total absence of subcutaneous fat cover, and (3) bi-annual shearing which adds to the amount of time when they are subject to exposure. Also the tendency of some animals to shed in the spring requires that they be sheared before summer temperatures have arrived. Death losses from exposure may take more than one form. The most dramatic is death from short-term exposure in which body heat is removed faster than it can be produced even at summit heat production. A second form is that due to prolonged exposure over a period of hours. In this case the animal was able to elevate heat production to equate initial losses, but muscle fatigue or exhausted stores of available energy results in the animal succumbing after prolonged exposure. Other types of loss can be due to infection or adrenal malfunction as a result of prolonged stress.

The various approaches taken by producers to evade these losses include shedding or confining to barns in time of inclement weather, shearing with combs which leave a greater amount of stubble hair on the animal, or the use of natural or artificial capes. The natural cape consists of leaving a strip of unshorn mohair down the back (Fig. 9) which can be removed later when a regrowth of hair at other points of the body has occurred. Artificial capes may be made of plastic or canvas ducking. None of these practices are totally satisfactory under all conditions. For instance, the presence of shelter in the pasture does not always insure protection as rapid changes in weather can chill the animal before they reach shelter. Only a few degrees drop in body temperature causes the animal to become immobile and cease shelter seeking activity. The author has revived chilled animals with body temperatures as

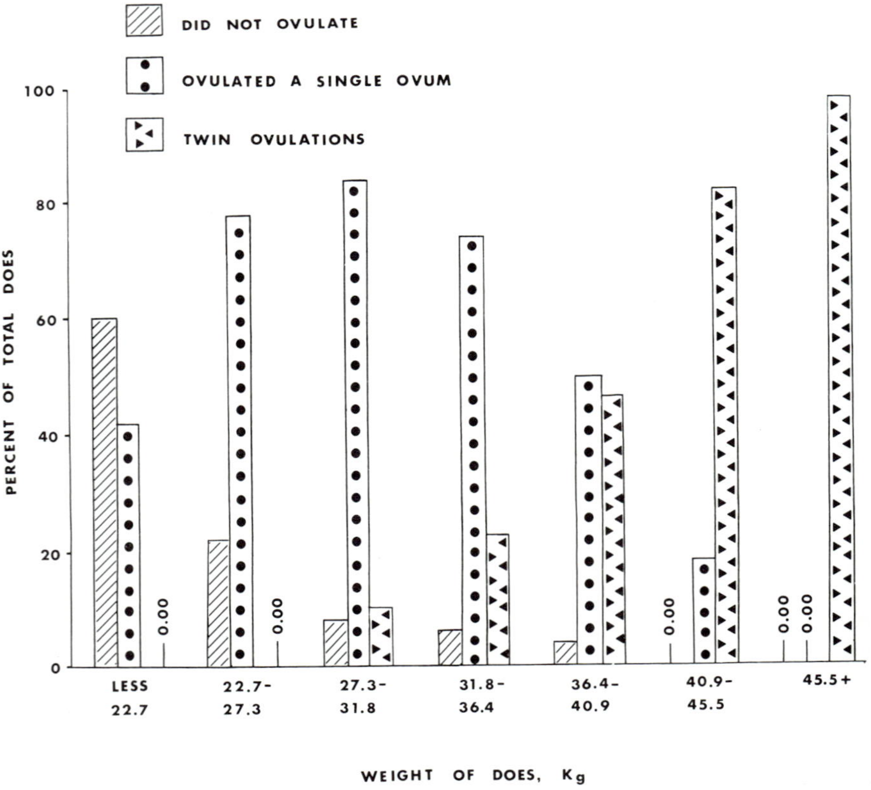

Fig. 10. The relation of doe weight to ovulation of the Angora doe.

Table VIII. Relation of size (breeding weight) of doe to various aspects of reproduction in Angora goats.

Body wt. (kg)	Ovulation rate				Conception data				Kidding data			
							No. embryos				No. kids	
	No.	% ovulating	Per 100 does ovulating	% of all does	No.	% pregnant	Per 100 pregnant does	% of all does	No.	% kidding	Per 100 does kidding	% of all does
< 22·7	17	41·2	100·0	41·2	9	44·4	100·0	44·4	8	50·0	100·0	50·0
22·8–27·3	54	77·8	100·0	77·8	27	51·9	100·0	51·9	19	68·4	107·7	73·7
27·4–31·7	56	92·9	109·6	101·8	41	82·9	100·0	82·9	28	67·9	100·0	67·9
31·8–36·4	69	95·7	122·7	117·4	53	81·1	120·0	98·1	27	85·2	108·7	92·6
36·5–40·9	35	97·1	147·1	142·9	21	95·2	155·0	147·6	21	81·0	122·7	104·8
41·0–45·4	11	100·0	181·8	181·8	5	100·0	180·0	180·0	6	100·0	166·7	166·7
> 45·5	4	100·0	200·0	200·0	3	100·0	200·0	200·0				
	246	87·8	122·7	107·7	159	77·4	122·0	94·3	109	75·2	113·4	85·3
Correlation coefficient, size and indicated variable				0·581				0·578				0·410
Regression on weight				0·053				0·053				0·040

low as 25°C provided they were in a good state of health before chilling and facilities were available to provide prolonged protection in a warm environment. Unfortunately few producers have these facilities available. The provisions of stimulants or readily available energy sources should be beneficial but their value has not been demonstrated in practice. Small size, poor condition and lack of a body fill of nutritious feed aggravates the problem of freeze loss and losses are sure to be greater under these conditions. A good supply of high quality roughage contributes to internal heat production and increases the body reserve of heat.

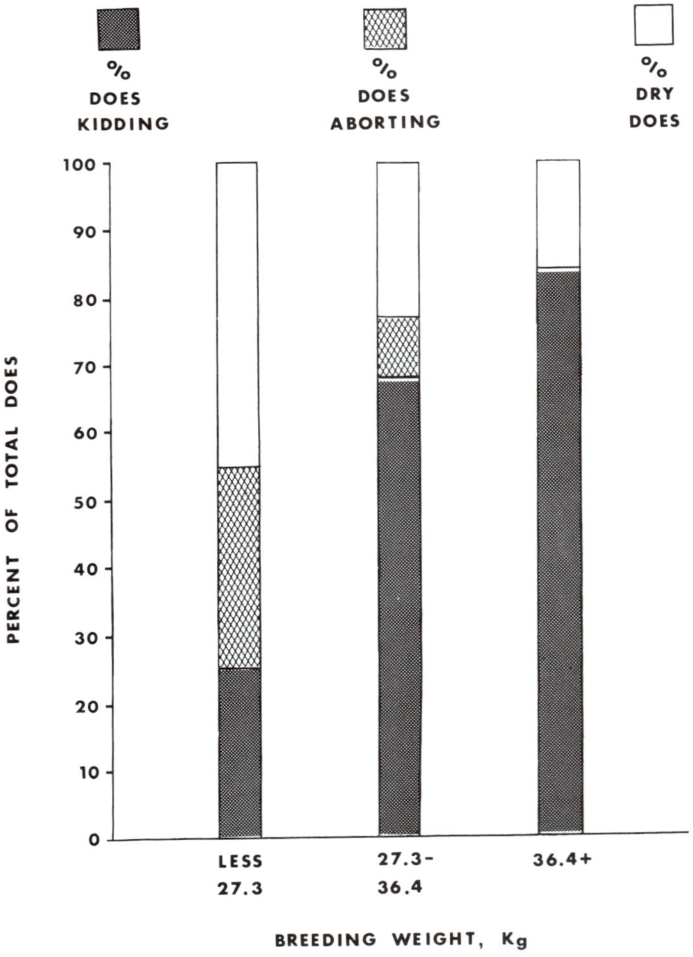

Fig. 11. The relation of breeding weight to various parameters of reproduction in the Angora female.

Poor reproductive efficiency is a major problem with the Angora (Van Heerden, 1964; Van Rensburg, 1970; Shelton and Groff, 1974). Although the Angora goat is valued primarily as a fiber producer reproductive rate is very important. To the individual producer, the most important reason for a good kid crop is the income to be obtained from the sale of surplus animals. Since surplus males are run as castrates for fiber production, there is generally a demand for the entire kid crop for replacement purposes. At times of expansion of the industry low reproductive rate is often a distinct limiting factor to increase in flock size. A good kid crop offers several additional advantages. When the population remains stable a high reproductive rate increases the amount of selection which can be practised, but more important it permits the reduction in the average age of the flock and the younger animals produce both more and better quality hair (Fig. 5). Also most goats are eventually slaughtered for meat production and thus the amount of meat produced is influenced by reproductive rate.

The overall cause of the poor reproduction is the high level of fiber production and the fact that in highly selected flocks this function tends to take precedence over other functions such as growth and reproduction. The physiological mechanism and preventative measures are more complex. Most noted is the effect of body size on ovulation rate (Fig. 10) and kid production (Table VIII). The small, underdeveloped female which does become pregnant is likely to abort (Fig. 11), especially if subjected to some type of stress (low temperature, driving for long distances in late pregnancy, inadequate nutrition, etc.). The problem of abortion has generally reached the most serious proportions in South Africa where the goats have been more highly selected for fine quality hair suggesting some relationship to this trait. Observed abortions start and are concentrated at approximately 100 days of gestation, but may have occurred unobserved prior to this and may continue up to normal parturition. However, losses clearly tend to peak around 100 days of gestation and it is significant that this coincides with a rapid acceleration in fetal development (Fig. 12). Adrenal malfunction (Van Rensburg, 1970) is almost certainly implicated in the mechanism of abortion. Abortion in young and undeveloped females appears to result largely from stress, but it can become habitual (Van der Westhuysen, 1974). Thus, unless aborters are removed from the flock, a high incidence can occur in older goats even in good condition as the percentage of habitual aborters builds up in the flock. In a broad sense, abortion is inherited since it is largely restricted to the Angora. However, it does not appear to be inherited independent of level of mohair production and adaptability to production conditions. The incidence of abortion distinctly increases as the level of mohair production relative to body size increases. It remains to be determined if it is possible to obtain a high level of production of fine mohair without

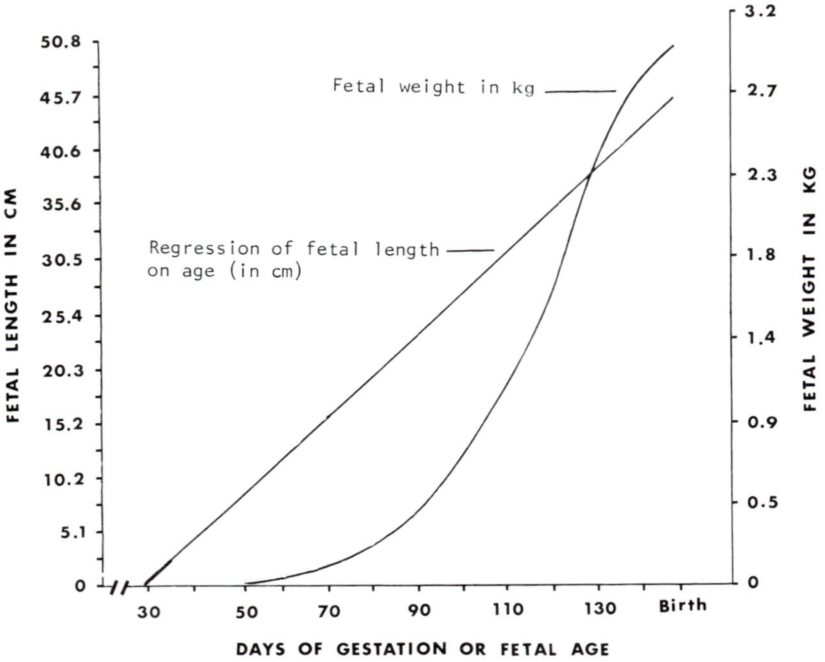

Fig. 12. The relation of stage of gestation to length or weight of the fetus. Fetal age can be approximated by the following formula:

age in days = 30 + crown–rump length in cm divided by 0·378.

incurring this problem. Corrective or preventative measures consist of selecting breeding males from does with a good reproductive history, managing to obtain good size and development, prevention of stress and culling of aborting females. The culling of aborting females can have a marked effect on reducing the number of aborters in the flock (repeatability), but must be practised over many generations to reduce or eliminate the problem.

Among does which conceive and carry kids to term a heavy death loss of kids is often encountered especially under extensive conditions. Some of the causes of these losses are predation, cold stress, starvation, etc. However, with the Angora, an added problem is poor mothering as expressed in abandonment of kids. If kidding occurs in large pastures the females will often leave the young in search of forage or water and not return to pick them up. This is primarily a problem under extensive conditions. Corrective measures consist of kidding in confinement or in very small pastures. If kidding occurs in larger pastures good nutrition should be provided with the does being disturbed as little as possible during kidding. When adequate labor is available this problem can largely be solved by kidding in barns. In the early years of the industry in the USA, the kids used to be staked (see Fig. 9) but this practice has largely been abandoned due to the high labor requirement.

Angora goats appear to be most susceptible to internal parasitism including both roundworms (Le Jambre and Royal, 1974) and coccidia (Shelton, 1969). Their susceptibility to roundworms appears to be related to their high protein requirement and frequent general undernutrition (Fig. 6). An improved level of nutrition will overcome much of the effect of internal parasitism. Goats in general, including but not restricted to the Angora, are highly susceptible to coccidiosis. However, coccidiosis tends to be more serious with the confined animal and excellent coccidiostats are available for use under these conditions.

Since the fiber produced is the major product of value from the Angora any condition which lowers the value of the fleece is serious. The mohair fleece readily picks up foreign material and the seed head of many plant species causes trouble (Bassett *et al.*, 1970; Bassett and Engdahl, 1971). For this reason, it is almost impossible to produce quality mohair in regions where certain plant species are present. On a world basis, there are many such species and no attempt will be made to identify them at this point. Mohair can be carbonized (removal of vegetable material by acid decomposition of foreign matter) as is often practised with wool, but the adverse effect on the fiber is greater than with wool. Many producers spray their goats with oil (oleic acid) in the belief that this helps to eliminate vegetable matter defect in the fiber (Bassett and Engdahl, 1971). However, the major effect of this practice is to increase grease fleece weight from which an advantage may accrue to the producer depending on the marketing system. Another practice is to shear immediately before release of the offending seed heads. This may have the effect of limiting the defect to one 6-month clip and this is the reason that some producers have at times sheared three times per year. Shearing three times per year invariably results in one or more clips with inadequate fiber length.

E. Acknowledgements

The author wishes to thank Dr J. H. Dreyer for permission to use Figs 1 and 2 which were originally published in *South African Journal of Agricultural Science* **10**, 477–500 reproduced under Copyright Authority 5917 of 25.5.1977 of the Government Printer of the Republic of South Africa. Figure 3 is used courtesy of Pringle of Scotland. Other figures were provided by the author or have occurred in various reports from the Texas Agricultural Experiment Station.

References

Al'Meev, I. A. (1973). *Anim. Breeding Abstr.* **43**, 136.

Al'Meev, I. A. (1975). *Anim. Breeding Abstr.* **41**, 610.

Bassett, J. W. (1966). Changes in mohair fleece characteristics as influenced by age and season. *Tex. Agric. Exp. Sta.* PR-2402.

Bassett, J. W. and Engdahl, G. R. (1971). Influence of vegetable matter defect on grease mohair value. *Tex. Agric. Exp. Sta.* PR-2935.

Bassett, J. W. and Engdahl, G. R. (1971). Removal of vegetable matter defect from mohair before shearing. *Tex. Agric. Exp. Sta.* PR-2936.

Basset, J. W., Engdahl, G. R. and Shelton, M. (1970). Vegetable matter defect in mohair. *Tex. Agric. Exp. Sta.* PR-2754.

Burns, M. (1965). *Trop. Agric. Trinidad* **42**, 243–259.

Burns, R. H., Von Bergen, W. and Young, S. S. (1962). *J. Textile Inst.* **53**, T45–T68.

Carter, H. B. (1939a). *J. Council Sci. Ind. Res. Aust.* **12**, 250–258.

Carter, H. B. (1939b). *Aust. Vet. J.* **15**, 210–213.

Carter, H. B. (1943). Studies in the biology of the skin and fleece of sheep. *Bull. Conc. Sci. Ind. Res. Aust.* No. 164.

Carter, H. B. and Hardy, M. H. (1947). Studies in the biology of the skin and fleece of sheep. *Bull. Conc. Sci. Inc. Res. Aust.* No. 215.

Davis, W. and Shelton, M. (1965). An analysis of some factors affecting weaning weight of Angora kids. *Tex. Agric. Exp. Sta.* PR-2341.

Devendra, C. and Burns, M. (1970). "Goat Production in the Tropics". Technical communication No. 19. Commonwealth Bureau of Animal Breeding and Genetics.

Dhororo, D. R. (1975). *Angora Goat Mohair J.* **17**, 17.

Dreyer, J. H. and Marincowitz, G. (1967). *S. Afr. J. Agric. Sci.* **10**, 477–500.

Duerden, J. E. and Ritchie, M. I. F. (1924). *S. Afr. J. Agric. Sci.* **21**, 480–497.

Forte, O. W. (1977). Personal communication.

Gallagher, J. R. (1972). Postnatal secondary follicle maturation in the Angora goat. *Tex. Agric. Exp. Sta.* PR-3028.

Gallagher, J. R. and Shelton, M. (1972). *J. Anim. Sci.* **34**, 319–321.

Gallagher, J. R. and Shelton, M. (1973). Fiber traits of primitive and improved sheep and goats. *Tex. Agric. Exp. Sta.* PR-3190.

Huston, J. E. and Shelton, M. (1967). The influence of level of protein on the performance of billy kids fed in dry lot. *Tex. Agric. Exp. Sta.* PR-2454.

Huston, J. E., Shelton, M. and Ellis, W. C. (1971). Nutritional requirements of the Angora goat. *Tex. Agric. Exp. Sta. Bull.* 1105.

Le Jambre, L. F. and Royal, W. M. (1974). *Proc. Aust. Soc. Anim. Prod.* **10**, 179.

Mamadaliev, F. K. H. (1974). *Anim. Breeding Abstr.* **43**, 48.

Margolena, L. A. (1966). *Virginia J. Sci.* **17**, 32–49.

Margolena, L. A. (1974). Mohair histogenesis, maturation and shedding in the Angora goat. *US Dept. Agric. Tech. Bull.* 1495.

Montagna, W. and Ellis, R. A. (1958). "The Biology of Hair Growth". Academic Press, New York and London.

Nikitenko, V. A. and Zaporozhtsev, E. B. (1972). *Anim. Breeding Abstr.* **41**, 74.

Pinkus, H. (1958). "Embryology of Hair. Biology of Hair Growth". Academic Press, New York and London.

Ryder, M. L. (1956). *J. Agric. Sci.* **47**, 6–11.

Ryder, M. L. (1966). *Anim. Prod.* **8**, 289–302.

Shatunov, V. A. (1973). *Anim. Breeding Abstr.* **42**, 130.
Shelton, M. (1960). *J. Anim. Sci.* **19**, 302–308.
Shelton, M. (1961). Factors affecting kid production of Angora does. *Tex. Agric. Exp. Sta.* MP-496.
Shelton, M. (1964). *J. Anim. Sci.* **23**, 355.
Shelton, M. (1969). Coccidiosis in Angora goats. *Tex. Agric. Exp. Sta.* PR-2635.
Shelton, J. M. and Huston, J. E. (1966). Influence of level of protein and other factors on the performance of yearling billies fed in dry lot. *Tex. Agric. Exp. Sta.* PR-2399.
Shelton, M. and Bassett, J. W. (1970). Estimate of certain genetic parameters relating to Angora goats. *Tex. Agric. Exp. Sta.* PR-2750.
Shelton, M. and Groff, J. (1974). Reproductive efficiency in Angora goats. *Tex. Agric. Exp. Sta. Bull.* 1136.
Shelton, M., Davis, S. P. and Bassett, J. W. (1965). A preliminary study of the importance of staple length in selecting Angora goats. *Tex. Agric. Exp. Sta.* PR-2343.
Shelton, M., Thompson, P. V. and Huston, J. E. (1976). Nutritional efficiency of energy use for wool production. *Tex. Agric. Exp. Sta.* PR-3403.
Short, B. F. (1955). *Aust. J. Agric. Res.* **6**, 683–872.
Smith, I. D., Clarke, W. H. and Turner, H. N. *J. Aust. Inst. Agric. Sci.* **39**, 128–131.
Spector, W. S. (1956). "Handbook of Biological Data", p. 60. Sanders, Philadelphia.
Stampa, S. (1975). *Angora Goat Mohair J.* **17**, 61.
Stapleton, D. L. (1976). *Proc. Aust. Soc. Anim. Prod.* **11**, 45–48.
Van Heerden, K. M. (1964). *J. South Afr. Med. Assn.* **35**, 19–20.
Von Bergen, W. and Mauersberger, H. R. (1947). "American Wool Handbook". Textile Book Publishers, New York.
Van der Westhuysen, J. M. (1974). *Angora Goat Mohair J.* **17**, 29.
Van Rensburg, S. J. (1970). Reproductive physiology and endocrinology of normal and habitually aborting Angora goats. A Thesis for Doctor of Veterinary Science. Dept. of Physiology, University of Pretoria, South Africa.
Van Rensburg, N. A. J. and Wessels, N. G. (1968). "Angora Goats. In the Small Stock Industry in South Africa". Government Printer, Pretoria, South Africa.
Wentzel, D. and Dreyer, J. H. (1967). *S. Afr. J. Agric. Sci.* **10**, 1049–1054.
Yusupov, Kh. (1974). *Anim. Breeding Abstr.* **43**, 48.

13. Husbandry

C. Gall

Little scientific work has been directed towards systems of managing goats and to their husbandry. This partly reflects the negative attitude towards goats in past decades. Strangely enough, since it had been widely accepted that goats were dangerous to vegetation, it was apparently not considered worth investigating these relationships and experimenting with husbandry of goats, where appropriate, using methods that would avoid negative effects. However, in many publications practical aspects of management are discussed, mainly for intensive conditions and on the basis of individual experience (Mackenzie, 1967; Devendra and Burns, 1970; French, 1970; Gall, 1971; Quittet, 1975; Guss, 1977). Furthermore, many publications of extension services and the like give recommendations for management of goats, indicating particular requirements of breeds and of local conditions.

A. Size and Structure of Herds

Due to their early maturity and fertility, goats have the potential to increase rapidly in number. This is one of their virtues, because herds can be quickly established and provide their owners with a self-generating source of protein food or income, and herds whose numbers are reduced by drought or epidemics increase quickly. The annual growth rate of goat herds may range between 30 and 40% (Dahl and Hjort, 1976) and exceed that of cattle and sheep. Rebuilding a herd, which has been reduced by one half, to its original size will take about 9 years in cattle but only about 2 years in goats. In herds mixed of cattle and goats with a ratio of 9:11 or 1:1, re-establishment

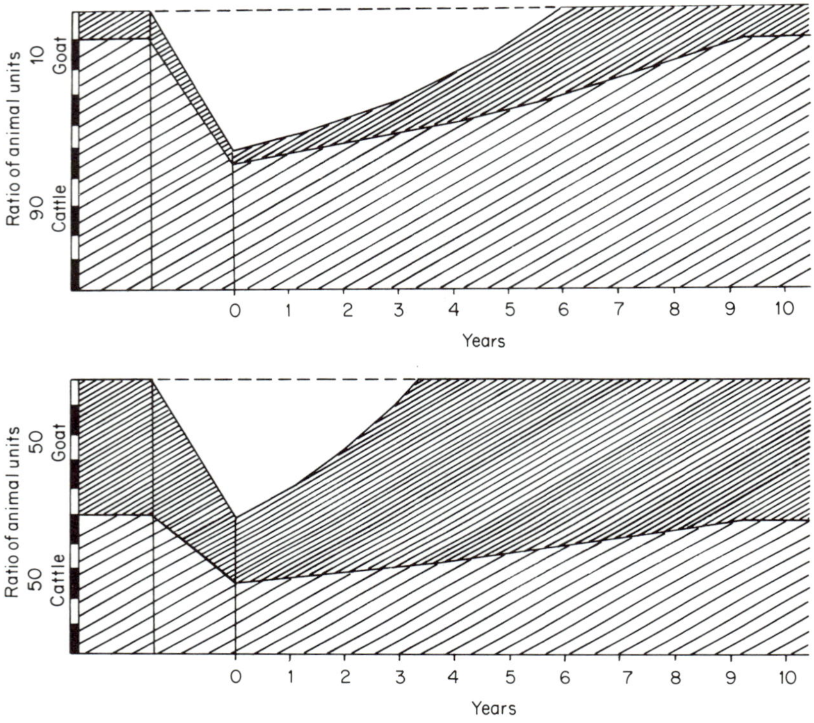

Fig. 1. Recovery of stock numbers after 50% reduction at different cattle:goat ratios (Peters and Horst, 1980).

Assumptions	Cattle	Goat
birth rate (%)	55	140
losses in first year (%)	15	20
losses to first breeding (%)	5	10
effective reproductive rate (%)	40	98
losses among breeding females (%)	12	15
age at first breeding (years)	4	2

of animal numbers after the same reduction by a half will take 6 or 3·5 years respectively, but the ratio will have changed to 2·25:1 and 0·5:1 respectively (Fig. 1) (Peters and Horst, 1980). However, rapid reproduction entails the danger of overstocking, and in most areas where goats are kept, overgrazing is apparent.

Large herds seem to be an almost universal goal of goat-keepers. Much has been written about the reasons for this attitude. The most conspicuous arguments for large herds are the insurance against recurrent draught or epidemics and the growth of wealth which may be easily capitalized when goats are sold in order to meet small cash needs. The notion of prestige is disputed. It is so closely related to wealth and economic possibilities that it is difficult to consider prestige as a particular quality derived from large herds. On the other hand, under conditions of communal land use, there is an incentive for the individual stock-owner to increase his stock numbers and try to maximize his individual share of the common grazing resources (Meyn, 1971).

If goats were to provide the whole food for a West African Nomad family, 74 to 86 goats would be necessary to supply the protein (depending on their live weight) but 205 to 223 to supply the energy (Dahl and Hjort, 1976). The productivity of a herd depends on its composition. Aged infertile does and adult males not needed for breeding are the unproductive part of the herd, which contribute unnecessarily to overstocking. In East Africa, many herds contain as much as 45% adult males (Allonby, 1976). Improved marketing conditions which ensure an adequate price at any time of the year, would encourage timely slaughter of these animals.

In commercial goat dairy production under intensive conditions, the size of herds is determined by the available surface and the family income which can be produced, as compared with alternative forms of land use, or by labour availability. This is discussed in Chapter 1 and 10.

While in some semi-arid areas families may live entirely on goats, mixed herds of sheep and goats (and cattle) are more common. Behaviour of sheep and goats is very similar in many respects. Sheep raised with goats behave like goats, and goats raised with sheep behave like sheep. But after 60 to 90 days association with their own species, normal affinity to their own kind is re-established. In general, goats are more social than sheep (Price and Tomlinson, 1979). Due to different feed preferences, sheep and goat are to a great extent complementary. The reproductive pattern of both species may be different too. For instance, in the Atlas mountains of Morocco it was observed that sheep were lambing once a year only, with a peak in January and February, about half of the parturitions of goats were between September and December but another third occurred in March and April (Bourbouze and Guessous, 1977). Hence, in a mixed flock, parturitions may be spread more evenly over the months of the year. Sometimes there is a tendency to exchange small stock for cattle (or other large stock) if certain numbers are exceeded. The higher prestige connected with owning cattle is involved as well as the intention of complementing the utilities of stock and to spread the risk, since the risk factors for small and large stock are somewhat different (Dahl and Hjort, 1976).

Goats (and sheep) are complementary to cattle in several respects: in grazing, goats make better use of browse and may live entirely on it; goats may lactate in different months of the year when cows are dry. When oestrus occurs shortly after the onset of the rainy season, goats will begin to lactate 5 months later, while it takes 9 months in cattle; a small herd of goats is less demanding on grazing and may be kept in the immediate surroundings of the household thus supplying daily milk, while cattle are grazing further away. Goats may be tended and milked by minors; goats may supply small quantities of meat for festive occasions or eating when a cattle carcass would exceed the immediate requirements.

B. Behaviour

An essential basis for successful animal management is knowledge of animal behaviour. Inventories of goat behaviour are still incomplete (Hafez and Scott, 1962; Sambraus, 1978). Many of the particular features of goats' behaviour may be understood in terms of adaptation to its original environment, mountainous Mediterranean country. Wild goats live in small groups within which, even in herds of considerable size, they recognize each other individually. Urine odour may be an important means of recognition. Females differentiate between urine of males better than between that of other females (Baldwin, 1977).

Goats select carefully individual plant foliage which enables them to maintain a nutritive balance under a wide range of feed supplies: leaves of woody plants, which may contain more than 20% crude protein; fruits which may be rich in both energy and protein; twigs and bark which are almost pure cellulose with a high lignin content.

Goats feed on a great variety of fodder plants and completely reject very few plants. Taste preference has been established (in decreasing order) as sweet—salty—acid—bitter (Goatcher and Church, 1970). Taste responses may to some extent be altered by taste modifiers like gymnemic acid, inosinic acid or monosodium glutamate (Mehren and Church, 1976), which may be useful for inducing goats to consume by-products or wastes. Goats are browsers by preference. As such, during evolution they may have had little opportunity to develop resistance to parasites whose larvae are less present in the higher and dry parts of plants. The adaptation to such a wide range of feeds during evolution must have led to a dependency on feed variety, which is evident from difficulties with one-sided diets, especially when nutrients are unbalanced. Goats are able to climb rocks and steep slopes reaching the most exposed sites in a mountainous environment. The hooves are particularly useful for these surfaces, with soft soles protected by a rim of very hard wall horn which restores abrasion by rapid growth. Wherever the goat has no need or desire to climb rocks, the abrasion of the horn will be insufficient and malformed hooves will result.

Some of the difficulties encountered in managing goats intensively and in confinement can be explained by these behavioural characteristics. Goats are not very gregarious, do not like to feed on grass, like to alternate between different feeds, they select the more nutritious parts of the plants leaving much waste and are sensitive to nutrient imbalances. They are susceptible to internal parasites notably when grazed on grass. Goats are alert and easily find gaps in the system of confinement, they are easily frightened but learn quickly. Awareness of the conditions of the original goat habitat will aid understanding of the goats' behaviour and adequate husbandry can compensate for the inadequacies of the artificial environment.

The hierarchy in goat herds depends on body weight, horns and age. In herds with polled animals and with many animals of the same age, the order is less firmly established (Sambraus, 1978). There seems to be a relation between rank order and productivity (Marincowitz, 1968, Pretorius, 1970). Faced with new and palatable feed offered with insufficient space at the feed rack, goats may try to break through the established order and the resulting fights may reduce feed intake (Syme *et al.*, 1974).

C. Reproductive Management

1. Breeding Males

Young males may be used for breeding when they are about 7 months of age. Live spermatozoa were present in all ejaculates of Damascus goats tested at this age (Louca *et al.*, 1977) (see also page 171). Generally one male is considered sufficient for about 50 females. When the breeding season is restricted, more males may be needed in order to ensure a high conception rate. During the breeding season, males, engaged in fighting other bucks and breeding does, may devote little time to feeding. They may draw heavily on body reserves accumulated during the summer months. It has been observed that bucks almost stopped feeding during the mating season (Louca *et al.*, 1977) and lost up to 17% of their body weight between August and October (Gall and Fautz, 1971). Breeding males may serve does up to 20 times a day, and about 350 services seem to be possible in a limited breeding season of the septentrional zones (Sambraus, 1973).

Where goats are kept in small numbers, the individual goat owner does not normally keep a male. One sire kept by an individual will serve goats of many surrounding owners who will bring does in oestrus to the male. In Korea, goats in oestrus are brought to markets where bucks are available for breeding for a small charge (Rumich, 1968). In countries like Germany, dwindling of goat numbers has been accelerated by the difficulty of keeping bucks. Faced with the inconvenience of keeping a male, people have often decided to not keep any goats at all. Artificial insemination has not been able to compensate because of unsatisfactory results and costs. De-odorizing bucks has been recommended as a means of overcoming one of the main causes of the inconvenience in keeping males. The odour is limited to the sexual season; bucks kept in clean premises or in the open do not have a strong odour in spring and summer.

In white goats like the Saanen, the hair of the head and neck gets stained during the mating season. It has been shown in England (McEwan Jenkinson *et al.*, 1967) that in entire males, sebaceous glands increase in size from April to October. While there was some increase in all sites sampled, it was most pronounced between the horns. The characteristic odour was analysed from the sebum of males. It was not apparent in females and male castrates, less pronounced in males out of the breeding season, and on body sites other than the head. The exact chemical composition of the smelling substance has not been analyzed but there were indications that these were C_{10} and C_{12} branched fatty acids. Neither sweat nor urine contributed to the odour produced.

The main localization of the scent glands behind the horns or the polls may be destroyed in the same way as the horns. In kids this could be done simply by extending the destroyed area (which is beneficial anyhow for successfully dehorning without formation of scurs) and including in the operation all male and female kids irrespective of visible horn buds (Ford, 1965).

It is generally claimed that the presence of bucks in goat premises will contaminate goat milk with the specific buck flavour. However, Korvald (1958) reported that keeping a buck together with milking does had no influence on milk flavour.

In areas where goats reproduce seasonally, there are often fluctuations of prices of goat products synchronous with the reproductive cycle. This would favour organizing reproduction of goats to produce milk and meat out of season.

The goat's ability to lactate for extended periods if not pregnant may be used to produce equal quantities of milk throughout the year by not serving one part of the herd in every service period, so that it will produce milk when the rest of the herd is dry.

Introducing males into the female herd shortly before the onset of the breeding season will induce cycles in the majority of does within about 8 days (Shelton, 1961). Earlier parturition during the season may not only be favourable with regard to product prices but, under conditions prevailing e.g. in Europe, will favour lactation yields too (see page 321). For herd management it may be practical to limit the mating season in order to have young kids and does who need special attention shortly before and after parturition, for a limited period only. In northern Mexico, for example, it is common practice to limit matings to a very short period of about 2 weeks

(Carrera, 1971). However, it has been shown that in order to have high reproductive rates in a herd, a mating period covering at least two cycles, i.e. about 6 weeks will be necessary (see page 540).

2. Kids

In free grazing herds of goats, where parturitions occur without any attendance by man, survival of kids depends on mothering ability and the firm establishment of mother–kid bonds. Kids are "labelled" by their mother during the first physical contacts. A kid which had no contact with its own dam may be accepted by another doe within a few hours after birth (Gubernick *et al.*, 1979). Hence, orphan kids or kids which are not nursed by their dam may be nursed by a foster mother. While olfactory stimuli probably mediate recognition initially, after about 4 days kids are recognized by their voice (Leonhardt, 1977). When goats are exclusively grazed on extensive range and no pen feeding is possible, dams usually have to leave with the herd for grazing immediately after parturition, while kids stay in the pen during daytime, for about 1 to 2 months. If the does kid early in the morning the time until leaving for grazing may not be sufficient for dam–kid relationships to be established and kids may not be able to suckle their dam without human assistance. Furthermore, the period from birth to the first separation may be too short for the kid to consume sufficient colostrum during the limited period when immunoglobolins are fully absorbed.

When a group of African dwarf does, pregnant after hormonal synchronization, kidded simultaneously in one paddock individual dam–kid relationships were not established; all kids suckled on most of the dams (Scheurmann, 1980). This could be used to set up nurseries with a limited number of more productive goats who stay on the premises and raise the kids.

Mothering instinct in primiparous goats often needs some time to develop fully. This may take so long that the kids are abandoned or labelled by other goats so that afterwards they will not be accepted by their dam. Mothering ability seems to differ between breeds. Poor mothering instinct is attributed to Angora does (Guthery and Beasom, 1978). If does have large pendulous udders with over large teats, i.e. mainly extended cisterns, kids may be unable to suckle and may starve to death with their dams' udders brim full of colostrum. In an experiment with Angora goats about 15% of the kids born alive would have died for this reason (Guthery and Beasom, 1978).

Under normal conditions, twins get about equal chance to nurse since the dam gives the weaker kid priority in nursing (Klopfer and Klopfer, 1977). However, if the vigour of kids differs beyond a certain limit this mechanism does not work and the weaker kid will be abandoned. This may be an efficient control mechanism to ensure sufficient milk for one kid, under harsh environment conditions.

From the observations on kidding behaviour it follows that in extensive operations, goats should for kidding be separated from the herd in supervized pens. Here, goats may be assisted at parturition (which is very seldom necessary) and at the kids' first suckling (if udder conformation is unfavourable). In order to ensure the establishment of firm dam–kid relationships this confinement should continue for about 4 days. Kids should remain on the site for at least 14 days, if the herd has to cover larger distances for grazing.

Except in intensive dairy operations, goats are frequently milked while still nursing their kids. In Chapter 10 the effect of this on kid growth and milk yield is discussed. The usual practice is to separate kids from their dams during the daytime, milk in the evening and then reunite the kids with their dams for suckling during the night. Alternatively, kids may be separated during the night and the does milked early in the morning. This is applied when milk is sold liquid and has to be delivered in the morning. The quantity of milk left for the kids may be geared by the time intervals between milking and suckling. Unless this is handled carefully, kids may not receive sufficient milk and the resulting kid losses may cancel out any gain from additional milk yield. No experiments have been reported on the influence on milk yield and growth performance of the interval between suckling and milking. It is doubtful, however, if milking more than once a day would be of any advantage, because of the additional labour requirements.

3. Length of Productive Life

Little is known about the length of productive life of goats. Under confined and housed conditions, life span may exceed 10 years. Since after mature age the salvage value of the does' carcass does not decrease markedly, the goat is normally kept as long as she reproduces. The span of life may be limited by wear of the teeth. Generally, the average age at death in harsh environments is estimated at 5 to 6 years but about 10 years with Yomut flocks in Iran (Irons, 1972).

D. Herding and Fencing

Under intensive conditions goats are successfully grazed on natural pastures. With adequate pasture management, good utilization of grass is obtained and refusal need not be greater than with cattle, but, as with cattle, it is difficult to combine highest feed intake with minimum pasture waste. Since it is necessary to graze grass while still short, the protein to energy ratio tends to be too narrow and supplementation with energy- and fibre-rich feeds is necessary. Furthermore, intensive grazing enhances the disease risk from parasites. Good fencing is indispensable, if good use is to be made of pastures; otherwise goats will get out in search of easier feeding. Secure fencing may not be easy if differences in feed availability and quality on both sides of the fence are marked. Eight strands may be required for a goat proof fence but for absolute security chain link fences may be necessary. Electric wires are useful for reinforcing fences and for interior subdivisions. They are never absolutely goat proof. At least two wires are required, but three or more are sometimes used.

On extensive range, goats are grazed without being herded both with and without fences. Costs may prohibit fencing as well as herding but losses may be high due to predators and theft. In Zulia, Venezuela for instance, theft has become one of the main limiting factors in goat production (Hernández, 1971). The lack of regulated land tenure precludes any control of trespassing as a preventive measure. Predators may take a heavy toll on goats. It may force goat raisers to refrain from grazing extensive bush pasture or force them completely out of business (Wade and Conolly, 1980). In a trial in Texas, coyotes (*Canis latrans*) took as much as 95% of the kid crop of an Angora herd (Guthrey and Beasom, 1978). Most of the kids were killed during the first few days of life. Adult nannies were not killed as long as there were prey kids available. When the last kid was lost, predation on adults began, it took 24% of the herd. Weak animals and those separating from the herd, made preferential prey.

In many parts of the world goats are herded on pastures mainly in order to graze where feed is available. However herdsmen frequently hold the herd back from travelling to remote grazing grounds in order to avoid physical strain, whereas an experienced and devoted shepherd may encourage the herd to use pasture far away from the pen and the water point. Free grazing goats have been observed to travel 10 km daily (Cory, 1927) but herded goats may cover regularly 15 km and more (Scheurmann, 1980). More invesitigations are required on the influence on feed intake and feed utilization of grazing time, periods for rest and rumination, distances travelled, open or confined grazing (i.e. with the herd dispersed or closely assembled).

The number of goats that may be herded by one person varies according to breed, type of terrain and vegetation. It may be several hundreds in plains and with equally spread forage such as aftermath. In arid areas with sparse bush vegetation, 20 to 30 goats may be the maximum because the animals move quickly and tend to disperse. They are said to be herded more easily together with sheep who are more gregarious and under the same circumstances about 250 to 300 mixed goats and sheep may be herded.

Fig. 2. Criollo goats grazing cotton aftermath, herdsman on horseback.

Herding by children is quite common especially of smaller herds. Children of different families tend to stay with their herds close together and during noon hours to rest together. This may add to overgrazing of pastures near the settlements. Dogs are used for herding in South America (Maubecín, 1979) and all other areas but in general not very frequently. The contention that dogs start killing kids or small adult animals with time is certainly not well founded. Advice on selection of useful dogs, their training and adequate management, including feeding, may be an easy way of improving herd management. Relatively little use is made of herding goats on horseback doubtless because goats predominantly graze hilly and broken terrain and the social environment is not favourable for keeping horses. However, herding of goats on horseback may be quite useful for grazing plains, if only because surveillance is much easier from the elevated position (Fig. 2). Tethering is common practice in agricultural areas. Similarly, two or three animals may be tied together thereby limiting their movements severely.

While goats are a permanent danger to cultivations, this may not necessarily be generalized. For example, in Yemen goats are grazed in wheat to feed on herbs (Fig. 3). This is a sophisticated technique which makes use of the grazing preferences of the goat. It is practised by owners of the fields only. Young fattening males, having grazed on the range for several hours are taken in the afternoon to wheat fields. These are planted in tussocks and not in rows, leaving much space between the groups of plants. The animals use this space and do not step on the plants. They follow their preference and carefully select the herbs growing between the wheat and do not touch

Fig. 3. Goats grazing wheat field in the A.R. Yemen.

the gramineas (personal observation of the author in the A. R. Yemen, October 1979).

Observations in South Africa indicate that in hot and dry climates goats feed a good deal during the night. Hence, better use is made of pasture if goats are allowed to graze at night (Kennan, 1969), but the risk of losses from predation and theft fully justifies the practice of penning goats at night.

Time available for grazing may influence feed intake, physical status of the animals and productivity. Carrera (1971), with criollo goats in Mexico, observed herds taken out for grazing at dawn and after 8 a.m. respectively. Herdsmen claimed that the soil and plants wet from dew had an adverse effect on goats. However, milk production and the general status was better in goats grazed for longer periods. Similarly, from observations of local Tunisian goats, it was concluded that the herdsmen's habit of rushing grazing goats, probably with the intention of keeping the herd together, is detrimental to feed utilization because it does not allow for the necessary periods of rumination between grazing (Scheurmann, 1980).

Goats very actively graze and browse. When taken for grazing in the morning, they may discontinue feeding within about an hour and then rest and ruminate. If they are not allowed to do so, e.g. when forced to continue walking, they seem to feel a desire to ruminate expressed by short attempts to chew an odd bolus (Scheurmann, 1980). Under these conditions goats spend about the whole night ruminating. Probably more normal is the partitioning of activities as observed by Carrera (1967) and Carrera and Aguirre (1971) with about one-third of the time spent feeding, one-third ruminating and the remaining one-third in search of feed or resting. For two-thirds of the night time, the animals did ruminate. In semi-arid areas with low forest or high bush vegetation, shepherds may lop trees or bushes, cut sprouts of yuccas, and pads, branches or fruits of cacti which are out of reach of the goats. Loppings may be the only available fodder during winter months, in Mediterranean countries as Corsica or Morocco. Grazing of goats in forests may produce severe problems for the production of timber. However, by complete elimination of grazing in forest areas a valuable resource will remain unused and cooperation should be aimed at instead of confrontation in forest areas. Undergrowth is undesirable in forests because it competes with trees and is fuel for wood fires. Use of fodder should be made possible by cutting during the early stages of plantation and later by controlled grazing. Since productivity of tropical woods is often quite limited some of them may be replaced with fodder trees like *Acacia albida* which gives excellent results in Africa or *Leucaena leucocephala* in Latin America, South Asia and in Africa too.

If goats are used for controlling bush and weeds on pastures grazed mainly by other livestock (cattle and sheep), as in Texas or South Africa, specific grazing techniques have to be applied. It requires skill and experience to time the grazing with goats ("put and take") so that undesirable vegetation is checked but grass is saved for those animals who are expected to make most use of the pastures.

Renewed interest has been reported from New Zealand in using goats for weed brush and control (created by increasing costs of mechanical and chemical weed control and environment protection). It is reported that goat-proof fencing was no more expensive than good standard hill-country fencing. Pregnant lactating and growing goats may have higher nutrient requirements than yielded by scrubland if stocking is sufficient to suppress scrubs and weed. During winter, forage production might be unsufficient, so careful management will be necessary in order to avoid goats damaging the grass cover and reducing cattle and sheep feed (Batten, 1979).

The use of grazing grounds in dry areas may be limited by water availability. Strategic localization of water points may increase pasture use. While it will be difficult to pay for the cost of a well by the additional production from goats alone, simple forms of water catchments or cisterns may be an economical way of bringing goats through the dry season (Fig. 4).

Fig. 4. Waterproofed run-off area drained into a water tank in Jamaica. With 300 mm annual rainfall, 300 litres can be collected from 1 m^2.

Fig. 5. Goat pen in Venezuela.

Fig. 6. Kid pen in Northern Mexico. Kids are pegged at a simple guard against wind, sun and rain.

E. Facilities and Care of Goats

The necessary goat facilities differ according to systems of management and climatic conditions which of course overlap to a good extent. Aspects of goat housing in cold climates are discussed in Chapter 15. Housing of goats in the humid tropics is discussed by Devendra and Burns (1970). In dry tropical countries goat installations normally include a fenced area for animal handling and a roofed pen for night shelter and protection against rain (Fig. 5). Very simple protection is sometimes provided for kids during their first week of life when they cannot yet leave with their dams for grazing (Fig. 6). Goats are often milked in a rather disorganized way where does with sufficient milk are picked from the herd and milked on the spot (Fig. 7). If larger herds with higher production are milked regularly, they may be driven through a corridor. Although milking parlours are very simple to construct and goats are easy to train, little use is made in tropical countries of such convenient facilities for improving the milking routine (Fig. 8).

Unless goats walk daily on hard ground and cover considerable distances or climb rocks, abrasion of hoof horn does not balance horn growth. If not

Fig. 7. Milking of criollo goat in Venezuela.

Fig. 8. Milking parlour in Mexico.

trimmed, horns tend to malform (Fig. 9). In goats kept in confinement or on soft pasture in the plains, hooves have to be trimmed at least every 6 months. Most of the work can be done with garden (rose) pruning shears (Fig. 10).

Supernumerary teats are quite inconvenient for milking. They are easily removed in kids. When kids are dehorned, the mammary gland may also be checked and extra teats cut off with curved scissors. Since they are of genetic origin, it should be registered that an animal had extra teats so that it may not be used for breeding in herds with a high occurrence of the defect.

Horns are generally considered undesirable in goats kept in confinement. The general contention that horns involve the risk of goats injuring each other by butting, is at variance with observations that among horned goats rank order is more stable than among polled goats, apparently because the horned animal emphasizes its dominance with more authority (Sambraus, 1978). Since breeding for polledness affects fertility (Chapter 4), dehorning is recommended. Growth of horns commences early and is very vigorous, in males in particular. Therefore, kids have to be disbudded during the first 2 weeks of life, male kids during the first week. It is a good practice to dehorn as soon as kids suckle strongly and frequently. Techniques applied are the hot iron, nitric acid or caustic paste. Nitric acid causes the least pain but takes more time;

Fig. 9. Deformed hooves of scavenger village goat in A.R. Yemen.

the cherry red iron destroys a ring of skin around the bud very rapidly and most effectively but the apparatus may be too expensive; caustic paste is difficult to control, because effect is not immediately visible, and deforming scars may result. Although the pain at dehorning is very short when the operation is done skilfully, kids may be easily anaesthetized.

Grown horns may be removed surgically. Elastic bands work with females but failures are frequent and deformed horns may continue to grow. This is almost invariably the case in males. Furthermore, since extended pain and distress are caused by the band, this is not a recommended technique.

Males intended for slaughter at higher body weights are often castrated. It has been shown in Beetal and their crossbreeds with Alpine and Saanen goats that growth rate until slaughter at 9 months of age was not affected by castration. There was no difference whether kids were castrated at 5 days, 3 or 6 months of age. Carcass weight was slightly higher in intact males. Dressing percentage was about 50% in all three breed groups (Chawla and Nath, 1979). Entire Damascus male goats grew faster than kids castrated at the age of 7 days, and at 7 months weighed about 14% more. Thereafter,

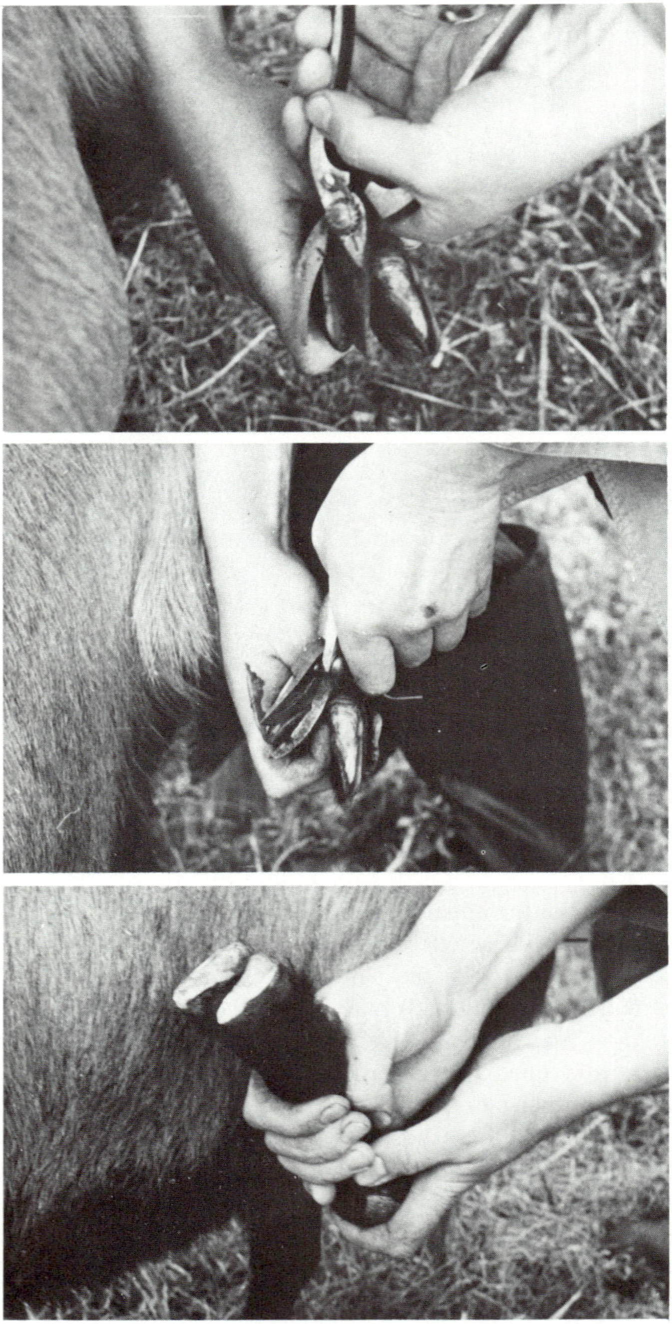

Fig. 10. Hoof trimming with garden shears: (*top*) wall horn tends to grow over the sole; (*middle*) most of the trimming is done quickly and safely with the shears; (*bottom*) trimmed hoof; note trimmed point.

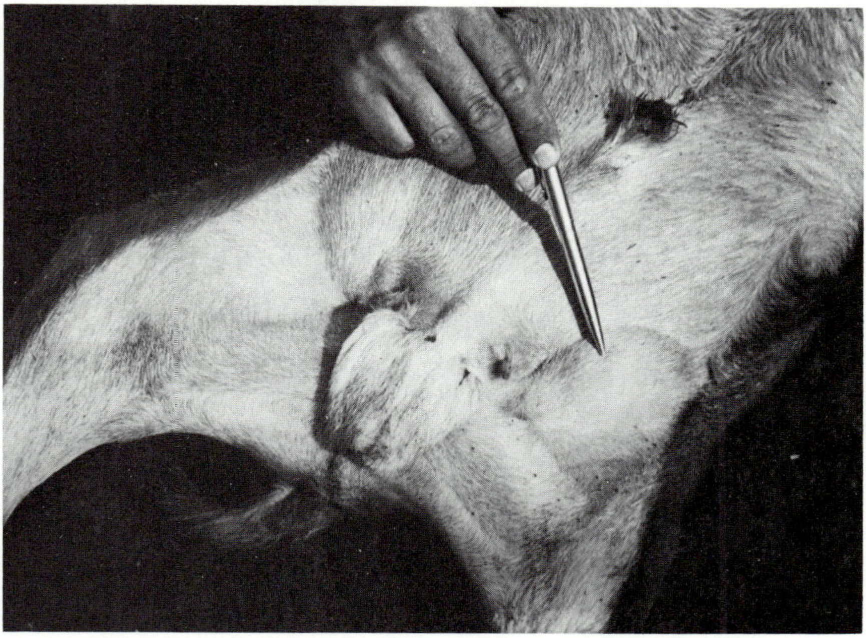

Fig. 11. Castration of a criollo goat in Venezuela. The testis is twisted around the spermatic cord whereby it is held back in the inguinal canal and atrophies.

with sexual activity beginning, they about ceased to grow. Carcass weight at 10½ months (318 days) was about the same in both groups, but muscle to bone ratio in the chops of entire males was 1·53 *vs* 1·08 in castrates. The meat of castrates was free of the flavour which was invariably present in entire males. Castration at 7½ months had the same effect on meat flavour as early castration (Louca *et al.*, 1977). Hence, castration is necessary if animals are to be kept beyond the onset of sexual activity, which may be earlier than the 9 months reported in the Damascus males, depending on breed and geographical latitude. Castration will be beneficial for meat flavour and produce more lean meat.

Castration methods are the same as in other species: surgical, rubber bands, emasculator. A traditional method is practised in Venezuela. The testis is twisted within the scrotum around the spermatic cord and pushed back into the inguinal canal (Fig. 11). It requires great skill but is very convenient because every risk from infection or flies is avoided.

Under extensive management, goats may be conveniently identified by neck straps with number tags. Ear tags tend to tear out; they will serve for a very limited time and then leave ugly slashes. For permanent identification goats

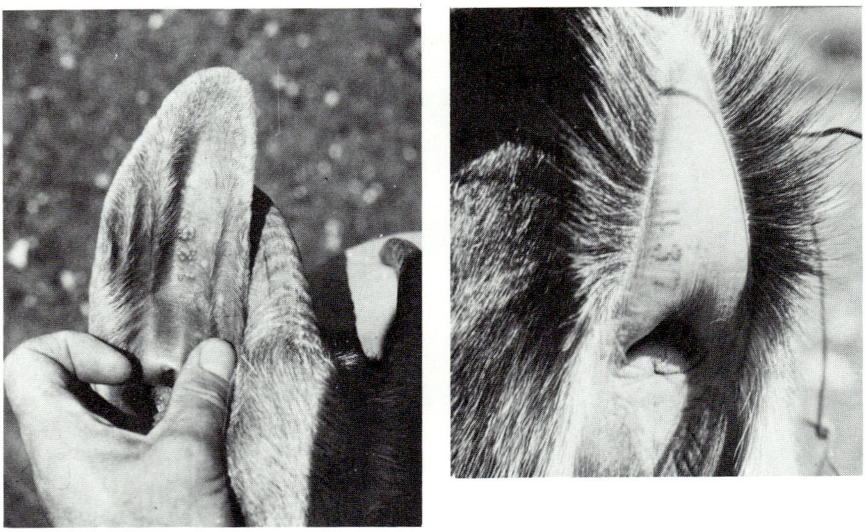

Fig. 12. Tattoos in ear (*left*) and under the tail (*right*).

Fig. 13. Halter for preventing weaned kid from suckling.

have to be tattooed. As with other species, the preferred site is inside the ear but the tail may be convenient for dairy goats (Fig. 12).

Various devices are used to restrain goats. Two legs may be hobbled, a yoke around the neck may prevent goats from leaping over fences or hedges, a loose suspended large udder may be retained by a bag fixed over the rump. In a similar way the udder may be protected from suckling, against the weaned kid, vicious companion goats or the goat herself. Otherwise, kids may be prevented from suckling by halters with a peg fixed in the mouth (Fig. 13).

References

Allonby, E. W. (1976). *In* "The Role of Sheep and Goats in Agricultural Development", pp. 30–31. Winrock International Center, Morrilton, Arkansas, USA.

Baldwin, B. A. (1977). *Appl. Anim. Ethol.* **3**, 145–150.

Batten, G. J. (1979) *In* "Proc. 32nd Weed and Pest Control Conf.", Dunedin, New Zealand, August 1979, pp. 292–296. New Zealand Weed and Pest Control Society, Palmerston North.

Bourbouze, A. and Guessous, F. (1977). In "Symposium on Goat Breeding in Mediterranean countries", Malaga-Grenade-Murcia, Spain, October 1977 pp.207–227. EAAP and Spanish National Committee Animal Production, Madrid.

Carrera, C. (1971). Algunas Observaciones del Ganado Caprino en México, con Especial Enfasis en el Noreste. Dept. Zoot., Div. Cienc. Agropec. Marit., ITESM Monterrey, Mexico.

Carrera, C. and Aguirre, E. (1970) XII Inf. Invest., pp. 166–168. ITESM Monterrey, Mexico.

Chawla, D. S. and Iqbal Nath (1979). *Indian J. Anim. Sci.* **49**, 406–408.

Cory, V. L. (1927). *Tex. Agr. Exp. Sta. Bull.* 367.

Dahl, G. and Hjort, A. (1979). "Having Herds. Pastoral Herd Growth and Household Economy". Dept. of Social Anthropology, Univ. Stockholm, 335 pp.

Devendra, C. and Burns, M. (1970). "Goat Production in the Tropics". Tech. Comm. No. 19, CBABG, Edinburgh, 184 pp. Commonwealth Agricultural Bureaux, Farnham Royal, Bucks, England.

Ford, R. S. (1965). How to deodorize bucks. *Dairy Goat J.*

French, M. H. (1970). "Observations on the Goat", FAO Agricultural Studies, No. 80 (English, French and Spanish editions), 204 pp. Food and Agricultural Organization of the United Nations, Rome.

Gall, C. (1971). "Producción Caprina y Ovina. Primera Parte: Producción Caprina" 88 pp. Monterrey, N. L., México, ITESM, 1971.

Gall, C. and Fautz, J. (1971). *In* "II. Int. Conf. Goat Breeding", Tours 1971, pp. 354–357. ITOVIC, Paris.

Goatcher, W. D. and Church, D. C. (1970). *Anim. Sci.* **31**, 364–372; 373–382.

Gubernick, D. J., Jones, K. C. and Klopfer, P. H. (1979). *Anim. Behav.* **27**, 314–315.

Guss, S. B. (1977). "Management and Diseases of Dairy Goats", 222 pp. Dairy Goat Publishing Corporation, Scottsdale, Arizona, USA.

Guthery, F. S. and Beasom, S. L. (1978). *J. Range Mgt.* **31**, 168–173.
Hafez, E. S. E. and Scott, J. P. (1962). *In* "The Behaviour of Domestic Animals" (E. S. E. Hafez, ed.), pp. 297–333. Ballière, Tindall and Cox, London.
Hernández, I. (1971). "Situación Actual de la Explotación Caprina en el Distrito Urdaneta—Estado Zulia", 54 pp. Univ. del Zulia, Fac. Agr. Maracaibo, Venezuela.
Irons, W. (1972). Cited by Dahl and Hjort (1976), p. 94.
Kennan, T. C. D. (1969). *Rhodes. Sci. News* **3**, 331–336.
Klopfer, P. and Klopfer, M. (1977). *Anim. Behav.* **25**, 286–291.
Korvald, T. (1958). Rpt. 68, Dairy Res. Inst. Agric. Univ. Norway, 16 pp.
Leonhardt, M. L. (1977). *Appl. Anim. Ethol.* **3**, 211–219.
Louca, A., Economides, S. and Hancock, J. (1977). *Anim. Prod.* **24**, 387–391.
Mackenzie, D. (1956). "Goat Husbandry" (2nd edn 1965), 368 pp. Faber and Faber, London.
Marincowitz, G. (1968). *Angora Goat Mohair J.* **10**, 25–26.
McEwan Jenkinson, D., Bluckburn, P. S. and Proudfoot, R. (1967). *Br. Vet. J.* **123**, 541–549.
Mehren, M. J. and Church, D. C. (1976). *Anim. Prod.* **22**, 255–260.
Meyn, K. (1971). World Bank, Washington. Personal communication.
Peters, K. J. and Horst, P. (1980). "Entwicklungspotential der Ziegen in den Tropen und Subtropen". Expert Consultation "Technical Assistance to Goat Production in the Tropics and Subtropics", Berlin, January 1980. German Foundation Int. Development.
Pretorius, P. S. (1970). *Agroanimalia* **2**, 161–164.
Price, E. O. and Tomlinson, K. A. (1979). *California Agric.* **2**, 4–6.
Rumich, B. (1968). FAO, Reginal Office, Bangkok, Mimeograph, 7 pp.
Sambraus, H. H. (1973). "Fortschr. d. Verhaltensforschg" 12. Paul Parey, Berlin, Hamburg.
Sambraus, H. H. (ed.) (1978). "Nutztierethologie", pp. 152–167. Paul Parey, Hamburg.
Scheurmann, E. (1980). Schadverhalten der Ziege am semiariden Standort, 84 pp. Institut für Tropische Veterinärmedizin, Univ. Gießen.
Shelton, M. (1960). *J. Anim. Sci.* **19**, 368–375.
Syme, G. J., Syme, L. A. and Jefferson, T. P. (1974). *Anim. Prod.* **18**, 309–312.
Wade, A. and Connolly, E. (1980). *Tex. Agric. Prog.* **26**, 12–16.

14. Diseases

Christine S. F. Williams (ed.)

This chapter reviews certain infectious and parasitic diseases which have been the subject of recent investigations. The individual reviews were prepared by several contributors who are particularly familiar with the respective diseases. These contributors are cited in brackets below the section headings, and their generous cooperation is gratefully acknowledged. The diseases are presented in groups according to the main symptoms which they produce in the animal. It is stressed, however, that in many diseases other symptoms may prevail, e.g. listeriosis may manifest itself mainly by abortions, salmonellosis may produce hardly any symptoms at all, paratuberculosis may produce unthriftiness, and some diseases may show a wide variety of symptoms.

A. Diseases which Cause Fever, Enteritis or Pneumonia

1. Bluetongue
(H. E. Metcalf)

Bluetongue is caused by an insect-transmitted virus which infects many species of domestic and wild ruminants. Goats can be readily infected with the virus (Luedke and Anakwenze, 1972). Infection in goats is primarily inapparent and little or no evidence of disease is usually seen. Clinical disease in goats has been reported from Israel (Komarov and Goldsmit, 1951) and India (Sapre, 1964). The disease was described as resembling the disease seen in sheep with swelling of the lips, salivation, and hyperemia of the oral mucous membranes. The disease has not been reported in goats from other areas of the world where the virus is enzootic. In the United States, sera from many goats have been found to have antibody to bluetongue virus but recovery of the virus has never been reported from a natural case.

There is no specific treatment for this disease, other than general nursing care. Because the infection is primarily inapparent, goats are not routinely vaccinated.

2. Foot and Mouth Disease
(J. W. McVicar)

Etiology

Foot-and-mouth disease (FMD) is a contagious viral disease of cloven-hoofed animals including cattle, sheep, swine and buffalo, as well as goats. Many wild species such as deer and antelope are also susceptible. The disease is widespread in Asia, Africa and South America. It is considered enzootic in Europe although largely controlled by vaccination. Australia, New Zealand, Denmark, Finland, the United Kingdom, and Central and North America are currently free of FMD.

Seven immunologically distinct types of the virus are recognized and numerous subtypes and strains have been identified. Infection results from direct contact with infected animals, from contact with a virus-contaminated environment, and most commonly by inhalation of aerosols containing virus from animals in the early stages of infection.

Diagnosis

The disease is characterized by vesiculation of the epithelium of the mouth, nostrils, muzzle, feet and teats. In goats, blister-like lesions on the tongue, lips, gums or dental pad rupture easily, leaving shallow erosions sometimes with small tags of epithelium around the periphery. Foot lesions may involve the coronary band or interdigital space. They start as blanched areas but soon rupture, revealing raw, reddened, extremely sensitive underlying tissues. Lameness may be the first sign of disease. Pregnant does may abort and nursing kids may die suddenly, with lesions confined to pale streaks in the heart muscle. The disease is frequently relatively benign in adults, and transient loss of appetite, lameness, or drop in milk production are the only signs; however, outbreaks with more severe symptoms have been reported.

Differential diagnosis should include bluetongue, contagious ecthyma, goat pox, foot rot, and vesicular stomatitis. The disease can be diagnosed positively only in a qualified laboratory by isolation of the virus from samples of vesicular material, saliva, or blood or, later in the course of infection, by identification of antibodies in the serum.

Treatment

There is no treatment for FMD.

Control

Vaccination can be used as a means of control, but eradication should be the goal. Direct monetary losses due to FMD in goats have not usually been severe except in those relatively few flocks where abortions, neonatal death, severe lameness, or cessation of lactation becomes excessive. Severe losses do result from eradication procedures, however, which require the slaughter of all susceptible animals on or near an infected premises. Additional economic impact is felt as a result of sanctions placed on infected countries by countries free of the disease.

In summary, FMD may have a relatively minor impact on the goat industry world-wide, but the political and economic implications of the disease are enormous. The extensive international movement of people and animals has increased the risk of accidental spread of the virus, and owners must be continually vigilant because early diagnosis is the most effective weapon (Honeker, 1938; McVicar and Sutmoller, 1968; 1971, 1972; Reppin, 1929).

3. Leptospirosis

(C. S. F. Williams)

Leptospirosis as a clinical disease of goats has been reported infrequently. The majority of the reports have been concerned with serological titers in goats which were apparently healthy. Rajesekhar and Nanjiah (1971) found 23·1% of 108 goats in Mysore, India, to have positive titers, with *L. pomona* being the most frequent. Sheep which were kept in close contact with the goats, had only 3·2% positives. Tripathy (1977) found 50% of 133 goats in Orissa, India, positive to the pool antigen, with *L. autumnalis* being the most common. Sebek (1961) in a survey of 373 goats in Czechoslovakia, found 43 reacting to *L. sejroe*, 26 reacting to *L. icterohaemorrhagiae*, 16 reacting to *L. grippotyphosa* and 25 reacting to other serotypes. Mailloux (1969) found 27% of 388 apparently healthy goats in Morocco reacting against *L. icterohaemorrhagiae* and *L. ballum*. Farina and Sobrero (1960) found 11/86 goats in Somalia to be positive. Van Riel and Van Riel (1956) found 12/353 goats in Central Africa to be positive.

Maronpot and Barsoum (1972) tested 195 goats, and many other domestic animals in Egypt. Forty-two per cent of the goats had titers of 1 : 128 or higher. This was almost as high as the percentage of high reacting pigs, 43%. The percentage of goats with titers of 1 : 32 or higher was 69%, which was the highest of all domestic animals tested, approached only by the camel, at 58%. However, clinical cases of leptospirosis in man and animals were not then being diagnosed or reported in Egypt.

Kharole and Rao (1968) found leptospires in 11/48 goats with white spots in the kidneys. They felt these goats had subclinical leptospirosis. Sahoo *et al.* (1972) found, in a slaughterhouse survey, a high incidence of white spots in caprine kidneys, associated with the presence of leptospires.

Morse and Langham (1958) infected goats with *L. pomona*, experimentally, and by contact with infected hogs. The infections were inapparent clinically, but the white spot lesions in the kidney were reproduced. Small gray white circumscribed areas in the kidney cortex were composed mostly of lymphocytes and plasma cells. Four out of five of the goats also had meningitis and encephalitis seen microscopically, but not clinically. One of the five goats was still shedding leptospires 31 days after its initial contact with infected material.

Morter and Morse (1956) were able to infect one of two goats placed in contact with calves which were shedding *L. pomona*. The affected goat was killed in the acute stage, after a short illness, with anorexia, pyrexia, lethargy, and acute nephritis.

The few reports of acute natural leptospirosis usually are associated with heavy losses. It is quite possible that subacute or chronic leptospirosis goes undiagnosed. Van der Hoeden (1953) reported a severe outbreak of icterohemoglobinuria in goats, with many fatalities, in Arab settlements in Israel. The disease in cattle, sheep and goats had been known for many years and it was termed "safra". Anaplasmosis also existed at a high level in the native goats but only caused sporadic hemoglobinuria.

The goats showing clinical signs of leptospirosis had illnesses of 2–4 days duration, with little evidence of pyrexia. Anorexia, lethargy and jaundice were common. Mortality was very high in those that developed symptoms. In one area, 44% of all the goats died. Many of the pregnant ones aborted. Many goats with inapparent illness developed high titers against *L. grippotyphosa*. Inoculation of isolated *L. grippotyphosa* into experimental goats, did not cause illness, but the grey-white pinpoint spots on the kidneys were seen at slaughter 46 days later. Little or no symptoms were seen in sheep and cattle in contact with affected goats in this outbreak.

Gayot and Cannepin (1954) in Tunisia considered leptospirosis to be a problem in the nomadic goat herds. In one flock of 200, 70 died over a 5-week period. Peracute leptospirosis resulted in the goats dying in less than 24 hours. Acute leptospirosis caused an illness of 1–3 days with inappetance, depression, incoordination and nervous cries prior to death. Chronic leptospirosis was characterized by an illness of a week or more, with diminished appetite, pale membranes, mild icterus and normal or subnormal temperatures. Animals which survived shed leptospires in the urine for at least a month. Many developed inapparent infections. Vaccination appeared to be helpful in reducing losses.

Hakioglu (1956) in Turkey found leptospires in the kidney of a jaundiced goat with hemoglobinuria. A similar disease occurred in cattle and sheep in the same area. Hartley and Hakioglu (1970) attributed cases of icterohemoglobinuria in goats in Turkey to *L. grippotyphosa*, with a possible concurrent *Mercurialis perennis* plant poisoning.

Burdin *et al.* (1958) in Kenya, reported one of the rare outbreaks of leptospirosis where several species were involved simultaneously, namely cattle, sheep and goats. The animals were acutely ill, anorexic, 30% became jaundiced, and many died. Urine was bright red at the onset. If it turned black, the animal usually died. In animals which had been infected for a long while, light colored patches were seen in the kidney cortex. It was very difficult to assess the effect of treatment, since many animals recovered spontaneously. Formolized whole culture vaccines did not stimulate titers. Control was achieved by moving the animals away from contaminated watering areas. Leptospirosis was considered to be latent in Kenyan native cattle.

Rarely has leptospirosis in goats been reported in association with abortions. Van der Hoeden (1953) reported on abortions occurring in the severe outbreak in Israel.

Pargoanker (1964) examined 15 goats that had aborted and found 8/15 with titers of 300–1000 against *L. pomona*. Brucellosis was ruled out as a causative agent on the basis of serology. Nigam *et al.* (1974) attributed an outbreak of 12 abortions among 70 goats to the use of two imported Alpine bucks, both of which had titers of 1:100 against *L. pyrogenes*. Of the 70 females, one had a titer of 1:100 against *L. pyrogenes* and many had 1:30 titers against several serotypes. All the females were negative to the *Brucella* test.

On the basis of published reports, it would appear that chronic subclinical leptospirosis is widespread in goats, and that the economic losses and public health significance have not been fully evaluated.

4. Paratuberculosis (Johne's Disease)
(A. B. Larsen)

Etiology

Paratuberculosis is a chronic infectious disease of the intestinal tract, characterized by a gradual loss of weight. The disease has a worldwide distribution. Death losses in infected herds of goats range from 2 to 10% of the adult animals annually, but usually only 1 or 2 goats show symptoms of the disease at any given time. The causative agent, *Mycobacterium paratuberculosis*, is a small rod-shaped organism that is found in typical clumps in smears or histological sections prepared from intestinal tissue and stained with an acid-fast stain.

Goats probably become infected in the first few weeks of life from the feces of infected goats.

The organism has been isolated from bull semen, but it is not known whether the disease has ever been spread by semen. Cattle fetuses can be infected *in-utero*. It is not known whether these two situations can occur in goats.

Diagnosis

Because the incubation period is a year or more, clinical signs of disease are usually observed only in mature animals. A gradual loss of flesh is the first clinical sign with intermittent diarrhea and soft rather than liquid feces. Milk production is diminished in milking goats. Sick animals gradually waste away, develop an unthrifty appearance and eventually die.

The disease may be diagnosed by culturing the organism from the feces, but this takes several months. Acid-fast intracellular organisms forming typical clumps in fecal smears indicate Johne's disease. The rectal biopsy technique used in cattle is not practical for goats. Microscopic and microbiologic examination of tissue from the ileo-cecal valve area, and the associated mesenteric lymph nodes can be used to confirm an infection. Serum or skin tests, if they were reliable, would be theoretically preferable to culture or histopathology for herd screening. The gel diffusion test is reportedly effective on a flock basis for sheep, but has not been studied extensively in goats. However, it does show promise as a herd test for goats. The intradermal test, once used in cattle, is not recommended now, and is not likely to be of value in goats. The intravenous johnin test, which will identify about 80% of infected cattle showing clinical signs, has not been extensively used or reported in goats. The complement fixation test is of limited value for diagnosing paratuberculosis in cattle, and with present knowledge would not be recommended for diagnosing the disease in goats.

Cattle can recover from light infections, and will retain their immunologic reactivity. Whether goats can do the same is not known.

It is important to understand that at the present time, there is no test available which can be used to guarantee that an apparently healthy goat does not have Johne's disease and is not incubating it.

Control

There is no satisfactory treatment for this disease but a veterinarian can recommend several control measures for infected herds. The recommendations would probably be as follows:

1. Maintain the herd on a ration adequate for good growth, reproduction and milk production because well-fed goats may be more resistant to infection.
2. Keep the premises clean and sanitary to reduce the spread of the disease. Disinfectants which are active against tuberculosis organisms are effective against *M. paratuberculosis*. Sodium orthophenol-phenate is particularly useful in dairy herds, since it has no odor, and will not taint the milk.
3. If possible, raise kids separate from mature goats because young animals are easily infected.
4. Eliminate at once any animals showing signs of Johne's disease because feces from diseased goats usually contain large numbers of organisms.
5. Manure from infected animals should not be spread on pastures.
6. Manure from diseased cattle and sheep will also infect goats. Therefore, goats should have their own quarters and pastures.
7. Once each year obtain a fecal sample from each adult goat in the flock and send it to a laboratory for fecal culture to determine if the goat is infected and shedding organisms. Infected goats should be removed from the flock at once and sold for slaughter. These diseased goats should never be sold for breeding.
8. Experimental vaccines may be available in some countries and a vaccine may be recommended if the herd is badly infected. The vaccine would only be used for replacement kids. Vaccinations should be used as an adjunct and not take the place of good management and sanitation (Goudswaard, 1971).

5. Pasteurellosis
(G. R. Carter)

Etiology

The form of pasteurellosis seen most commonly in goats is that manifested as an enzootic pneumonia. This bronchopneumonia which resembles clinically and pathologically enzootic pneumonia of sheep is most often chronic and limited in the extent of lung involvement. Borgman and Wilson (1955) have reported several outbreaks of enzootic pneumonia in the United

States in which *P. multocida* was implicated as the principal cause. A large, destructive outbreak of *Pasteurella* (species unspecified) pneumonia in goats transported from Switzerland to Austria was described by Jiřina (1953). Other outbreaks of acute pneumonic disease attributed to *P. haemolytica* were reported from Kenya by Mugera and Kramer (1967) and from India (Mishra *et al.*, 1970) and from Uganda by Gourlay and Barber (1960). Pande (1943) described an enzootic "pleuropneumonia" of goats in India caused by an organism that fitted the description of *P. multocida.* In the author's laboratory during the period 1974 to 1978 *P. multocida* was recovered from the lungs of seven goats with pneumonia; *P. haemolytica* and *P. multocida*-like organisms were isolated from the same source and disease in one and in four instances respectively. Cultures with the basic characteristics of *P. multocida* but having one or more aberrant reactions were identified as *P. multocida*-like.

As in sheep, cattle and swine, it seems likely that the *Pasteurella* species in goat infections are most often secondary invaders to such primary agents as viruses and mycoplasmas. That such predisposing factors as crowding, parasitisms, shipment and exposure to elements are important in the development of severe pneumonia is evident from a review of available reports. It also seems likely that in goats as in other species the pneumonia may be very limited and only become fulminant under the influence of stresses of the kind just referred to.

Septicemia of neonatal lambs due to *P. haemolytica* is well known. A similar type septicemia caused by this species has been described by Gourlay and Barber (1960) in three neonatal goats. *P. haemolytica* has been recovered from septicemia in neonatal goats in Michigan in the author's laboratory. Although both *P. multocida* and *P. haemolytica* are important causes of mastitis in ewes, there are few reports of mastitis due to *Pasteurella* spp. in the goat. Smith and Harnden (1943) described pasteurella mastitis in sheep and goats. The organisms they recovered, apparently from several cases in both sheep and goats, fit the description of *P. multocida. P. haemolytica* has been recovered in the author's laboratory from the milk of a goat with severe mastitis.

It seems likely as more laboratory examinations are carried out on diseased goats, that *Pasteurella* species will be found associated with as many and as varied infectious processes as are encountered in sheep and cattle. In addition to pneumonia *P. multocida*-like organisms have been isolated from a draining fistula, a septic joint and an abscess.

Diagnosis

Clinical signs of pneumonia and septicemia are not sufficient to make a definitive diagnosis of pasteurellosis since many other agents can cause similar disease manifestations. The laboratory diagnosis depends on the microbiological examination of such clinical specimens as trachea washings, milk, blood and the tissues of necropsied goats.

Treatment

The results of *in vitro* susceptibility tests performed on clinical isolates of *P. multocida* and *P. haemolytica* from various domestic animals have shown the following drugs to have considerable activity against both species: penicillin, ampicillin, streptomycin, various tetracyclines, chloramphenicol, sulfonamides, novobiocin, erythromycin, neomycin, polymyxin, cephalothin, tylosin and spectinomycin. Perhaps the most widely used drugs are penicillin–streptomycin combinations, sulfonamides, the tetracyclines and chloramphenicol. The fact the multiple drug resistance has been observed in clinical isolates of *P. multocida* and *P. haemolytica* (Chang and Carter, 1976) from cattle and pigs underlines the need for the susceptibility testing of all isolates obtained from goats.

Control

Bacterins containing *P. multocida* and/or *P. haemolytica* have been used with irregular results to prevent pasteurellosis in goats (Jiřina, 1953; Smith and Harnden, 1943). Until more is known about the occurrence of the serotypes and immunotypes of these two species *vis-à-vis* disease in goats it is unlikely that an effective bacterin or vaccine will be developed. It is possible, however, that autogenous bacterins containing sufficient "antigen" prepared from freshly isolated smooth or mucoid cultures may be of some value in reducing the losses in extensive outbreaks.

6. Pest of Small Ruminants
(F. M. Hamdy)

Etiology

Peste des petits ruminants, pseudo rinderpest of goats and sheep, stomatitis-pneumoenteritis complex of sheep and goats and Kata are synonyms. Pest of small ruminants (PSR) is an acute or subacute viral disease of goats and sheep characterized by fever, necrotic stomatitis, gastroenteritis, and pneumonia.

The infectious agent of PSR is a member of the morbillivirus group of the family Paramyxoviridae and is closely related to rinderpest, canine distemper, and measles viruses. PSR virus grows in various cell cultures of ovine or caprine origin as well as in a green monkey kidney cell line (VERO). PSR virus is heat labile but maintains viability for several weeks in the cold and for several months in frozen animal tissues. The virus is inactivated at pH 3 and by ether and chloroform.

The disease has been reported in West Africa; Ivory Coast, Dahomey, Senegal and Nigeria. Sheep are less susceptible than goats, and cattle are only subclinically infected. Surviving animals develop a dual immunity, to PSR and to rinderpest.

Diagnosis

The acute form of PSR is characterized by a sudden rise of temperature to 40–42°C. Affected animals appear ill and restless and have a dull coat, dry muzzle, and depressed appetite. The mucous membranes are congested. Early in the disease, a serous nasal discharge is observed; it later becomes mucopurulent and causes a fetid odor to the breath. Diarrhea develops during the later stages of the disease, and the animals become emaciated. The animal dies 5–7 days after onset of fever. In the subacute form of PSR, the disease may linger for several days and the animal dies in 14–21 days or occasionally survives. The disease is more severe in young animals than in adults and morbidity and mortality rates are higher. Secondary latent infections are activated and may complicate the clinical picture of the disease.

Post-mortem lesions in PSR are similar to those of rinderpest, mainly erosive stomatitis, abomasal congestion, and enteritis. Bronchopneumonia is a constant finding.

A presumptive diagnosis can be made on clinical, pathological, and epizootiological findings; but because of the similarity of PSR to other diseases (e.g. rinderpest), virus isolation and identification is essential for laboratory confirmation. Detection of virus-neutralizing antibodies at a rising titer in surviving animals is also diagnostic.

Control

Cell-cultured rinderpest vaccines are used to protect goats and sheep against PSR and true rinderpest. Sanitary measures should be applied in case of outbreak of either disease (Dardiri *et al.*, 1977; Gargadennec and LaLanne, 1942; Hamdy *et al.*, 1976a, b; Mornet *et al.*, 1956; Rowland *et al.*, 1969; Whitney *et al.*, 1967).

B. Diseases which Cause Nervous Symptoms

1. Listeriosis
(C. S. F. Williams)

Listeriosis is caused by *Listeria monocytogenes* and the disease occurs in three forms, encephalitis, fetal septicemia with abortion and neonatal septicemia. As a general rule, it is unusual for simultaneous abortion and encephalitis to occur in the same flock. The disease has been reported in goats in the USA (King, 1940), Norway (Naerland, 1950), Japan (Tajima, 1950), USSR (Amanzhulov, 1953), Bulgaria (Ivanov, 1957), Germany (Gall, 1962), the Sudan (Pillai, 1962), Yugoslavia (Berberovic, 1964) and Canada (Wood, 1972). Usually the disease is sporadic with isolated cases, although major outbreaks occur occasionally. An animal which aborts a listeric fetus usually does not develop systemic or encephalitic listeriosis. If the goat does not abort cleanly, she may die of listeric metritis. Sandbu (1956) found that, in three herds numbering 91 goats, 14 goats aborted during the last month of pregnancy and five of those died of metritis. Four goats in one herd died of *Listeria* encephalitis. Amanzhulov (1953) found a wave of abortions and detected *Listeria* organisms in the fetal stomach contents. Some of the kids born alive later developed systemic listeriosis and died. The outbreak was brought under control by moving the goats to a new environment.

Animals that are suffering from the disease shed the organisms in all the body secretions, including milk, and from all the orifices. Carrier animals also occur (Khalimbekov, 1957). The organism may also be isolated from soil, bedding, feed and water. The encephalitic form of the disease may be extremely rapid in onset and death can occur in 2–3 days. Other animals may linger on showing neurological signs for weeks. In the German outbreak, some control was achieved by giving all the goats tetracycline at 3 mg/kg i.m. for 3 days followed by 20 mg aureomycin/animal orally for 7 days (Gall, 1962). Berberovic (1964) used 250 mg chlortetracycline orally four times daily for 3 days.

The effectiveness of vaccination for the control of listeriosis in goats is not known.

2. Scrapie
(J. Hourrigan, A. Klingsporn, W. W. Clark and M. de Camp)

Etiology

Scrapie is a naturally occurring fatal disease of sheep and goats caused by a filterable, transmissible, self-replicating agent considered by most American workers to be a small virus. The virus is unusually resistant to heat and other physical and chemical abuse, including formaldehyde, and to ultraviolet and ionizing radiation.

Early signs of natural scrapie in the goat are usually subtle changes in behavior or temperament, causing the animal to appear apprehensive or excited when confronted with stimuli of noise or motion. Tremors of the head and neck are also seen. Signs shown by affected goats become progressively more obvious and include in most animals loss of weight and condition, bilateral pruritus, rubbing against fixed objects, and varying degrees of biting at feet, limbs, rump, and sides, resulting in hair loss and skin abrasions. Rubbing of the animal's rump or back may stimulate a nibbling or lip reflex. Some affected goats masticate more slowly than normal, move the lower jaw back and forth, chewing crossways, and drool saliva, resulting in damp chin hairs. As the clinical course of the disease progresses, weakness, incoordination, inability to balance and walk or run normally become more pronounced until the goat is unable to stand. The course of the clinical disease varies from 1 to 3 or 4 months or longer ending in death. We have never observed an obvious clinical case that recovered. Goats develop natural clinical disease at an average age of about 40 months, rarely less than 30 or more than 60 months of age. Scrapie virus is widely disseminated in various tissues of affected goats as demonstrated when these tissues were inoculated into mice.

Diagnosis

The clinical diagnosis of scrapie is confirmed by demonstrating typical histological changes in the brain of the affected animal (McDaniel and Morehouse, 1966).

Transmission by Inoculation

Forty years ago French workers transmitted scrapie by inoculation from sheep to sheep and from sheep to goats (Cuille and Chellé, 1939). Subsequently, the disease was passed serially in both sheep and dairy goats and from goats back to sheep and to mice (Stamp, 1962).

In 1954, British workers inoculated 12 dairy goats intracerebrally with scrapie material, resulting in all developing the disease in 15 to 22 months (Gordon, 1966).

One of the authors (W. W. Clark) observed that 17 of 20 dairy goats inoculated subcutaneously with scrapie virus developed clinical disease in 10 to 46 months, and that all of 12 Angora goats dosed orally or inoculated subcutaneously or intracerebrally with scrapie virus developed clinical disease in 12 to 37 months.

Natural Scrapie

Scrapie was recognized in sheep some 250 years ago and has been reported in England, Scotland, Wales, Continental Europe, Iceland, India, Kenya, Norway, Ireland, Northern Ireland, Columbia, Canada, and the United States. The disease was also reported in Australia, New Zealand, and the Republic of South Africa, but is believed to have been eradicated from these three countries.

Evidence of natural scrapie in goats is accumulating. The natural disease was reported in 1942 in a goat reared with a flock of scrapied sheep in France (Chellé, 1942). In Scotland, four out of six dairy goats maintained in a stall in contact, from the time they were less than 1 week old, with a succession of natural cases (mainly Suffolks) developed scrapie at 39, 39, 39, and 47 months of age. In a second similar experiment, 55% of 11 exposed goats developed scrapie at 29, 32, 32, 32, and 37 months of age (Brotherston *et al.*, 1968). Three goats developed scrapie when held in contact from 1 day of age with scrapied sheep for 40 months (Stamp, 1962). In Britain, scrapie was reported in a 4-year-old and an 8-year-old Saanan goat which had very little contact with any sheep (Harcourt and Anderson, 1974). Scrapie was diagnosed in an uninoculated control goat purchased at 3 years of age from a dealer (MacKay and Smith, 1961). In Canada, scrapie was diagnosed in a 7- or 8-year-old goat born in a flock where scrapie had occurred in sheep (Stemshorn, 1975). In the United States scrapie was diagnosed in a 5-year-old goat reared in a scrapied flock of Suffolk sheep in Missouri (Hourrigan *et al.*, 1969).

In an experiment conducted by the authors, previously unexposed goats were placed in direct contact with an infected sheep flock. The unexposed goats included 11 Angora goats and 6 Nubian or Toggenburg dairy goats. Scrapie was not detected clinically or histologically in any of the previously unexposed dairy or Angora goats. However, natural scrapie occurred in 6 (17%) of the Angora goats and 44 (61%) of the dairy goat progeny born on the infected premises and at risk. The data were prepared so as to show scrapie incidence in the progeny of dairy goat dams and sires of known scrapie status. The incidence was 58% in the progeny (24 at risk) of Free X Free parents. Scrapied sires did not increase the incidence in goat progeny (it was 44%), scrapied dams increased in the incidence to 71%. When both sire and dam were scrapied, 89% of 9 progeny at risk developed clinical scrapie. In another experiment scrapie continued to occur (6 cases have occurred) in goats born and reared apart from any sheep, but exposed to infected goats. Later certain dairy goats were removed from exposure at birth and at 6 or 20 months of age and placed in isolation pens for long-term observations. Those removed from scrapie exposure at 6 months of age or older developed scrapie, however, those removed at birth did not (Hourrigan *et al.*, 1979).

Treatment

No successful treatment has been found although efforts to demonstrate useful treatments have included drugs, chemicals, extracts and other materials.

Control

In the absence of useful treatment of individual animals, control of scrapie is on a flock or herd basis. Disease control is based on preventing exposure of susceptible animals and/or increasing their resistance. Efforts to produce a vaccine to increase resistance have failed.

Dairy and Angora goats are very susceptible to natural scrapie or inoculation or oral dosing with scrapie virus. Preventing exposure of goats to affected goats or sheep is the only presently known method of avoiding outbreaks among goats.

Scrapie spreads from farm to farm, state to state, and country to country by the movement of animals which are affected but not yet showing clinical signs of the disease. Since the scrapie agent has been demonstrated to be present in many tissues of affected animals, it is clear that these animals must be destroyed. There is no diagnostic test for the disease in the live animal not showing signs; consequently, there is no way of knowing how many exposed animals may harbor the infectious agent. This factor, linked with the long incubation period, has necessitated the slaughter of exposed animals. The consensus has been that the immediate progeny of exposed animals are also dangerous and should be slaughtered.

C. Diseases which Cause Alterations of the Skin

1. Poxes
(G. R. Scott)

Etiology

Several poxes, some benign and some malignant, affect goats. Two, perhaps, three, appear to be species-specific.

(i) *Benign Poxes.* The benign poxes include an unclassified goat-pox virus circulating in herds in America, Australia and Europe, and contagious ecthyma virus which has a world-wide distribution. Both viruses affect man (Newsom and Cross, 1934; Bakos and Brag, 1957) and in addition, contagious ecthyma virus attacks sheep frequently and dogs rarely (Wilkinson *et al.*, 1970). The two viruses differ antigenically and exposure or vaccination with one does not induce immunity to the other (Renshaw and Dodd, 1978).

The disease induced by the benign goat-pox is characterized by the development of papulopustular pocks on the lips, in the mouth and on the udder and teats (Bakos and Brag, 1957). The surfaces of the pocks erode and exude serum forming crusts. The course lasts about six weeks. Deaths are rare. The lesions of contagious ecthyma are similar but typically, they affect the lips and mouth (Zeller, 1920; Glover, 1928). Lesions will also develop at any site that has been abraded. Nannies suckling affected kids develop proliferative lesions on the teats. Some, thereafter, resent suckling and mastitis ensues.

(ii) *Malignant Poxes.* The malignant poxes are enzootic in Africa and Asia. They are characterized by fever and the development of visceral pocks in addition to generalized skin lesions. Case mortality rates range from less than 10% to more than 50% being highest in young animals. The common African virus attacks goats and sheep indiscriminately and is related to the virus of lumpy skin disease of cattle (Capstick, 1959). In Asia, the two best studied diseases are loosely designated goat-pox (Sen and Datt, 1968) and goat dermatitis (Haddow and Idnani, 1948). Both causal viruses are alleged to be host-specific but a recent report from India describes the transfer of goat-pox virus to three animal attendants (Sawhney *et al.*, 1972). The Asian poxes do not naturally affect sheep.

The skin lesions of Asian goat-pox and African goat- and sheep-pox are characteristically nodular in character ranging in diameter from 0·5 to 2·0 cm, are strictly cutaneous and seldom erode superficially. If erosion does occur the lesion becomes encrusted. In contrast, the centres of mucosal nodules are usually eroded leaving ulcer-like lesions with a raised white periphery. The nodules in viscera are shaped like pigeon eggs, are white and have a rubber-like consistency. They are found most often in the lungs but may occur in other viscera. Animals developing pneumonia die within a few days.

The skin lesions of goat dermatitis are widely distributed and consist of numerous circular nodules 4–12 mm in diameter (Haddow and Idnani, 1946). Most affected goats die following the onset of pneumonic symptoms. In the few survivors the lesions necrose to form sitfasts which when shed leave shallow red ulcers with clear-cut, punched-out rims. Scabs develop and healing occurs slowly from the edges. Visceral pocks occur as grayish-white nodules in the alimentary tract, kidneys and lungs.

Transmission

Transmission of all the pox viruses except that of goat dermatitis readily occurs when there is direct and indirect contact between sick and healthy goats provided the animals at risk have fresh abrasions. Insect vectors were once considered to be the main mode of transmission of African goat- and sheep-pox virus (Plowright and Ferris, 1958). Davies (1976), however, suspected that infection usually followed inhalation.

Treatment

Antiviral drugs have not yet proved to be economic or efficacious. Nevertheless distress should be relieved by symptomatic treatment. In addition, affected animals should be segregated from healthy animals. The administration of attenuated virus vaccines to scarified skin is claimed to limit the spread of infection but protocols have yet to be published.

Control

The management of goats in closed herds strictly segregated from other goats and sheep is the optimal prophylactic technique for most poxes affecting goats. In many countries where the diseases are endemic, communal browsing and grazing are practised, together with close herding at night in kraals. In addition, transhumance is an essential part of the livestock system. Vaccination is therefore the only feasible prophylactic; both live and killed vaccines are used.

D. Mastitis

(M. Roguinsky and Mary C. Smith)

Etiology

A great variety of organisms has been associated with clinical and subclinical mastitis in the goat (Plommet, 1974). Infection is typically via the streak canal and, thus, may be related to improper milking or inadequate hygiene. The number of does infected with major pathogens varies from 3·6 to 25·9%; 21·7 to 53% are infected by non-hemolytic staphylococci and 39 to 52·4% harbour no bacteria (Bryan, 1942; Roguinsky *et al.*, 1971; Smith and Roguinsky, 1977). In some commercial herds, all of the does are free of pathogens (Poutrel and Roguinsky, 1978).

Hemolytic *Staphylococcus aureus* is the most important pathogen. Caprine strains of staphylococcus belong to biotype C, along with, but serologically distinct from, bovine and ovine strains (Roguinsky and Grandremy, 1978). Seven out of eight caprine strains tested by Holbrook* were not enterotoxin producers. The severity of the infection depends more on the animal's resistance and antitoxin titer than on the strain involved. Gangrene of the udder is the most severe form which staphylococcus infection can take. Usually, there is preceding insult to the udder such as trauma, improper milking, or udder dermatitis caused by contagious ecthyma or staphylococcal infection. Non-hemolytic staphylococci can cause severe irritation of the udder, as measured by increased somatic cell counts and decreased production, but do not often cause obvious clinical mastitis (Plommet, 1974; Roguinsky *et al.*, 1971).

Many different species of streptococci have been isolated from the udder of the goat. These include *Streptococcus agalactiae*, *S. uberis* and *S. dysgalactiae*; *S. faecalis* is frequently encountered (Poutrel and Roguinsky, 1978). *S. zooepidemicus* has caused a herd problem with chronic mastitis atrophy, induration, and abscess formation (Nesbakken, 1975). *Pseudomonas* has been reported to cause acute purulent mastitis, sometimes progressing to gangrene and death of the goat (Toidze, 1954); it also has been the cause of subclinical infections (Poutrel and Roguinsky, 1978).

Pasteurella haemolytica (*P. mastitidis*) can cause clinical mastitis, with marked swelling of the gland and the supramammary lymph node (Bagadi and Razig, 1976; Schroter, 1954). A yellowish gray watery secretion, possibly blood tinged, with clots and a leukocyte count higher than 2 million/ml, occurs. The organism is normally in the upper respiratory tract and may be acquired from suckling kids.

Corynebacterium pyogenes and *C. ovis* (C. pseudotuberculosis) were isolated from goat mastitis in Egypt (Farrag and Oof, 1966). With *C. Pyogenes*, udders show a severe progressive reaction with swelling, then fibrosis of the gland and creamy pus. In herds where Caseous lymphadenitis occurs, *C. ovis* may invade the udder and cause typical greenish caseated pus there or in the supramammary lymph node.

Proteus, *Escherichia coli*, and *Klebsiella* have been isolated infrequently.

Mycoplasmal mastitis is a serious entity in some parts of the world. Most of the research has been conducted in Europe, but clinical outbreaks have appeared in the United States. Strains first classified as *Mycoplasma mycoides* subsp. *capri*, then recently reclassified as *M. capricolum*, are a major cause of contagious agalactiae in goats in France (Perreau, 1977). This form of

* The authors wish to thank Dr R. Holbrook, Unilever Research, UK, for testing the strains.

mastitis spreads rapidly through the herd and may be accompanied by arthritis, keratitis or abortion. Other species involved include *M. agalactiae* subsp. *agalactiae* (Miege, 1978), *M. putrefaciens* (Perreau, 1977), *M. mycoides* subsp. *mycoides* (Perreau, 1977) and true *M. mycoides* subsp. *capri* (Littlejohn and Cottew, 1977).

Diagnosis

Mastitis in the goat is not always easy to diagnose. Consequently, a variety of tests has been applied, the proper interpretation of which awaits further research. Body temperature (normal 38·5 to 40°C, (101–104°F), pulse (normal 75 to 90/min) and respirations rates (normal 25 to 35/min, with panting common in healthy animals), should be determined. Palpation of the udder may reveal inflammation, fibrosis or necrosis. A strip-plate examination and bacteriological culture of the mammary secretion should be performed.

The California Mastitis Test (CMT) is used when results of the strip-plate examination are negative or inconclusive; however, the CMT must be interpreted cautiously when applied to goats. The reaction depends on the release of DNA from epithelial cells and leukocytes (mainly neutrophils). One study gives an average of 750 000/ml total cells, 69·3% are neutrophils, and 0·4% are epithelial cells (Okada, 1960). Schalm *et al.* (1971) have interpreted the CMT reaction on goats' milk in terms of numbers of neutrophils. From these data, and from the authors' experience, only gel reactions (2+ and 3+ in Schalm's description) indicate an infection of the gland by a major pathogen, except in late lactation. Differences between the two halves are especially important.

Other methods, either microscopic or electronic, are available for a more direct determination of cell count. Coulter (Cardoen, 1977; Nesbakken, 1976; Roguinsky, 1974) and Fossomatic (Cardoen, 1977) electronic counters have been used. For goat milk, the correlation between them is very good ($r = +0{\cdot}97$). A study of 34 herds sampled three times shows that the correlation is lower for herd milk ($r = +0{\cdot}79$); Coulter counts are more correlated to CMT and bacteriological results than Fossomatic (Poutrel and Roguinsky, 1978). Further research is needed for the types of cells involved in these counts (Claveau *et al.*, 1978). With Coulter TA, which classifies cells counted by volume into 16 groups or channels, histograms may be obtained which allow differentiation of mastitic and nonmastitic samples (Smith and Roguinsky, 1977).

Counts less than 1 million/ml represent healthy glands or some weak irritation, such as improper milking; 500 000 to 2 million cells/ml suggests the presence of non-pathogenic or weakly pathogenic organisms such as most non-hemolytic staphylococci. Counts above 1·5 million/ml signal a pathogenic organism, most often a *Staphylococcus*. It should be observed that very high counts have been obtained, up to 50 million (Roguinsky, 1977). From a study of 118 herds, the distribution of cell numbers was found to be log-normal, with a geometric mean of 1·1 million. Herds with more than 1 million cells are often infected, as determined by isolation of staphylococci on selective media (Poutrel and Roguinsky, 1978). The cell numbers increase dramatically, up to 12 million, at the end of lactation (Roguinsky, 1977), when less than half of the goats are milked, even in non-infected herds.

Treatment

Ideally, the treatment of caprine mastitis should depend on isolation of the agent and determination of antibiotic sensitivities. In practice, however, except for the goat that is detected by routine screening tests or that has a chronic mastitis, treatment cannot be delayed until laboratory results are available.

Commercial preparations for goats are not generally available. This leads to two problems: milk withdrawal times are uncertain, and often the tip of the mastitis syringe is too large to penetrate the goat's teat sphincter. One half of the contents of a tube of bovine mastitis medication/treatment every 12/24 hours is sufficient in most instances. Cephalosporins, penicillin–streptomycin, sulfathiazole, and tetracycline are all acceptable for initial treatments before laboratory results are available. A male cat catheter or an over the needle IV catheter can be attached to the tube if the teat orifice is too small. The milk withdrawal time specified for cows should be considered a minimum and, if possible, a sample should be checked for bacterial inhibition before the treated doe's milk is sold or used for cheese or yoghurt production.

Animals that are systemically ill or that have chronic, non-responding infections may benefit from intramuscular antibiotics; udder amputation, corticosteroids (to non-pregnant animals) and other ancillary treatments may be indicated in severe cases.

Control

Prevention of mastitis begins with close attention to husbandry and sanitation. The barn, sleeping area, and exercise yard should be well drained and ventilated. All animals should be dehorned, and feet should be trimmed once a month to reduce the risk of udder trauma. Animals with draining abscesses should be isolated or culled. Therapy during the non-lactating period has been shown to eliminate approximately two-thirds of the staphyloccal infections that have not progressed to gangrene (Roguinsky, 1977). One half tube of a non-lactating cow preparation in each side of the udder seems to be a reasonable dose. Treatment should be used for problem herds or animals that have experienced mastitis during the preceding lactation. Reducing feed and water for a few days at the end of lactation will reduce the danger of new infections.

Teat dipping with iodophors is effective, giving the same protection against new infections as is afforded to cows (Bouffet, 1973; Cardoen, 1977). Unfortunately, bovine preparations tend to be sold in large volumes; a pH change that will cause teat irritation often occurs in the remainder of the teat dip before it can be used in a small goat herd. Chlorhexidine, with or without glycerol, also seems to be a satisfactory teat dip for most goats. Teats should be dipped after every milking and also twice a day for about 1 week after the end of lactation and again for a week before parturition.

The vaccination of goats against staphylococcal mastitis deserves serious attention and controlled research. Two doses of a staphylococcal toxoid vaccine* (50 IU α toxoid and 50 IU β toxoid) subcutaneously at 15-day intervals have given good clinical results, but these trials lacked controls.

E. Diseases which Cause Abortion

1. Chlamydiosis
(Annie Rodolakis)

Etiology

Chlamydiosis is an infectious disease caused by the microorganism *Chlamydia psittaci* and characterized in most cases by abortion, usually in the last month of gestation; in some cases pneumonia, keratitis and polyarthritis occur. The initial report involving *Ch. psittaci* associated with abortion in goats was from Germany (Staub, 1959). Although the agent was not isolated, the author observed elementary bodies in the mucus adhering to an aborted fetus and demonstrated chlamydial antibodies in aborted goats. The disease has been since reported in Bulgaria (Semerdziev *et al.*, 1966), Spain (Blanco, 1967), United States (McCauley and Tieken, 1968), France (Faye *et al.*, 1971) and India (Jain *et al.*, 1975). Pneumonia due to *Chlamydia* in goats was reported in Japan (Omori *et al.*, 1953). In the United States, keratitis and polyarthritis associated with abortion were described by McCauley and Tieken (1968). Clinically healthy goats can harbor the organism in their intestines (Omori *et al.*, 1957).

The causative agent is a basophilic bacteria, obligate intracellular parasite;

* Capristaph, Roger Bellon, Nevilly, France.

it multiplies in the cytoplasm of the host from cytoplasmic inclusions. It enters the cells after being phagocytosed, and a complex developmental cycle follows. The small infectious forms (elementary bodies, diameter 250 to 450 nm) change into large homogeneous forms (initial bodies, up to 1000 nm), which multiply by binary fission and differentiate into small forms. The large forms are not infectious. These organisms possess RNA and DNA, and a limiting membrane of complex chemical composition similar to the cell walls of gram negative bacteria.

The disease is introduced in healthy herds by purchasing infected pregnant goats. At the time of abortion, *Chlamydia* are discharged profusely in uterine fluids, placenta and fetuses. The first aborting goats may thus become an important source of infection for the other goats in an earlier stage of gestation. Goats at their third month of pregnancy seem to be the most susceptible. A rapid spread of infection occurs within such goats in the same herd, leading to important economic losses. The infection is maintained in the herd by young goats born from infected does; they abort at their first pregnancy. These young goats create a reservoir of the infectious agent, but they cannot be detected by the complement fixation test.

Diagnosis

Chlamydial infection of pregnant goats results in abortion, usually in the last month of pregnancy, stillbirth or delivery of weak kids. The abortion rate usually does not exceed 5%, except when the infection is first introduced in the herd; in this case it can rise as high as 30%. Specific clinical signs are not observed until abortion is imminent. The incidence of retained placenta is not increased. Goats may be sick for some days after abortion and a brown discharge from the vagina may be observed. The mortality of infected goats is very low, but the milk production can be greatly reduced after abortion (McCauley and Tieken, 1968). Experimental inoculation of *Chlamydia* into the mammary gland produced a severe mastitis (Omori *et al.*, 1957) but this organism has not been isolated from natural cases. Chlamydial abortions can be associated with a high level of non-productive goats in the herd.

The diagnosis may be made directly, by isolation or demonstration of *Chlamydia* in samples, or indirectly by serological or allergic tests. Demonstration of chlamydial elementary bodies is possible in placental impression smears. These bodies appear as red dots when stained by Stamp's method; however, chlamydial agents are so small that the demonstration of *Chlamydia* by this method is quite difficult. The organism can be confused with Q-fever (*Coxiella burneti*) and *Brucella* organisms. Isolation of *Chlamydia* by culture in developing chicken embryo or better on human cells gives a definitive etiologic answer. Isolation on cell cultures is performed by plaque assay (Rodolakis and Chancerelle, 1977) or by staining of cytoplasmic inclusions, in McCoy or Hela cells. Vagina swabs taken as close as possible after abortion are the best available samples for the isolation of *Chlamydia.* In case of pneumonia, chalmydial agents can be isolated from lung, nasal discharge or milk (Omori *et al.*, 1957).

The complement fixation (CF) test is a valuable diagnostic aid but the antibodies detected are directed against the Chlamydial group (including antigen *Ch. trachomatis*). The study of the serological pattern of infected herds shows that the CF test cannot be used for a definitive individual diagnosis (Rodolakis *et al.*, 1977).

An allergic diagnosis can also be used, by means of a skin test. The results observed are in agreement with those obtained by the CF test (Rodolakis *et al.*, 1977).

Control

Drugs known to affect the replication of *Chlamydia* such as tetracyclines can be effective in preventing abortions. For example, 20 mg/kg of oxytetracycline given by intramuscular route each 15 days to pregnant goats can stop the abortions in an infected herd.

Killes vaccines prepared from *Chlamydia* grown in embryonated eggs or tissue culture are used before breeding, but the protection obtained is very variable.

2. Salmonellosis
(R. Sanchis)

At the present time, the incidence of *Salmonella* in goats is not well documented. The isolation of various serotypes from apparently healthy goats has been reported from many countries (Doutre and Bouche, 1976; Kapur *et al.*, 1973; and Tadjebakhche and Namin, 1974).

There are few reports of clinical disease, and the pathological consequences of infection seem to be minor, when compared to the losses due to salmonellosis in cattle and sheep.

S. typhimurium, *S. abortus ovis* and *S. dublin* are the three serotypes which have been isolated from goat salmonellosis (Buxton, 1959). Although *S. abortus ovis* is traditionally considered as pathogenic only for sheep (Blood and Henderson, 1968), serologic evidence of the infection in goats has been described (Tadjebakhche and Gatel, 1972) and about 2% of the sera from aborting goats show significant levels of specific seroagglutinating antibodies (Sanchis, 1978). Bacteriological confirmation of the disease is rarer. The organism has been isolated from goat abortions in France (Yalcin and Gane, 1970) but the disease does not show the enzootic characteristics that it has in the ewe. *S. abortus ovis* was isolated from a fatal case of septicemia in a buck, but such cases are extremely uncommon (Sanchis, 1978). Although abortion has been produced by experimental inoculation of *S. abortus ovis* (Tadjebakhche *et al.*, 1974), the goat appears to be an apparently healthy carrier; the epidemiologic significance of this should not be overlooked.

3. Vibriosis
(C. S. F. Williams)

Vibrio infections in the goat are reported rarely. Dobbs and McIntyre (1951) isolated *V. fetus* from the only fetus submitted for examination during an abortion outbreak in the USA where at least 19 out of 203 dairy goats aborted. Krylov and Shakhimina (1972) reported on an outbreak of vibriosis, in the USSR, where 39 (8·8%) goats aborted in the third and fourth month of gestation, 57 (12·9%) of the kids were stillborn and 33 (7·4%) were weakly and died after a week or two. The goats were Orenburg down-goats and some of the management practices were considered to be suboptimal. Bergeron (1976) isolated a vibrio-like agent from goat abortions in the USA. Newsam (1925) showed that *V. fetus* of bovine origin could become established in the goat reproductive tract. Webster and Thorp (1952) were able to induce abortions in pregnant goats by intravenous or intrauterine inoculation of *V. fetus*.

F. Diseases with Complex Symptoms

1. Caseous Lymphadenitis
(M. K. Ashfaq and S. G. Campbell)

Etiology

Caseous lymphadenitis is a chronic, contagious disease of goats caused by the gram-positive bacterium *Corynebacterium pseudotuberculosis* (*C. ovis*). The organism produces a soluble exotoxin, however, the lymph node abscesses are caused by the pyogenic factor present in the cell wall of the bacterium (Cameron, 1964; Zaki, 1976) which attracts phagocytic cells to the site of infection, thereby producing an abscess. *C. pseudotuberculosis* has been associated with 70% of the superficial abscesses in goats in the USA. However, other corynebacteria, staphylococci, and streptococci, as well as mixed infections account for occasional abscesses in goats. These occasional abscesses may not be associated with lymph nodes and do not usually spread from goat to goat.

Natural infections of caseous lymphadenitis in goats have been reported from the USA (Shaw and Seghetti, 1939; Stafseth and Bryan, 1945; Ayers, 1977), India (Dhanda and Singh, 1955; Sarkar and Bhattacharyya, 1975), Pakistan (personal observation), Egypt (Abdel-Hamid, 1973), Venezuela (Gallo and Morris, 1965) and Sicily (Cortella, 1963).

In the USA the disease has been observed in all parts of the country and it has now emerged as one of the more important diseases of goats (Turner, 1973). While careful statistics have not been compiled on the incidence of the disease, in a recent survey in the USA, 8% of the goats reported had current abscesses. In herds which are being assembled or in herds with a rapid turnover rate, it is not unusual to have as many as 50% of the animals affected, while in small closed herds, no abscesses may occur.

There seems to be no difference in the sex or breed distribution of the disease. Goats ranging from 3 months of age to 10 years have been reported to be infected. However, the incidence is higher in the older goats, perhaps because of repeated exposure to infection or because a goat once infected rarely eliminates the infection completely.

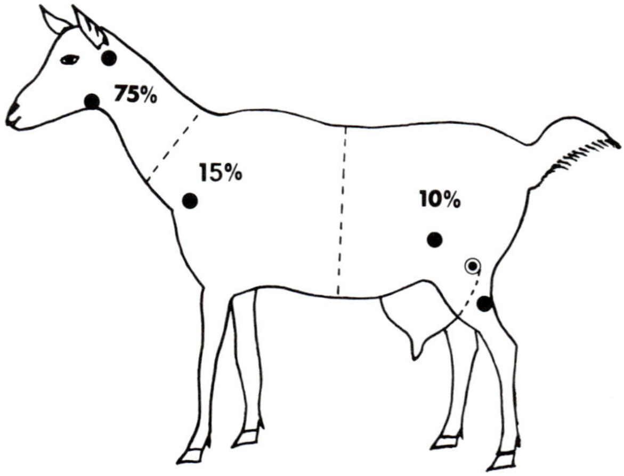

Fig. 1. The body distribution of caseous lymphadenitis in the goat. The location of superficial lymph nodes where abscesses commonly occur is marked thus ● and the percent distribution of superficial abscesses is noted in three areas of the goat's body.

Diagnosis

The clinical diagnosis of caseous lymphadenitis is usually based on the visible lesions, that is abscesses in the superficial lymph nodes. The location of these lymph nodes is shown in Fig. 1. It will be noted from this diagram that 75% of the abscesses occur in the head and upper neck region.

Abscesses most frequently occur in the mandibular lymph nodes, followed by parotid, superficial cervical, subiliac and the supramammary lymph nodes (see Fig. 2). In heavily infected goats, a chain of abscesses may be seen running down their necks. Some goats have deep abscesses in their thoracic and abdominal cavities and many show associated signs such as coughing, respiratory distress or digestive disorders, or they may become emaciated and a few die. After an incubation period ranging from 2–4 months or even longer, infected lymph node(s) become enlarged. As the abscess matures, the

Fig. 2. Naturally occurring caseous lymphadenitis due to infection with *C. pseudotuberculosis* in the right sub-mandibular lymph node of a goat.

lymph node may increase 4–10 times its normal size but it still remains painless. The skin over the abscessed lymph node becomes denuded of hair and finally the abscess bursts, discharging an odorless, thick, greenish-yellow pus. Bacteriological examination usually reveals the presence of *C. pseudotuberculosis*, frequently in pure culture.

In contradistinction to this, abcesses which are caused by other bacteria (other cornyebacteria, streptococci, staphylococci, etc.) may occur anywhere on the body, that is they may not be associated with superficial lymph nodes. They may be soft, fluid filled, and painful and contain thin, smelly pus. The site of entry may be apparent (e.g. dirty needle, scar, etc.) and such sporadic abscesses seldom spread and involve other animals.

Several serological tests, a toxin neutralization test (Doty *et al.*, 1964) and antihemolysin inhibition test (Zaki, 1968), and an indirect hemagglutination test (Shigidi, 1978) originally used for the diagnosis of caseous lymphadenitis in sheep, can be used for the detection of *C. pseudotuberculosis* infection in goats. But, the complexity and general unreliability of these tests has rendered them of very limited general value.

Transmission

The introduction of an infected goat into a susceptible herd is the most usual method of starting a herd outbreak. A goat with an enlarged abscessed superficial lymph node poses a real threat to a herd, because the abscess is likely to rupture and the discharged pus contains large numbers of bacteria (1 000 000 bacteria/g). Other goats in the herd may contact this directly by licking, butting, or other close contact. Alternatively, the pus may contaminate the environment, milking stands, feed racks, clippers, etc., and thus other animals in the flock are exposed indirectly. While it is theoretically possible for animals with deep abscesses to shed bacteria or for does to excrete them in milk or colostrum, these methods of spread seem to be rare events in contrast to pus contamination from a superficial abscess.

The bacteria enter the body by two main routes, through a wound, or abrasion of the skin, or by mouth. Wounds can be caused by butting, ear chewing, barbed wire, sharp hay racks, nails, and by surgical manipulations such as ear tagging, inoculations, clipping, etc. Animals can become infected by mouth after licking pus. Abrasions of the mucous membranes or tooth shedding may facilitate entry. Once the organism has penetrated the skin or mucous membranes, it lodges in the regional lymph nodes which become enlarged and filled with pus. The abscess eventually bursts and discharges pus, thus completing the cycle of infection.

Treatment

The commonest method of treating caseous lymphadenitis consists of simply opening and draining the abscess. The infected goat should be isolated and the hair clipped around the abscess, which is opened by a large vertical incision from top to bottom. The pus is removed, and the abscess flushed out with disinfectant to ensure cleaning. The cavity may be packed with swabs and disinfectant and cleaned mechanically each day until it heals and new skin forms over the site. A tranquilizer may be given for restraint and prophylactic penicillin may be administered concurrently. In addition, discharged pus, contaminated swabs, etc., should be removed so that the environment and other animals will not be infected. The common practice of opening abscesses and immediately returning the patient to the herd perpetuates the disease and is to be deprecated.

Alternatively, the entire abscessed lymph node may be removed surgically. This requires a general anaesthetic and the services of a veterinarian. It is more expensive, but the abscesses seldom recur—a problem with simple drainage—and it is worthwhile in selected animals. While *C. pseudotuberculosis* is sensitive to various antibiotics *in vitro*, antibiotic therapy for animals with caseous lymphadenitis using a wide variety of antibiotics has generally proved to be unsuccessful.

Control

The following measures are recommended for the control of caseous lymphadenitis. In the first place, do not introduce infected animals into a herd. This can be done by inspecting the superficial lymph nodes, particularly those of the head and neck of new animals. Should abscesses appear on a goat, isolate it and either open the abscess or have it removed surgically. Ensure that such animals are cured before returning them to the herd. Because wound contamination is a likely route of infection, common sense measures should be taken to prevent wounds in goats. Thus new goats should be carefully observed for bullying, and sharp corners on stalls and feed bins, barbed wire, etc. should be removed. Surgical procedures should be undertaken only with sterile equipment. If wounds are observed on goats, they should be clipped and cleaned and treated with a disinfectant. Goats which are very thin and unproductive should be considered for culling since they may have deep abscesses, and goats with recurrent superficial abscesses should also be culled, as they are a source of infection for the rest of the herd and they too may develop deep abscesses.

Various commercial and autogenous vaccines have been tried and are available for the prevention of caseous lymphadenitis in goats (Cameron *et al.*, 1972). At best, they are of unproven value and at worst they have been known to cause abscesses at the site of inoculation. While an efficacious vaccine would be most welcome, results to date have been disappointing and undocumented.

Caseous lymphadenitis is an insidious, chronic and contagious disease of goats. It does not kill many goats, but it does spread through a susceptible herd causing unsightly abscesses and affected goats are often barred from shows or sales. The abscesses, if severe, or if they involve a vital organ, may cause clinical signs, loss of productivity, and even death. They can be controlled by carefully observing good principles of hygiene and husbandry as suggested above. In the absence of a suitable vaccine and diagnostic test for inapparent carriers, constant vigilance, treatment and culling are the price which has to be paid for maintaining a herd of goats free of caseous lymphadenitis.

2. Caprine Arthritis–Encephalitis
(T. B. Crawford)

Etiology

Caprine Arthritis–Encephalitis (CAE) is a newly-discovered disease of goats in the USA that is characterized by paralysis of young kids (Cork *et al.*, 1974) and arthritis in adults. It is caused by a virus that is a member of the same group as the leukemia viruses (Retroviruses), which cause tumors in many species of animals. The virus is spread to kids from older, infected goats early in life, by routes that are not yet defined. As far as is known, once infected, goats remain carriers for life.

Diagnosis

Typically, a low percentage of infected kids experience central nervous system disease between 2 and 4 months of age. This may vary in severity from a barely noticeable unsteadiness of gait (ataxia) to rapidly advancing fatal paralysis. The hind legs are often affected earlier and worse than the front legs, but control of both is usually lost within a period of 1 to 3 weeks. There is no fever. The kids remain alert and will usually continue to eat and drink with assistance until near the time of death. Most but not all affected kids die; some progress only to a certain level of paralysis and stabilize. Though diligent nursing may maintain paralyzed goats for months, the effort is ultimately fruitless. The damage to the spinal cord is permanent, and use of the limbs is never recovered once paralysis is advanced.

Most infected kids (estimated at 80 to 90%) escape the acute paralytic form of this syndrome. However, many of the escapees have low-grade persistent disease of the brain that may flare up in later life.

The most prevalent manifestation of this virus in adult goats is a chronic progressive arthritis. An as yet undefined proportion of goats infected as kids develop, by the age of 1 or 2 years, generalized disease of the connective tissue characterized by unthriftiness, rough hair coat, loss of condition, lameness, and "big knees". Although swollen joints and lameness are the most noticeable signs, the disease process actually affects all synovial-lined structures, including tendon sheaths and bursae; it occasionally causes lesions in other organs such as kidneys, blood vessels and skeletal muscle as well. The beginning of the arthritis may be so subtle as to go unnoticed, or it may show up suddenly as acute lameness. Generally, the condition progressively worsens over a period of years and goats die of the arthritis or its secondary complications at an early age (e.g. 5 to 8 years).

The arthritis is non-suppurative, no bacteria or chlamydia can be isolated from the joint fluid or tissues and antibiotics have no effect.

Treatment

At present, there is no effective therapy available. Non-steroidal anti-inflammatory and analgesic drugs such as phenylbutazone give some temporary relief of joint pain, but tend to lose effectiveness after prolonged administration. About all that can be done is to maintain a high nutritional plane to help delay the wasting and, when possible, avoid excessive exposure to cold weather, as this seems to worsen the joint stiffness and pain.

Control

Only a minority of infected goats develop sufficiently severe joint changes to come to the attention of the breeder. Like the acute paralytic form, most infected goats, by means not yet understood, escape severe disease and show no, or only negligible, arthritis. Nevertheless, it is believed that they serve as reservoirs of transmission to newborn kids.

Current data suggests a surprisingly high prevalence of the virus among goats. In the states of Washington and Idaho for example, at least 79% of goats carry the virus. No breed or sex predilection has been found. This newly recognized virus poses a difficult problem for goat breeders. More information on the natural history of the virus and how it is transmitted is needed before effective control measures can be designed.

3. Caprine Mycoplasmosis
(H. E. Adler)

Etiology

Ten serotypes of mycoplasma have been isolated from goats with various pathologic manifestations. These strains have been characterized by biochemical, serological and biological means and designated as distinct species. The mycoplasmas with potential effect to the livestock industry are: *Mycoplasma agalactiae* subsp. *agalactiae*; *M. mycoides* subsp. *capri*; *M. capricolum*; *M. conjunctivae* and *M. mycoides* subsp. *mycoides*. There are five mycoplasmas of uncertain pathogenicity; T-strain mycoplasmas, *M. arginini*, *M. dispar*, *Acholeplasma oculusi* and *M. putrefaciens*.

M. agalactiae subsp. *agalactiae* is the cause of contagious agalactia in goats and sheep. The infection has been found in the Mediterranean countries, other parts of Europe, Africa, Asia and the USA (one isolate). Transmission occurs through aerosols of exudate, milk and excreta. Evidently this is one of the few mycoplasmal infections acquired by ingestion. The disease is acute, subacute or chronic and progresses from septicemia to localized lesions in the mammary glands, uterus, joints and eyes. Primary pathological changes of the udder are interstitial mastitis with eventual fibrosis. Mastitis is associated with cellular and chemical changes in the milk manifested by increased cellularity and pH. Arthritis is a common feature of this infection most frequently affecting the carpal and tarsal joints with primary damage to connective tissue and synovial layers. Eye lesions are keratitis, iriditis and purulent conjunctivitis (Zavagli, 1951).

M. mycoides subsp. *capri* is the major cause of contagious caprine pleuropneumonia (CCPP). It is uncertain if this infection is present in the USA. This disease has been observed in parts of Europe, Central Asia, Africa and India. Transmission occurs mainly by inhalation of contaminated respiratory discharges. Experimental and natural disease is an acute sero-fibrinous pleurisy and pneumonia usually involving an entire lobe. Aside from the classical hemorrhagic infarction and the red and gray variegation observed in the lungs, a common sequela of septicemia is severe arthritis. In a few instances there may be edema of the lower jaw and lumbar muscles (Longley, 1951).

Polyarthritis seems to be the lesion most commonly associated with mycoplasmosis of goats. Joint involvement following septicemia is always seen with *M. capricolum* infection. The mycoplasma has been described in Australia, Turkey and in the USA from only one state. However, unpublished reports indicate the presence of this infection in other parts of the country.

Transmission might be by inhalation of exudate and excreta but never demonstrated experimentally. During acute outbreaks during the kidding season, an entire kid crop can be lost from septicemia and debilitating arthritis. Mortality as much as 20% has been observed in yearling and adult animals (Cordy *et al.*, 1955). *M. capricolum* has been isolated from a few cases of mastitis in the absence of other bacteria.

Several mycoplasmas have been isolated from "pink-eye" among goats. *M. conjunctivae* was recovered from epidemic caprine keratoconjunctivitis in Australia, Canada, Switzerland and USA. Goats inoculated subconjunctively with purified clones of *M. conjunctivae* propagated in tissue cultures developed mild conjunctivitis to severe keratoconjunctivitis. The latter disease syndrome was characterized by early marked lacrimation and injection of blood vessels in the conjunctiva that progressed to corneal opacity in about 2 weeks. From this point, clinical signs became less severe and recovery was apparent by 6 weeks without therapy (Trotter *et al.*, 1977).

M. mycoides subsp. *mycoides* infections of goats has been associated with arthritis, peritonitis, pneumonia and conjunctivitis. Within the last 10 years bilogical evidence revealed *M. mycoides* subsp. *mycoides* infection has been present among infected herds in Europe, Asia, Africa, Australia and the USA. The infection had not been identified previously or was reported to be *M. mycoides* subsp. *capri*.

Experimental infection with a goat-adapted organism induced unilateral disease involving the right lung. Affected lobes had gray and red hepatization, dilation of interlobular septae, edema and fibrinous pleurisy. The severe respiratory lesions resulted in a mean mortality rate exceeding 74% (Ojo, 1976).

Five additional mycoplasmas have been recovered from normal and diseased goats. T-strain mycoplasmas found in the urogenital tracts of goats have not been shown to be pathogens (Livingston and Gauer, 1975). *M. arginini* was isolated from arthritis and pneumonia of affected goats (Al-Aubaidi *et al.*, 1972). Culture of goats with enzootic pneumonia yielded *M. dispar* (Perreau, 1973). *Acholeplasma oculusi* (*oculi*) has been cultured from goats with keratoconjuctivitis (Al-Aubaidi *et al.*, 1973). Recently *M. putrefaciens* was found to cause subacute and chronic mastitis accompanied by blood elements in the milk without gross changes.

Diagnosis

Ordinarily mycoplasmas are mainly transmitted via the respiratory route. The nasal mucosa of healthy animals is much more resistant to infection than the trachea and lungs. Many times infection limited to the upper respiratory tract does not produce clinically apparent disease. Further, at necropsy the anatomical sites of the various isolations may not be suspect for mycoplasma since the expected pathognomonic lesions caused by a specific mycoplasma are absent. The lack of clinical signs and the failure to recover mycoplasma from infected tissues using routine bacteriologic procedures is a major problem for the diagnostician.

Mycoplasma have fastidious nutritional requirements. Usually a basal medium prepared from meat infusion, peptone and yeast product supplemented with 15% inactivated horse serum is adequate for isolation of many mycoplasmas from infected tissues. There are a number of commercially available media that satisfy these requirements. Several mycoplasmas from goats will not grow on this standard medium and other formulations are required (Stalheim, 1976).

The most reliable diagnosis of mycoplasmal infections is by isolation and identification of the organism. Sterile swabs are inserted into affected tissues and placed in a selective broth "transport" medium, described below. For mastitis diagnosis, foremilk samples taken several hours after regular milking are preferred. If immediate examination is not feasible, specimens should be kept cool and not frozen. As soon as possible, a smear of the tissue, exudate or milk is made; fixed in Bouin's solution or methyl alcohol and stained in diluted Giemsa. Mycoplasma are barely visible coccoid bodies. Contaminated samples are seeded into selective broth and on agar media containing 1000 U/ml penicillin G potassium. Agar and tubes of broth are incubated in a moist chamber for 3 to 6 days at 37°C since many mycoplasmas will not colonize on solid media of low water content. Moisture may be maintained by sealing the plates with masking tape. After 3 and 6 days of incubation, the agar is observed for colonies with a dissecting microscope at 10 to 50 magnifications using obliquely transmitted light. Characteristic colonies have a dark central papilla and a light periphery. Colonies range in size between 20 and 80 μm. For further identification, an agar block colony-side upward is placed on a slide and covered with a stained coverslip. More detailed studies of colony morphology are made by placing agar blocks colony-side downward on a slide; fixed with Bouin's solution and stained with diluted Giemsa. The broth is fixed and stained in the same manner. With the usual methods of handling mycoplasmas, coccoid bodies will predominate in broth preparations. Fine filaments may be present, but difficult to demonstrate.

Mycoplasmas can be isolated from infected tissues by yolk sac inoculation of five to seven day old chicken embryos. Growth can occur in eggs without killing the embryo. Tissue culture is an ideal culture medium, but growth is influenced by the cell type, suspending fluid and the serum source.

Serologic measurements used for the indirect diagnosis of mycoplasmal infections in goats with varying success include: slide and tube agglutination, complement-fixation, agar-gel double diffusion and surface inhibition tests.

Control

Mycoplasmal infections are mainly transmitted by cohabitation. Healthy kids should be isolated to avoid exposure to droplet infection from the "healthy carrier" animals. Indirect transmission of mycoplasma by contaminated feeders, waterers and other fomites does not occur readily with these organisms due to their osmotic fragility. Lacking a cell wall, mycoplasmas can be destroyed by soap, anionic and cationic detergents, and a variety of other commonly available disinfectants. A particularly effective disinfectant is alkalinized quaternary ammonium compounds.

Inactivated vaccines in general appear to offer only modest protection against mycoplasmal infections (Whittlestone, 1976). The failure to produce adequate immunity may be related to selection of antigenic types, adjuvants, and inoculation routes. Satisfactory immunizing products against *M. agalactiae* subsp. *agalactiae* have been prepared from infected tissue and a broth culture of the organism (Lopez and Lopez, 1952). Formalin-killed, chicken embryo-propagated organism seemed to offer protection from this infection. *M. agalactiae* subsp. *agalactiae* attenuated by 40 passages in embryonated chicken eggs protected goats against a lethal challenge (Sharmir, 1954). Possibly the effectiveness of the tissue and egg products was related to lymphocyte sensitizing factors and adjuvant effect from yolk. Other than killed *M. agalactiae* subsp. *agalactiae*, there have been no reports of effective vaccines for the other mycoplasmas isolated from goats. If *M. mycoides* subsp. *mycoides* infection becomes a major problem, it is doubtful that attenuated vaccines used in cattle will be applied due to possible reversion to pathogenicity.

Treatment

A large number of chemotherapeutic agents have been used to treat mycoplasmas of goat origin. *In vitro* tests showed erythromycin to be effective against *M. capricolum* infection (Adler *et al.*, 1956), although field evidence indicated tylosin to be most effective. Tylosin tartrate at 5 mg/lb body weight given intramuscularly to goats infected with *M. Mycoides* subsp. *capri* was effective after a single dose. Fever subsided and animals were free of lesions at necropsy. Treatment with oxytetracycline at 7 mg/lb was also effective but fever subsided only after three daily injections and two of four animals had lesions when killed. Neither penicillin plus streptomycin nor chloramphenicol was effective (Onoviran, 1974). Tylosin has been assessed in the treatment of pleuropneumonia in goats. Animals that received tylosin intramuscularly at either 3 or 9 mg/lb daily for 5 days recovered from the infection and the organism was not present in the tissues at necropsy; whereas, untreated goats had severe lung lesions and viable mycoplasma (El Nasri, 1964).

G. Parasites

(T. W. Schillhorn van Veen)

1. Blood Parasites

Blood-borne parasitism is rare in goats kept in temperate zones but can be of considerable importance in (sub)tropical climates. Heartwater, trypanosomiasis, babesiosis and malignant theileriosis can occur as fatal disease outbreaks in Africa and, in the case of the latter two infections, in other parts of the world as well. Losses due to these diseases are rarely seen in areas where the parasites and their intermediate invertebrate vectors are endemic but the diseases become important at the edges of endemic zones, during periods of lowered resistance in exposed animals due to stress and concomitant diseases, or in situations of a high infection-challenge.

Heartwater is a rickettsial disease transmitted by species of *Amblyomma* ticks. It is characterized by fever, nervous signs and lymphopenia. The course of the disease is sometimes so rapid that animals may die within a few hours after the onset of the clinical signs. Excessive fluid in the pericardium ("heartwater"), thorax and lungs is commonly observed at necropsy but the diagnosis can only be confirmed by finding the causative agent, *Cowdria ruminantium*, in brain smears or in histological sections of the blood vessel wall. Treatment, with tetracyclines, is only successful if it is given before the nervous signs develop. It is possible, however, to vaccinate susceptible animals by exposure to *Cowdria* followed by drug treatment (Ilemobade and Blotkamp, 1978).

Two other blood parasites, *Eperythrozoon* and *Anaplasma ovis* occur in goats and are rarely associated with unthriftiness.

Trypanosomiasis is transmitted by tsetse flies and is only found in goats in the humid parts of Africa. The savanna-type tsetse flies prefer to take their blood meal from other domestic animals, rather than goats (Boyt *et al.*, 1978), which may explain the low prevalence of trypanosomiasis in goats in that ecozone. The major parasite associated with the disease is *Trypanosoma congolense*, but there are other trypanosomes to which goats are susceptible (Griffin, 1978). Unthriftiness, anemia, opacity of the eyes and occasionally abortion are the major clinical signs during the acute stage of the infection; chronic infections are rarely apparent. The diagnosis can be made by demonstrating the trypanosomes in fresh blood preparations or Giemsa stained bloodsmears. Most animals respond very well to treatment with diminazene-aceturate or homidium-chloride, but drug resistance does occur in certain strains of *T. congolense* as well as in other species. The disease can be prevented by housing the animals in fly-proof pens, especially in daytime when the vectors are active.

Clinical babesiosis due to *Babesia motasi* or *B. ovis* is transmitted by various tick species and occasionally occurs as a clinical entity, in the tropics, the Middle East and in southern Europe (Hoffman *et al.*, 1971). In endemic areas, however, nearly all animals are immune and rarely show detectable disease. Although susceptible animals sometimes die without previous signs, most affected goats show fever, standing haircoat, jaundice and, occasionally, hemoglobinuria ("redwater"). The diagnosis is made by demonstrating the parasite in the red blood cells of the affected animals. Treatment with diminazene-aceturate or quinuronium-sulphate often results in a rapid improvement. The disease is prevented by regular tick control which may be combined with vaccination by exposure to infected blood. Immunity is maintained by occasional exposure to infected blood or, under natural conditions, to infected ticks. A too rigid tick-control program may be contra-indicated in endemic zones as the animals can lose their resistance due to insufficient exposure to the parasite, so that there is no booster effect.

Malignant theileriosis, also a tickborne disease, is mainly found in Eastern Europe, the Middle East and in the northern parts of Africa. The major clinical signs are fever, unthriftiness and enlargement of the lymphnodes. Since there is no effective treatment, the disease is often fatal. Tick control and vaccination are the main preventive measures taken.

2. Ectoparasites

Although most animals are always infected with some ectoparasites, the occurrence of clinically detectable infestations is rare and depends on various factors among which are the climate and the management system as well as the nutrition and immunological status of the animals.

The bloodsucking blue lice (*Linognathus* spp.) as well as the biting red lice (*Bovicola* spp.) spend their entire life cycle on the goat and infestations may become manifest when undernourished animals are crowded in sheds and stables. The highest infection rates are found at the end of the winter or dry season, even in animals kept outdoors. Heavy lice infestations do cause irritation and, in case of sucking lice, may lead to severe anemia in young animals.

Fleas (*Ctenocephalides* spp.) are rarely found on livestock in temperate climates but some African species can cause severe blood loss in young kids born in flea infested barns (Schillhorn van Veen and Mohammed, 1976). Both lice and flea infestations can be controlled by spraying, dipping or dusting with insecticide; also neck collars impregnated with insecticides appear useful in the control of these parasites (Darrow, 1973).

Demodectic mange, a widespread skin disease of goats, is of considerable economic importance in areas where goat skins are highly priced, as in Somalia or in the West African savanna. The elongated *Demodex* mites live in the glands deep in the skin causing pustules, especially around the neck, front legs and shoulder but occasionally also on the head, eyelids and other parts of the body (Euzeby *et al.*, 1976). The pustules can easily be squeezed out for diagnostic purposes and contain yellowish wax-like material in which the various stages of the parasite can be demonstrated. The condition does not appear to cause irritation to the infected animals. It is frequently overlooked in live goats, especially in Africa where the pustules are usually less than 5 mm in diameter. In the USA, the pustules often reach 1 cm in diameter.

Sarcoptic mange or scabies, however, may cause considerable itching and can be found in young kids as well as in adult goats. The infection is mainly observed on the head, ears and on the legs. Animals with prolonged infection show a typical thick crusty dermatitis on the ears in which the parasites are very difficult to demonstrate. Multiple deep skin scrapings at the edge of affected sites may be necessary before a few mites or mite eggs can be found.

Other mites which can be found on the ears or head, besides *Sarcoptes*, are *Chorioptes* spp. (which are more common on the legs), *Psoroptes* spp. and *Raillietia auris*. The latter two are often overlooked as they rarely cause detectable gross lesions (Cook, 1981). *Psoroptes*, however, can be a cause of irritation, and scratching and rubbing are the most common signs (Littlejohn, 1968). Mange is a fairly persistent condition and often requires repeated treatment like dipping, spraying or washing with insecticides, (Chotchaev, 1974; Das and Misra, 1972) and temporary change of housing. Occasionally, protozoa and helminths are associated with skin lesions. *Besnoitia* cysts have been observed in the skin, blood vessels and scrotum of goats in East Africa and Iran (Cheema and Toofanian, 1979) and *Stephanofilaria* worms have been incriminated as the cause of dermatitis on the legs of goats in S.E. Asia (Patnaik and Roy, 1968).

The five most important tick genera are *Rhipicephalus*, *Boophilus*, *Ixodes*, *Amblyomma* and *Haemaphysalis*. The first one is mainly found on the head and ears, the latter on the feet and the other two on the body. Ticks are of major importance as the vectors of tickborne protozoal and viral diseases, but they are also a cause of irritation ("tickworry") and occasionally of tick paralysis. Tick worry is easily recognized when caused by adult ticks but also the very small larvae can be a cause of irritation. *Amblyomma* spp. in particular are sometimes deeply embedded in the skin around or between the hoofs and, besides causing severe pain and lameness, they introduce bacterial infections which may lead to foot abscesses. Goats have been considered to be a reservoir-host of *Haemaphysalis* ticks (New Zealand, 1978).

Among the many other ticks found on goats is the spinose-ear tick (*Octobius megnini*), which is a soft tick occurring in hot arid regions. These ear ticks are a cause of irritation and, when secondary bacterial infections occur, of ear canker.

Regular dipping and spraying is the most commonly applied method of tick control. Some consideration should, however, be given to the frequency of application: too frequent dipping may eliminate the ticks on the animals but may also eliminate the low challenge of tickborne blood-parasites necessary to maintain immunity, and a too low frequency of dipping or spraying may allow a too high challenge of these parasites causing the immunity to break down.

Blowfly and screwworm-fly infestations are rarely a problem in short-hair goats. Botflies (*Oestrus ovis*), however, do attack goats and may cause some disturbance in the herd during the flying season, which is generally during the early summer. Their larvae migrate in the nose and sinuses and may, rarely, affect the brain (Schillhorn van Veen, 1978). The most common signs of nosebot infestation are nasal discharge and sneezing, but many infected animals do not show any signs at all. The infection rate is similar to that in sheep although the number of bots per animal is generally lower, probably related to the greater effort of goats to get rid of the larvae in the nose by sneezing and shaking (Horak and Butt, 1977). If necessary the animals can be treated with rafoxanide or ruelene.

Warbleflies (*Przhevalskiana* spp.) may cause a skin condition which clinically resembles *Hypoderma* infections in cattle. The condition is mainly found in the Mediterranean region and Orient and appears to be more common in dry, hilly areas than in the more humid lowlands (Sayin *et al.*, 1972). The flies emerge from the pupae in spring and deposit their eggs on goats. The larvae enter through the skin and stay in the subcutaneous tissue during summer and autumn, causing nodules which are mainly located on the back of the animal. In winter time the larvae drop off, pupate and survive on the ground till the next spring. Application of organophosphorous insecticides appears to be the most common and successful treatment (Klement'eva and Drummiev, 1976).

The larvae of the ocular fly of sheep (*Gedoelstia* spp.) may also affect goats, causing conjunctivitis, intra-ocular hemorrhage, bulging eyeballs, and finally blindness (Basson, 1962). This conditions, however, is only found in those parts of Africa where small ruminants are grazing in areas where blue wildebeests (the prime host of the parasite) are present.

3. Helminth Endoparasites

Intestinal tapeworms, mainly *Moniezia*, *Avitellina* and *Thysaniezia* spp., are very common in goats, but rarely cause disease. The fringed tapeworm, *Thysanosoma actinioides*, occurs in the intestine near the bile duct and in the bile ducts themselves but is also of minor importance. The whitish proglottids of tapeworms are frequently found in the droppings of young goats in summer but infections rarely cause detectable clinical signs. Treatment with niclosamide, bunamidine, lead arsenate or older drugs of botanical origin is often only applied for esthetic reasons as owners do not like to see evidence of tapeworm infections in the droppings of their animals, especially in those used for exhibition.

Cysticercus tenuicollis, the larval stage of a dog tapeworm (*Taenia hydatigena*), is very common in the abdominal cavity of goats at slaughter. The cysts look like thin walled waterbags and may reach the size of a tennis ball. They rarely cause detectable clinical signs in live animals; very heavy infections may cause some liver damage. *Cysticercus ovis*, the larval stage of the canine *T. ovis* is rare in goats, although the cysts appear to persist longer than in sheep (Anon, 1973).

Cysts in the liver are often caused by the larval stages of another canine tapeworm, *Echinococcus* spp. This parasite is of public health importance since man may acquire similar larvae when in contact with dog feces. The major control method for both conditions is to prevent dogs from feeding on the viscera of slaughtered ruminants as this ultimately reduces the chances of infection in goats and other mammals.

Dogs and wild carnivores are also carriers of *Multiceps* spp., a tapeworm of which the larval stages, the so-called "coenurus bladders", sometimes occur in the brain of goats. Coenuriasis in goats may cause the same typical "staggers" syndrome as in sheep. Since there is no cure other than surgical extirpation, the condition is controlled by preventing contact with dogs as well as regular treatment of infected dogs.

The two major flukes affecting goats are *Fasciola hepatica* in the temperate zones and *F. gigantica* in the tropics. Mixed infections occur in the subtropics and in some tropical highlands. Both parasites spend part of their life cycle in *Lymnaea* snails: *F. hepatica* in mudsnails and *F. gigantica* in watersnails. Disease outbreaks are generally associated with grazing on poorly drained lowlands, floodplains, irrigated pastures or ricefields. The infection, which is acquired by ingesting the small encysted fluke larvae on the grass, shows two distinct clinical syndromes. Firstly, acute fascioliasis due to the approximately 2-months-old fluke larvae migrating through the liver and causing massive destruction of liver tissue. Secondly, chronic fascioliasis associated with the irritation and bloodsucking of adult flukes in the bile ducts (Hughes *et al.*, 1974).

Animals suffering from acute fascioliasis may die suddenly without obvious clinical signs before death. The destruction, hemorrhages and beginning fibrosis in the liver are easily detectable at post mortem.

Chronic fascioliasis is characterized by unthriftiness, reduced milk or meat production and slight anemia. Casualties are rare in well fed animals but undernourishment, which is fairly common in many arid goatbreeding areas in the (sub)tropics, aggravates the course of the disease. The most commonly used drugs are, at present, rafoxanide, nitroxynil, niclofolan, oxyclozanide, albendazole and hexachloroaethane but older drugs like hexachlorophene, bithionol and carbontetrachloride are still used (Gibson, 1975).

The control of fascioliasis is mainly based on pasture management (drainage, fencing-off flood plains, building drinking troughs away from natural waters), grazing management (avoiding infested pastures), snail control, and regular anthelmintic treatment to prevent pasture and water contamination with fluke eggs.

Paramphistomiasis is also a snail-borne disease. The adult paramphistomes (barrel flukes) in the rumen or reticulum do not appear to be very harmful to the host, but massive infections with juveniles in the intestine may cause severe enteritis with watery diarrhea and heavy mortality. Treatment with niclosamide, resorantel or bithionol generally results in a rapid improvement. Disease outbreaks are associated with flood plain pastures previously grazed by cattle and are as such very uncommon in goats.

Other flukes in goats are the ubiquitous small liverfluke (*Dicrocoelium* spp.), the tropical bloodfluke (*Schistosoma* spp.) (Hurter and Potgieter, 1967) and the Asian pancreas fluke (*Eurytrema pancreaticum*). These flukes, however, have rarely been incriminated as a cause of clinical disease.

Lungworms are common but, apart from *Dictyocaulus filaria*, are not a major cause of disease. Lungworm infections mainly occur in goats in temperate and subtropical climates. *D. filaria* larvae are acquired from infested pastures. After ingestion the larvae penetrate the gut-wall and migrate via the bloodstream to the lungs where they settle in the air passages. The eggs produced by the adults are carried out of the respiratory system by coughing and are then swallowed. They hatch in the gut system and the first stage larvae can be found in the animals' droppings. Typical clinical signs of this disease, which primarily affects kids of 6–9 months of age, are nasal discharge and spasmodic cough. Heavy infections may lead to pneumonia and death, especially on younger animals. There is some evidence that goats are more susceptible than sheep (Kassai *et al.*, 1973).

Anthelminthic treatment, with tetramisole, levamisole or diethylcarbamazine may prevent casualties. *Dictyocaulus* infections are among the few parasitic diseases which can be prevented by vaccination.

Protostrongylus spp. in the air passages, and *Muellerius* and *Cystocaulus* spp. in small nodules in the lungs are usually of minor importance but may occasionally cause a chronic pneumonia. They have an indirect life cycle with small landsnails as intermediate hosts. A related snail-transmitted worm of American whitetail deer, *Parelaphostrongylus tenuis* may cause fatal cerebrospinal damage in unnatural hosts like the goat, but is only found in situations when goats and deer are utilizing the same grazing area (Guthery and Beason, 1979).

Internal parasitism due to helminths and coccidia is, world-wide, probably the most common problem in goat husbandry (Fabiyi, 1973; Van Tonder, 1975), either as an easily detectable "acute" disease, or as a subclinical chronic condition due to a variety of helminths. Although the mortality of an outbreak of, for instance, acute hemonchosis can sometimes be spectacular, the economic loss is often higher in chronic helminthiasis when all animals in a flock gradually lose weight and many die without particular signs of disease. There is some evidence that subclinical parasitism may reduce milk yield in goats (Farizy, 1970).

Helminths do not multiply within the host and the larval infection has to be acquired while grazing. The basic epidemiological and pathophysiological concepts of internal parasitism in goats are similar to those in sheep. There is no evidence that goats are more resistant to worms than sheep, although certain breeds are better able to resist the effects of an infection than others (Edwards and Wilson, 1958). The major difference derives from the browsing behavior of goats which may reduce the intake of infective larvae. Moreover goats, apart from Angoras, are rarely kept in large herds in fenced pastures. Most animals are kept by small-holders either tethered or freely roaming, and this husbandry method usually prevents a rapid build-up of pasture contamination with worm eggs and larvae.

The diseases due to worms are also not very different from those in sheep. Generally they can be grouped as follows:

(a) those due to specific helminth species, i.e. hemonchosis, ostertagiasis, hookworm infections, trichuriasis etc.

(b) those due to a mixed infection of various strongylids and other worms, often called "trichostrongylosis".

Hemonchosis is due to the bloodsucking activity of *Haemonchus contortus* (the barber pole worm) in the abomasum. In summer rainfall areas, this disease may cause severe losses in young kids. Kids may die before showing any signs. The major clinical sign is anemia, but occasionally the animals also show edema of the lower jaw ("bottle jaw"). The egg output in the feces is very high in acute infections which may occur within the first 3 months of grazing. A similar condition occurring in small ruminants in south-eastern Asia is caused by *Mecistocirrus* spp.

Ostertagiasis is mainly due to the brown stomach worm, *Ostertagia circumcincta*, although other trichostrongylids are often also involved. The syndrome is more common in animals in cooler climatological regions. The signs in goats are similar to those in sheep: unthriftiness, diarrhea, and weight loss mainly occurring at the end of the grazing season.

Hookworm infections are common in the tropics and are due to a voracious bloodsucker *Gaigeria pachycelis*. Only very few worms, in the intestine, are needed to cause severe anemia. The disease is not well studied and needs urgent attention as it does cause severe losses in sheep and goats in the humid tropics.

Trichuriasis is a widespread but not very common disease of animals kept in paddocks or even in stables, caused by *Trichuris* spp. The disease is characterized by intermittent diarrhea and is fairly persistent as the parasites are not affected by the common anthelminthics and the eggs are very resistant to adverse environmental conditions. *Strongyloides* (the white bankrupt worm) infections may affect kids of only 4 weeks of age. The condition is often aggravated by, or coincident with, coccidial infections and as such may cause considerable losses (Anon, 1973).

Although other helminth species like *Oesophagostomum* (the nodular worm), *Trichostrongylus* and *Chabertia* sometimes also cause a specific disease, they are generally involved in multiple infections with the previously listed helminth species. The resulting chronic helminthiasis syndrome is characterized by ill thrift, poor growth or milk production, rough haircoat, varying degrees of anemia and diarrhea. It is sometimes difficult to differentiate this syndrome from other common causes of ill thrift, like undernourishment, deficiencies and fascioliasis. The diagnosis in such cases is based on experience, fecal and blood examination, or diagnostic treatment. It should be realized, however, that recovery of animals which suffered from clinically detectable helminth infections takes a long time, especially under circumstances of marginal nutrition. Curative anthelmintic treatment then shows often disappointing results and should be replaced by preventive anthelmintic treatment combined with husbandry methods aimed to prevent a build-up of the infections in the animals as well as on pasture.

Control programs should always include the complete herd and should be based on a sound knowledge of the epidemiology of the most common parasites in the area. Specific instructions are difficult to give but treatments at the end of a grazing season to reduce the worm burden, and at the beginning of the grazing season to prevent pasture contamination, are generally advised as a minimum requirement. In young animals, however, it may be necessary to increase the frequency of deworming during warm humid summers to once every month. Deworming of does before parturition is beneficial to the animals themselves as well as to their offspring. The choice of drugs depends on the local helminth-spectrum, price, availability and on the possible methods of application. Commonly used drugs at present are the benzimidazole derivatives, the tetramisole derivatives, pyrantel, organic phosphates, rafoxanide, and phenothiazine (Gibson, 1975). All have their advantages and restrictions and there are no "cure-alls". the use of anthelmintics without supporting changes in husbandry or management often shows disappointing results. Especially in the tropics some consideration should be given to local knowledge on helminth spectrum and epidemiology, as the information in textbooks is mainly based on studies in temperate areas with well-fed animals kept under sophisticated husbandry systems, for example (Baker, 1975).

4. Coccidiosis

(P. Yvore)

Etiology

Coccidia are the intestinal parasites most frequently found in goats. Nevertheless, the species involved and the effects on the host are not well documented. All of the species described so far belong to the genus *Eimeria* and are located in the intestine (Pellérdy, 1974). It is difficult to differentiate all of the species and infections often involve several species of coccidia. The most common species is *E. arloingi*. Nine of the twelve species are considered to occur in both goats and sheep but in most cases attempts to transmit parasites from one host species to the other are unsuccessful (Levine and Ivens, 1970). Consequently, there is little or no risk that kids will be infected by contact with sheep.

In all management systems, the kids become infected very early on in life. We have noted that kids separated from their mothers on the second day of life, and fed artificially, will start to excrete oocysts at the end of the third week. During the fourth and fifth week, all the kids in a group will be excreting oocysts. During the first year of life a large number of oocysts may be present in the feces (up to 60 000/g of feces) without causing any apparent problem to the host.

After 6 months of age, the levels begin to diminish. Adults may be chronic excretors, shedding up to a few thousand oocysts/g of feces.

Diagnosis

Kids kept in close confinement under intensive management are particularly susceptible. Clinical signs are loss of weight, diarrhea and unthriftiness, but not all animals in a group will show these signs. Stress of weaning, hot weather or unsanitary conditions may precipitate an outbreak. At the onset of the disease, very large numbers of oocysts may be found in the feces (more than 100 000 oocysts/g) which facilitates the diagnosis. In France, mortality is rare and the symptoms disappear after 5–6 days, but the animals will show retarded growth rates. Adult goats may excrete low numbers of oocysts, but usually show no symptoms.

In the USA, peracute coccidiosis in kids is uncommon but generally fatal and chronic coccidiosis can cause high morbidity and variable mortality in kids (Williams, 1979). In the United Kingdom, clinical coccidiosis is a considerable hazard for hand-reared kids under intensive indoor conditions (Fitzsimmons, 1967).

Treatment

There is no treatment which will completely eliminate coccidial organisms permanently. However, treatment will suppress the oocyst count and the symptoms and shift the host–parasite relationship in favor of the host. Since the environment is contaminated with oocysts, the animals are continuously exposed to the organism. Where kids are concerned, stresses should be reduced as much as possible. The building should be kept dry, well ventilated and clean. Reducing the animal density may be helpful. Treatments currently available are sulphonamides, in particular sulphadimethoxine at 75 mg/kg for 1–5 days. Treatment with amprolium 25–50 mg/kg orally for 3 to 5 days gives satisfactory results in France.

In France, systematic preventive treatment before weaning is of no benefit. Therapy is only begun after the disease in kids has been diagnosed, on the basis of clinical signs and the presence of high oocyst counts (more than 100 000/g of feces). The whole group of kids should be treated even if only a few individuals show signs of the disease.

Preliminary observations, by the author, in France, suggest that treatment of lactating does with sulphadimethoxine at 75 mg/kg for 1–3 days will result in a 5–10% increase in milk yield for 3–4 days after treatment, and this coincides with a reduction in the number of oocysts excreted.

5. Other Sporozoan Parasites
(T. W. Schillhorn van Veen)

Toxoplasmosis is caused by the asexual stages of *Isospora* (*Toxoplasma*) *gondii*, a coccidium of cats. The disease is widespread and affects nearly all mammals including man. The latter acquires the infection by consuming raw meat or any other food soiled with cat feces. As such the infection in goats is also of potential public health importance.

The knowledge of toxoplasmosis in goats is scanty. Serological studies revealed the presence of toxoplasma antibodies in goats throughout the world, but there exists little evidence of detectable disease. Abortion, stillbirth and weak offspring, the most common signs of the infection in sheep, do occur in goats after experimental infection (Dubey *et al.*, 1979) but there are surprisingly few reports of naturally occurring outbreaks. Incoordination, circling, torticollis and muscular rigidity can occur in rare cases of toxoplasmosis (Fatser, 1974) but the diagnosis is difficult as many other diseases cause similar nervous signs.

Goats probably acquire the infection through contact with cat feces or, by carnivorism (eating infected tissues such as placenta). Larval stages of the parasite invade the gut wall and eventually enter the cells of various organs and muscles. The parasites multiply within the hosts cells forming cysts which may break and release the tachyzoites. These enter other cells, and more organs become infected. There is some evidence that *Toxoplasma* may be found in the milk (Tizard *et al.*, 1977). Treatment in goats has not been attempted but in other hosts, it has little effect once the signs have become apparent and the best way to prevent the infection is by avoiding contact with cat feces. Although cats excrete the infective stages for only a few weeks in their lifetime, these oocysts may survive in the soil for over one year.

Sarcosporidiosis caused by *Sarcocystis* spp. shows, as far as the life cycle of the parasite is concerned, some similarities with toxoplasmosis. The sexual development of *Sarcocystis* probably takes place in carnivores and the cysts are found in goats and other animals. These ellipsoidal cysts can be fairly large and are observed in the striated muscle mainly of heart and esophagus. Sarcosporidiosis is mainly a slaughter observation and is rarely incriminated as a cause of clinical signs (Collins and Charleston, 1979). The affected organs are condemned at slaughter to prevent further dissemination.

A slightly different type of coccidiosis, globidiosis, is caused by *Eimeria* (*Globidium*) *gilruthi*. In contrast with the other *Eimeria* spp. this parasite invades not only the intestine, but especially the abomasum where the gray-white pinhead size nodules are visible with the naked eye. These cysts can be crushed on a slide and examination under a microscope reveals the numerous merozoites. Severe infections are characterized by foetid diarrhea, weakness and weight loss (Mugera and Bitakaramire, 1968). The diagnosis can be made by examination of the feces for merozoites, rather than oocysts. Treatment has rarely been attempted but anticoccidial drugs could probably be used.

H. Anthropo-zoonoses

1. Brucellosis
(Lois M. Jones)

Etiology

Brucellosis is one of the most important bacterial diseases of goats not only because of the economic loss it causes the goat industry but because it can cause serious illness in man. Brucellosis is a true zoonosis in that virtually all human infections are acquired from animals.

Brucella melitensis is the species of *Brucella* which infects goats causing a generalized infection which may persist for years. In pregnant goats the bacteria multiply in the uterus and mammary gland. This usually causes abortion in second half of pregnancy or weak kids and decreased milk production. Large numbers of bacteria are discharged with the placenta, and in uterine fluids, urine and milk. The excretion of bacteria may continue for several weeks from the vagina or for months or years from the udder. The goat may produce a normal kid at the next parturition although excretion can occur again at that time. In the chronic form of the disease the bacteria are localized in the udder and lymph nodes and there are no clinical signs and little effect on the milk yield. Thus, a flock of goats may appear to be free of disease, but if susceptible young goats of breeding age are introduced, an abortion storm may result.

Transmission of infection can occur by mouth, by inhalation, by the conjunctiva or through the intact skin. At the time of kidding large numbers of brucella may be present in aerosols in goat stables. Fresh cheeses and other unheated milk products made from milk of infected goats may contain large numbers of *B. melitensis*.

Persons who are in daily contact with infected goats, such as herdsmen, children tending flocks and slaughter house workers are most likely to contract the disease. Infection in city-dwellers is most frequently traced to consumption of fresh cheese.

Brucellosis was first recognized in British troops on the island of Malta and the causative organism was isolated from a human case by Bruce in 1886. In 1905 Zammit discovered that the Maltese goat was the reservoir of infection. *B. melitensis* still exists in Maltese goats as well as in goats in most of the countries surrounding the Mediterranean. Goat brucellosis extends eastward to Iran, India and Pakistan and the southern and eastern parts of the USSR and Mongolia. It is also prevalent in Africa and throughout Latin America. Sheep are also susceptible to *B. melitensis* infection although there are marked differences in susceptibility between breeds, e.g. Maltese sheep are highly resistant whereas the fat-tailed Awassi sheep of the Middle East are very susceptible. In general it is found wherever goats and sheep are the principal animals of a peasant economy, but the morbidity rates, where known, vary greatly from country to country and area to area.

The most common cause of brucellosis in cattle is *Brucella abortus*. This species causes natural infection in goats or sheep very rarely. This means that in countries where *B. abortus* is the only endemic brucella species, brucellosis rarely, if ever, occurs in goats. Therefore, when abortions occur in goats in these countries, they should not be attributed to brucellosis without absolute proof such as isolation of the organism.

Diagnosis

In endemic areas there may be little or no clinical evidence of brucellosis in goats. Few abortions occur and these are often attributed to other causes such as malnutrition or parasites. The occurrence of human infection is sometimes the first indication that goats are infected.

Since brucellosis is not the only cause of abortion in goats and not all infected goats abort, laboratory methods must be employed to diagnose infection. In general, these are the same as those used for bovine brucellosis and details are available in the WHO Monograph No. 55 (Alton *et al.*, 1975).

The tube agglutination test employs the standardized *B. abortus* antigen, but the dilutent should be 5% sodium chloride instead of 0·85% solution. The rose bengal plate agglutination test (or card test) is a rapid test which can be performed with a minimum of equipment. However, where adequate laboratory facilities exist, the complement-fixation test is the most reliable test for serum.

The milk ring test is more reliable when performed on samples from individual goats rather than on pooled samples. With goat milk, visible agglutination of the stained antigen indicates a positive reaction irrespective of whether the clumps of agglutinated antigen rise to the surface of the milk column (as in the bovine milk ring test), remain in suspension, or are deposited on the bottom of the tube. The addition of several drops of saline solution, sufficiently concentrated to provide a final concentration of 5% sodium chloride in the milk sample, will improve the efficiency of the test.

Where serological surveys of animal populations are performed with the tube agglutination test, a low incidence of 1–2% positive reactions in goats should not be ignored as this test is not reliable in detecting all infected animals. Goats may be serologically negative during the incubation period of the disease, titers rise after an abortion or infected parturition, then recede to negative again even though the animals are still infected and capable of spreading the disease. It is recommended that tests be repeated at monthly intervals, and that more than one serological method be applied. Brucellosis is a herd disease and the presence of even one reacting animal suggests that others may be harboring the disease.

Isolation of brucella is the definitive method of diagnosis. The use of selective bacteriological media makes it possible to isolate from aborted fetuses and placentae. If these are unobtainable, a visit to the flock should be made within a week of the reported abortion, and vaginal swab and milk samples should be examined bacteriologically.

Control

In areas where the incidence of disease is very low or non-existent, continued surveillance is required if raw goats' milk is sold for human consumption. Flocks should be checked at least once a year by serum or milk tests. Goats brought into a disease-free area should be quarantined before they are allowed to mingle with other animals and should be obtained from flocks in disease-free areas.

In areas where the numbers of goats and sheep are very large and the incidence is not known, a skin test is sometimes used to determine which flocks are infected. It should be emphasized that both goats and sheep should be skin tested. Positive tests in either goats or sheep are indicative of infection in the flock. All animals should then be tested serologically. The number of skin test reactors and serological reactors may not correlate as the two reactions may appear independently at different stages of infection.

Eradication may be attempted by slaughter of reacting animals and repeated serological testing every few months to detect new infections. This can be costly and unless replacements are known to be disease-free, and goats are isolated from the flock at the time of kidding, the disease usually continues.

If there are a limited number of reacting flocks in a given geographical area and disease-free replacements are available, eradication may be accomplished sooner and more economically by slaughtering the entire infected flock.

Common grazing is frequently practised in areas where brucellosis occurs among goats and sheep. Under these conditions the epidemiological unit becomes the village or geographical area, rather than the individual herd. Where the incidence of brucellosis is high, vaccination may be the only method of control. The most practical policy (for the government or the individual owner) is to vaccinate all goats for a number of years until the incidence is reduced. Where contact between flocks is the rule, even disease-free flocks must be vaccinated as a means of protection.

Controlled laboratory studies as well as field experience in a number of different countries in the world have shown that the living attenuated vaccine *B. melitensis* Rev 1* is the most satisfactory vaccine for use in goats and sheep (Alton and Elberg, 1967). Injection of young goats before breeding age will protect them for their entire life. If administered to adult goats the full dose of Rev 1 may cause abortion or excretion in the milk. Reduced doses may be given to adult goats without risk of abortion or excretion and without diminishing the protective effect (Alton *et al.*, 1972).

Goats given the full dose of Rev 1 will develop serum agglutinins and skin test reactions which cannot be distinguished from those in infected goats. The complement fixation test can be used provided it is not carried out within 6 months of vaccination. The reduced dose of Rev 1 does not cause a serological response which interferes with diagnosis.

Experience in Italy over a 5-year period has shown that vaccination of kids and lambs with Rev 1 has resulted in greatly reduced numbers of abortions due to brucella, an improvement in the cooperation of the shepherds and a reduction in the number of cases of human brucellosis (Farina and Bellani, 1974). Rev 1 vaccination has been employed on a large scale in Iran (Arshadi, 1974), Israel (Landau, 1974) and in the USSR (Ulavevich, 1974) with satisfactory results.

2. Q-Fever
(C. S. F. Williams)

Q-fever is a zoonosis, caused by *Coxiella burnetti*. This rickettsial organism is thought to be normally non-pathogenic for domestic stock, but it has been implicated in goat abortions. It is a public health problem more than an animal problem.

Caminopetros (1948) infected goats with the blood of humans suffering from Q-fever. Intrapulmonary or intranasal inoculation caused febrile bronchopneumonia, subcutaneous inoculation caused a febrile reaction and local inflammation, whereas intraocular inoculation produced ophthalmia. He found that infected goats can shed the organisms in the milk for at least 3 months, and that kids become infected by suckling. Goats could also become infected by being housed with goats suffering from Q-fever pneumonia.

In California, USA, Lennette *et al.* (1949) found that 43·6% of the goats in contact with human cases, had positive titers against Q-fever. The organism was demonstrated in the milk of goats, by guinea pig inoculation and seroconversion. Ruppaner *et al.* (1978) found 24% of 1054 goats in California, USA, positive to Q-fever, and 2/29 were shedding the organism in the milk.

* Method of production of Rev 1 vaccine is given in WHO Monograph No. 55.

They found that the risk of infection increased with herd size. Hummel (1976) found 13·6% of 575 goats in Tanzania had positive titers, but this was of no economic concern. Haumesser and Poutrel (1973) found that 54% of Red Maradi goats tested in Niger, were positive to *Chlamydia* and 40% had positive titers against Q-fever. They found no correlation between abortions and titers, and thought abortions were nutritional, although brucellosis existed at a low level in the same flocks. Gidel (1965) found 13·75% of 134 goats in Chad to have a positive titer against Q-fever. Kaplan and Bertagna (1955) found 11 of 31 goats positive in Cyprus, using the complement fixation test. Dade *et al.* (1973) tested goats in Egypt and found 47% with titers of 1 : 8 or more. Crowther and Spicer (1976) reported on a serological survey in Cyprus among goat flocks in which Q-fever organisms had been isolated from placentas. Seventy-two per cent of the aborting goats were positive, 44% of the non-aborters were positive. Those that had aborted had significantly higher titers than the non-aborters. The abortions were not due to *Chlamydia*, *Brucella*, *Salmonella* or *Vibrio*. In their experience, rickettsial abortions occur in the second half of kidding season, during the last month of pregnancy, with the dam showing no signs of illness. Waldhalm *et al.* (1978) in the USA isolated Q-fever organisms from an epizootic of abortions in a herd of 39 adult goats where 20 aborted. No other microorganisms were isolated and antibody titers to *Leptospira* spp., *Campylobacter* (*Vibrio*) *fetus*, *Brucella* spp., *Chlamydia* spp. and *Toxoplasma gondii* were all negative. Genchev *et al.* (1973) found 51% of 201 goats in Bulgaria had positive titers. They also isolated four strains of the Q-fever organism from aborting goats and sheep.

References

Abdel-Hamid, Y. N. (1973). *J. Egypt. Vet. Med. Ass.* **33**, 45–53.

Adler, H. E., Yamamoto, R. and Cordy, D. R. (1956). *Cornell Vet.* **46**, 206–216.

Al-Aubaidi, J. M., Dardiri, A. H., Muscoplatt, C. C. and McCauley, E. H. (1973). *Cornell Vet.* **63**, 177–179.

Al-Aubaidi, J. M., Taylor, W. D., Bubash, G. R. and Dardiri, A. H. (1972). *Am. J. Vet. Res.* **23**, 87–90.

Alton, G. G. and Elberg, S. S. (1967). *Vet. Bull.* **37**, 793–800.

Alton, G. G., Jones, L. J., Garcia-Carrillo, C. and Trenchi, A. (1972). *Am. J. Vet. Res.* **33**, 1747–1751.

Alton, G. G., Jones, L. M. and Pietz, D. E. (1975). "Laboratory Techniques in Brucellosis" (2nd ed). World Health Organization Monograph Series No. 55, Geneva.

Amanzhulov, S. A. (1953). *Veterinarya* (*Moscow*) **30**, 22–23.

Anon. (1973). *J. S. Afr. Vet. Med. Ass.* **44**, 321–330.

Arshadi, M. (1974). *Bull. Off. Int. Epizoot.* **82**, 73–82.

Ayers, J. L. (1977). *J. Am. Vet. Med. Ass.* **171**, 1251–1254.

Bagadi, H. O. and Razig, S. E. (1976). *Vet. Rec.* **99**, 13.
Baker, N. F. (1975). *J. Am. Vet. Med. Ass.* **167**, 1069–1075.
Bakos, K. and Brag, S. (1957). *Nord. Veterinaermed.* **9**, 431–449.
Basson, P. A. (1962). *Onderstepoort J. Vet. Res.* **29**, 203–204.
Berberovic, M. (1964). *Vet Glas.* **18**, 1101–1104.
Bergeron, J. A. (1976). Personal communication.
Blanco, A. (1967). *Rev. patron. Biol. Anim.* **11**, 35–58.
Blood, D. C. and Henderson, J. A. (1968). "Veterinary Medicine". Bailliere, London.
Borgman, R. F. and Wilson, C. E. (1955). *J. Am. Vet. Med. Ass.* **126**, 198–205.
Bouffet, P. (1973). Mammites de la chèvre. Thesis Doct. Vet., Lyon.
Boyt, W. P., MacKenzie, P. K. I. and Pilson, R. D. (1978). *Bull. Entomol. Res.* **68**, 497–500.
Brotherston, J. C., Renwick, C. C., Stamp, J. T., Zlotnik, I. and Pattison, I. H. (1968). *J. Comp. Pathol. Ther.* **78**, 9–17.
Bryan, C. S. (1942). *Am. J. Vet. Res.* **3**, 92–95.
Burdin, M. L., Froyd, G. and Ashford, W. A. (1958). *Vet. Rec.* **70**, 830–835.
Buxton, A. (1959). *In* "Infectious Diseases of Animals—Diseases due to Bacteria" (A. W. Stableforth and I. A. Galloway, eds), Vol. II, pp. 481–528. Butterworths, London.
Cameron, C. M. (1964). *Onderstepoort J. Vet. Res.* **31**, 119–132.
Cameron, C. M., Minnar, J. L., Engelbrecht, M. M. and Purdom, M. R. (1972). *Onderstepoort J. Vet. Res.* **39**, 11–24.
Caminopetros, J. P. (1948). *Ann. Parasitol. Hum. Comp.* **23**, 107–118.
Capstick, P. B. (1959). *Bull. Epizoot. Dis. Afr.* **7**, 51–62.
Cardoen, J. M. (1977). Les mammites de la chèvre. Thesis Doct. Vet., Alfort.
Chang, W. H. and Carter, G. R. (1976). *J. Am. Vet. Med. Ass.* **169**, 710–712.
Cheema, A. H. and Toofanian, F. (1979). *Cornell Vet.* **69**, 159–168.
Chellé, P.-L. (1942). *Bull. Acad. Vt. Fr.* **15**, 294–295.
Chotchaev, A. (1974). *Veterinariya* (*Moscow*) **3**, 66–67 (from *Rev. Appl. Entomol.* **13**, 2958, 1974).
Claveau, R., Grappin, R. and Roguinsky, M. (1978). Etudes préliminaires des méthodes de numération des cellules du lait appliquées au lait de chèvre. XXe Int. Dairy Congress, Paris.
Collins, G. H. and Charleston, W. A. G. (1979). *N.Z. Vet. J.* **27**, 260–262.
Cook, R. W. (1981). *Austr. Vet. J.* **57** (in press).
Cordy, D. R., Adler, H. E. and Yamamoto, R. (1955). *Cornell Vet.* **43**, 50–68.
Cork, L. D., Hadlow, W. J., Crawford, T. B., Gorham, J. R. and Piper, R. C. (1974). *J. Infect. Dis.* **129**, 134–141.
Cortella, I. (1963). *Nouva Vet.* **39**, 11–24.
Crowther, R. W. and Spicer, A. J. (1976). *Vet. Rec.* **99**, 29–30.
Cuille, J. and Chellé, P.-L. (1938). *C. R. Séances Acad. Sci.* **206**, 78–79.
Cuille, J. and Chellé, P.-L. (1939). *C. R. Séances Acad. Sci.* **208**, 1058–1060.
Dade, J. E. M., Zaklama, N. S. and Iman, I. Z. E. (1973). *J. Egypt. Public Health. Ass.* **48**, 101–108.
Dardiri, A. H., DeBoer, C. J. and Hamdy, F. M. (1977). *Proc. U.S. Anim. Health Ass.* **81**, 337–344.
Darrow, D. I. (1973). *J. Econ. Ent.* **66**, 133–135.
Das, D. M. and Misra, S. C. (1972). *Indian Vet. J.* **49**, 96–101.
Davies, F. G. (1976). *J. Hyg.* **76**, 163–171.
Dhanda, M. R. and Singh, M. M. (1955). *Indian Vet. J.* **32**, 43–46.

Dobbs, E. M. and McIntyre, R. W. (1951). *Calif. Vet.* **4**, 19.
Doty, R. B., Dunne, H. W., Hokanson, J. F. and Redi, J. J. (1964). *Am. J. Vet. Res.* **25**, 1679–1685.
Doutre, M. P. and Bouche, R. (1976). *Rev. Elev. Med. Vet. Pays Trop.* **29**, 205–209.
Dubey, J. P., Sharma, S. P., Lopes, C., Williams, J. F., Williams, C. S. F. and Weisbrode, S. (1980). *Am. J. Vet. Res.* **41**, 1072–1076.
Edwards, E. E. and Wilson, A. S. B. (1958). *J. Helminthol.* **32**, 195–210.
El Nasri, M. (1964). *Vet. Rec.* **76**, 876–877.
Euzeby, J., Chermette, R. and Gevrey, J. (1976). *Bull. Acad. Vet. Fr.* **49**, 423–430.
Fabiyi, J. P. (1973). *Bull. Epizoot. Dis. Afr.* **21**, 177–184.
Farina, R. and Bellani, L. (1974). *Bull. Off. Int. Epizoot.* **82**, 41–44.
Farina, R. and Sobrero, L. (1960). *Zooprofilassi* **15**, 925–936.
Farizy, P. (1970). *Recl. Med. Vet. Ec. Alfort* **146**, 251–260.
Farrag, H. and Oof, F. (1966). *Indian Vet. J.* **43**, 771–777.
Fatser, R. (1974). *Schweiz. Arch. Tierheilkd.* **116**, 219–224.
Faye, P. Lecoanet, J., Charton, A., Delahaye, J. and LeLayec, C. (1971). *Bull. Acad. Vet. Fr.* **44**, 61–64.
Fitzsimmons, W. M. (1967). *Vet. Rec.* **80**, 24–26.
Gall, C. (1962). *Münch. Tierärztl. Wschr.* **75**, 365–367.
Gallo, P. and Morris, H. (1965). *Rev. Med. Vet. Parasitol.* **19**, 91–94.
Gargadennec, L. and LaLanne, A. (1942). *Bull. Serv. Zootechn. Epizoot. Afr. Occident. Fr.* **5**, 16–21.
Gayot, G. and Cannepin, R. (1954). *Rec. Med. Vet.* **130**, 367–370.
Genchev, G., Kolev, V. and Ognyanov, D. (1973). *Vet. Med. Nauki* **10**, 37–42.
Gibson, T. E. (1975). "Veterinary Anthelmintic Medication" (3rd edn), pp. 210–218. Techn. Comm. No. 33, Commonwealth Agricultural Bureaux.
Gidel, R. (1965). *Rev. Elev. Med. Vet. Pays Trop.* **18**, 127–136.
Glover, R. E. (1928). *J. Comp. Pathol. Ther.* **41**, 318–340.
Gordon, W. S. (1966). Report of Scrapie Seminar, ARS 91–53, U.S. Department of Agriculture, Washington, DC, 19–40.
Goudswaard, J. (1971). "Ziekte van Johne bij de geit" [Johne's Disease in Goats]. Thesis, Inst. Vet. Bact. Fac. Diergenees. Rijksuniv. Utrecht, Netherlands.
Gourlay, R. N. and Barber, L. (1960). *J. Comp. Path. Ther.* **70**, 211–216.
Griffin, L. (1978). *Vet. Bull.* **48**, 819–825.
Guthery, F. S. and Beason, S. L. (1979). *J. Wildl. Dis.* **15**, 37–42.
Haddow, J. R. and Idnani, J. A. (1946). *Indian J. Vet. Sci.* **16**, 181–184.
Hakioglu, F. (1956). *Zentralbl. Allg. Path. Anat.* **95**, 525–527.
Hamdy, F. M., Dardiri, A. H., Breese, Jr., S. S. and DeBoer, C. J. (1976a). *Proc. U.S. Anim. Health Ass.* **80**, 168–179.
Hamdy, F. M., Dardiri, A. H., Nduaka, O., Breese, Jr., S. S. and Ihemelandu, E. C. (1976b). *Can. J. Comp. Med.* **40**, 276–284.
Harcourt, R. A. and Anderson, M. A. (1974). *Vet. Rec.* **94**, 504.
Hartley, W. J. and Hakioglu, F. (1970). *Pendik Veteriner Kontrol Arastirma Enstitusu Dergisi* **2**, 67–68.
Haumesser, J. B. and Poutrel, B. (1973). *Rev. Elev. Med. Vet. Pays Trop.* **26**, 293–298.
Hoffman, G., Hörchner, F., Schein, E. and Gerber, H. (1971). *Münchn. Tierärtzl. Wschr.* **84**, 152–156.
Honeker, A. (1938). *Berl Münch. Tieräztl. Wschr.* **54**, 17–20.
Horak, J. G. and Butt, M. J. (1977). *Onderstepoort J. Vet. Res.* **44**, 65–68.

Hourrigan, J. L., Klingsporn, A. L., McDaniel, H. A. and Riemenschneider, M. N. (1969). *J. Am. Vet. Med. Ass.* **154**, 538–539.
Hourrigan, J. L., Klingsporn, A., Clark, W. W. and de Camp, M. (1979) *In* "Slow Transmissible Diseases of the Nervous System" (W. J. Hadlow and S. B. Prusiner, eds.). Vol. I, pp. 331–356, Academic Press, New York and London.
Hughes, D. L., Treacher, R. J. and Harness, E. (1974). *Res. Vet. Sci.* **17**, 302–311.
Hummel, P. H. (1976). *Vet. Rec.* **98**, 501–505.
Hurter, L. R. and Potgieter, L. N. D. (1967). *J. S. Afr. Vet. Med. Ass.* **38**, 444–446.
Ilemobade, A. A. and Blotkamp, J. (1978). *Tropenmed. Parasitol.* **29**, 71–76.
Ivanov, I. (1957). *Bull. Off. Int. Epizoot.* **48**, 571–583.
Jain, S. K., Rajya, B. S., Mohanty, G. C., Paliwal, P. O., Mehrotra, M. L. and Sah, R. L. (1975). *Curr. Sci.* **44**, 209–210.
Jiřina, K. (1953). *Monatsh. Veterinärmed.* **8**, 271–272.
Joint FAO/WHO Expert Committee on Brucellosis, Fifth Report, Wld Hlth Org. Techn. Rep. Ser., No. 464, 1971.
Kaplan, M. M. and Bertagna, P. (1955). *Bull. WHO.* **13**, 829–860.
Kapur, M. P., Kalra, D. S. and Randhawa, A. S. (1973). *Indian Vet. J.* **50**, 859–862.
Kassai, T., Altaif, K. T., Kadhim, J. K., Al-Jabbir, M. H. and Bayyar, H. A. (1973). "Isotopes and Irradiation in Parasitology" Vol. III, p. 51. I.A.E.A.
Khalimbekov, M. M. (1957). *Veterinarya* (*Moscow*) **34**, 17–20.
Kharole, M. U. and Rao, U. R. K. (1968). *Indian Vet. J.* **45**, 14.
King, L. S. (1940). *Am. J. Path.* **16**, 467–478.
Klement'eva, E. V. and Drummieu, K. T. (1976). *Problemy Vet. Sanitarri* **52**, 175–181.
Komarov, A. and Goldsmit, L. (1951). *Refu. Vet.* **8**, 96–100.
Krylov, A. I. A. and Shakhimina, R. S. (1972). *Veterinarya* (*Moscow*) **8**, 104–105.
Landau, M. (1974). *Bull. Off. Int. Epizoot.* **82**, 61–68.
Levine, N. D. and Ivens, V. (1970). "The Coccidian Parasites (Protozoa, Sporozoa) of Ruminants". Illinois Biological Monograph No. 44, University of Illinois, Urbana.
Littlejohn, A. E. (1968). *Vet. Rec.* **82**, 148–155.
Littlejohn, I. R. and Cottew, G. S. (1977). *Aust. Vet. J.* **53**, 297–298.
Livingston, Jr., C. W. and Gauer, B. B. (1975). *Am. J. Vet. Res.* **36**, 313–314.
Longley, E. O. (1951). "Contagious Caprine Pleuropneumonia. A Study of the Disease in Nigeria". Col. Res. Publ. No. 7, H. M. Stationery Office, London.
Lopez, C. and Lopez, F. (1952). *Bull. Acad. Vet. Fr.* **25**, 23–25.
Luedke, A. J. and Anakwenze, E. I. (1972). *Am. J. Vet. Res.* **33**, 1739–1745.
MacKay, J. M. K. and Smith, W. (1961). *Vet. Rec.* **73**, 394–396.
Mailloux, M. (1969). *Bull. Soc. Pathol. Exot.* **62**, 508–516.
Maronpot, R. R. and Barsoum, I. S. (1972). *Am. J. Trop. Med. Hyg.* **21**, 467–472.
McCauley, E. H. and Tieken, E. L. (1968). *J. Am. Vet. Med. Ass.* **152**, 1758–1765.
McDaniel, H. A. and Morehouse, L. G. (1966). Report of Scrapie Seminar, ARS. 91-53, US Department of Agriculture, Washington, DC, 41–52.
McVicar, J. W. and Sutmoller, P. (1968). *Proc. US Livest. Sanit. Ass.* **72**, 400–406.
McVicar, J. W. and Sutmoller, P. (1971). *Proc. US Anim. Health. Ass. 75th Proc.* 194–199.
McVicar, J. W. and Sutmoller, P. (1972). *Arch. Gesamte Virusforsch.* **38**, 85–96.
Miege, R. (1978). *Rev. Med. Vet.* **129**, 247–259.
Mishra, H. N., Pandurangarao, G. C. and Kera, S. S. (1970). *Indian Vet. J.* **47**, 808–809.

Mornet, P., Orue, J., Gilbert, Y., Thiery, G. and Mamadon, S. (1956). *Rev. Elev. Med. Vet. Pays Trop.* **9**, 313–342.
Morse, E. V. and Langham, R. F. (1958). *Am. J. Vet. Res.* **19**, 139–144.
Morter, R. L. and Morse, E. V. (1956). *J. Am. Vet. Med. Ass.* **128**, 408–413.
Mugera, G. M. and Bitakaramire, P. K. (1968). *Vet. Rec.* **82**, 595–597.
Mugera, G. M. and Kramer, T. T. (1967). *Bull. Epizoot. Dis. Afr.* **15**, 125–131.
Naerland, G. (1950). *Nord. Veterinaermed.* **2**, 1–18.
Nesbakken, T. (1975). *Nor. Vet. Tidsskr.* **87**, 188–191.
Nesbakken, T. (1976). *Nord. Veterinaermed.* **28**, 550–556.
Newsam, I. D. B. (1925). Unpublished observation, cited by J. R. Lawson, 1959, *in* "Infectious Diseases of Animals" (A. W. Stableforth and I. A. Galloway, eds.). Butterworths, London.
Newsom, I. E. and Cross, F. (1934). *J. Am. Vet. Med. Ass.* **84**, 799–801.
New Zealand, (1978). *Ann. Rep. Min. Agric. Fish.* (1978), 24–25.
Nigam, S. K., Shukla, V. and Ranat, R. (1974). *Indian J. Anim. Health* **13**, 93–94.
Ojo, M. O. (1976). *J. Comp. Path.* **86**, 519–529.
Okada, M. (1960). *Tohoku J. Agric. Res.* **11**, 31–44.
Omori, T., Ishii, S., Harada, K., Ichikawa, O., Murase, N., Katada, M. and Araumi, W. (1953). Exp. Rep. No. 27 of Govt. Experimental Station for Animal Hygiene, Tokyo, pp. 109–120.
Omori, T., Morimoto, T., Harada, K., Inaba, Y. and Matumoto, M. (1957). *Jap. J. Exp. Med.* **27**, 131–143.
Onoviran, O. (1974). *Vet. Rec.* **94**, 418–420.
Pande, P. G. (1943). *Indian J. Vet. Sci.* **13**, 44–58.
Pargoanker, V. N. (1964). *Indian Vet. J.* **41**, 441.
Patnaik, B. and Roy, S. P. (1968). *Indian J. Vet. Sci. Anim. Husb.* **38**, 450–462.
Pellérdy, L. P. (1974). "Coccidia and Coccidiosis" (2nd edn). Paul Parey, Berlin.
Perreau, P. (1973). *Rev. Elev. Med. Vet. Pays Trop.* **26**, 13–25.
Perreau, P. (1974). *Bull. Acad. Vet.* **47**, 179–188.
Perreau, P. (1977). "Les Mycoplasmoses des Petits Ruminants en France: Données actuelles", "3èmes Journées de la Recherche Ovine et Caprine", pp. 228–237. INRA, ITOVIC, Paris.
Pillai, C. P. (1962). *Vet. Rec.* **74**, 15–17.
Plommet, M. (1974). *Ann. Zootech.* numéro hors série, 87–95.
Plowright, W. and Ferris, R. D. (1958). *Br. J. Exp. Path.* **39**, 424–435.
Poutrel, B. and Roguinsky, M. (1978). Unpublished results.
Rajesekhar, M., Nanjiah, R. D. (1971). *Indian Vet. J.* **48**, 1087–1095.
Reppin, K. (1929). *Arch. Wiss. Prakt. Tierheilk.* **60**, 166–190.
Renshaw, H. W. and Dodd, A. G. (1978). *Arch. Virol.* **56**, 201–210.
Rodolakis, A. and Chancerelle, L. (1977). *Ann. Microbiol.* **128B**, 81–85.
Rodolakis, A., Dufrenoy, J. and Souriau, A. (1977). *Ann. Rech. Vet.* **8**, 213–219.
Roguinsky, M. (1974). *Ann. Zootech.* numéro hors série, 117–118.
Roguinsky, M. (1977). "Les mammites des Petits Ruminants", "3ème Journées de la Recherche Ovine et Caprine", pp. 65–77. INRA, ITOVIC, Paris.
Roguinsky, M. and Grandremy, T. (1978). *La Chèvre* **108**, 25–26.
Roguinsky, M., Redon, J. F., Lemens, P., Gendron, H. and Allard, P. (1971). *La Chèvre* **68**, 4–5.
Rowland, A. C., Scott, G. R. and Hill, D. H. (1969). *J. Path.* **98**, 83–87.
Ruppaner, R., Riemann, H. P., Farver, T. B., West, G., Behymer, D. E. and Wijayasinghe, C. (1978). *Am. J. Vet. Res.* **39**, 867–870.

Sahoo, S. K., Rao, A. T., Misra, M. S. and Murty, T. L. (1972). *Indian Vet. J.* **49**, 1107–1109.
Sanchis, R. (1978). Unpublished results.
Sandbu, H. (1956). *Nord. Veterinaermed.* **8**, 585–594.
Sapre, S. N. (1964). *Vet. Rev.* **15**, 69–71.
Sarkar, P. and Bhattacharyya, H. M. (1975). *Indian J. Anim. Health* **14**, 5–6.
Sayin, F., Meric, I., Koseoglu, H., Sincer, N. and Ayabakau, S. (1972). *Ankara Univ. Vet. Fak. Derg.* **19**, 338–343.
Schalm, O. W., Carroll, E. J. and Jain, N. C. (1971). "Bovine Mastitis". Lea and Febiger, Philadelphia, PA.
Schillhorn, van Veen, T. W. (1978). Personal observation.
Schillhorn, van Veen, T. W. and Mohammed, A. N. (1976). *J. Niger. Vet. Med. Ass.* **4**, 93–96.
Schröter, A. (1954). *Berl. Münch Tierärztl. Wschr.* **67**, 3–5.
Sebek, Z. (1961). *Vet. Cas.* **10**, 271–281.
Semerdzhiev, B., Ognyanov, D. and Makaveeva-Simova, E. (1966). *Vet. Med. Nauki* **3**, 115–121.
Sen, K. C. and Datt, N. S. (1968). *Indian J. Vet. Sci.* **38**, 388–393.
Sharmir, A. (1954). *Refu. Vet.* **11**, 241–243.
Shaw, J. N. and Seghetti, L. (1939). *N. Am. Vet.* **20**, 39–40.
Shigidi, M. T. A. (1978). *Res. Vet. Sci.* **24**, 57–60.
Smith, H. C. and Harnden, E. E. (1943). *Vet. Med.* 299–301.
Smith, M. C. and Roguinsky, M. (1977). *J. Am. Vet. Med. Ass.* **171**, 1241–1248.
Stafseth, H. J. and Bryan, C. S. (1945). *N. Am. Vet.* **27**, 22–25.
Stalheim, O. H. V. (1976) (Ed.). "Laboratory Diagnosis of Mycoplasmosis in Food Animals". *Proc. 19th Ann. Meet. Am. Assoc. Vet. Lab. Diagnosticians*, November, pp. 1–156.
Stamp, J. T. (1962). *Vet. Rec.* **74**, 357–362.
Staub, H. (1959). *Dtsch. Tierärztl. Wochenschr.* **66**, 98–99.
Stemshorn, B. W. (1975). *Can. Vet. J.* **16**, 84–86.
Tadjebakhche, H. and Gatel, A. (1972). *Recl. Med. Vet.* **148**, 1027–1029.
Tadjebakhche, H. and Namin, P. (1974). *Rev. Med. Vet.* **125**, 1469–1474.
Tadjebakhche, H., Hosseinioun, M. and Nadalian, M. (1974). *Rev. Med. Vet.* **125**, 711–718.
Tajima, M. (1950). *Jpn. J. Vet. Sci.* **12**, 241–246.
Tizard, I. R., Carrington M. and Lai, C. H. (1977). *Can. Vet. J.* **18**, 274–277.
Tripathy, S. B. (1977). *Ind. Vet. J.* **54**, 584–585.
Trotter, S. L., Franklin Baas, R. M. and Barile, M. F. (1977). *Infect. Immun.* **18**, 816–822.
Toidze, A. I. (1954). *Veterinariya* (*Moscow*) **31**, 24–29. (1955) Abstract in *Vet. Bull.* **25**, 150.
Turner, L. W. (1973). *Proc. U.S. Animal Health Ass.* **77**, 650–656.
Ulasevich, P. S. (1974). *Tr. Vses. Inst. Exsp. Vet.* **42**, 257–261.
Van der Hoeden (1953). *J. Comp. Path.* **63**, 101–111.
Van der Werve, G. F. (1949). *J. S. Afr. Vet. Med. Ass.* **20**, 93–98.
Van Riel, J. and Van Riel, M. (1956). *Doc. Med. Geogr. Trop.* **8**, 55–60.
Van Tonder, E. M. (1975). *Vet. Med. Rev.* **1/2**, 109–158.
Waldhalm, D. G., Stoenner, H. G., Simmons, R. E. and Thomas, L. A. (1978). *J. Am. Vet. Med. Ass.* **173**, 1580–1581.
Webster, H. D. and Thorp Jr., F. (1952). *M.S.C. Vet.* **12**, 152–156.

Whitney, J. C., Scott, G. R. and Hill, D. H. (1967). *Bull. Epizoot. Dis. Afr.* **15**, 31–41.
Whittlestone, P. (1976). *Adv. Vet. Sci. Comp. Med.* **20**, 277–307.
Wilkinson, G. T., Prydie, J. and Scarnell, J. (1970). *Vet. Rec.* **87**, 766–767.
Williams, C. S. F. (1979). Personal observation.
Wood, J. S. (1972). *Can. Vet. J.* **13**, 80–82.
Yalcin, N. and Gane, P. (1970). *Recl. Med. Vet.* **146**, 705–709.
Zaki, M. M. (1968). *Res. Vet. Sci.* **9**, 489–493.
Zaki, M. M. (1976). *Res. Vet. Sci.* **20**, 197–200.
Zavagli, V. (1951). *Bull. Off. Int. Epizoot.* **36**, 336–362.
Zeller, H. (1920). *Arb. Reichsgesundheitsamt* (*Berlin*) **52**, 501–537.

15. Goats in Cold and Temperate Climates

S. Trodahl, Trygve Skjevdal and Torstein A. Steine

A. Goats in the Nordic Countries

1. Structure of the Goat Industry

The number of goats has decreased considerably in all the Nordic countries over recent decades. In Norway, where goats are still included in the official census of domestic animals, goat numbers went down from about 200 000 between 1907 and 1919 to 74 000 in 1979. Over the years between 20 and 35% of the total were less than 1 year old. In Sweden the number of goats was 66 257 in 1927, and decreased to 3856 in 1956. According to Sjödin (1976), there is now a growing interest in goat-keeping. The number of goats is known to have increased in certain localities since 1960, and is now estimated at approximately 10 000 head. Finland has 2000, Iceland 500, and Denmark 3000–5000 head. The goats in Norway are distributed over a wide range of natural and climatic conditions. One is the south-eastern region where farms are located in the valleys and mountainous areas. Here winters are severe and long. The south-west has farms mainly in the fjords with shorter winters and rather a humid climate. In the northern region beyond the Arctic Circle goats are kept right up to the Arctic Ocean at 71° north. The farms in this region are also located in the fjords and on islands along the coast. About 50% of the goat population in Norway is in this region. The average milk yield per doe does not differ much from one region to the other in spite of very different climatic conditions (Table I). In Sweden goats are mainly kept in the northern part of the country where conditions are rather similar to the eastern region in Norway.

Table 1. Milk production in Norway (Steine, 1975a).

Region	No. of lactations	Milk (kg) $\bar{x}$	Milk (kg) σ	Fat percentage $\bar{x}$	Fat percentage σ	Milk fat (kg) $\bar{x}$	Milk fat (kg) σ
Eastern	9061	575	141	3·67	0·40	21·1	5·24
Western	9456	573	142	3·61	0·38	20·6	5·03
Northern	3521	586	153	3·31	0·39	19·3	4·95
Total	22038	576	143	3·59	0·39	20·6	5·10

About 90% of the goat milk produced in Norway is delivered to cooperative creameries for production of brown cheese. This is a special product of whey mixed with milk and heated until most of the moisture is evaporated. It contains the lactose, fat and minerals of the milk, and part of the protein. This traditional cheese is very much in demand at the local market. Formerly it was made by local farmers, but today very few farmers engage in its production. The situation is the opposite in Sweden. There most of the farmers make their own cheese, and the main products are different types of white cheese.

In recent years, the goat industry in Norway has tended towards more specialization. Formerly the farmers used to keep some cattle and sheep along with goats. Also it was common to rear some kids for meat production. These were usually slaughtered after one grazing season. This practice has been omitted after the switch to earlier kidding and more specialized production. Today there are a number of farms which keep only goats or goats and sheep. The average size of herds is slightly above 40 does, and specialized herds keep as a rule 40–100 does. Most of the goats are milked by machine, but goat-keeping is still a rather labour intensive enterprise. An economic survey (Mytting, 1976) showed that in herds with about 60 does and 12–15 sheep the average labour requirement was about 3500 man hours a year.

2. Climatic Conditions

The ground may be covered with snow for 5–7 months in several of the areas where goats are kept. Therefore a long period of indoor feeding and good housing facilities are required. Even on mountain pastures during the summer there is a need for shelter against cold rain and winds. The goats are out on pasture for $4\frac{1}{2}$ months. However, there are considerable variations. In the far north and at high altitudes only 80–90 days may be available for pasturage, while in the south-western part of Norway goats may be kept on pasture for about 7 months. In these areas there are also occasional flocks of domestic goats that have gone astray and formed feral populations that have sustained and propagated out in wilderness for years. One such population can be seen at Stavenes in the district of Sogn og Fjordane, where they feed partly on seaweed at the beaches during the winter.

3. Goat Types

Goats are known to have played an important role in the household in Nordic countries. Tales in mythology and historic records tell of the usefulness of goats in everyday life for the common man. In more recent times we have records on steps taken to improve goat-keeping. Sporadic imports

Fig. 1. Norwegian buck, Ruff 2434, 2 years.

of Saanen and Angora goats were tried but, as far as the records can be traced, these imports did not have any significant advantage over the local types. Thus it is very likely that the goats found in the Nordic countries can be classified as an old landrace, with very little influence from foreign blood. In recent years bucks have been sold from Norway to Sweden and Finland, and the goats in these countries today are more or less of the same type.

An exception may be Denmark where the population consists mainly of goats of the Saanen and Toggenburg types together with a number of dwarf types.

The Nordic goats are rather short legged (Figs 1 and 2). Height over the croup is generally 65–70 cm. Average body weight of mature bucks is 75–80 kg, but some individuals may attain well above 100 kg. Average weight of does is 48·5 kg, varying from 39 kg in yearlings to 51 kg in 5-year-old does (Table II). Nordic goats have a long hair coat and often a dense woolly undercoat. This trait is frequently found in Icelandic goats where the flocks have been outwintered for several generations. Colours vary (Berge, 1966) and goats have not been selected for uniform colour. Emphasis is put on improving the teat and udder conformation. Both horned and polled individuals are found, but polled goats are often preferred because they damage each other less by butting. But polledness is generally connected with hermaphroditism. According to Haugen (1960), 8% of the offspring were hermaphrodites when both parents were polled, whereas this abnormality did not occur when both parents were horned, see Chapter 4.

Fig. 2. Norwegian doe, Gullhøna, 4 years. Yield 1975: 796 kg milk, 3·9% fat; yield 1976: 929 kg milk 3·7% fat.

Table II. Effect of age on body weight within herds when season of kidding is kept constant (Steine, 1975a).

Age at kidding years	Least square constants for age groups	
	kg	Standard error
mean	48·5	±0·1
1	−9·7	±0·2
2	−2·8	±0·2
3	−0·7	±0·2
4	2·1	±0·2
5	2·4	±0·2
6	2·5	±0·2
7	1·9	±0·3
8	1·6	±0·4
9	1·2	±0·5

4. Reproduction and Fertility

(a) *General Conditions*

It is common practice to let goats have their first kid at an age of approximately 1 year. In districts where goats formerly had to search for their food during the better part of the year on a poor level of nutrition, it was common to wait until the goats had reached 2 years for the first kidding. Traditionally the kidding season lasted from March to May. This has gradually been pushed forward to earlier dates. Today the average date of kidding is in the first part of February with some differences between age groups and regions (Table III).

Table III. Average date of kidding in Norway within age groups and regions 1968–1972 (Steine, 1975a).

Age group	Eastern region	Northern region	Western region
1 year	24 Feb.	25 Feb.	18 March
2 years	31 Jan.	14 Feb.	22 Feb.
3 years	24 Jan.	11 Feb.	14 Feb.
>3 years	23 Jan.	4 Feb.	8 Feb.

Kidding in all age groups occurs earliest in the eastern region followed by the north and western region. In 30 herds in the Swedish districts of Vesternorrland and Jemtland the kidding season lasted from November to March with the highest frequency in January–February (Rejkestam and Cornell, 1977).

Goats may show false heat without ovulation. It occurs mainly during the first part of the mating season and can be distinguished from ordinary heat only by being shorter. A normal oestrus may follow within 4 to 7 days. Frequency of false heats was 7·1 and 10% (Skjevdal, 1975c; Fougner, 1979). In natural service an average of 85·5 ± 1·8% out of 5702 does did show repeated heats after mating.

In Norway frequencies of singles, twins and triplets were 50·7%, 47·8% and 1·5% respectively. The average number of kids born was 1·5 with a standard deviation of 0·5. It increased with age of dam up to 5 years. The difference between 1- and 5-year-old goats was 0·48 kids born. Kidding season also had a small effect. Does kidding in April had 0·08 kids more than goats kidding in December (Steine, 1975a).

(b) *Artificial Insemination*

The first experiments with fresh semen in Norway in the 1950s yielded quite encouraging results. However, due to the sparse distribution of the herds and long routes of transportation, it soon became evident that only deep frozen semen would have any prospective use. In 1972 a new technique for intra-uterine insemination was developed and it was discovered that centrifugation of the semen improved sperm survival (Fougner, 1979). In 1978/79 1026 goats were inseminated in 130 herds by 26 technicians. The average kidding percentage was 60·9 but results vary considerably between technicians and herds. Some areas have a pregnancy percentage above 80. This variation indicates that the training and experience of the technical staff is important. Also, better heat detection by the farmers will improve results.

A breeding scheme based on AI and without heat synchronization is more expensive than a scheme based on natural mating in buck circles. Yet, in recent years AI is used as a supplement to the buck circles. Eighty per cent of the cost of AI in this scheme is covered by a government grant.

B. Milk Production

1. Milk Recording

(a) *Schemes*

An official goat recording scheme was started early in the 1930s along the same lines as for dairy cows. Today a minimum of six weighings during the lactation is required. The first of three of these are taken with 1 month intervals, the rest with 2 months intervals. Fat tests are required for does in their first and second lactation only and are taken three times during lactation. Participation in the scheme was very modest at the beginning, but increased when modern data processing was adopted with the possibility of giving the breeders better information on progeny tests of bucks and goat indices. Table IV shows the development since 1952. The increase in numbers of recorded goats is partly due to an increase in the average size of the recorded herds which was 17 does in 1952 and 48·9 in 1978. The decrease in fat percentage is partly due to the fact that no special emphasis has been put on this trait. Creameries processing goat milk maintain that about 3·2% fat is sufficient for the kind of products they make, and no decrease in fat percentage has been noticed in milk delivered to the creameries during the last five years.

Table IV. Participation and yield in the Norwegian goat recording scheme (yearlings included).

Year	No. of goats recorded[a]	Total no. of goats older than 1 year	Average milk yield (kg)	Fat (%)	Milk fat (kg)
1952	4107	83992	421	3·53	14·9
1955	5551	86944	387	3·61	14·3
1960	7321	76408	441	3·56	15·7
1965	12256	82367	515	3·55	18·3
1970	10369	66231	590	3·47	20·5
1975	15385	50882	573	3·28	19·0
1978	19428	52251	592	3·24	20·2

[a] The number of goats is calculated on full year basis, i.e. goats with 365 days in the herds.

The yields in Table IV are averages for the whole scheme, several goats are yielding up to 1200–1500 kg, and the better herds have average yields of about 800–900 kg milk (Table V).

Figure 3 shows the average yield for the total goat population.

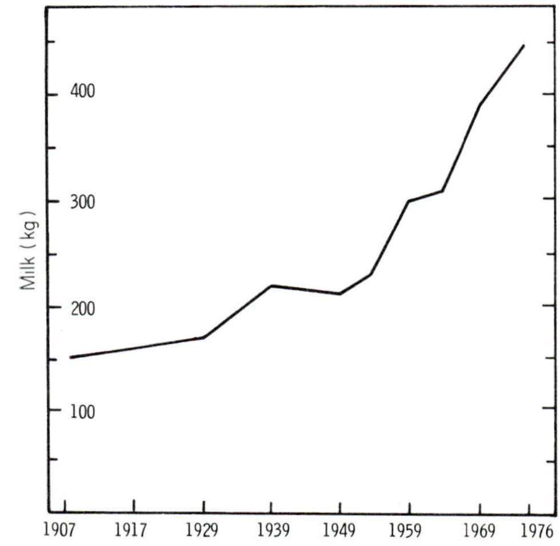

Fig. 3. Average milk yield of the goat population in Norway, 1907–1976.

Table V. Highest yielding herds and goats in the Norwegian goat recording scheme (yearlings included).

	The highest yielding herd[a]			The goat with highest yield[a]		
Year	Milk (kg)	Fat (%)	Milk fat (kg)	Milk (kg)	Fat (%)	Milk fat (kg)
1960	956	3·75	35·5	1257	3·91	49·2
1965	893	3·65	31·8	1504	3·10	45·8
1970	868	4·04	35·1	1624	2·80	46·2
1975	1103	3·80	42·0	1171	4·90	58·4
1976	1291	3·60	46·0	1729	3·00	53·1
1977	873	3·60	32·0	1960	2·80	56·0
1978	884	3·70	33·0	1836	3·20	60·4

[a] Ranked on the basis of milk fat production.

In Sweden two different systems of recording are in operation. One is recording milk and fat percentage, and the results from this system are given in Table VI.

Table VI. Results from goat recording with fat analysis in Sweden (yearlings included) (Rejkestam and Cornell, 1977).

Year	No. herds	No. goats	Goats per herd	Average milk yield (kg)	Fat (%)	Milk fat (kg)
1969/70	16	266	16·6	536	3·4	18·2
1970/71	14	151	15·1	514	3·6	18·5
1971/72	7	210	30·1	445	3·2	14·2
1972/73	7	175	25·0	495	3·2	15·9
1973/74	5	96	19·2	420	3·3	13·8
1974/75	5	82	16·5	466	3·1	14·6

In the other system the dry matter content is calculated instead of fat percentage. The analysis according to Sjödin (1976) is carried out by the farmers themselves three times a year with one conformatory check by an unchallangeable person once a year on the day after one of the three ordinary recordings. The equipment necessary for this analysis is a balance capable of registration down to 0·01 g, and an electric oven. The procedure is simple to follow and has caused no difficulty to the farmers. Results from this scheme are given in Table VII

For the conformatory analysis a modified gravimetric method is applied. The relation between the field test and the laboratory analysis shows a correlation of 0·85. The correlation between fat and dry matter content has been calculated at $r = 0{\cdot}5836$ (Sjödin, 1976).

Table VII. Results from goat recording with dry matter analysis in Sweden (Rejkestam and Cornell, 1977).

Year	No. herds	No. goats	Goats per herd	Average milk yield (kg)	Dry matter (%)	Dry matter (kg)
1971/72	15	270	18·0	556	11·6	63·9
1972/73	17	396	23·3	593	11·3	67·3
1973/74	22	483	21·9	576	11·4	65·8
1974/75	18	426	23·6	686	11·3	77·8

(b) *Utilization of Data*

In Sweden each farm with over 15 goats receives a recommendation at the end of the year for planned matings. The best third of the goats is selected to breed herd replacements. In Norway data from the recording scheme are used for calculation of indices for the evaluation of each individual doe. The procedure is described in Chapter 4. The best 5% of the does in the population are regarded as élite goats and are eligible for planned matings with élite bucks. The indices are calculated once a year and are presented to the farmers together with the annual accounts of the recording scheme. Thus the farmers can use indices in their herd selection. The indices are also utilized in connection with judging of goats at local shows. In this context scores for body conformation and for udder and teats are added to the index and the sum determines the grade awarded to the individual goat. Furthermore, research workers at scientific institutions at the Agricultural University of Norway (A.U.N.) have access to the data from the milk recording scheme, and several of the papers referred to in this chapter are based on these.

C. Progeny Testing of Bucks

Bucks are progeny tested on field data from the milk recording scheme (Trodahl, 1976b). With a scheme well developed in all regions it is possible to have the daughters tested under conditions prevailing in the different areas. Thus, not only the daughters' milking capacity is estimated but also their adaptation to the local environment is reflected. The first bucks were tested in 1963, based on daughters in one or two herds only. Therefore it was difficult to obtain reliable results from the progeny tests. In order to overcome this difficulty, buck-circles were organized along the same lines as had been done previously in the sheep industry for progeny testing of rams (Gjedrem, 1969). Each buck-circle must have at least 3 herds and 500 does, and a capacity to test at least 5 bucks. In the mating season each buck is circulated to all the herds in the ring so that offspring may be obtained in each herd. In 1976 there were 23 circles with 10 817 goats in operation. In recent years AI has been used as a supplement to the buck circles. In this way, semen from 4 to 6 of the best progeny tested bucks is used on élite goats. In 1977 this was utilized by 19 out of 24 buck-circles. Very strict regulations are imposed on the use of AI in the buck-circles in order not to impair the progeny-testing scheme. No more than 15% of the does may be mated to élite bucks, the rest of the goats being mated to young test bucks. Bucks must be approved by a goat breeding commission appointed in each county before they can be put on test. In this procedure the indices of the buck's sire and dam as well as the dam's udder and teat score are considered together with the buck's own score for body conformation. All approved bucks are ear-tagged and entered in the official buck register which also serves as a herd-book. For a valid test a minimum of 15 daughters in 3 herds are required. Each buck is tested for 2 consecutive years. The number of new bucks included in the test each year increased from about 20 in 1967 to 91 in 1978.

The result of the progeny test is presented as an index, giving the buck's relative breeding value in percentage of the population average. Bucks with indices above 110 are considered as "élite bucks" and used for planned matings with élite does. Bucks with indices in the range from 103 to 109 are considered as positive and may be sold to private herds or breeders associations, which get a subsidy from the government for their purchase, while bucks with indices below 103 have to be slaughtered according to the regulations. Normally, young bucks are selected when they are 3–4 months old and put on test service the same year. The first daughters complete their first lactation when the test bucks are 3 years of age. Then selected sires will be available for service in the following season when 3½ years old. The generation interval may be shortened and promising bucks used at 2½ years if the milk yield of the daughters can be preliminarily assessed before the beginning of the mating season. This is done on the basis of the average highest daily milk yield of the progeny group which has a correlation with the final index of $r = 0{\cdot}57$ to 0·74 or on the basis of a lactation record projected from part lactation yield (Steine, 1975b).

D. Specific Flavour of Goats' Milk

1. Products from Goats' Milk

Most of the goats' milk in Norway is processed to whey cheese (*geitost*). A similar product is made from cows' milk (*mysost*) and another from a mixture of goats' and cows' milk (*Gudbrandsdalsost*). Consumers enjoy cheese made from goats' milk because of the specific and strong flavour. Therefore the price paid to the producers is 40–50% higher than for cows' milk, and it is important to maintain the goat flavour of the milk. At A.U.N., selection experiments have been started in order to clarify the variation and relationship between goat flavour and various components of the milk

2. Components of Milk Flavour

Korvald (1958) found that the flavour was most prominent in cream and butter-milk. In skim milk and whey it was weaker than in whole milk, while casein was without specific flavour. A significantly higher content of volatile fatty acids was found in milk with strong flavour than in milk with weak flavour. Fat from genuine goat cheese had a higher content of volatile fatty acids than cheese made of mixed cows' and goats' milk. Volatile C_6–C_{10} acids could not be found in significant quantity in whey cheese from cow or mixed milk, while they were found in genuine goat cheese (Brandsaeter and Abate, 1959). The highest content of volatile fatty acids was found in cheese with medium flavour. It is mainly volatile fatty acids with a short structure, up to C_{10}, which are found in cheese with prominent flavour. Also a positive correlation with lipase-activity and flavour score has been found (Bjørke, 1974).

Other experiments have shown that milk with a weak flavour has a significantly higher content of polyenic acids than milk with a stronger flavour (Skjevdal, 1974a; Astrup *et al.*, 1972). The only components found in goats' milk which are not present in cows' milk are ortho- and metacresol (Svensen, 1968).

3. Factors affecting Flavour

Norwegian trials have shown that insanitary conditions reduce the goat flavour (Funder, 1933), and that keeping a buck together with the does has no influence on flavour (Korvald, 1958). Storing the milk for up to 42 hours increased the flavour of both milk (Skjevdal, 1974a) and cheese (Fyksen and Steinsholt, 1974).

The influence of nutrition on flavour has been investigated in several experiments. Formerly it was claimed that feeds with strong flavour like some leaves and pine needles created a strong flavour. Korvald (1958) observed that milk from herds fed such feeds had a rather weak flavour and in experiments it was found that these feedstuffs did not create a strong flavour (Opstvedt, 1967). There was a tendency towards a stronger flavour in milk from goats fed hay in comparison to a mixed ration of hay and grass silage. When feed rations of hay and silage were compared to silage only, there was a tendency towards stronger flavour in milk from goats fed mixed rations (Skjevdal, 1972a). Feeding of swedes has also caused a somewhat stronger flavour, although this was not significant (Skjevdal, 1974a). Several experiments with intensive feeding, often with a high content of concentrates, have given a strong flavour, especially when concentrates were fed on pasture (Opstvedt, 1967; Skjevdal, 1972c, 1974c). The effects of feeding on pasture and of stall feeding are not clear. In some experiments cheese flavour was stronger during the pasture period (Opstvedt, 1963; Korvald, 1958) but in others it was stronger during the stall period (Rønningen, 1965b). However, the differences were either small or inconsistent. These effects may be partly confounded with an effect of the stage of lactation since flavour is weak at the beginning and the end of the lactation (Rønningen, 1965a) and milk with high fat content gave cheese with low flavour score (Opstvedt, 1963).

4. Genetics of Milk Flavour

An analysis of milk from a few Saanen goats which had been brought to Norway from Denmark showed a volatile fatty acid content of 1·10 mEq/litre as for cows' milk, and its taste could hardly be distinguished from cows' milk. In Norwegian goats Rønningen (1965b) calculated a heritability coefficient for goat flavour of 0·25. Genetic correlations of flavour with milk yield and fat content were 0·77 and 0·28 respectively. Positive phenotype correlations between yield and flavour were calculated by Opstvedt (1963), Rønningen (1965a) and Skjevdal (1976c). Table VIII shows the correlations between flavour score and other components of goat milk found in three different surveys.

Table VIII. Correlation between flavour score and different components of milk.

Component	Opstvedt (1967)	Rønningen (1965a)	Skjevdal (1974a)
Dry matter (%)	−0·46[a]	—	−0·23
Fat (%)	−0·34[a]	−0·20[a]	−0·04 and −0·35[a]
Protein (%)	−0·31[a]	−0·35[a]	−0·28
Lactose (%)	−0·21[a]	−0·20[a]	−0·13
Ash (%)	−0·02	—	+0·10

[a] Significant figures.

In a selection experiment at the A.U.N. a herd of approximately 90 goats was split into two lines, one selected for high flavour (HF), the other for low flavour (LF). The flavour was recorded according to a scale where 1 indicates no goat flavour and 5 very strong goat flavour. In approximately five generations the selection response was almost linear (Fig. 4).

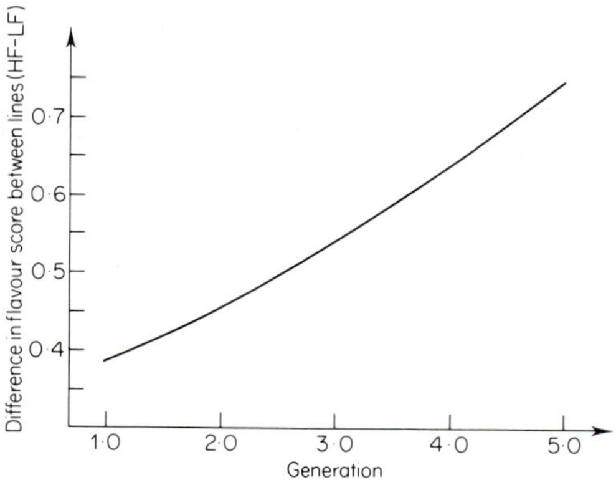

Fig. 4. Selection response in flavour score estimated as difference between selection lines (Steine, 1978).

Table IX. Heritability estimates for flavour score[a] (Steine, 1978).

Generation interval	Number of				$h^2 \pm$ s.e.
	tests	lactations	goats	sires	
0	1761	282	105	37	$0·37 \pm 0·21$
1·0–2·5	2004	283	104	42	$0·10 \pm 0·22$
>2·5	1573	228	115	42	$0·07 \pm 0·15$

[a] The flavour scores are adjusted for significant effects of year, age and stage of lactation.

Steine (1978) calculated heritability estimates for three generation groups in order to account for the effect of selection (Table IX). The standard errors are large, but there is a clear trend towards a decrease in the genetic part of the variation as the selection goes on. By using linear regressions Steine found a decrease in daily milk yield of 0·14 kg and 0·07 kg per generation in the HF and LF line respectively. The analysis also indicated a negative genetic correlation between daily milk yield and flavour score which is inconsistent with the positive correlation of $0·77 \pm 0·15$ found by Rønningen (1965b). Differences between HF and LF in milk yield were small (Steine, 1978) but protein, lactose and total solids content was higher in LF (Table X).

Steine (1978) suspects a genetic relationship between flavour intensity and milk components which makes it questionable to select for flavour intensity as the only measure of quality.

Milk from HF does gave stronger flavoured brown cheese (Bakkene and Steinsholt, 1975) and ordinary white cheese (Fyksen and Steinsholt, 1974).

Table X. Least squares constants ± s.e. for selection lines for some milk components (Steine, 1978).

Milk component	HF	LF	$\bar{x}$
Protein (%)	−0·11 ± 0·013	0·11 ± 0·013	2·85 ± 0·01
Lactose (%)	−0·06 ± 0·020	0·06 ± 0·020	4·41 ± 0·02
Total solids (%)	−0·26 ± 0·057	0·26 ± 0·057	12·72 ± 0·06
Fatty acids as % of milk fat:			
Lauric acid	−0·23 ± 0·05	0·23 ± 0·05	4·03 ± 0·05
Palmitic acid	0·54 ± 0·16	−0·54 ± 0·16	22·41 ± 0·10
Free fatty acids (mEq/litre)	2·07 ± 0·24	−2·07 ± 0·24	4·33 ± 0·28

All differences are highly significant ($p < 0{\cdot}01$).

This milk also had a much higher content of volatile fatty acids (5·65 mEq/litre). Similar differences were found in cheese.

E. Feeding of Goats

1. General Conditions

Usually the goat farms are rather small. Out of the 1500 farms with goats in Norway, two-thirds have less than 5 hectares arable land, and very few have more than 10 hectares. The land is mainly used for fodder production. The goats therefore have to be pastured on uncultivated areas like rough ranges and mountain pastures. Only 6% of the goat population in Norway is pastured in paddocks, although this system is rather common in Sweden. Zero grazing is not used to any extent except for feeding of green forage in the autumn.

During the summer, a great many (40%) of the goats are moved to mountain pastures. These are mainly located on out-farms above the timber-line at 600–1000 m altitude. Some 40 cooperative pastures have been established, mainly in order to keep the goats from grazing in the woods during the summer. The conflict between the goat and forest industry is especially evident in areas with conifer woods. Therefore goats are mainly concentrated in the northern regions and other areas where such woods are of less economic importance.

Some years ago goats were kept on pasture for a longer period, especially in the autumn. Since more emphasis is now put on high yield, it has been necessary to increase the period of indoor feeding. According to Skjevdal (1974c) twenty years ago the average period of indoor feeding was 210 days, compared to 230 in 1974. The feeding regime also varies according to the climatic conditions. Rønningen and Gjedrem (1966) found that in the eastern, western and northern regions the average percentage of feed from pasture and the average number of days on indoor feeding were 30, 41, 39 and

239, 177, 235 respectively. Because of this, research work on goat nutrition has been directed towards roughage quality and utilization of roughage in combination with concentrates.

2. Feed Stuffs

(a) *Types of Grass*

Hay and silage from grass are the main fodder during the winter in the northern areas. Timothy (*Phleum pratense*) and meadow fescue (*Festuca elatior*) are the main types used for leys. Older meadows are more dominated by red fescue (*Festuca rubra*), common bent (*Agrostis tenuis*) and smooth stalked meadowgrass (*Poa pratensis*). Goats prefer the finer and softer hay from natural swards on old meadows to the more mature and wiry hay from timothy as it is commonly harvested nowadays.

(b) *Grass Silage*

Climatic conditions with high rainfall make it very difficult to cure hay in the field. Fence drying of hay which was used formerly in some areas demands too much labour and it is difficult to cut the grass at a sufficiently early stage of development. These difficulties can be overcome by silage conservation. In Norway grass silage is a major constituent of the winter ration for ruminants. In dairy cattle silage amounted to three-quarters of the harvested home-grown fodder (Trodahl, 1976a) and most farms today have grass harvesters equipped for spraying additives (formic acid) when cutting (Saue and Breirem, 1967; Ekern *et al.*, 1976). Since the encouraging results of feeding grass silage to cattle, it has also become the principal part of the goat rations. In 1976 it accounted for 62% of the home grown winter ration. In a trial, silage, hay and potatoes fed at a daily rate of 0·5–1·3, 0·4–0·9 and 1·0–1·5 kg per head were compared with a simple hay ration (Opstvedt, 1967). All rations were balanced in nutrient content and concentrates were supplemented according to milk yield. Goats fed both hay and silage had 7·3% higher F.C.M. yield than goats fed hay only. The results of a feeding trial with silage as a sole roughage are given in Table XI.

Because of a somewhat variable goat herd and due to rather late harvesting of the hay, the result seems to favour silage too much. Trials with cattle and sheep where grass silage was compared to hay have given similar favourable results for silage (Presthegge, 1959; Nedkvitne, 1977). By cutting grass early and using preservatives a fodder rich in energy and proteins is ensured. A final product with pH below 4·2, less than 0·1% butyric acid and ammonia nitrogen less than 8% of total nitrogen (Breirem and Homb, 1970) indicates successful fermentation. Preliminary results of feeding trials with silage of

Table XI. Milk yield and feed consumption of two groups fed grass silage and a mixed ration of hay and silage for 180 days (Skjevdal, 1972a).

	Silage only	Hay and silage
Milk yield (kg/day)	2·30	1·92
Fat (%)	4·09	3·85
Kg 4% F.C.M.	2·33	1·88
Consumption of roughage		
Hay (kg)	—	0·50
Silage (kg)	3·10	1·90
D.M. (kg)	0·70	0·83
F.U.	0·50	0·53

grass cut at varying stages of development (Robstad, 1975) indicate that with early harvest, feed consumption and milk were increased by 15 and 10% over late cut grass. Fat content was not affected. Average D.M. consumption of silage and milk yield were 0·5–1·0 kg and 2·5–3·2 kg F.C.M. respectively. Several farmers are using silage as the sole roughage apparently with good results. Nevertheless some hay is recommended if the fermentation is not successful or during cold spells when the silage is frozen. Poorly fermented silage has led to some cases of listeriosis. Also soiling of pens due to higher moisture content of the faeces may be a problem. However, this may be overcome with slatted floors (see p. 511).

Silage has also been fed as the sole roughage to kids. They grew and thrived at least as well as the ones fed hay only (Snaebjørnsson and Nedkvitne, 1968).

(c) *Hay*

Because of the great interest in grass silage, relatively little has been done to evaluate hay in the ration of goats. In ten Swedish farms Rejkestam and Cornell, (1977) found that, in addition to conentrates, goats consume on average 1·0 kg hay per day, with a variation from 0·5 to 1·4 kg, while Sjödin (1970) gives 1·0–1·3 kg as a normal intake. In trials conducted by Opstvedt (1967) daily consumption of hay was 0·7–0·8 kg in addition to 1·5 kg potatoes and concentrates. According to practical experience goats fed hay and concentrates only will seldom eat more than 1·2 kg hay per day (0·56 F.U./kg D.M.). By feeding hay of good quality 3–4 times per day combined with a restrictive use of concentrates (less than 0·4 kg/kg 4% F.C.M.), goats may consume 1·5–2·0 kg hay per day.

(d) *Barn-Cured Hay*

The method of barn curing has gained some popularity, particularly in

Sweden. However, with the unstable weather conditions prevailing in many of the areas where goats are kept the necessary wilting in the field is not always possible. Accordingly farmers have to be equipped for different systems of hay making. In most of the trials at A.U.N. quoted above the hay was barn cured.

(e) *Dehydrated Grass*

In northern Norway goats have been fed 0·3 kg of grass pellets a day together with 1·0–1·5 kg grass silage and concentrates. Some herds have used grass pellets as the sole roughage for periods of up to 1–2 months without any sign of digestive disturbances, except some cases of enterotoxemia. Preliminary results of trials at the A.U.N. indicate that 0·6 kg grass pellets (0·4 F.U.) per day can be fed to goats in addition to hay, grass silage and concentrates without any detrimental effect on milk yield.

(f) *Treated straw*

Straw is available in only a few areas where goats are kept, but in these areas alkali- or ammonia-treated straw is used as fodder for goats. While in the average recorded goat herd in Norway in 1976 treated straw was only 2% of the harvested roughage, in some herds it amounted to 25% (Trodahl and Jacobsen, 1977). Treated straw holds 17·5% D.M. with 0·13 F.U. per kg; a usual ration is 1·0 to 1·5 kg per day.

(g) *Potatoes and Roots*

In the Nordic countries potatoes and roots are more or less regarded as concentrates with a low dry matter content, because of the high digestibility of the dry matter. These crops may be grown in all areas where goats are kept and in case of limited supply of concentrates, they can be used as a substitute. In Norway potatoes and roots account for 2% of the total food consumption of all farm animals and only 0·5% for goats.

Raw and ensiled potatoes together with swedes (*Brassica napus* L. *napobrassica* L.) have been fed experimentally to dairy goats (Skjevdal, 1974a). Three kg swedes and 1·5 kg potatoes could substitute about 0·3 F.U. in ground barley without any loss of milk yield but with reduced fat content. There was a tendency of energy in potatoes being better utilized than in swedes. A feed test with five goats for 3 weeks showed that they had no difficulty in consuming 3 kg raw potatoes per day. Thus it may be concluded that potatoes can form a large part of the concentrate ration for dairy goats.

(h) *Green Fodder*

In recent years it has been the practice to feed goats some rape in the autumn when grass is getting scarce and the quality of pasture is deteriorating. Goats are given access to a field of late rape kale (*Brassica napus* L. *oleifera*) in addition to ordinary grass. No detrimental effects have been encountered, provided that sufficient quantities of grass are consumed together with the rape. Also leaves of swedes and turnips (*Brassica rapa* L. *rapifera*) are used as green fodder after the goats are put in stables for the winter.

Trials with rape have been conducted for 2 years with goats on mountain pasture at an altitude of 1000 m (Robstad, 1977a). About 0·01 hectare per goat was planted by the end of June. Experimental goats had access to the rape on about 20 August when the grass growth in the mountains was diminishing; a control group continued on pasture only. Daily milk yield was 1·4–1·7 kg. For 5 weeks in the first year and 3 weeks in the second year the milk production of the experimental group exceeded that of the controls by 20 and 10% respectively. The difference was greater in the high yielding individuals. A supplement of 0·3 kg concentrates was not sufficient to maintain milk yield of the controls at the initial level.

For 2 months following the trial positive after-effects were observed in the rape group. Blood haemoglobin and haematocrit did not differ between groups, and cheese production was not negatively affected by the rape supplement.

(i) *By-products from the Dairy Industry*

Farmers consider skim milk to be of specific value for dairy goats and may feed 1–2 kg daily. Also kids may be fed skim milk until the age of 4–5 months. The growth rate of kids experimentally fed 1·5–1·8 kg skim milk per day did not differ from those fed hay and concentrates (Skjevdal, 1974a). When offered whey (6% D.M.) *ad libitum*, lactating does may take 4–7 litres daily. The faeces become wetter, but this has no detrimental effects on slatted floors. Preliminary results of feeding trials at the demonstration farm at Valbjør indicate that whey and acidified skim milk can replace equal amounts of F.U. in concentrates. The average consumption of whey fed *ad libitum* was 6 litres per day (Robstad, 1979).

Milk replacers formulated for calves or lambs have been used to a large extent for rearing of kids in Sweden and Norway. Following trials in Norway,

Skjevdal (1974b) and Robstad and Skrøvseth (1978) recommended using 175 g of powder for each litre of liquid feed. Daily rations larger than 1·0 litre per kid divided into not more than two feeds may cause gastrointestinal irritation with stomach and duodenal ulcers. So far there are no milk replacers available which ensure fast growth rate without digestive troubles. Trials with milk replacers where part of the protein component consisted of fishmeal and soybean have not been successful (Skjevdal, 1976).

(j) *Concentrates*

The average proportion of concentrates fed to goats in Norway is estimated to be 40% of the total feed intake. It was about 33% in 1974 (Skjevdal, 1974e), Since then, milk yield in the whole population has increased from 400 to 450 kg (Fig. 3). In milk recorded herds the concentrate percentage in the ration is about 45%, i.e. 220 kg per year (Trodahl and Jacobsen, 1977).

3. Feeding Regimes

(a) *Feeding Requirements*

Feeding standards based on feeding experiments (Breirem, 1975; Opstvedt, 1967; Skjevdal, 1974a, d and 1975a) are recommended as follows:

Maintainance requirement in F.U. $= 0{\cdot}6\left(\frac{\text{live weight}}{50}\right)^{0\cdot75}$, with 80 g digestible crude protein per F.U. (0·1 Mcal M.E. per kg metabolic body weight)

Milk production = 0·4 F.U. per kg 4% F.C.M. with 150 g digestible crude protein per F.U. (1·2 Mcal M.E. per kg 4% F.C.M.)

The average farmer, however, seems to give more liberal feeds than these recommended allowances (Skjevdal, 1974a).

(b) *Pasture Grazing*

Most of the goats are out on pasture in the first part of June. The first crop of grass is very rich, with about 90 F.U. per 100 kg D.M. and 170–180 g digestible crude protein per F.U. Therefore it is common practice to feed roughages during the first days on pasture. Generally milk fat content at this time of the year will be at a very low level. Goats put on pasture at low

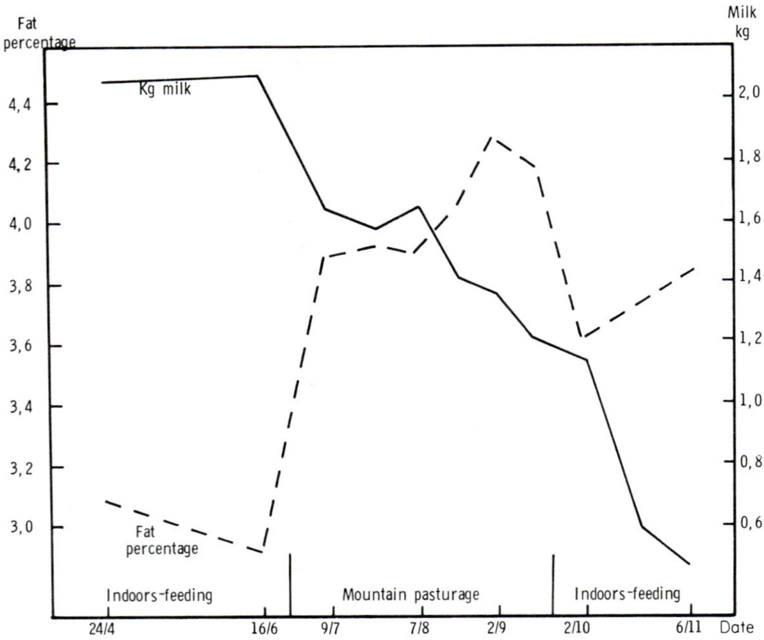

Fig. 5. Effect of indoor feeding and mountain pasture on milk yield and fat percentage during lactation from 24 April to 6 November. 140 goats on mountain pasture, 1000 m above sea level (Skjevdal, 1972c).

altitudes will maintain the fat content they had when they were stall-fed, but goats put on pasture in higher regions will often have an increased fat percentage and a drop in yield (Fig. 5). During cold and rainy periods fat content may increase. The nutritive value of pasture decreases somewhat later in summer, but goats like sheep are able to select the most energy-rich plants even later in the season (Olafsson, 1972). Young grass at high altitudes will usually be available as late as July if the herds graze continuously higher areas as the snow melts. In the northern and coastal districts of Norway where the grass growth is more lush, daily yields of 2·0 kg per doe are reported for the whole grazing season without any supplementary feeding. In recent years with increased yield, a drop of 0·3–0·5 kg milk per day is frequently experienced when goats are put on mountain pastures. On the other hand fat content increases by 1·0%. In an experiment (Skjevdal, 1972c) on ranges at 1000 m altitude, different levels of concentrates were fed to goats with an annual milk yield of 500–700 kg milk, which during the experiment gave between 1·1 and 2·2 kg milk daily. The intake in F.U. from grazing was calculated from nutrient requirements for maintainance, weight gain and lactation as 1·0, 0·9, 0·8 and 0·7 for goats receiving 0·1, 0·3, 0·5 and 0·7 kg concentrates per day respectively. At 0·1 kg of concentrate, both milk yield and fat content decreased. At each increment of 0·2 kg concentrates the goats gained an average of 0·5 kg body weight during the experiment and had a higher milk yield after the experiment and in the following year. Feeding concentrates at a constant level during the whole season gave better results than adjusting rations to the decreasing grass growth, probably because the goats had reached a rather late stage of lactation by the time grains were fed. Feeding higher levels of concentrates had no effect on the onset of oestrus, but the number of kids was positively affected.

Following these experiments recommendations for concentrate feeding to goats on mountain pasture are given in Table XII. Recommendations are similar for goats yielding 700 to 850 kg per year. In northern Norway feeding more than 0·1 kg to goats producing about 1·5 kg daily is not economical even in years with poor grazing. For the highest yielding goats 0·6 kg will be sufficient in good years but in bad years 1·0 kg will be necessary. Kids grazing on mountain pasture were fed 0·1 and 0·3 kg concentrates per day which was about the limit of voluntary intake (Skjevdal, 1975a). Growth rates were 80 and 94 g per day for the two groups. Body weight at the end of the grazing season and at parturition and first lactation milk yield in the 0·3 kg group were only slightly higher, namely 1·0, 0·5 and 5 kg respectively. Furthermore, the larger concentrate ration appeared to reduce grass consumption and therefore feeding kids concentrates during the grazing season is not recommended if the average daily weight gain is in the region of 80 g.

Table XII. Recommended rations of concentrates to goats on mountain pasture (Skjevdal, 1975a).

Yield per doe per day when being put on pasture	kg concentrates per doe per day during grazing season
Less than 2 kg	0·3
2–3 kg	0·5
More than 3 kg	0·7 or more

4. Rearing of Kids

The range of birth weight of kids in Norway is 2 to 4 kg. The average female kid born as twin by adult goats or as single by yearlings is 2·9 kg; male kids are 10–15% heavier than females. Kids with birth weights less than 2·5 kg are difficult to rear. Variations in birth weight are largely due to the nutritional status of their dams. Goats are expected to drop their first kid as yearlings, and they should attain 80% of adult weight at 18 months of age. In Norway the average yearling weight in 1968–1972 was 38·8 kg (Steine, 1975a); weight at kidding varied between 25 and 45 kg in feeding experiments.

There is a relationship between weight at 1½ years of age and future milk yield (Table XIII) which is mainly due to environment. It allows prediction of herd yield but predictions within herds will be less reliable (Steine 1976).

Table XIII. Predicted yield of goats related to body development at 1½ years.

kg live weight at 1½ years (average of herd)	Expected yield in 1st lactation (kg milk)	Expected yield when mature (kg milk)
25	250	300–350
30	310	450
35	380	550
40	450	650
45	550	650–900

In Norway feeding of kids is deficient mainly from the age of 8 months onwards. In feeding trials, therefore, 0·25 and 0·5 kg concentrates have been fed between 8 months of age and parturition (Skjevdal, 1975a). Roughage intake was 0·43 and 0·33 F.U., and average daily weight gain was 55 and 70 g for the two groups respectively. More animals were later culled in the group that received 0·5 kg concentrates (Table XIV). Although this effect was not confirmed in other experiments (Skjevdal and Nedkvitne, 1972), it seems that care has to be taken in feeding high concentrate rations to growing kids. In this connection it should be mentioned that few ruminants are more sensitive with respect to roughage quality than kids, and their feed intake varies considerably with the concentration of F.U. in the foodstuffs.

Table XIV. Concentrate allotments in relation to viability. Number of animals at start and with completed first and second lactations (Skjevdal 1975a).

Concentrates fed:	0·25 kg/day	0·5 kg/day
At start	145	141
At 1st kidding	126	120
1st lactation	113	102
2nd lactation	90	66

In Norway the recommended nutrient allowances for rearing of kids are as given in Table XV (Snaebjørnsson and Nedkvitne, 1968; Skjevdal and Nedkvitne, 1972; Skjevdal 1972b, 1974b, 1975a).

F. Houses for Goats

In winter, the temperature in Nordic countries may fall below − 20 to − 30°C for short periods. Therefore houses for goats must be of high standard with good insulation. Because of ground-frost in some areas foundations have to

Table XV. Recommended nutrient allowance for rearing of kids.

Age (weeks)	Live weight (kg) at end of period	Minimum daily gain (g)	Nutritive requirements a day	
			F.U.	digestible protein (g)
0–6	8	110	0·30	40
6–12	12	100	0·40	60
12–18	17	100	0·55	80
18–34	27	80	0·75	100
34–44	35	100	0·85	90
44–kidding	45	110	0·90	90
After kidding	38–40			

be as deep as 2 m. Different designs for byres are being tried (Hilling, 1974; Berge 1968), and in recent years 25–30 new byres for goats have been erected each year. Most of these have room for 70–100 lactating does, and generally the manure is collected in a basement under the pens. In order to reduce not only costs of building but also the volume and area of outside walls (transmission heat loss) per housed goat, the area of the pens has been limited to 0·63–0·88 m^2 per goat. In the areas where goats are kept it is usually difficult to procure sufficient material for bedding, so most of the byres are equipped with slatted floors. This is also an advantage where grass silage is used extensively because of higher water content of the faeces. The most common types of slats are 3·8 cm boards 6–10 cm wide which are laid with slots of 25 mm. It is easier to keep floors clean if boards are narrow and slots wide, but such floors do not have the desired durability. Also expanded metal has been used for dairy goats, and results of 9 years experience with this at A.U.N. are quite satisfactory (Skjevdal, 1977).

In most of the byres there is a raised platform, 30–50 cm high, in connection with the manger where the goats are tied during feeding and milking. Formerly the structures were mainly of wood, but in recent years iron structures have been more common. Indoor climate in most goat-houses in Norway is unsatisfactory with temperatures, relative humidity and NH_3 content of about 10–15°C, 88% and 20 p.p.m. respectively (Hilling, 1974). Swedish surveys (Rejkestam and Cornell, 1977) show similar results.

Calorie production in a goat byre is not as constant throughout the year as it is for cows. In the dry period before parturition, it may be nearly half of that of the lactation period. Calculations on the heat balance made by Hilling (1974) show that a byre with 100 goats and good insulation and with indoor and outdoor temperatures of + 12°C and – 15°C respectively, would need an addition of 2643 kcal/h during the dry period in order to keep the relative humidity below 80%. This may be reduced to 1078 kcal/h after the does have started lactating. So far extra heating in the byres has, however, not been practised, but it may be an advantage because some herds have had problems with pasteurellosis, probably caused by poor indoor climate.

Recently milking machines and milk tanks have been installed. Separate milking parlours have been used in very few instances but they do not seem to imply any significant improvement. Two operators with four clusters each in a herd of 200 lactating does and a separate milking parlour (Skjevdal, 1975b) did not manage to milk faster than in pens where each milker operates only two clusters (Hilling, 1974). In both systems the labour requirement per lactating doe was 2 man minutes per day. The time for the teat-cups to be attached to the udder during milking seems to be more dependent on milk yield than in cows. In data from 230 goats, it was found that the time of teat-cups attached increased by 1·5 min when the amount of milk increased by 1 kg. In this experiment average time of teat cups attached was 2 min with goats 2 years and older.

References

Astrup, H. N., Nedkvitne, J. J., Skjevdal, T., Fauske, R., Lindstad, P. and Bakke, F. (1972). Report 396, Dept. Anim. Nutr., Agr. Univ. Norway, 11 pp.

Bakkene, G. and Steinsholt, K. (1975). Report 189, Dairy Research Institute, Agr. Univ. Norway, 11 pp.

Berge, E. (1968). *In* "Geitehald", pp. 129–149. Bønd, Forlag, Oslo.

Berge, S. (1966). Report 226, Dept. Anim. Genetics and Breeding, Agr. Univ. Norway, 24 pp.

Bjørke, K. (1974). Hovedoppgave, Dairy Research Institute, Agr. Univ. Norway, 45 pp.

Brandsaeter, E. and Abate, V. (1959). *Meld. Norg. Landbr. Hogsk.* **38** (10), 11 pp.

Breirem, K. (1975). *In* "K. K. Heje Lommehåndbok for jordbrukere, skogbrukere, meierister og hagebrukere", pp. 118–123. Steenballes Forlag, Oslo.

Breirem, K. and Homb, T. (1970). "Fôrmidler og fôrkonservering". Forlag, Buskap og Avdrått A/S, Gjøvik.

Ekern, A., Saue, O. and Vik-Mo, L. (1976). "Proc. 2nd International Silage Research Conference", Chicago, 1975. Mimeogr. paper 68, Dept. Anim. Nutr., Agr. Univ. Norway.

Fougner, J. A. (1979). *Zuchthyg.* **14**, 104–110.

Funder, L. (1933). Report 26, Dairy Research Institute, Agr. Univ. Norway.

Fyksen, S. and Steinsholt, K. (1974). Report 183, Dairy Research Institute, Agr. Univ. Norway, 16 pp.

Gjedrem, T. (1969). *Acta Agr. Scand.* 116–126.

Haugen, E. (1960). *Meld. Norg. Landbr. Høgsk.* **39** (10), 33 pp.

Hilling, V. J. (1974). Mimeogr. paper 122 Dep. Agr. Structures, Agr. Univ. Norway, 80 pp.

Korvald, T. (1958). Report 68, Dairy Res. Institute, Agr. Univ. Norway, 16 pp.

Mytting, H. E. (1976). "Økonomien i Geiteholdet". Norges Landbruksøk. Inst., Oslo.

Nedkvitne, J. J. (1977). Rep. EAAP 28th Annual Meeting. Brussels, 6 pp.

Olafsson, G. (1972). Lic. thesis. Dept. Anim. Nutr., Agr. Univ. Norway, 114 pp.

Opstvedt, H. (1963). Report 237, Dept. Anim. Nutr., Agr. Univ. Norway, 20 pp.
Opstvedt, J. (1967). Tech. Bull. 134, Dept. Anim. Nutr., Agr. Univ. Norway, 114 pp.
Presthegge, K. (1959). Tech. Bull. 93, Dept. Anim. Nutr., Agr. Univ. Norway, 237 pp.
Rejkestam, Å, and Cornell, B. (1977). Rapport 47, Inst. för Husdjurens utfodring och vård. Lantbrukshögskolan, Ultuna, Sweden, 99 pp.
Robstad, A. M. (1975). "Husdyrforsøksmøtet, Norges Landbr. Høgsk", pp. 203–208 Rådet for husdyrforsøk, Oslo.
Robstad, A. M. (1977a). "Husdyrforsøksmøtet, Norges Landbr. Høgsk", pp. 232–237 LOT, Ås.
Robstad, A. M. (1977b). *Sau og Geit.* **30**, 176–177.
Robstad, A. M. (1979). *Sau og Geit.* **32**, 116–117.
Robstad, A. M. and Skrøvseth O. J. (1978). *Sau og Geit.* **31**, 175–179.
Rønningen, K. (1965a). Report 213, Dept. Anim. Genetics and Breeding, Agr. Univ. Norway, 19 pp.
Rønningen, K. (1965b). *Acta Agr. Scand.* Report 219, Dept. Anim. Genetics and Breeding, Agr. Univ. Norway, 42 pp.
Rønningen, K. and Gjedrem, T. (1966). Report 229, Dept. Anim. Genetics and Breeding, Agr. Univ. Norway, 17 pp.
Saue, O. and Breirem, K. (1969). "Proc. 3rd Gen. Meet. Europ. Grassl. Fed.", Braunschweig, pp. 161–172.
Sjödin, E. (1970). "Getter, Avel, Utfodring, Skötsel, Ekonomi". LTs förlag, Stockholm.
Sjödin, E. (1976). Report EAAP 27th Annual Meeting Zürich, 6 pp.
Skjevdal, T. (1972a). Report 288, Dept. Anim. Nutr., Agr. Univ. Norway, 10 pp.
Skjevdal, T. (1972b). Report 392, Dept. Anim. Nutr., Agr. Univ. Norway, 7 pp.
Skjevdal, T. (1972c). Report 407, Dept. Anim. Nutr., Agr. Univ. Norway, 10 pp.
Skjevdal, T. (1974a). Tech. Bull. 169, Dept. Anim. Nutr., Agr. Univ. Norway, 42 pp.
Skjevdal, T. (1974b). Tech. Bull. 173, Dept. Anim. Nutr., Agr. Univ. Norway, 19 pp.
Skjevdal, T. (1974c). Mim. paper 35, Dept. Anim. Nutr., Agr. Univ. Norway, 30 pp.
Skjevdal, T. (1975a). "Husdyrforsøksmøtet Norges Landbr. Høgsk.", pp. 195–202. Rådet for husdyrforsøk, Oslo.
Skjevdal, T. (1975b). *Sau og Geit.* **28**, p. 12.
Skjevdal, T. (1975c). *Sau og Geit.* **28**, p. 72.
Skjevdal, T. (1976). Mim. paper, Dept. Anim. Nutr., Agr. Univ. Norway, 5 pp.
Skjevdal, T. (1977).*Sau og Geit.* **30**, 121.
Skjevdal, T. and Nedkvitne, J. J. (1972). Report 397, Dept. Anim. Nutr., Agr. Univ. Norway, 7 pp.
Snaebjørnsson, J. and Nedkvitne, J. J. (1968). "Husdyrforsøksmøtet. Norges Landbr. Høgsk.", pp. 105–115. Rådet for husdyrforsøk, Oslo.
Steine, T. A. (1975a). Report 369 Dept. Anim. Genetics and Breeding, Agr. Univ. Norway, 30 pp.
Steine, T. A. (1975b). Report 385, Dept. Anim. Genetics and Breeding, Agr. Univ. Norway, 25 pp.
Steine, T. A. (1976). Report 398, Dept. Anim. Genetics and Breeding, Agr. Univ. Norway, 19 pp.
Steine, T. A. (1978). Rep. EAAP 29th Annual Meeting, Stockholm.
Svensen, A. (1968). "Meieriteknikk". Dairy Research Institute, Agr. Univ. Norway, 8 pp.
Trodahl, S. (1976a). *Buskap og Avdrått.* 188–189.
Trodahl, S. (1976b). Report EAAP 27th Annual Meeting Zürich, 6 pp.
Trodahl, S. and Jacobsen, S. (1977). *Sau og Geit.* **30**, 97–99.

16. Goats in the Dry Tropics

O. García and C. Gall

A. Dry Belts of the Globe

Before discussing the specific aspects of goat production in the dry tropics, we must provide a definition of these zones comprising arid and semi-arid areas and an idea of their extent. Wallen (1966) defines an arid and semi-arid area as one where water losses through run-off, evaporation and plant transpiration exceed water available from precipitation, soil moisture and subterranean water.

While in the past there has been some discussion about the different degrees of aridity, it is now generally accepted that any area with less than 250 mm annual rainfall is arid or at least semi-arid (UNESCO, 1958). Often areas with 600 to 400 mm are considered semi-arid, with 400 to 100 mm arid and below 100 mm desert. But other factors than precipitation itself are important in determining aridity, such as monthly distribution of rains, variability, air humidity and air movement which all determine water availability to plant growth and water balance.

There are in the world four main arid areas (Fig. 1):

(1) the area stretching from northern Africa, the Arabian Peninsula, Pakistan and India to the Desert of Gobi in China;
(2) the belt lining the pacific coast of North and South America;
(3) the centre of Australia;
(4) the Atlantic coast of South Africa.

These areas occupy 29 000 km of coast lines and 4900 million hectares, that is about one third of the terrestrial area of the globe (UNESCO, 1958). As shown in Fig. 1, a high proportion of these areas, and with which this chapter

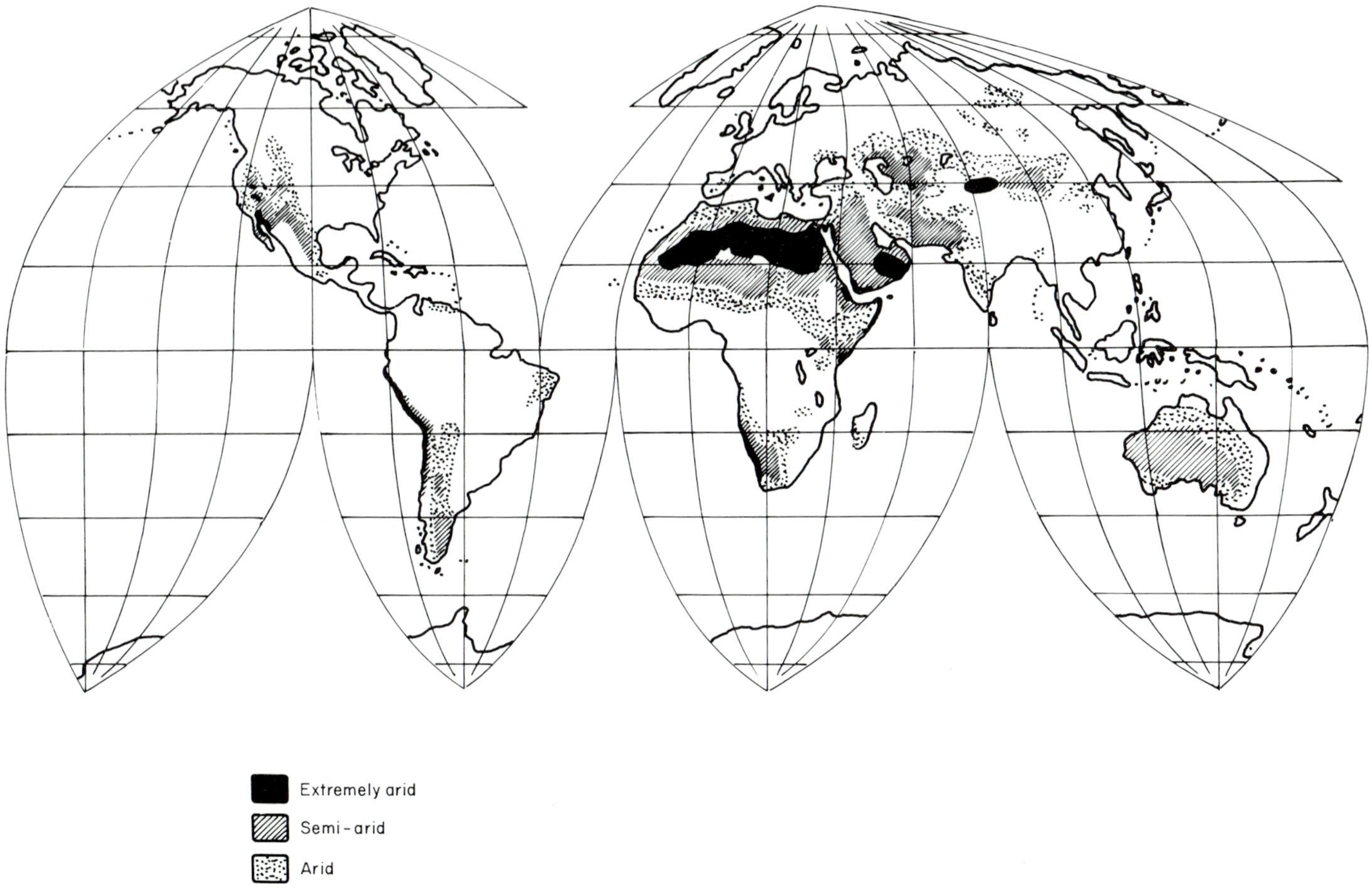

Fig. 1. Distribution of dry areas (UNESCO, 9958).

deals, is situated in the tropics (i.e. between the tropics of cancer and capricorn). We shall, however, to some extent discuss goat production in areas which may be considered sub-tropical dry, such as north and south Africa, or semi-humid, such as Mediterranean and Near-East countries, because the situation there is, at least for part of the year, very similar to the dry tropics (Hills, 1966). The Mediterranean environment, in addition to extended dry periods in summer, is characterized by cold and wet winters; in both periods, much of the vegetation is dormant (Bourbouze and Guessous, 1977).

B. Distribution of Goats

Reliable statistics on the goat population in the areas concerned are difficult to obtain. This difficulty reflects the disinterest in the development of goat production, of public and private institutions. Devendra and Burns (1970), based on FAO statistics, assume the goat population in the tropics to be 262 million, corresponding to two-thirds of the world total. The latest FAO statistic (1979) lists a world goat population of about 446 million of which 225 million in tropical or sub-tropical countries that comprise dry areas (UNESCO, Fig. 1).

The goat is the best ally of the inhabitants of these areas which is illustrated by the fact that about 52% of the world goat population lives in countries comprising dry tropical areas, as compared with about 39 and 25% respectively of the cattle and sheep population (FAO, 1979).

The highest concentration of goats may be seen on the Asian continent and particularly in India, with 16% of the world goat population. Of course not all the goats in these countries do live in dry areas. But it is safe to assume that it will be a high proportion, since larger herds are usually kept in dry areas only. In Venezuela 85% of the goats are found in dry areas (Castillo, 1971) and in Brazil 72·5% of the national goat herds live in the north-east, the only dry area of the country (Jardin, 1971). Goats are almost exclusively found in the dry coastal area of Peru (Nolte and Arauzo, 1973) and similarly in the Peninsula Guajira of Columbia (Colombia, 1963). In Mexico, the majority of goats are found in the northern dry areas (Agraz, 1971) and similarly in Argentina (Maubecin, 1979). In Iran 13 million goats are being kept mainly by nomads in the vast arid areas of the country; they rank second as livestock (Sattari, 1971).

C. Economic Importance

From the FAO statistics (1979) it may be deduced that goats' contribution to the world milk and meat production is 1·7 and 1·4% respectively. If we consider only those countries with dry areas this contribution is about 10 and 7% respectively. This illustrates the importance of goat products in the dry tropics where they replace meat and milk from cattle and sheep to a certain extent.

In India there are 71 million goats, comprising one fifth of all livestock. They produce an estimated 0·7 million tons of milk, 278 tons of meat (FAO) and 23 million skins (Sarup, 1971). In Peru, 7·5% of the total milk and 8·3% of meat is produced by goats (Nolte and Arauzo, 1973). Agraz (1971) has estimated that in Mexico some 200 000 families derive their livelihood from goats who produce US $64 millions worth of milk and US $58 millions worth of meat (Juárez, 1979).

In Brazil, the production of goat meat is estimated at 23000 tons per year. Skins at a value of about US $700 000 are exported, while milk is mainly used for home consumption (Jardin, 1971). In Venezuela, the total value of goat products is estimated at US $4·85 million (Venezuela, 1975). While in the humid tropics many of the goats are kept as a sideline in small family holdings, fed mainly on by-products and forage resources between agricultural land, in the dry tropics goats are kept in larger herds on range-land which does not yield anything but grazing, often so scarce that only goats can utilize it. The owners may keep other livestock as well as goats or else keep only goats which may provide their sole means of economic survival.

Goats (and sheep) in Mali are favoured because the several animals that may be kept instead of one cow will provide some milk at all times of the year, while the cow will be dry for many months. Mixed herds of sheep, goats and cattle are thought to be complementary in their grazing and thus make better use of vegetation than one species alone. However in some ways, goats are looked upon as an intermediate solution. Immediately after a disaster they may have to ensure the family's subsistence, but later on surplus small stock may be sold in order to purchase cattle (Braun, 1979).

D. Social Problems

Goat production in the dry tropics is characterized by its marginal position and the low rentability of holdings. A study in a dry area of Venezuela revealed that goats are being kept for subsistence mainly and that very few herds are producing for the market (Carrasco, 1976). Under present conditions, improving goat production would not improve the living conditions of rural families; these are limited by the lack of infrastructure, communications, schools, health care, electricity etc. Economic analyses of goat production in Venezuela and in Peru are presented in Chapter 18. Returns were low and in some instances did not equal the minimum official wage. The situation of goat production will be very similar in other countries, since goats are mainly being used where other domestic species will not survive and are receiving little, if any, assistance for development. The ranges used for grazing goats are usually public land, owned by governments or communities. As a rule, they are overstocked and heavily damaged by overgrazing. In many areas formerly there have been strict regulations for the use of ranges and wells by stock-owners which effectively limited stock numbers. Tolls were levied for passing certain areas, pasture reserves were kept for dry periods and there were agreements with agriculturists for grazing their fields during the dry period. Some of these regulations go back to Islamic traditions, like the Hema System of the Arabic Peninsula (Draz, 1969). With the decline of feudal systems, most of these restrictions have been eliminated, and nowadays most of the common land is grazed without any restrictions or regulations. Furthermore, with population increases much former grazing land has been cultivated and this has added to the destruction of traditional grazing systems (Bourbouze and Guessous, 1977). The lack of regulated land tenure is usually limiting development in dry tropical areas. A man who does not own the land cannot have a sense of responsibility for the land resources he is utilizing. If he held a deed to his grazing grounds, he would feel the need to conserve them and their water resources and ensure adequate feeding of his herd year after year.

Nevertheless, it would still be difficult to develop goat production in areas of very low productivity and where herds are scattered. By contrast, in areas where feed resources are more abundant and where there are possibilities for production of supplementary forage through mechanization and possibly irrigation on some small plots, development may be envisaged, with services and technical assistance provided by public organizations. This could be achieved best and quickest if producers were organized in cooperatives, both with communal grazing and individually owned grounds.

In African Mali, after the great Sahel draught, cooperatives which regulate transhumant grazing were revived. While moving north with their herds in June/July they put certain grass areas in reserve. On their return in September/October, these reserves will be available for grazing. By that time, new areas are being assessed for reserve suitability in the period between February/March and June/July. Movement into the grazing lands where the waters of the Niger have receded is controlled by similar cooperative groups. Observance of the code is ensured by patrols of tribesmen aided by government officials. Offenders are warned first and fined at the second attempt. Since the great draught has reduced stock numbers, excessive herd size is not the main problem at present. However, the cooperative is trying to build up a collective herd which may serve as a stock reserve, so that tribesmen will not have to keep excessive stock numbers as an insurance against draught and disease (Braun, 1979). In a development project for range improvement and small stock management in Syria, an attempt is being made to revive the traditional system of range conservation through cooperatives (Draz, 1969).

E. Production Systems

Goats in the dry tropics are generally managed extensively on unfenced range, sometimes herded by men but often left grazing alone. It is this management system which has led to the goat being accused of causing all the damage to vegetation observed in areas of goat grazing. The main reason for uncontrolled grazing is the marginal socio-economic situation of the people keeping goats in dry tropical areas. They depend on goats for milk and meat but are often in a vicious circle: because of the low productivity of pasture and animals they do not seem able to apply better methods of management; but due to poor management, productivity cannot be increased and people tend to keep large numbers of animals, in order to ensure a minimum of products.

With few exceptions, goats are kept virtually without management. Animals graze free on communal range, which excludes any possibility of rational management and range improvement. Females and males graze together, so that controlled matings and a breed improvement programme through selective breeding are impossible. By the same token, young female goats are mated at a very young age which impairs growth and future productivity. Since different herds are in constant contact, diseases are easily spread and any disease control programme through preventive measures or treatments which individual progressive farmers may apply, are jeopardized from the beginning.

Migratory systems of goat production—nomadism and transhumance—suffer from the same deficiencies as the sedentary systems, in the dry tropics, but in a much more accentuated form. While they are an adaptation to the extreme seasonal fluctuation in forage and water availability, any deviation from the normal pattern of seasonality is likely to strike them hardest. In addition, they are faced with a host of social and political problems including those concerning grazing rights.

There is little place in the dry tropics for intensive systems of goat production. Apart from goats being kept at the house for supplying milk for the family, we find small herds for educational purposes, as in Israel, or larger experimental herds for research. Animals kept permanently in feed lots or houses, under intensive conditions, will have to have a high productive potential and products must fetch high prices in order to pay for the high cost of management.

An intermediate form of semi-intensive management, where animals are stabled during the night or on rainy days, is gaining popularity. It allows for some control over animals and pastures and would meet the minimum requirements for extending technical assistance. For example, in Venezuela a programme has been started under which rural enterprises for goat breeding are established. They are formed as cooperatives by peasants originating from extensive grazing areas and resettled in areas which have been fenced and equipped with the necessary facilities for a more controlled goat operation.

Juárez (1974) has proposed a stratified system of goat production, where in an irrigated zone of Northern Mexico a breeding centre would serve as nucleus, with a number of small herds integrated in mixed agricultural operations, which would multiply improved stock for the wider population. This system is based on forage availability in the intensive strata, large demand and high prices for goat dairy products.

Low productivity of goats makes herding a rather inefficient occupation. In many countries this and the need to cover large distances daily makes herding a job for children. Where only small numbers of animals are kept by each family, herding is sometimes left to youngsters who are paid on a doe per year basis, e.g. in Yemen Y.R 20, which in 1978 was a tenth of the value of a 3-month-old kid. Some ranchers in Latin American countries let their cattle hands graze a limited number of goats for their own meat supply while with the herds (Maubecin, 1979).

There is no general pattern of herd size in tropical dry areas. We find single goats kept within settlements, small groups of 10 or 20 goats herded by children and large herds of several hundreds or over a thousand goats kept in range areas. Excessive numbers and overstocking are a widespread problem. Due to the high reproductive rate of the goat, in periods of favourable environment conditions, goat numbers may increase rapidly. With experience of animal losses from recurrent draught and disease, there is a tendency to keep larger goat numbers as a kind of insurance. Quartermain (1975) suggested that in Zambia there was a self-regulation of individual herd size and thus a limitation of the goat population to what the environment can support. The natural increase of goat herds by high fertility rates was counterbalanced by losses due to neo-natal mortality, disease, straying, predation and consumption of meat. Losses increased with herd size, and labour, worry or social pressure which are associated with managing larger herds may effectively limit goat numbers to the observed 30 and below.

While the large herds in the typical semi-arid goat areas are of goats only, in transitional areas herds of goats and sheep, often mixed with cattle, are quite common. The aspects of common grazing are discussed in Chapter 7.

F. Breeds

The various breeds of goats are described in detail in Chapter 3.

Breeds encountered in the dry tropics are manifold, but three types predominate: long-legged, shallow-bodied, loop-eared of the Sudanese-Nubian type; small (but not of the achondroplast dwarf type) goats like the North African black Bedouin and Arab goat; and the medium-sized goats of multi-breed origin like the different Latin-American Criollos.

While meat is probably the more important product of goats in the dry tropics, most of the goats are eventually milked and may yield appreciable quantities. Goat and kid skins are by-products which do not yield much income to the producer, because processing costs are high. They may, however, be important assets to the national economy (Joubert, 1973; Robinet, 1967). There are considerable differences in skin quality both between breeds, e.g. the Sokoto Red excelling in quality, and within breeds, short-haired fine skins yielding higher qualities. However, these breed differences are not generally accepted and some tanners believe that feeding has a much greater influence on quality than breed.

Inbreeding seems to be a very common feature in herds of goats in the dry tropics. This is due to management practices: sires are selected from within the herd, and purchase from other herds is unusual. Striking uniformity in some breeds might be a consequence, as well as small uneconomic sizes in others.

G. Physiological Factors

Goats are undoubtedly well adapted to hot environments as long as humidity is low. If the high temperature is combined with high humidity the effect on the goat is adverse, and large herds are seldom found in such climates. Studies on the mechanisms of heat tolerance in goats have given some conflicting results. It seems, however, that some of the most important features are the ability to maintain feed intake under heat stress and reduce metabolic rate and water turnover, as well as adaptive behaviour patterns (Maloiy, 1973; Khan *et al.*, 1979). Apleman and Delouche (1958) in (Anglo-)Nubian goats found the limit of heat tolerance to be between 35 and 40°C. Heat exposure reduced physical activity and feed consumption, increased body temperature and respiration rate while pulse rate was little changed. Plasma and whole blood specific gravity were unchanged, haematocrit and plasma chlorides were lowered, plasma protein was increased. Joshi *et al.* (1977) reported similar results in Jamnapari males. During exposure to 45°C rectal temperature in goats rose 1·96°C as compared with 0·85°C in Merino sheep (Johnson, 1971).

Feed consumption of East African goats in closed cages (Schoen, 1968) or cages exposed to the sun (Quartermain and Broadbent, 1974) was not affected by heat exposure. These authors attribute the heat tolerance of goats largely to the ability to maintain feed consumption, increase body temperature, increase ventilation rate and reduce water loss. Evaporative heat loss differs considerably between breeds and animals (Robertshaw, 1968). Sweating rate seems to increase with age in kids (Rai *et al.*, 1979). Saanen goats at 40°C lost 26 ml moisture per m^2 per hour (Jenkinson and Robertshaw, 1971). In crosses of Beetal × Sirohi goats 209–216 g water loss/m^2/h has been reported (Rai *et al.*, 1979). Sweating rate in black Bedouin goats was more than twice as much at the side of the head than at the thorax (Table I). This difference was not due to sweat gland density but rather the activity of individual glands (Dmi'el *et al.*, 1979) which was up to 10 times the average value observed in other bovides. Regulation of sweat gland function is similar as in cattle (Robertshaw, 1968). The contribution of cutaneous evaporative heat loss to total heat loss seems to differ between goat breeds. Johnson (1976) found respiratory evaporation to be more important than cutaneous evaporation while Borut *et al.* (1979) found the reverse to be true in the black Bedouin goat. Passive diffusion remained about constant at 32–35 ml/m^2/h during heat exposure and did, therefore, not contribute to the high moisture loss (Dmi'el *et al.*, 1979).

Black coat colour may be an essential feature of adaptation to desert conditions where both cold and heat, often at short intervals, put a heavy strain on the animals (Shkolnik *et al.*, 1972; Borut *et al.*, 1979). When exposed to solar radiation, black Bedouin goats took an additional heat load about equal to the heat produced by the resting metabolism. Evaporative cutaneous heat loss increased at about the same rate. Thus, black animals make better use of solar radiation for regulating their body temperature at low environmental temperatures. The extra heat load during hot conditions may be compensated for by more intensive sweating (Finch *et al.*, 1980). The very high sweating rate at the side of the head, observed by Dmi'el *et al.* (1979) (Table I), may cool the blood supplying the brain. A similar function of

Table I. Sweat gland density and function in black Bedouin goats, $n = 4$ (Dmi'el *et al.*, 1979).

	Maximal sweat rate (ml/m^2/h)	Sweat gland density (glands/mm^2)	Sweat rate (ml/gland/h)
Thorax	137	2·14	66
Head	363	3·38	114

the horns and their vascular system has been suggested by Taylor (1966).

Goats seem to be able to survive desert conditions with very low water intake. Fed exclusively on dry feeds, goats survived for 38 days (Ohya, 1964).

Where water content of feed plants is high, goats are known to be practically independent of drinking water and Dunson (1974) found goats under arid conditions, with feed plants of 52–67% water content and no drinking water, to be able to maintain their water balance. Maloiy and Taylor (1971) suggest that, like some wild ungulates, goats might be able to survive desert conditions without drinking water, if they were allowed to graze at night. When goats were deprived of water for 5 to 8 days, feed intake, body weight and milk yield were reduced to 74–92%, 66–78% and 66–94% respectively. Body temperature, pulse and respiratory rate were slightly affected. Physiological parameters changed faster when feed was withheld, but returned earlier to normal values (Tulenbekov, 1961). Shkolnik *et al.* (1972) kept (black Bedouin) Mountain goats and Sinai goats for two weeks on dry feed without water, at 30°C and 30% relative humidity. The goats lost about 30% of their body weight. While the Mountain goats ceased eating after a week of water deprivation, the Sinai goats did not lose their appetite. Metabolic rate fell to 40% below the value predicted according to body weight, and water loss was reduced. Barmer goats, after 4 days without water, had lost only 1·5% of their body weight per day but feed consumption was reduced by 40% (Khan *et al.*, 1978). At the end of the water deprivation period, Sinai goats drank 30–40% of their body weight within 2 minutes (Shkolnik *et al.*, 1972). Osmotic stability of the erythrocytes, which is not high in goats, is not challenged by this rapid re-hydration; but the rumen plays an important role in this osmotic regulation (Chosniak and Shkolnik, 1978). The different behaviour of very similar breeds indicates specific breed adaptation to arid conditions. Milk yield of black Bedouin goats deprived of water at 30°C and 30% relative humidity was not affected until 2 days following the last drink. After 4 days, it decreased to 20% of the initial yield, but after drinking it was completely restored within 3 days (Maltz and Shkolnik, 1980). Goats on some arid islands have been observed to live without any fresh drinking water. Under such conditions they may drink seawater. Dunson (1974) has shown that feral goats voluntarily drank seawater. Consumption of less than 4% of the body weight apparently helped to reduce weight loss during dehydration.

H. Feeding

1. Pasture

On the range goats depend on grazing and browsing on available plants. Their ability to ingest practically all available plants, except toxic ones, accounts for their great adaptability. In the dry tropics, the most important typical goat ranges are the deciduous matorral (or monte, i.e. scrubland)

and the espinar or thorn forest (Castillo, 1974). Both are forms of frutescent perennial vegetation of desert and semi-desert habitats. The matorral typically consists of small trees, bushes and ligneous forbs. They are all characterized by small size and the absence of a distinct layer. Most of them lose their leaves during a certain period of the year. Related to the matorral is the espinar, which presents the typical aspects of xerophytes, scattered perennial plants of asperous appearance, with sparse, small and sharp leaves. Thorn-bearing plants are either hardwood or succulents able to store water in their tissue. Some sink deep roots to find water in order to survive in semi-desert or desert zones.

Goat ranges in East African Zambia have been described as Acacia Tree-Grassland with *Acacia* sp., *Ficus*, *Strychnos* and other small trees and *Hyparrhenia*-dominated grassland (Quartermain, 1975).

Pasture production depends on many environmental (climate, soil quality, vegetation type) and management factors (stocking rate, grazing pattern). In the dry tropics, annual rainfall seems to be by far the most important among these factors. Le Houérou and Hoste (1977), on the basis of numerous

Fig. 2. Browsing horizon on oaks marking the height goats can reach on these preferred trees (Texas).

observations in eight countries, found close relationships between average annual rainfall and pasture production, in the Sahelian zone of Africa with average annual rainfall between 100 and 1500 mm. They calculated regression equations with $r = 0{\cdot}83$:

$$\text{kg consumable dry matter/ha/year} = 42{\cdot}17 + 1{\cdot}03 \text{ annual rainfall (mm)}$$
$$\text{tropical livestock units/ha/year} = 0{\cdot}0169 + 0{\cdot}0004 \text{ annual rainfall (mm)}$$

The vegetative stratum utilized by goats ranges from the soil surface to the level they can reach standing erect on their hind feet, i.e. according to their size up to 2 m. A browsing horizon often indicates this limit (Fig. 2). They may even climb trees whose branches permit doing so like fir, thuriferous-, argan-, and juniper tree.

For adequate forage utilization in high bush and low forest, it may be desirable to reduce the height of plants. This may be done through partial destruction by fire and chemicals or by mechanically knocking down with heavy chains or bulldozers. In both cases regrowth will be within the reach of the animals (see pp. 243–245).

Huss (1972) reported a grass production of 452 kg/ha in a natural bush vegetation in northern Mexico with about 700 mm annual rainfall. Production increased to 1146 and 1628 kg respectively, after roller cutting and burning of the bushes. Stocking rates in the same experiment were 7 and 14 goats per 5 ha.

In the same region of northern Mexico stocking rates per ha were analysed in two different areas with vegetation typed as *"matorral mediano subperennifolio"* with *Acacia–Cordia–Cercidium* as the dominating species, and compared with the carrying capacity estimated from green matter production. Observed and recommended ha per goat were 1·22 and 3·4, 3·75 and 6–7 for the two sites respectively, indicating severe overstocking (Carrera, 1971). Malechek and Leinweber (1972) considered 1 goat per 5 ha a heavy, 1 goat per 16 ha a light stocking rate at the Sonora research station in Texas.

The carrying capacity of a tropical savanna pasture with average annual rainfall of 1269 mm was estimated at 2·3 breeding does per ha. By improving the pasture through brush clearing, seeding with grass and legumes and fertilizing, the carrying capacity was increased to 7·5 does per ha (Wilson *et al.*, 1980).

The ability to consume and digest coarse feed seems to differ between breeds. When fed exclusively on wheat straw, black Bedouin goats maintained their body weight and consumed 24% more energy than Saanen goats, who lost weight rapidly on the same feed. Energy digested was closely related to water turnover:

$$\text{digestible energy intake (kcal. kg}^{-1}) = 1{\cdot}28 + 1{\cdot}097 \times \text{water turnover (ml. kg}^{-1})\quad r = 0{\cdot}996$$

This equation may be used to estimate energy intake in desert conditions (Silanikove *et al.*, 1979).

Cacti are widespread in many of the semi-arid areas of America. They have invaded the destroyed dry forests of Latin America (Hueck, 1966) and may play an important role in preventing soil erosion. At the same time, cacti are an important feed for ruminants, and goats in particular. Goats may gnaw at the base of tall growing species until they fall. Depending on the degree of spining, the fallen plant may be eaten immediately or after some degree of wilting (Fig. 3). Goats ably nibble between spines to get access to the succulent flesh or the fruits, without injuring lips or tongue. When sheep try to eat the fruits of *Cylindropuntia*, their lips get pricked easily. Subsequent swelling and infection may lead to starvation. Some of the species of *Opuntia* readily shed their pads when touched, e.g. by grazing goats (Fig. 4). Fallen and wilted pads are scattered on the ground and eaten by goats. We have seen goats in Northern Venezuela carefully picking pads by the free end and then rubbing them against a stone, deliberately breaking off the spines. Although we could not measure the intake, quite considerable quantities seemed to be ingested.

Cacti are a valuable feed reserve for drought periods. Cultivation of spineless *Opuntia* in northern Africa has not proven successful since herdsmen tended to graze them continuously and not leave them as reserve. Experiments are being carried out in Mexico for improving suitability of *Opuntia* as animal feed.

Closely related to the discussion of possible damage to vegetation by goats is their relative utilization of gramineas, forbs and bushes. Quite conflicting results have been published. Carrera (1970) and Carrera and Aguirre (1970) in northern Mexico found that goat diets consisted of 16·8% gramineas. In the experiment reported by Huss (1972), goats preferred browse although plenty of gramineas were available. Likewise Edwards (1948), Jordan (1957), Staples and Hornby (1962) and Wilson (1957) had concluded from their studies, that gramineas contribute little to the diet of goats and were taken only if there was not sufficient browse available. On the other hand Martínez *et al.* (1972) in Northern Venezuela found gramineas to be the major constituent of the diet. Knight (1965), in a dry area of Kenya, differentiated selective feed intake by counting feeding stops according to plant species. His results are reported in Table II. There was a clear preference for gramineas. In experiments in Texas, goats appeared to be grazers rather than browsers (Malechek and Leinweber, 1972). These seemingly contradictory results may be explained by plant availability and their palatability and nutritive value, mainly determined by stage of growth. This is discussed in detail in Chapter 7. Apparently the goats' particularity is that they effectively use a much wider range of feed plants than sheep and cattle. Their capacity

Fig. 3. Cactus fallen to the ground with flesh eaten between needle-sharp spines.

Fig. 4. Criollo goat with cactus pads sticking to the skin on jaw, neck and flank.

Table II. Feeding behaviour of goats on pasture (Knight, 1965).

Season	Grass (*n*)	Grass (%)	Bush (*n*)	Bush (%)	Herbs (*n*)	Herbs (%)	Total
	Feeding stops in 1 hour						
Rainy	236	65·4	98	27·2	27	7·4	361
Dry	402	82·5	58	11·9	27	5·6	487

to take advantage of forages that are not consumed by other herbivores enables them to utilize marginal range areas in the dry tropics whose best forage resources have been exhausted by cattle and sheep. By their preference for browse they exert less grazing pressure on the grass cover. Bourbouze and Guessous (1977) report that goats in southern Tunisia consume herbs in spring and afterwards pass rapidly to browse. This change is delayed if rains are abundant.

Browsing goats may ingest, in low forest-brush vegetation of the dry tropics, diets that after the rainy season may contain up to about 25% crude protein (Carrera, 1968; Castillo, 1970; Martínez *et al.*, 1972). However, during the dry season, protein content of grasses is reduced by 50% (Gonzales *et al.*, 1966) and young sprouts and other green forage from browse are severely diminished, although their nutrient content hardly varies.

Crude protein content (CP) of fresh leaves of *Acacia nilotica* and *Colophospermum mopane* in Zambia in mid-rainy season was 14–27 and 10·5% respectively. Seed heads of *Urochloa trichopus* which were selectively eaten had 12% CP in the rainy season, but annual dry grass remnant during the dry season had only 4% CP. However, weights of goats recorded in mid-dry and mid-rainy season did not differ, indicating a well balanced nutrient supply (Quartermain, 1975). Until recently, in Zambia's Valley Tonga the agro-pastoral system, which includes goats, seemed to make the greatest possible use of the land without causing degradation. With increasing land pressure, this equilibrium may be endangered.

More investigations are necessary in different ecosystems of animal–plant interactions, where not only feeding habits of goats are studied but also the reaction of plants to defoliation under various grazing pressures and possibilities of cultivating forage plants for these environments.

2. Effects on Vegetation

In Chapter 7 the interaction of goats with vegetation is discussed in detail. Any adverse effect of the goat on vegetation will be more marked in a dry climate, since plant recovery is slow or even impossible. Disastrous results are well known in Mediterranean, African and Asian countries and are visible in many others. A fact often overlooked is that in the dry tropics there is often an abundant supply of feed after the rainy season. Stockmen are tempted to keep stock in numbers related to forage availability during the lush season, which then starves through the dry season. Overgrazing regularly occurs, which may initiate disaster if the rains fail and plant growth is stopped during extended periods of drought. If such events cumulate, all ruminants but the goat may gradually lose the bases for even meeting their maintenance requirements. In the past, the goat, as the only remaining domestic animal, has been blamed in a very unilateral way for the destruction which in the end faces it. Now, however, after a more differentiating analysis, it is generally accepted that the goat is only one link in a chain of factors that have been badly managed by man (Huss, 1972; Le Houérou, 1975; Bourbouze and Guessous, 1977; Huston, 1978). These include:

—deforestation or agricultural use of marginal lands in low rainfall areas, where the land, void of its natural vegetational cover between two crops, is exposed to the forces of erosion;

—the use of fire, for preparing land for cultivation, and opening of new pastures in forest areas. This would not by itself be serious if the fire was well controlled; but it often happens that man loses control over the fire which year after year causes the loss of vast forest areas. Unfortunately government authorities often remain rather indifferent to this. Even today in a country like Italy the forest area annually destroyed by fire exceeds the area afforested;

—collection of firewood. In the coastal plains of Nueva Esparta, Venezuela, Castillo (1966) studied the changes in the vegetation cover after the goat eradication campaign in the 1950s. With a sample of 1376 families and 14 026 goats, he found that browsing by goats caused less damage to the vegetation than cutting firewood. Deprived of the income from goats, small as it was, the families were compelled to make direct use of the sparse vegetation;

—keeping excessive stock numbers, with cattle and sheep mostly preceding goats as long as the vegetation will support them.

3. Supplementary Feeding

While from a nutritional point of view, supplementary feeding of range goats during the critical periods of the reproductive cycle is desirable in the dry tropics, it is very difficult and in most herds even impossible to practise. Low productivity of goat herds does not seem to justify the high investment in concentrates or the location of farms may simply prevent concentrates from being transported to them. Still, some progressive farmers do supplement their goats, using mainly by-products of cereals, molasses, sorghum, seeds of bush legumes, e.g. *Prosopis* sp., *Caesalpina cassia* sp., *Acacia* sp., and minerals. Often, supplementary feeding is preceded by extended periods of underfeeding so that the additional nutrients ingested are used to replenish the depleted organism, not leaving any nutrients for production of milk or for growth. This is how some of the disappointing results of grain feeding may be explained. Trials with supplementary feeding of goats have been carried out in a semi-arid area of northern Mexico by Carrera and his group. Results were quite varied. Feeding phosphorus and cobalt did not significantly affect condition and milk yield in range goats (Carrera *et al.*, 1969). Sorghum or maize grain and cotton-seed meal partly supplemented with urea, fed to pregnant goats before kidding (Carrera and Flores, 1970) and to milking goats on pasture (Carrera and Robles, 1970) resulted in increased production over the control group. Similar supplementation of goats prior to being served did not, however, result in an improvement of reproductive performance which was about 150 kids per 100 does kidding, in both supplemented and unsupplemented control animals, although weight gains in does were higher with supplementation (Carrera and Flores, 1970; Diego, 1969). Leigh (1969) in Peru supplemented rations for fattening goats with cottonseed cakes. He concluded that this kind of feeding was not economical, under normal production conditions.

European dairy breeds, under intensive feeding conditions in Mexico, received about 1·8 kg of alfalfa hay, and an experimental group was given ground sorghum grain *ad libitum* from 2 weeks prior to parturition and during early lactation. Consumption was 1 kg. Birth weight, average daily weight gains of kids, and milk production of the treatment group exceeded those of controls by 380 g, 70 g and 1·27 kg respectively. Net return was about 45% through sales of milk alone (Juárez and Vázquez, 1972).

Thus direct economic returns from supplementation may be expected only if the basic nutrient requirements are met by the available forage, and if goats are genetically fit to convert additional nutrients into saleable products. Level of production and relations between prices of concentrates and of products are further preconditions for successful supplementation.

I. Reproductive Performance

Knowledge of the reproductive performance of goats in the dry tropics is of vital importance for the planning of breed improvement programmes and for recommending herd management practices. Doubtless, the environment and fluctuations in environmental conditions will be an important cause of variation in breeding efficiency. This is documented by experiments in India, where it was shown (Sachdeva *et al.*, 1973) that insufficient energy supply seriously affected reproductive performance, notably the percentage of

Table III. Reproductive performance of goats of local breeds in semi-arid areas.

Breed	Country	% twins	Kids/litter	Source
Criollo	Mexico	56	1·53[a]	Diego (1969)
Criollo	Venezuela	58	1·6	García *et al.* (1972b)
Criollo	Venezuela	53·1	—	García *et al.* (1977)
Criollo	Peru	—	1·74	Vélez-Nauer and Callacná (1979)
Criollo	West India	—	2·11[b]	Devendra and Chenost (1973)
Créole	Guadeloupe	—	2·33	Cognie *et al.* (1971)
Baladi	Egypt	76[c]	—	Tantaway and Ahmed (1960)
Zaraibi	Egypt	—	1·84	Shalash *et al.* (1970)
Zaraibi	Morocco	—	0·6–0·9	Bourbouze and Guessous (1977)
Mubende	Uganda	32	—	Sacker and Trail (1966)
Sokoto Red	Niger	—	1·47	Haumesser (1975)
Small East African (Gwembe)	Zambia	13	—	Quartermain (1975)
West African Dwarf	Ghana	—	1·9	Vohradský and Šáda (1973)
West African Dwarf	Nigeria	—	1·7	Kirkpatrick and Akindele (1974)
Boer	South Africa	55·6	—	Skinner (1972)
Maltese	Israel	—	1·59	Maule (1966)
Damascus	Israel	—	1·76	Maule (1966)
Mamber	Israel, south	—	0·7	Israel (1962)
Mamber	Israel, north	84	1·3	Israel (1962)
Damascus	Lebanon	53	1·65	Choueiri (1973)
Malabari	India	—	1·64	Mukundan and Rajagopalan (1971)
Malabari	India	—	1·44	Mukundan (1976)
Malabari	India	53[c]	—	Mukundan and Rajagopalan (1973), Prasad *et al.* (1972)
Beetal	India	—	1·70	Bhatnagar *et al.* (1976)
Local	India	53–33	—	Singh *et al.* (1977)
Barbari	India	34·38	—	Singh *et al.* (1977)
Jamnapari	India	—	1·45	Singh and Singh (1974)
Jamnapari × Local	India	20·0	—	Singh *et al.* (1977)

[a] after a service period of 12 days.
[b] up to 6 kids in 2 years.
[c] increasing with age of the dam.

multiple births, kidding interval and number of kiddings. There is no reason why the effect of plane of nutrition on semen quality in bucks found by Arbeiter (1963) and Hiroe and Tomizuka (1965) in temperate zones should not apply to the dry tropics as well.

The importance of nutrition for fertility is reflected in relationships between body weight and kidding percentage, as reported in Angora does by Menzies (1968). Malnutrition may affect fertility rates through abortions which often remain unobserved (van Rensburg, 1971).

Table IV. Reproductive performance of breeds of temperate areas and their crosses in the dry tropics.

Breed	Country	Multiple births(%)	Kids/litter	Source
Saanen	Venezuela	45·2	—	García *et al.* (1972c)
Saanen	Egypt	—	1·01	Shalash *et al.* (1970)
Saanen	Israel	—	1·9	Epstein and Herz (1964)
Saanen	South Africa	61·3	1·73	Hofmeyr (1962)
Saanen × Barbari	India	14·3	—	Singh *et al.* (1977)
Saanen × Jamnapari	India	12·5	—	Singh *et al.* (1977)
Saanen × Jamnapari	India	33·33[a]	—	Singh *et al.* (1977)
Israeli Saanen × Ethiopian Highland	Ethiopia	—	1·67	Galal and Beyene Kebede (1977)
Toggenburg	Venezuela	41·2	—	García *et al.* (1972c)
Toggenburg	South Africa	60·3	1·63	Hofmeyr (1962)
Toggenburg ($\frac{1}{2}$)[d]	Venezuela	42·1	—	García *et al.* (1977)
French Alpine	Venezuela	42·2	—	García *et al.* (1972c)
Alpine	South Africa	53·8	1·91	Hofmeyr (1962)
Alpine	India	—	1·49	Bhatnagar *et al.* (1976)
French Alpine	India	52·07	—	Gill and Dev. (1972)
Alpine ($\frac{1}{2}$)[d]	Venezuela	55·6	—	García *et al.* (1977)
Alpine × Beetal	India	—	1·6	Bhatnagar *et al.* (1976)
Anglo-Nubian	Venezuela	45·1	—	García *et al.* (1972c)
Anglo-Nubian	Egypt	—	1·91	Shalash *et al.* (1970)
Anglo-Nubian[d]	Peru	43·6	1·48[b]	Nolte (1977)
Anglo-Nubian	Israel	—	1·75	Epstein and Herz (1964)
Anglo-Nubian	India	57·68	—	Gill and Dev. (1972)
Anglo-Nubian ($\frac{1}{2}$)[d]	Peru	—	1·84	Vélez-Nauer and Callacná (1979)
Anglo-Nubian ($\frac{1}{2}$)[d]	Venezuela	64·9	—	García *et al.* (1977)
Anglo-Nubian ($\frac{3}{4}$)[d]	Venezuela	57·8	—	García *et al.* (1977)
Granadina	Mexico	—	1·5[c]	Hernandez (1969)

[a] $\frac{3}{4}$ Saanen.
[b] maximum at 7 to 8 months interval between parturitions.
[c] after a mating period of 12 days.
[d] produced by crossbreeding with criollo.

1. Fertility and Prolificity

The well known high reproductive efficiency of goats is evident in the dry tropics as well. Tables III and IV give observations by various authors on autochthonous breeds, breeds of European origin and crosses of the two.

There are very fertile breeds with high twinning rate in both temperate and tropical areas. Crossing with some of the bigger and more productive temperate breeds seems to have the potential of increasing fertility as well. However, if environmental conditions are not adequate, this advantage may not be realized. The data also indicate that environmental conditions, in general, are as important as breed differences in fertility variation. Seasonal differences will reflect environmental conditions, mainly with respect to rainfall pattern and the resulting feed availability.

Table V. Reproductive performance of criollo and crossbred goats in Venezuela (Garcia *et al.*, 1977).

Type of cross	Weight at parturition	Singles	Twins	Triplets	Quadruplets	Multiples	Abortions
Criollo	(376) 39·3 ± 6·2	(195) 44·0%	(235) 53·1%	(11) 2·5%	(2) 0·5%	(248) 56·0%	(22) 4·7%
ND	(251) 42·0 ± 8·5	(87) 35·1%	(142) 57·3%	(18) 7·3%	(1) 0·4%	(161) 64·9%	(26) 9·5%
NND	(86) 40·2 ± 6·9	(38) 42·2%	(48) 53·3%	(4) 4·4%	— —	(52) 57·8%	(10) 10·0%
AD	(58) 43·0 ± 9·3	(32) 44·4%	(37) 51·4%	(3) 4·2%	— —	(40) 55·6%	(8) 10·0%
TD	(62) 38·7 ± 8·6	(33) 57·9%	(24) 42·1%	— —	— —	(24) 42·1%	(7) 12·3%

( ) = number of observations.
ND = $\frac{1}{2}$ Nubian, $\frac{1}{2}$ Criollo.
NND = $\frac{3}{4}$ Nubian, $\frac{1}{4}$ Criollo.
AD = $\frac{1}{2}$ Alpine, $\frac{1}{2}$ Criollo.
TD = $\frac{1}{2}$ Toggenburg, $\frac{1}{2}$ Criollo.

Table VI. Duration of sexual cycle in goats in dry tropical areas and duration of oestrus as reported in the literature.

Breed	Country	Cycle duration (days)	Oestrus duration (h)	Source
Criollo	Venezuela	20·3 ± 5·6 (66%) 6·3 ± 1·2 (33%)		Garcia *et al.* (1972b)
			33·2 ± 5·8	Gonzales (1972)
Anglo-Nubian × Criollo[a]	Venezuela		20·4 + 6·37 (7–32)	Fuenmayor *et al.* (1972)
Anglo-Nubian	Chile	22 ± 5		Santisteban *et al.* (1977)
Granadina[b]	Mexico	20·9 ± 1·9	34·8 ± 8·1	Carrera and Juárez (1968)
Baladi	Egypt	19·45		Sorror (1973)
West African Dwarf	Ghana	24 ± 9·9	17 ± 9·7	Otchero and Nimo (1975)
West African Dwarf	Nigeria	20·4	24–40	Kirkpatrick and Akindele (1974)
Boer	South Africa	22 ± 0·3[c]	19 ± 8[d] (8–48)	Hofmeyr *et al.* (1966) Skinner (1972)
Barbari	India	17–22 (57%) 5–10 (28%)	26 ± 7[e]	van Rensburg (1964) Sahni and Roy (1967)

[a] after synchronization with fluorogestone-acetate.
[b] penned.
[c] many short cycles included.
[d] spring and early summer.
[e] August–January.

Reproductive performances of local goats and crossbreds with European breeds in Venezuela are given in Table V. Some of the crosses are more prolific than the criollos, but not all. Kid losses are higher in all crosses so that there was little increase in the number of kids weaned. The performance of European breeds is generally reduced in dry tropical areas because of the environment stress and, more important, inadequate nutrition.

Table VI provides data on the length of the oestrus cycle and duration of oestrus as reported in the literature. The average duration is between 20 and 21 days, as in septentrional countries, but the occurrence of short oestrus cycles seems to be quite frequent. They are linked to infertile cycles and apparently are a symptom of a hormonal imbalance. This syndrome needs to be further investigated.

In Table VII some observations are recorded on the duration of pregnancies of imported and crossbred goats in tropical countries. There is a trend toward increased length of gestation from small to large breeds.

2. Age at First Kidding

While puberty in the goat begins at about 6 months of age, pregnancies at such an early age would affect growth and irreversibly impair future performance (García *et al.*, 1972a). This is of particular importance with goats of autochthonous breeds of the dry tropics raised under extensive range conditions and which have very low weight at their first oestrus. Based on their observations at the experiment Station El Cují in the arid zone of

Table VII. Duration of pregnancy in goats in dry tropical areas, as reported in the literature.

Breed	Country	Duration	Source
Criollo	Mexico	148·2	Hernandez (1969)
Criollo	Venezuela	149·7 ± 2·7	Garcia *et al.* (1977)
Saanen	Venezuela	149·7 ± 2·4	Garcia *et al.* (1972d)
Toggenburg	Venezuela	149·0 ± 2·8	Garcia *et al.* (1972d)
Toggenburg × Criollo	Venezuela	149·5 ± 3·4	Garcia *et al.* (1977)
French Alpine	Venezuela	152·2 ± 3·7	Garcia *et al.* (1972d)
French Alpine × Criollo	Venezuela	149·7 ± 3·4	Garcia *et al.* (1977)
Anglo-Nubian	Venezuela	148·8 ± 2·8	Garcia *et al.* (1972d)
Anglo-Nubian × Criollo	Venezuela	149·0 ± 3·1	Garcia *et al.* (1977)
West African Dwarf	Ghana	141·3 ± 4·6	Otchere and Nimo (1978)
West African Dwarf	Nigeria	143·5	Kirkpatrick and Akindele (1974)
Sirohi	India	146·4 ± 0·24[a]	Mishra *et al.* (1979)

[a] Standard error of mean.

Venezuela, García *et al.* (1976a) recorded weights at first service of 25 kg and 28 kg for criollos and crosses with European breeds respectively. These weights are attained at about 1 year of age and first kiddings would result at $1\frac{1}{2}$ years approximately. Goats of breeds which in temperate zones usually have their first parturition at an age of 14 to 16 months in tropical conditions will kid much later (Gill and Dev, 1972; Gonzales *et al.*, 1976). This may partly be due to lower growth rate, as does which were born as twins, with lighter birth weights and slower growth rate, kidded about 1 month later

Table VIII. Age and weight at first service or first kidding of different goat breeds in the dry tropics.

		Age at first kidding	Weight (kg) at first kidding	Age at first service	Weight (kg) at first service	
Breed	Country					Source
Europ. breeds[a]	Venezuela	866 d	—	—	—	Gonzales *et al.* (1976)
Europ. breeds[a] × Criollo	Venezuela	639 d	—	—	—	Gonzales *et al.* (1976)
Anglo-Nubian	Peru	1·51 y	—	—	—	Nolte (1977)
Anglo-Nubian	India	25·4 m	—	—	—	Gill and Dev (1972)
Anglo-Nubian × Criollo	Peru	1·52 y	—	—	—	Nolte (1970)
French Alpine	India	20·8 m	—	—	—	Gill and Dev (1972)
Alpine	India	556 d	—	—	—	Khatter *et al.* (1977)
Alpine × Beetal	India	498 d	—	—	—	Khatter *et al.* (1977)
Baladi	Egypt	1·6 y	21·4	—	—	Ahmed and Tantawy (1960)
Mubende	East Africa	19 m	—	—	—	Sacker *et al.* (1966)
Maradi Red	Niger	427 d	—	—	—	Haumesser (1975)
Damascus	Israel	24 m	—	—	—	Devendra and Burns (1970)
Jamnapari	India	2·0 y	—	—	—	Singh and Singh (1974)
Jamnapari × Baladi	India	534 d	—	—	—	Raja and Mukundan (1974)
Beetal	Israel	22 m	—	—	—	Devendra and Burns (1970)
Beetal	India	545 d	—	—	—	Khatter *et al.* (1977)
Malabari	India	495 d	—	—	—	Raja and Mukundan (1973)
Malabari	India	596 d	—	447 d	—	Mukundan (1976)
Saanen × Malabari	India	—	—	314 d	—	Mukundan (1976)
Alpine × Malabari	India	—	—	257 d	—	Mukundan (1976)
Beetal	India	526 d	—	—	—	Bhatnagar *et al.* (1976)
Alpine	India	542 d	—	—	—	Bhatnagar *et al.* (1976)
Alpine × Beetal	India	461 d	—	—	—	Bhatnagar *et al.* (1976)
Black Bengal	Bangladesh	—	—	10 m	15·0	Ali *et al.* (1973)
Barbari	India	—	—	7–12 m[b]	9–16	Prasad and Bhattacharya (1979)

[a] Anglo-Nubian, French Alpine, Toggenburg, Saanen.
[b] first oestrus observed.

than their single born sisters (Vohradský and Šáda, 1973; Ali *et al.*, 1973). Crossbreds with indigenous breeds may kid earlier than their purebred dams. Poor feeding may delay age at first conception, as well as the following breeding interval (Singh and Sengar, 1970). Some observations on age and weights at first breeding are compiled in Table VIII.

Table IX. Occurrence of oestrus in goats in countries of the dry tropics.

Breed	Country	Season[a]	% Oestrus	Source
Northern Hemisphere				
Criollo	Mexico	VIII–XII (VI–II)		Carrera and Juárez (1971)
European breeds[b]	Venezuela	VII–XII,	79·5	Gonzales *et al.* (1974)
		III–VI	19·4	
Criollo	Guadeloupe	VII–X[b]	62·6	Devendra and Chenost (1973)
Maradi Red	Niger-Maradi	IX–XI[b]	51·7	Haumesser (1975)
West African Dwarf	Ghana	XI[b]	67	Vohradski and Sada (1973)
Desi	India	X, VI, III		Mishra and Biswas (1966)
Beetal	India	X–XI, V–VI		Amble *et al.* (1964)
French Alpine Anglo-Nubian	India	VIII–X[b]	74·4	Gill and Dev (1972)
Southern Hemisphere				
Criollo	Brazil	I–III	44·3	Jardin (1965)
Anglo-Nubian	Brazil	I–V	76	Jardin (1965)
Anglo-Nubian	Peru	II	19·0	Nolte (1977)
		III[b]	17·4	
Boer	South Africa	II–VIII, X–XI		Hofmeyr *et al.* (1966)

[a] Number I–XII correspond to the months of the year (January–December).
[b] Calculated from kidding data.

3. Seasonality of Reproduction

After a period of sexual inactivity, oestrus cycles in septentrional countries start in autumn. First cycles, in which does conceived regularly, occurred in the first week of September in Munich (Gall, 1979) and between 15 and 25 March in Cape Province, South Africa (Marincowitz, 1962). It is well established that this seasonality is governed by photoperiodicity. As a consequence, goats should breed all through the year in the tropics, where changes in the daily duration of sunlight in the course of the year are minimal. Indeed in the tropics, kiddings are observed throughout the year. Horst (1976) and Emady (1976) found a high frequency of pregnancies in all months of the year and as far north as Shiraz/Iran. There are, however, months with clearly reduced sexual activity. These months differ from region to region and are apparently linked to the temperature regime and to feed availability (Table IX). It is a common observation in the tropics that oestrus occurs shortly after the onset of the rainy season, or even after sporadic rains during the dry season. These relationships have not been studied in depth but some reports have been published. Ashmawy (1978/79) studied the seasonality of sexual behaviour in male Baladi goats in Egypt. He found a significant effect of season, mid-summer (July/August) being the most efficient mating season. In Niger, 52% of parturitions of the Maradi Red goat were between mid-February and mid-April. The corresponding oestruses in September, October, November were favoured by the end of the rains and a change from deficient feeding to better feed supply at stubble grazing (Haumesser, 1975). In India, the majority of parturitions occur during or after the monsoon season (Amble *et al.*, 1964; Moulick *et al.*, 1966; Prasad *et al.*, 1972; Singh and Singh, (1974), although some occur throughout the year. García *et al.* (1972b), from their work with criollo goats in semi-arid Venezuela, report a rather uniform distribution of parturitions over the year. However, there are two prominent seasons from August to October and from January to April, where 75% of the parturitions occur, resulting in a certain seasonality of reproduction in this breed of the very tropics. In semi-arid northern Mexico, Carrera and Juárez (1968) observed about equal occurrence of oestrus between June and February (60 to 100% of possible oestrus) and very low occurrence during

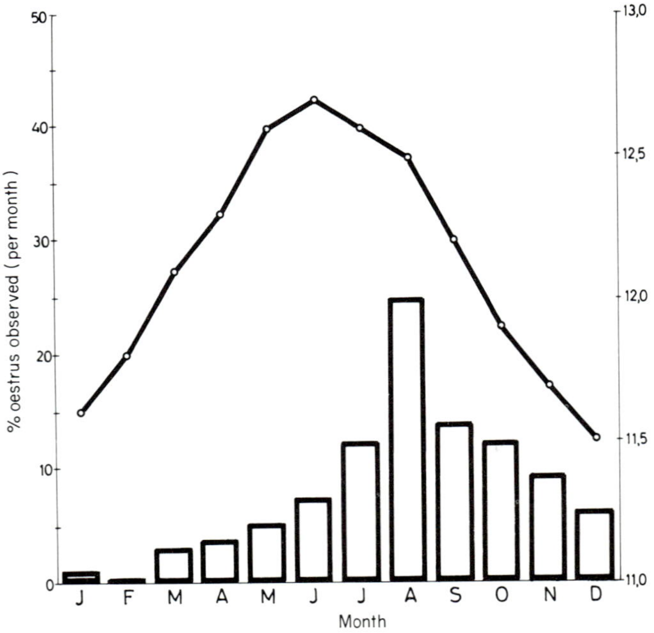

Fig. 5. Occurrence of oestrus in relation with daily light hours in the dry tropics in Venezuela, Long. 69° 18′, Lat 10° 09′ N, Altitude: 580 m above sea level (Gonzales *et al.*, 1974).

March, April and May (4 to 54% of possible oestrus). In the same environment the average kidding percentage of these breeds was 177. In an analysis of a herd of about 100 criollo does in northern arid Peru with annual rainfall averaging 100 mm, Vélez-Nauer and Callacná (1979) found an accumulation of kiddings in May, June and August, with 12·7, 14·5 and 20·7% respectively, but with a marked annual fluctuation; in 3 out of 7 years the lowest kidding percentage was in May, June or July. The seasonal accumulation of kiddings is explained by forage availability which is best during February to May.

This seasonality is even more marked with European breeds implanted in the tropics, or with their crossbreds with autochthonous breeds. Gonzales *et al.* (1974) analysed the rhythm of sexual activity of four European breeds imported to Venezuela (Anglo-Nubian, French Alpine, Toggenburg and Saanen), during 1966 to 1970 (Figs 5 and 6). Two different seasons were

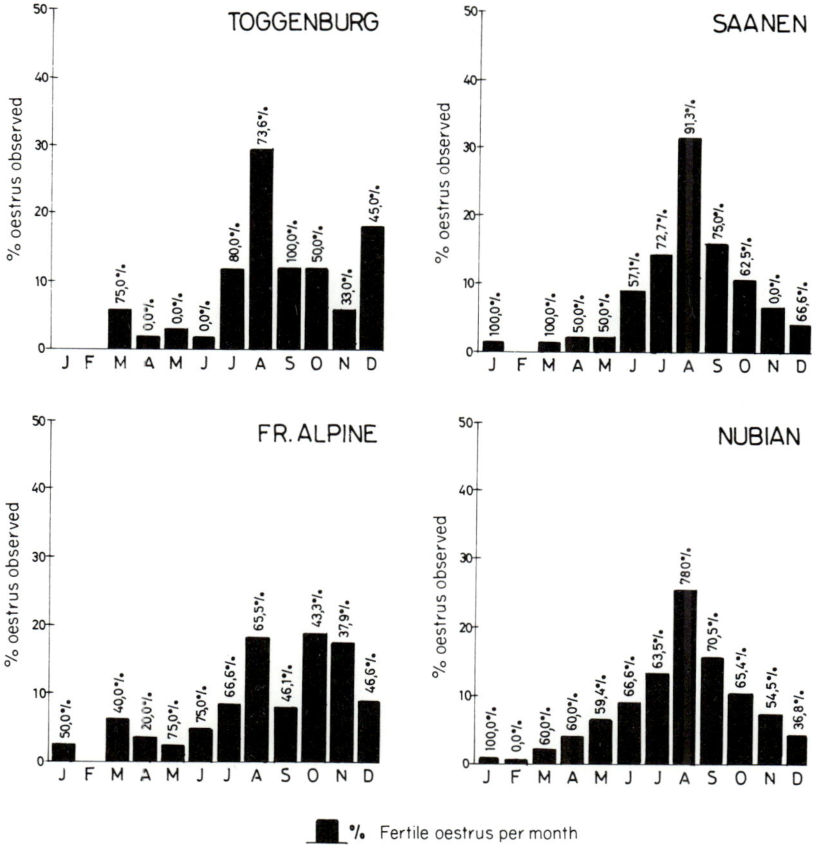

Fig. 6. Oestrus distribution in four breeds imported to a dry tropical area in Venezuela, 1966–1970 (Gonzales *et al.*, 1974).

observed. During June to November 81% of the oestrus occurred and 11·9% during March to May. The latter were apparently induced by the beginning of the rainy season. The authors concluded that contrary to what would be expected in a tropical environment, the goats exhibited a marked seasonality, which incidentally corresponded to the same rhythm in extensive goat herds of criollo goats in the country. The highest fertility was observed in August (77%) and the lowest during November–December (43%); of the does whose oestrus came between June and December, 89% became pregnant. Of all does served during a 5-year period of observation, 63·7% became pregnant (García and Gonzales, 1972). There are, however, clear breed differences. In 1975 parturitions of goats of European breeds managed intensively in the Tlahualilo Breeding Centre in northern Mexico, Juárez (1979) recorded the proportion occurring between July and December. These were 4·5, 6·9, 6·2, 27·4 and 15·4 in Saanen, Toggenburg, Alpine, Granadina and Anglo-Nubian respectively.

Carrera and Hernandez (1969/70) studied the relation between the length of the mating season and conception rate. In northern Mexico it is traditional to limit matings to a short period of 10 to 12 days in order to facilitate management of the herd of does and of kids as well. Percentage of adult does pregnant after 10, 15, 40 and 45 days exposure was 45, 50, 60, 70 respectively. Per cent pregnancy in yearlings after 10, 15, 25, 35 and 65 days exposure was 10·5, 21, 36·8, 63·2 and 73·7 respectively. In Venezuela Gonzales (1976) has shown that ovarian activity was stimulated in spring by introducing bucks to criollo does who had had no contact with males for the 2 preceding months.

4. Interval between Parturitions

The number of parturitions during the lifetime of a goat determined by the longevity and the interval between kiddings is of great importance for the economics of production. The variation of this characteristic principally depends on the seasonality of reproductive activity (see above). Intra-breed differences at a given geographical latitude would reflect differences in management. In Table X there are data on intervals between kiddings as reported in the literature. It may be noted that intervals range mainly between 240 and 390 days. However, some very short intervals indicate complete aseasonality of oestrus occurrence. On the other hand, imported animals of European breeds seem to have much longer intervals (+ 100 days according to García *et al.*, 1977). When data on frequency of kiddings are evaluated care should be taken to consider lactational anoestrus. High kid mortality leaving does without kids and not lactating may lead to much shorter kidding intervals than those normally observed with does raising their kids.

Table X. Interval between parturitions in goats in dry tropical areas, as reported in the literature.

Breed	Country	Interval (days)	Source
Local breeds			
Criollo	Venezuela	272·4 ± 87·3	Garcia *et al.* (1977)
Criollo	Venezuela	281 ± 81·5	Castillo *et al.* (1972c)
Criollo	Guadeloupe	237 (202–272)	Devendra and Chenost (1973)
Mubende	East Africa	296·7	Sacker and Trail (1966)
Small East African (Gwembe)	Zambia	223 180[a]	Quartermain (1975)
Maradi Red	Niger	332·4 ± 6·5[b]	Haumesser (1975)
West African Dwarf	Ghana	258	Vohradský and Šáda (1973)
Barbari	India	280	Singh and Sengar (1970)
Jamnapari	India	340	Singh and Sengar (1970)
Malabari	India	299	Raja and Mukundan (1973/74)
Malabari	India	307	Mukundan (1976)
Malabari × Jamnapari	India	285	Raja and Mukundan (1973/74)
Beetal	South India	328 ± 8·5	Wijeratne (1968)
	India	308·9	Bhatnagar *et al.* (1976)
European breeds			
Saanen	Venezuela	390·6 ± 77	Garcia *et al.* (1972c)
Toggenburg	Venezuela	407·2 ± 123·3	Garcia *et al.* (1972c)
French Alpine	Venezuela	360 ± 102	Garcia *et al.* (1972c)
French Alpine	India	349·58	Gill and Dev. (1972)
Alpine	India	382·4	Bhatnagar *et al.* (1976)
Alpine × Beetal	India	357·2	Bhatnagar *et al.* (1976)
Anglo-Nubian	Venezuela	385·3 ± 122·3	Garcia *et al.* (1972c)
Anglo-Nubian	India	335·0	Gill and Dev (1972)
Nubian[c]	Peru	283·2	Nolte (1977)
Anglo-Nubian × Criollo	Venezuela	282·5 ± 58·8	Garcia *et al.* (1977)
French Alpine × Criollo	Venezuela	295·7 ± 42·8	Garcia *et al.* (1977)

[a] in does whose kids died after birth.
[b] 266, 302, 343 in different seasons, 219 after abortions.
[c] bred by upgrading criollos.

5. Oestrus Synchronization and Artificial Insemination

In the dry tropics field application of these techniques on a large scale will be quite difficult because of the low level of production of the extensively managed herds and the lack of infrastructure. However, there is interest in applying the techniques in breeding nuclei for breed improvement work and for rapid expansion of genetic progress. Gonzales (1977) tested oestrus synchronization and artificial insemination extensively in Venezuela. Of does treated in early lactation with fluorogestone acetate (FGA) and pregnant mare serum (PMS), 90% came into oestrus; 67% conceived during the first, and 81·5% during the second cycle; of does synchronized with the same treatment and inseminated without oestrus detection with fresh semen, 74% conceived. Up to 86·6% of does conceived after one double insemination with frozen semen.

These results show that semen conservation, oestrus synchronization and artificial insemination with goats is possible in the dry tropics. Although application in the field will not be justified because of high costs involved, these techniques might be useful at specific points in government executed breed improvement programmes.

J. Kid Mortality

Kid mortality in the dry tropics is one of the main factors adversely affecting goat production. While it is difficult to assess exactly the extent of these losses because producers tend to give very subjective information, it is beyond doubt that losses are high and may occasionally take 50% of the kid crop. The causes are closely related to the system of production and therefore difficult to control. They include birth weight, milk production of the dam, predators, diseases and accidents. They are aggravated by poor management, as is illustrated by the following examples:

—if does are kidding unguarded on the range, kids are exposed to predators, and flies causing myasis of the umbelicus; or kids may simply be abandoned by their dams, which happens with primiparous does in particular;

—if kids remain unattended for some time, infections through the navel may cause disease, which might be easily prevented if the kids were subjected to preventive treatment in time;

—if kids are allowed to join their mothers grazing on the range and travel large distances in their first days of life, they may suffer dehydration;

—if there is no shelter to protect young kids from the extremes of weather, they may suffer from heat or cold, especially in combination with humidity;

—lack of basic veterinary measures such as vaccination and antiparasitic treatments causes early infection to be fatal.

There are few reports on kid mortality in the dry tropics, and those published are mainly on experimental stations where there is the possibility of registering data.

Bravo *et al.* (1976) analysed 307 losses out of 963 kids born at the Experimental Station Loma de León, Barquisimeto, Venezuela. The main factors influencing losses were:

—*type of birth*: mortality up to 1 year of age of kids born as singles, twins or triplets was 18·8%, 32·4% and 33·3% respectively. These differences were partly explained by differences of body weight and milk available from the dam;

—*birth weight*: of kids with birth weights between 1·5 and 2 kg, 57·9% died during the first 3 months of life and another 12·2% during the following 3 months; the percentage of losses was less with higher birth weights;
—*breed of sire*: does in this experiment were criollo, and losses to 1 year of age were 42% among kids sired by French Alpine, against 27% among kids sired by Anglo-Nubians. In another experimental station, García *et al.* (1976b) recorded losses to 1 year of age among pure bred European breeds. Losses were 57%, 65·6%, 36·7% and 55·0% with Anglo-Nubian, French Alpine, Toggenburg and Saanen respectively.

Among 5300 kids born in the breeding centre Tlahualilo, in northern Mexico, mortality in Saanen, Toggenburg, Alpine, Granadina and Anglo-Nubian was 14·5, 15·4, 12·5, 6·7 and 9·3% respectively (Juárez, 1979). In a herd of about 1000 Criollo does in Northern Peru, over 6 years, Vélez-Nauer and Callacná (1979) recorded among 11 364 kids born, losses up to 3 months of age of 23·9% and a further 17·2% until about 1 year of age. Replacement rate of adult goats varied between years from 14 to 30%, that of sires from 2 to 21%. In a herd of about 140 Anglo-Nubian in Peru over 9 years, Nolte (1977) recorded kid losses up to 18·5 and 10·8%, in July and August respectively. Pneumo-enteritis was the cause in 44·2% of the losses (Vélez-Nauer *et al.*, 1977).

Ahmed and Tantaway (1960) analysed records covering 14 years of the flock of Baladi goats of the University of Alexandria. Losses until weaning at 4 months of age were 33·7%; there was a relation between birth weight and losses, mortality being lower in heavier kids. Similarly losses were higher with does of low live weight and kids born as twins and triplets. In local Tswana goats, Boer goats and their crosses in Botswana, mortality to weaning at 4 month of age was 18·9, 34·6 and 9·7% respectively (Botswana, 1977).

At the Agricultural University of Bangladesh, Abdur *et al.* (1976) recorded 82·2% losses of kids until the age of 6 months. In villages losses were only 30%. Causes of the losses were: 43·8% respiratory diseases, 25% gastro-intestinal parasites, 18·8% ecthyma contagiosum, 2·3% digestive disorders and the rest unidentified. In India, Minnet (1960) analysed 540 cases of kid losses in government farms and with individual goat keepers. Fifty-five per cent had died by 3 months of age of which the majority died during the first 2 weeks of life. Ranatunga (1971) reported kid losses in two government farms in Sri Lanka at a level of 23·7% and 32·2%. Pleuropneumonia was the main cause of high kid mortality of 45 to 53% in French Alpine and Nubian imported to India (Gill and Dev, 1972). Birth weight had a direct bearing on early kid mortality in India (Mittal, 1976). Bautista and Berriz-beitia (1974) enquired from breeders in eastern Venezuela and in different localities found losses between 18 and 27%. In the Small East African goat,

21% kid losses were recorded in the Gwembe Valley in Zambia (Quartermain, 1975) and in Ghana (Vohradský and Šáda, 1973).

Table XI. Milk production of goats of local breeds in dry tropical areas.

Breed	Country	kg/day	kg/lact.	Days	Source
Criollo	Mexico	0·2–0·5		210	Carrera and Sevilla (1971)
Criollo	Venezuela		40–60	200	García *et al.* (1972a)
					Hernandez (1971)
					García and Peraza (1976)
Criollo	Puerto Rico		154·2	232·3	Sanfiorenzo (1957)
Criollo	Puerto Rico	1·0[a]			Sanfiorenzo (1957)
Barbados	Puerto Rico		273·5	270·8	Sanfiorenzo (1957)
Criollo × Barbados (F_1)	Puerto Rico		242·4	296·0	Sanfiorenzo (1957)
Barbados (F_2)	Puerto Rico		215·3	289·9	Sanfiorenzo (1957)
Nubian	Sudan	1–2			Devendra and Burns (1970)
Maradi		0·3–0·5[d]		120	Robinet (1967)
(Sokoto Red)		0·8–1·2[e]			
Boer	South Africa		228·1[c]	120	Skinner (1972)
			360·8[b]		
Damascus	Lebanon		392 (−500)		Choueiri (1973)
Mamber South type	Israel		60		Israel (1962)
Mamber (N)	Israel		150		Laor (1971)
Mamber	Syria	1·5	> 500		Devendra *et al.* (1970)
Israeli-Saanen	Israel		500[f]		Israel (1962)
Beetal	India		183·5	189·2	Bhatnagar *et al.* (1976)
Beetal	India		137	147	Prakesh *et al.* (1971)
Jamnapari	India		202	191	Singh and Singh (1974)
			222·5	224	
			150·3	252	Maule (1966)
Malabari	India		38·6	117·5	Mukundan (1976)

[a] with improved management.
[b] with singles.
[c] with twins.
[d] in dry season.
[e] in rainy season.
[f] record: 2293 kg/277 days.

All these observations indicate severe defects in the management practices of kid raising. This is obvious particularly when goats are confined and have to be managed in a way which is not common in the dry tropics. Furthermore, these observations reveal the need for better knowledge of the pathological conditions affecting kids in particular.

K. Productive Performance

1. Milk Production

Generally there is no clear distinction, in the dry tropics, between dairy and meat operations, except in those rare intensive or semi-intensive herds of specialized breeds. Production with autochthonous breeds is mainly for subsistence. Surplus beyond the needs of the family is processed by rustic techniques to cheese or other dairy products for local markets. Much of the goats' milk is marketed together with milk from cows or buffaloes, but in some instances it may fetch a higher price. In Mexico, where goats' milk is used for sweets the price in 1978 was the equivalent of US $0·24 per litre, 21·5% higher than that for cows' milk (Juárez, 1979).

Maule (1966) reviewed the literature on dairy performance of goats in

Table XII. Milk production of goats of European breeds in dry tropical areas.

Breed	Country	kg/day	kg/lact.	Days	Source
Saanen	Venezuela		292·7	268·8	García *et al.* (1977)
Saanen	Puerto Rico		704	344	Sanfiorenzo (1957)
Saanen	Mexico	2·01	502	250	Juárez (1979)
Saanen	Puerto Rico		291·5	—	Sanfiorenzo (1962)
French Alpine	Venezuela		233·0	233	García *et al.* (1977)
Alpine	Mexico	2·00	468	235	Juárez (1979)
Toggenburg	Venezuela		282·9	275	García *et al.* (1977)
Toggenburg	Mexico	2·05	491	238	Juárez (1979)
Granadina	Mexico	1·70	376	224	Juárez (1979)
Anglo-Nubian	Venezuela		137	218·8	García *et al.* (1977)
Anglo-Nubian	Mexico	1·70	338	203	Juárez (1979)
Anglo-Nubian	India		289	—	Gill and Dev (1972)
Saanen × Criollo	Puerto Rico		275·6	326·8	Sanfiorenzo (1957)
Saanen × Barbados	Puerto Rico		382·8	316·8	Sanfiorenzo (1957)
$\frac{1}{2}$ Saanen × $\frac{1}{4}$ Native	Puerto Rico Barbados		297·4	270	Sanfiorenzo (1962)[a]
$\frac{3}{4}$ Saanen × $\frac{1}{4}$ Native	Puerto Rico		284·2	—	Sanfiorenzo (1962)
$\frac{7}{8}$ Saanen × $\frac{1}{8}$ Native	Puerto Rico		272·4	—	Sanfiorenzo (1962)
Criollo × French Alpine	Venezuela		268·2	232·5	García and Peraza (1976)
Criollo × Anglo-Nubian	Venezuela		157·7	183	García and Peraza (1976)
Alpine	India		314·9	217·2	Bhatnagar *et al.* (1976)
French Alpine	India		310	—	Gill and Dev. (1972)
Alpine × Beetal	India		313·4	239·9	Bhatnagar *et al.* (1976)

[a] maximum production at 4 to 6 years of age.

Table XIII. Lactation performance in criollo and crossbred goats in Venezuela (Garcia *et al.*, 1977).

Type of cross	Lactation number I No.	I (kg)	I days	II No.	II (kg)	II days	III No.	III (kg)	III days	IV No.	IV (kg)	IV days	V No.	V (kg)	V days	VI No.	VI (kg)	VI days
Criollo	116	61·0 ± 29·6	166·3 ± 72·7	67	68·3 ± 33·0	159·7 ± 49·7	33	97·6 ± 35·9	170·8 ± 32·3	32	103·3 ± 40·9	168·1 ± 41·4	19	112·1 ± 49·1	171·2 ± 57·2	11	82·3 ± 29·1	149·5 ± 36·6
ND	35	84·69 ± 42·3	175·9 ± 54·4	45	113·8 ± 38·6	175·4 ± 45·8	39	140·0 ± 37·8	195·7 ± 54·3	21	141·7 ± 29·2	205·0 ± 39·2	12	134·6 ± 57·3	184·0 ± 92·7	4	112·1 ± 37·9	157·0 ± 43·1
NND	27	86·05 ± 30·4	196·9 ± 53·4	19	83·6 ± 34·1	165·7 ± 49·7	5	83·8 ± 23·3	127·2 ± 13·1		— —	— —	— —	— —	— —		— —	— —
AD	19	153·7 ± 63·8	194·7 ± 64·4	13	206·1 ± 71·7	213·6 ± 41·6	9	204·2 ± 76·5	197·5 ± 59·0	6	227·3 ± 186·8	199·8 ± 83·3	— —	— —	— —		— —	— —
TD	10	121·3 ± 84·2	206·8 ± 47·8	13	120 ± 54·7	188·9 ± 52·4	5	123·4 ± 37·9	143·6 ± 29·0	2	138·3 ± 10·9	167·0 ± 15·6		— —	— —		— —	— —
AAD	3	104 ± 57·9	175 ± 55·4		— —	— —		— —	— —		— —	— —	— —	— —	— —	— —	— —	— —

ND = ½ Nubian, ½ Criollo.
NND = ¾ Nubian, ¼ Criollo.
TD = ½ Toggenburg, ½ Criollo.
AD = ½ Alpine, ½ Criollo.
AAD = ¾ Alpine, ¼ Criollo.

the tropics. While production of about 250 kg in 210 days of lactation seems to be an average, much higher production of up to 1800 kg has been reported under very good management. Tables XI and XII show results of some more recent trials with autochthonous and imported breeds in the dry tropics. Production of criollos and crosses with European breeds by lactation number as registered in the Venezuela experiments is given in Table XIII. The experiments show that dairy breeds from other geographical latitudes can adapt to conditions in the dry tropics and that milk production increases in crossbreds with native breeds. Application of these possibilities in large-scale breed improvement programmes requires government action. Examples of this kind of action may be the programme initiated in Venezuela in 1965 and where at present, in four centres, sires mainly of the Anglo-Nubian and French Alpine breed are multiplied and sold to individual breeders. In northern Mexico, there is one big breeding centre supplying stock to surrounding goat keepers and nine large dairy operations with a total of 7000 does (Juárez, 1972).

2. Meat Production

Goat meat in many tropical areas is highly valued, and in some areas its contribution to the daily diet is higher than that of beef or mutton (Devendra, 1971). Preference for goat meat has been reported from Ghana (Jollans, 1959; Bose, 1963), India and Jamaica (Devendra and Chenost, 1973). The small size of the goat is an advantage in tropical areas, since the carcass may be used by the family or any surplus readily sold without need of refrigeration.

In Latin America most of the surplus kids are marketed at a very young age as *"cabrito"*. Raised on milk only, until 30 to 60 days of age in Argentina, they reach a weight of 6 to 8 kg with 3 to 5 kg carcass. Offtake is estimated at about 20%. Farm gate producer price per head in 1979 was about the equivalent of US \$10–12. Consumer prices were US \$3·10 to 3·80 per kg. Skins sold at about US \$2·0 and the abomasus (for rennet) at 1·50 (Maubecín, 1979). In northern Mexico *cabritos* are slaughtered at 30 to 40 days of age. Prices vary between the equivalent of US \$3·60 and 9·60, the highest prices being obtained between September and November when there is little on offer, because between March and May reproductive activity of goats is low (Carrera, 1971). Goat carcass prices in 1978 were the equivalent of US \$2·03 per kg, 13% higher than that of beef. Offtake rate was 27% (Juárez, 1979). In Brazil, goats are slaughtered at about 15 months of age and a live weight of 20 kg with 60% carcass yield (Nogueira Padilha, 1979). Birth weight of kids is important both for sales of young kids and meat production from older kids. McDowell and Bove (1977) have compiled data from tropical and subtropical areas. Small breeds, like the West African or

Table XIV. Least square means of birth weights of five European breeds in the Breeding Centre Tlahualilo, Mexico (Montaldo *et al.*, 1979).

	(*n*)	Birth weight (*kg*)	Weight of dam (*kg*)
Breeds			
Saanen	89	3·59	50·1
French Alpine	241	3·50	49·3
Toggenburg	87	3·43	47·7
Anglo-Nubian	62	3·14	53·1
Granadina	98	2·64	47·8
Sex			
males	289	3·42	
females	288	3·10	
Type of birth			
twins	477	3·37	
triplets and more	100	3·15	

Table XV. Growth rate of kids in dry tropical countries.

Breed	Country	Weight of dam (kg)	Average daily gain (g)	Until age	Source
Criollo	Mexico	39·6	65	30 d	Carrera (1971)
			68	90 d	
Criollo	Guadeloupe	20	48	1 y	Devendra and Chenost (1973)
Criollo	Guadeloupe	25	70/55	2 m	Cognie *et al.* (1971)
Anglo-Nubian	Peru	—	150/132[b]	12 wk	Vélez-Nauer *et al.* (1977)
East African	Zambia	24	64–66	12 wk	Quartermain (1975)
			32–46	24 wk	
Boer	South Africa	—	132	3 m[a]	South Africa, 1977
Boer	Botswana	—	90	4 m[a]	Botswana, 1977
Boer × Tswana	Botswana	—	88	4 m[a]	Botswana, 1977
Tswana	Botswana	—	85	4 m[a]	Botswana, 1977
Barbari	India	—	88/79[b]	2 m	Mittal and Pandey (1974)
Jamnapari	India	—	82/94[b]	2 m	Mittal and Pandey (1974)
Barbari, Jamnapari	India	—	45	14 m	Singh and Sengar (1970)
Malabari	India	—	38	4 m	Mukundan (1976)
Alpine × Malabari	India	—	66	4 m	Mukundan (1976)
Saanen × Malabari	India	—	41	4 m	Mukundan (1976)
West African Dwarf	Ghana	24·8	31	1 y	Šáda and Vohradský (1973)
West African Dwarf	Nigeria	—	84	12 wk	Kirkpatrick and Akindele (1974)
Feral goats	Australia	25–30	168	3 m[a]	Restall, 1976

[a] at weaning.
[b] males and females respectively.

Black Bengal, average at birth less than 1·5 kg while large meaty breeds like the Jamnapari or Boer may average up to 4 kg. Deviations of birth weight due to litter size, sex and age of dam may be about 0·3 and 1·0 kg respectively. Breed crosses may show about 5% heterosis (Montemurro, 1966; Carrera and de la Parra, 1971; Castillo *et al.*, 1972a). Juárez and Montaldo (1979) reported variations of birth weight by breed, sex and type of birth in five breeds kept under intensive conditions in northern Mexico (Table XIV). They estimated h^2 of 0·53 ± 0·24. Weights of kids born during the more favourable season may exceed those born during or after the difficult seasons (Singh *et al.*, 1977; Mittal, 1979). Mittal (1979) calculated correlation coefficients between birth weight of kids and weight of the dam ranging from 0·699 to 0·998. He observed a lower birth weight in dams kidding below 2 years of age which was probably due to the weight relation, but weights did not decrease with does up to 6 years of age.

Growth rates of goats in tropical countries have been compiled by McDowell and Bove (1977). In the dry tropics growth rate is generally low, ranging between 45 and 70 g, with body weight of dams between 24 and 40 kg (Table XV).

Table XVI. Meat production in Boer goats (Ueckerman, 1969).

	Roughage 100%		Roughage: conc. 60 : 40		Roughage: conc. 40 : 60	
Slaughter weight (lb)	70	100	70	100	70	100
n	231	336	182	308	168	301
weeks to slaughter weight	33	48	26	44	24	43
feed utilization	10·3	13·2	8·1	9·5	8·6	9·5
average daily weight gain	122	124	155	136	168	139
dressing %	49·4	52·4	51·8	56·3	54·6	56·5

Carcass yield of goats varies between 43 and 53% which is in the order of that of other ruminants. In criollo goats of northern Peru, Vélez-Nauer and Callacná (1979) recorded carcass weights in males of 6–10 months, females of 8–12 months, cull goats and cull bucks as 10·3 kg, 11·2 kg, 16·0 kg and 24 kg respectively.

While goats of all breeds are used in some way for meat production, there are only a few specialized meat breeds. The Boer goat of South Africa is a fairly big breed, with does weighing 50 to 60 kg, and males 75–90 kg. Data from an experiment on growth rate and carcass yield are given in Table XVI. A prerequisite for rapid growth of the kids is high milk yield in the dams, which has been reported in the Boer goat as up to 1·8 kg per day, during the first 120 days of lactation (Skinner, 1972). Another breed with good meat production aptitude is the Jamnapari of India. Devendra and Burns (1970) report carcass yields of 44 to 45% in 12-month-old males, and Johri and Talapatra (1971) 46·9% in males and females of between 9 and 18 months. Jamnapari have been used for crossing with native goats in the West Indies, Malaysia and Indonesia. Live weight at 1 year of age of these crossbreds exceeded those of Anglo-Nubian crosses with native goats (Devendra and Burns, 1970). Anglo-Nubians have been successfully used in Venezuela for improving meat production. Castillo *et al.* (1972a) report that live weights of crossbreds until 1 year of age exceeded criollos under the same conditions. However, the same results were obtained when crossing criollos with Toggenburg (Castillo *et al.*, 1972) and French Alpine (Castillo *et al.*, 1972b).

Meat production of goats in the dry tropics has some potential and there is urgent need for investigations on the technical aspects, like possibilities of intensive fattening, reaction to supplementary feeding, the effect of castration on growth and carcass composition, stocking rates of pastures for adequate weight gains, growth potential of breeds, breed improvement, and on socio-economic aspects such as market demands, marketing, consumer preferences in urban areas, prices and rentability of specialized fattening operations.

L. Conclusions

Goats are among the most important means of production in the dry tropics. Due to their ability to selectively use sparse and coarse feeds, often unpalatable to other livestock, and their tolerance to extreme environmental conditions, they may enable man to make productive use of some of the most uninhabitable areas of the world. The importance of goats in these areas is to be seen mainly in their contribution at the subsistence level. However, goat production in dry tropical areas, by applying rational production techniques, has the potential of developing into an industry, which may contribute considerable quantities of basic foods for the market. For developing goat production, governments have to carry out programmes of investigation, extension and investment. Their aim would be to utilize better and at the same time conserve the resources in semi-arid and arid areas, improve the socio-economic situation of the people living in these areas and contribute to the output of food. While improvement of management will have first priority, these measures will have to go hand in hand with deliberate breed improvement, considering that improving management cannot be economic if animals cannot respond, but that genetically productive animals may perform worse than natives, unless managed adequately.

References

Abdur, R., Muzahid, U. A. and Abdus, S. M. (1976). *Trop. Anim. Hlth Prod.* **8**, 90.

Agraz, G. A. A. (1971). "2nd Int. Conf. Goat Breeding", Tours, France, pp. 117–125. ITOVIC, Paris.

Ahmed, I. A. and Tantawy, A. O. (1960). *Emp. J. Exp. Agric.* **28**, 104–108.

Ali, S. Z., Hoque, M. M. and Hasnath, M. A. (1973). *Indian Vet. J.* **50**, 438–440.

Amble, V. N., Khandekar, N. C. and Garg, J. N. (1964). *I.C.A.R. Res. Ser.*, No. 38, New Delhi.

Apleman, R. D. and Delouche, J. C. (1958). *J. Anim. Sci.* **17**, 326–335.

Arbeiter, K. (1963). *Zuchthyg. Fort. Pfl. Stör. Besam. Haustiere* **7**, 349–362.

Ashmawy, G. M. (1978/79). *Z. Tierzücht. ZüchtBiol.* **95**, 269–276.

Bautista, E. O. and Berrizbeitia, H. (1974). *Esc. Zoot. Jusepin*—Venezuela, 84 pp.

Bhatnagar, D. S., Sharma, R. C. and Mishra, R. (1976). Crossbreeding of Indian goats with exotic breeds. *In* "India, National Dairy Research Institute", pp. 137–139, Karnal, Annual report, 1974.

Borut, A., Dm'iel, R. and Shkolnik, A. (1979). *Physiol. Zool.* **52**, 105–112.

Bose, A. B. (1963). *Agric. Res. New Delhi* **3**, 56–57.

Botswana (1977). Goats, pp. 104–105. *In* Ministry of Agriculture Animal Production Research Unit. An integrated programme of beef cattle and range research in Botswana, 1970–1977, Gaborone.

Bourbouze, A. and Guessous, F. (1977). "Symposium on Goat Breeding in Mediterranean Countries", Malaga-Grenade-Murcia, Spain, Oct. 1977, pp. 207–227. EAAP and Spanish National Committee Animal Production, Madrid.

Braun, Armelle (1979). *Ceres* No. 69, 19–24.

Bravo, J., García, B. O., Osal, N. and Arangú, M. (1976). *Est. Exp. El Cuji*, FONAIAP, Barquisimeto, Venezuela, 10 pp.

Carrasco, M. (1976). "IV Sem. Nac. Ovinos Caprinos Coro", Venezuela, 15 pp.

Carrera, C. (1968). "Int. Symp. Increasing Food Prod. Arid Lands", Monterrey, Mexico, pp. 219–224.

Carrera, C. (1970). "XII Inf. Invest". ITESM, Monterrey, Mexico, pp. 168–169.

Carrera, C. (1971). Dept. Zoot., Div. Cienc. Agropec. Marit, ITESM Monterrey, Mexico.

Carrera, C. and Aguirre, E. (1970). "XII Inf. Invest". ITESM, Monterrey, Mexico, pp. 166–168.

Carrera, C. and Flores, R. (1970). "XII Inf. Invest". ITESM, Monterrey, Mexico, pp. 157–158.

Carrera, C. and Hernandez, E. (1969/70). "XII Inf. Invest". ITESM, Monterrey, Mexico, p. 162.

Carrera, C. and Juárez, J. L. (1968). II World Conf. Anim. Prod. Univ. Maryland, College Park, Washington, D.C. USA.

Carrera, C. and de la Parra, L. (1971). "XI Inf. Invest". Escuela Agric. Ganaderia, ITESM, Monterrey, Mexico, p. 154.

Carrera, C. and Sevilla, A. (1971). "XII Inf. Invest". Escuela Agric. Ganaderia, ITESM, Monterrey, Mexico, 1969–70, pp. 152–155.

Carrera, C. and Robles, L. (1970). "XII Inf. Invest". Escuela Agric. Ganaderia, ITESM, Monterrey, Mexico, pp. 158–159.

Carrera, C., Mendizabal, F. and Maltos, J. (1969). *J. Anim. Sci.* **29**, 154.
Castillo, J. (1966). Min. Agric. Cria, Caracas, Venezuela, 130 pp.
Castillo, J. (1970). *Est. Exp. El Cuji*, FONAIAP, Barquisimeto, Venezuela, 11 pp.
Castillo, J. (1971). "II Int. Conf. Goat Breeding", Tours, France, pp. 163–167.
Castillo, J. (1974). Prog. Nac. Inv. Ovinos Caprinos Min. Agric. Cría, Venezuela, Bol. Inf. No. 5, 34–42.
Castillo, J., García, O. and Osal, N. (1972a). *Agric. Trop.* **22**, 251–259.
Castillo, J., García, O., Osal, N. and Arangú, M. (1972b). Prog. Nac. Inv. Ovinos Caprinos Min. Agric. Cría, Venezuela, Bol. Inf. No. 2, 8–14.
Castillo, J., González, S. C. and García, B. O. (1972c). "II Sem. Nac. Ovinos Caprinos", Maracaibo, Venezuela.
Castillo, J., García, O. and Figueroa, L. (1973). Prog. Nac. Inv. Ovinos Caprinos Min. Agric. Cría, Venezuela, Bol. Inf. No. 3, 1–7.
Chosniak, I. Shkolnik, A. (1978). *In* "Osmotic and Volume Regulation", Alfred Benzon Symp. XI, Minksgaard, pp. 344–359.
Choueiri, E. (1973). III World Conf. Anim. Prod. Pre-Conf. Vol. No. 1., Short Contributions, Melbourne, Australia.
Cognie, Y., Houix, Y. and Logeay, B. (1971). "II. Int. Conf. Goat Breeding" Tours, France, pp. 345–350. ITOVIC, Paris.
Colombia, Republica de (1963). Conferencia Latinoamericana para el estudio de las zonas aridas. Informe nacional. Min. de Gobierno, Div. Terr. Nac. Bogotá, Colombia, 111 pp.
Devendra, C. (1971). *Z. Tierzücht. ZüchtBiol.* **88**, 69–78.
Devendra, C. and Burns, M. (1970). "Goat Production in the Tropics". Tech. Comm. No. 19, CBABG, Edinburgh, Commonwealth Agricultural Bureaux Farnham Royal, Bucks, England, 184 pp.
Devendra, C. and Chenost, M. (1973). *Z. Tierzücht. ZüchtBiol.* **90**, 83–93.
Diego, R. (1969). Thesis, Esc. Agric. Ganaderia, ITESM, Monterrey, Mexico.
Dm'iel, R., Robertshaw, D. and Chosniak, I. (1979). *Physiol. Zool.* **52**, 558–564.
Draz, O. (1969). FAO Document PL:PFC/13, 10 June 1969.
Dunson, W. A. (1974). *Am. J. Physiol.* **226**, 662–669.
Edwards, L. C. (1948). *E. Afr. Agric. J.* **13**, 221–223.
Epstein, H. and Herz, A. (1964). *J. Agric. Sci.* **61**, 237–244.
FAO (1980). 1979 FAO Production Yearbook Vol. 33, FAO Statistics Series No. 28, 309 pp.
Finch, V. A., Dmi'el, R., Boxman, R., Shkolnik, A. and Taylor, C. R. (1980). *Physiol. Zool.* **53**, 19–25.
Fuenmayor, C., Garcıa, C., Mazzarri, G. and Castillo, J. (1972). Prog. Nac. Invest. Ovinos Caprinos Min. Agric. Cría, Venezuela, Bol. Inf. No. 2, 1–7.
Galal, E. S. E. and Beyene Kebede (1977). Inst. Agr. Res., Sheep and Goat Studies, Addis Ababa, *Anim. Prod. Bull.* **3**.
Gall, C. (1979). Vet. Fac. Univ. Munich, unpublished data.
García, O. and González, S. C. (1972). "II Sem. Nac. Ovinos Caprinos", Maracaibo, Venezuela.
García, O. and Peraza, F. (1976). "IV Sem. Nac. Ovinos Caprinos". Coro, Venezuela.
García, O., Castillo, J. and Gado, C. (1972a). *Agric. Trop.* **22**, 239–250.
García, O., Castillo, M. J., Peraza, F. and Osal, N. (1972b). Prog. Nac. Invest. Ovinos Caprinos Min. Agric. Cría, Venezuela, Bol. Inf. No. 3, 15–23.
García, O., Castillo, J. J., Osal, N. and Peraza, F. (1972c). Prog. Nac. Invest. Ovinos Caprinos Min. Agric. Cría, Venezuela, Bol. Inf. No. 1, 23–39.

García, O., González, S. C. and Castillo, M. J. (1972d) "II Sem. Nac. Ovinos Caprinos." Maracaibo, Venezuela.
García, O., Castillo, M. J., Verde, O. and Osal, N. (1976b). *Asoc. Latinoam. Prod. Anim.* Vol. II, pp. 27.
García, O., García, B. E., Arangú, M. and Camacaro, A. (1977). "Symposium on Goat Breeding in Mediterranean Countries", Malaga-Grenade-Murcia, Spain, Oct. 1977. EAAP and Spanish National Committee Animal Production, Madrid, pp. 80–89.
Gill, G. S. and Dev, D. S. (1972). *Ind. J. Anim. Prod.* **3**, 173–178.
González, J. E., Capó, E. and Trujillo, B. (1966). "VI Jornadas Agronómicas." Maracaibo, Venezuela.
González, S. C. (1972). "II Sem. Nac. Ovinos Caprinos." Maracaibo, Venezuela.
González, S. C. (1975). "V Reunión Latinoam. Prod. Anim." Maracay, Venezuela, 30 Nov.–5 Dec.
González, S. C. (1977). "Symposium on Goat Breeding Mediterranean Countries", Malaga-Grenade-Murcia, Spain, Oct. 1977. EAAP and Spanish National Committee Animal Production, Madrid, pp. 322–326.
González, S. C., García, O. and Castillo, M. J. (1974). *Cienc. Vet.* **4**, 223–248.
González, S. C., García, O. and Castillo, M. J. (1976). "IV Sem. Nac. Ovinos Caprinos." Coro, Venezuela.
Haumesser, J. B. (1975). *Rev. Elev. Med. Vet. Pays Trop.* **28**, 225–234.
Hernandez, E. (1969). Thesis, Esc. Agric. Ganaderia, ITESM, Monterrey, Mexico.
Hills, E. S. (1966). *In* "Arid Lands, a Geographical Appraisal" (E. S. Hills, ed.). Methuen, London; UNESCO, Paris.
Hiroe, K. and Tomizuka, T. (1965). *Bull. Nat. Inst. Anim. Ind.* No. 8, 17–24.
Hofmeyr, H. S. (1962). M.Sc. (Agric.) Thesis, Univ. Pretoria.
Hofmeyr, H. S., Joubert, D. M., Badenhorst, F. J. G., D. Steyn and G. J. van (1966). *Proc. S. Afr. Soc. Anim. Prod.* **5**, 134–136.
Horst, P. (1976). *Anim. Res. Develop.* **4**, 70–86.
Hueck, K. (1966). "Die Wälder Südamerikas." G. Fischer, Stuttgart.
Huss, D. L. (1972). *In* "Wildland Shrubs, their Biology and Utilization", Int. Symp. Utah State Univ. Logan, Utah, July 1971. USDA Forest Service, General Tech. Report INT-1, 331–342.
Huston, J. E. (1978). *J. Dairy Sci.* **61**, 988–993.
Israel (1962). Min. Agric. Hakirya, B.P. 7011, Tel-Aviv.
Jardin, W. R. (1965). *Revta. Agric. Piracicaba* **40**, 131–147.
Jardin, W. R. (1971). "II Int. Conf. Goat Breeding", Tours, France, pp. 163–167. ITOVIC, Paris.
Jenkinson, D.Mc. and Robertshaw, D. (1971). *J. Physiol. Lond.* **212**, 455–465.
Johnson, K. G. (1971). *J. Agric. Sci. Camb.* **77**, 267–272.
Johnson, K. G. (1976). *In* "Effect of Temperature on animals: Progress in Biometereology" (H. Johnson, ed.), Vol. I, pp. 140–147. Swets and Zeitlinger, Amsterdam.
Johri, G. B. and Talapatra, S. K. (1971). *Indian Vet. J.* **48**, 602–604.
Jollans, J. L. (1959). *J. W. Afr. Sci. Ass.* **5**, 64–78.
Jordan, S. M. (1957). *Kenya E. Afr. Agric. J.* **23**, 84–88.
Joshi, B. C., Aravindan, M., Singh, K. and Bahattacharyya, N. K. (1977). *Indian J. Anim. Sci.* **47**, 200–203.
Joubert, D. M. (1973). *Z. Tierzücht. Zücht Biol.* **90**, 245–262.
Juárez, L. A. (1972). IV Congr. Nac. Med. Vet. Zootéc. Cuernavaca, Mexico.

Juárez, L. A. (1974). *México Ganadero* No. 194.
Juárez, L. A. (1979). "VII Reunión Latinoam. Prod. Anim., Sem. Prod. Caprina en Medios Dificiles de America Latina". Panama, 25 September, 12 pp.
Juárez, L. A. and Vázquez, S. E. (1972). IV Congr. Nac. Med. Vet. Zootéc., Cuernavaca, México.
Khan, M., Ghosh, P. K. and Sasidharan, T. O. (1978) *J. Agric. Sci.* **91**, 395–398.
Khan, M. S., Sasidharan, T. O. and Ghosh, P. K. (1979). *J. Agric. Sci.* **93**, 247–248.
Khatter, S. S. and Mishra, R. R. (1977). *Indian J. Anim. Sci.* **47**, 846–847.
Kirkpatrick, R. L. and Akindele, Z. T. (1974). *J. Anim. Sci.* **39**, 163.
Knight, J. (1965). *E. Afr. Agric. Forest J.* **30**, 182–188.
Laor, M. (1971). "II. Int. Conf. Goat Breeding". Tours, France, pp. 114–116.
Le Houérou, H. N. (1975). *Org. Rech. Agron. Cent. Volcani.* Bet-Dagan. Israel. Public. Spec. **39**, 45–57.
Le Houérou, H. N. and Hoste, C. H. (1977). *J. Range Mgt* **30**, 181–189.
Leight, E. G. (1965). Thesis, Univ. Católica Lima, Perú.
Malechek, J. C. and Leinweber, C. L. (1972). *J. Range Mgt* **25**, 105–111.
Maloiy, G. M. O. (1973). *Z. Tierzücht. Zücht Biol.* **90**, 219–228.
Maloiy, G. M. O. and Taylor, C. R. (1971). *J. Agric. Sci. Camb.* **77**, 203–208.
Maltz, E. and Shkolnik, A. (1980). *Physiol. Zool.* **53**, 12–18.
Marincowitz, G. (1962). *S. Afr. Tyd. Landbouwet.* **5**, 211–223.
Martínez, J., Durand, G., Castillo, J. and García, O. (1972). Prog. Nac. Inv. Ovinos Caprinos Min. Agric. Cría, Venezuela, Bol. Inf. No. 2, 24–32.
Maubecin, R. A. (1979). "VII Reunión Latinoam. Prod. Anim., Sem. Prod. Caprina en Medios Dificiles de America Latina". ALPA, Panama.
Maule, J. P. (1966). *Anim. Breed. Abstr.* **34**, 153–158.
McDowell, R. E. and Lea Bove (1977). Cornell Int. Mimeograph 56. Dept. Anim. Sci. New York State Coll. Agric. Life Sci. Cornell Univ., Ithaca, New York, 40 pp.
Mena, L. H. and Escamilla, H. R. J. (1977). "XV INF. Invest". (1975/76), ITESM, Monterrey-México, pp. 130–131.
Minett, F. C. (1960). *Indian J. Vet. Sci.* **20**, 69–103.
Mishra, H. R. and Biswas, S. C. (1966). *J. Dairy Sci.* **19**, 132–134.
Mishra, R. K., Nivsarkar, A. E. and Arora, C. L. (1979). *Indian J. Anim. Sci.* **49**, 967–968.
Mittal, J. P. (1976). *Indian Vet. J.* **53**, 681–684.
Mittal, J. P. (1979). *Indian J. Anim. Sci.* **49**, 45–47.
Mittal, J. P. and Pandey, M. D. (1975). *Agra Univ. J. Res.* **23**, 67–70.
Montaldo, H., Juárez, A., Forat, M., Berruecos, J. M. and Villareal, M. (1979). 70th Annual Meeting, American Soc. Animal Science. Michigan State University, July 1978.
Montemurro, O. (1966). *Annali Fac. Agr. Univ. Bari* **20**, 1–17.
Moulick, S. K., Guha, H., Gupta, S., Mitra, D. K., Bhattacharya, S. (1966). *Indian J. Vet. Sci.* **36**, 154.
Mukundan, G. (1976). "Proc. 2nd Workshop on All India Coordinated Research Project on Goat Breeding". NDRI Karnal, 22–23 March.
Mukundan, G. and Rajagopalan, T. G. (1971). *Kerala J. Vet. Sci.* **2**, 95–98.
Nogueira Padilha, Terezinha (1979). "VII Reunión Latinoam. Prod. Anim., Sem. Prod. Caprina en Medios Dificiles de America Latina". ALPA, Panama.
Nolte, E. (1977). "Symposium on Goat Breeding in Mediterranean Countries", Malaga-Grenade-Murcia, Spain, Oct. 1977, EAAP and Spanish National Committee Animal Production, Madrid, pp. 327–328.

Nolte, M. E. and Arauzo B., O. R. (1973). Univ. Nac. Agr. La Molina. Lima, Perú, 167 pp.
Ohya, M. (1964). *Jap. J. Vet. Res.* **12**, 23.
Otchere, E. O. and Nimo, M. C. (1978) *Ghana J. Agric. Sci* **8**, 187–190.
Prakesh, C., Acharya, R. M. and Dhillon, J. S. (1971). *Indian J. Anim. Sci.* **41**, 356–360.
Prasad, S. P. and Bhattacharyya, N. K. (1979). *Indian J. Anim. Sci.* **49**, 969–970.
Prasad, S. P., Roy, A. and Pandey, M. (1972). *Agra Univ. J. Res. Sci.* **20**, 31–39.
Quartermain, A. R. (1975). "Proc. III World Conf. Anim. Prod.", pp. 102–106. Sidney University Press, Melbourne.
Quartermain, A. R. and Broadbent, M. P. (1974). *E. Afr. Agric. Forest. J.* **40**, 115–124.
Rai, A. K., Mehta, B. S., Devesh Gour and Manohar Singh (1979). *Indian J. Anim. Sci.* **49**, 546–548.
Raja, C. A. R. and Mukundan, G. (1974). *Kerala J. Vet. Sci.* **4**, 165–169.
Ranatunga, P. (1971). *Trop. Anim. Hlth Prod.* **3**, 52–61.
Restall, B. J. (1976). "2èmes Journèes de la Recherche Ovine et Caprine", 1–2 December, pp. 155–165. INRA, ITOVIC, Paris, France.
Robertshaw, D. (1968). *J. Physiol.* **198**, 531–539.
Robinet, A. H. (1967). *Rev. Elev. Med. Vet. Pays Trop.* **20**, 129–186.
Sachdeva, K. K., Sengar, O. P. S., Singh, S. N. and Lindahl, I. L. (1973). *J. Agric. Sci., U.K.*, **80** (3), 375–379.
Sacker, G. and Trail, J. (1966). *Trop. Agric. Trin.* **43**, 43–51.
Šáda, I. and Vohrandský, F. (1973). *Sbornik Vysoké Školy Zemědělske v Praze*, Institut Tropického a Subtropického Zemédélstvi No. 6, 173–185.
Sahni, K. L. and Roy, A. (1967). *Indian J. Vet. Sci.* **37**, 269–276.
Sanfiorenzo, J. H. (1957). *Univ. Pto Rico Agric. Sta. Bull.* **139**, 37 pp.
Sanfiorenzo, J. H. (1962). *Univ. Pto Rico Agric. Sta. Bull.* **46**, 205–212.
Santisteban, M. E., Morales, M. A., Hernandez, A. and Naus (1977). "Symposium on Goat Breeding in Mediterranean Countries", Malaga-Grenade-Murcia, Spain, Oct. 1977. EAAP and Spanish National Committee Animal Production, Madrid, pp. 310–314.
Sarup, R. (1971). "II. Int. Conf. Goat Breeding", Tours, France, pp. 97–104.
Sattari, M. (1971). "II. Int. Conf. Goat Breeding", Tours, France, pp. 105–113.
Schoen, A. (1968). *E. Afr. Agric. Forest. J.* **34**, 256–262.
Shalash, M., Mousa, A., Nawito, M., Farra, J. F., Off, F., Selim, M. K. and Tawfik, M. A. (1970). *Egypt. Vet. Med. J.* **18**, 295–312.
Shkolnik, A., Borut, A. and Choshniak, J. (1972). *Symp. Zool. Soc. Lond.* No. 31, 229–242.
Silanikove, N., Shkolnik, A. and Tagari, H. (1979). *Israel J. Zool.* **28**, 60.
Singh, S. N. and Sengar, O. P. S. (1970). RBS College, Bichpuri, Agra, India. Dept. of Animal Husbandry and Dairying. Technical Rep.
Singh, B. B. and Singh, B. P. (1974). *Indian Vet. J.* **51**, 316–322.
Singh, C. S. P., Mishra, H. R., Singh, S. K., Mukherji, D. K. and Prasad, B. (1977). *Indian J. Anim. Sci.* **47**, 592–594.
Skinner, J. D. (1972). *Trop. Anim. Hlth Prod.* **4**, 120–127.
Staples, R. and Hornby, R. M. (1962). *E. Afr. Agric. J.* **8**, 62–70.
Sorror, B. H. M. (1973). Vet. Thesis, Cairo Univ., cited by Ashmawy (1978/79).
South Africa (1977). Production potential of Boer goats. *In* "Ann. Rpt. Secr. Agr. Tech. Serv.", 1975–1976, p. 96.

Tantawy, A. O. and Ahmed, I. A. (1960). *Emp. J. Exp. Agric. 28*, 74–82.
Taylor, C. R. (1966). *Physiol. Zool.* **39**, 127–139.
Tulenbekov, I. (1961). Izvestija Akademii nauk Kazachskoj SSR. *Serija mediciny i fiziologii*, Alma-Ata Nr. 1, 83–90.
Ueckermann, L. (1969). M. Sc. Agr. Thesis, Univ. Pretoria.
UNESCO (1958). El programa de la Unesco, para las tierras áridas. *Man. Inf. Unesco* No. 3, 32 pp.
Rensburg, S. J. van (1964). *Onderstepoort J. Vet. Res.* **31**, 97–106.
Rensburg, S. J. van (1971). *J.S. Afr. Vet. Med. Ass.* **42**, 305–308.
Vélez-Nauer, M. and Callacná, M. A. (1979). *Entw. ländl. Raum* **13** (1), 14–16.
Vélez-Nauer, M., Cordero Bonilla, T. and Callacná Custodio, M. (1977). "Symposium on Goat Breeding in Mediterranean Countries", Malaga-Grenade-Murcia, Spain, Oct. 1977, EAAP and Spanish National Committee Animal Production, Madrid, pp. 349–353.
Venezuela (1975). Ministerio de Agricultura y Cría de Venezuela, Anuario Estadistico Agropecuario M.A.C., Caracas.
Vohradský, F. and Šáda, I. (1973). *Sbornik Vysoké Skoly Zemedelske v Praze*, Institut Tropického a Subtropického Zemedelstvi, No. 6, 161–172.
Wallen, C. C. (1966). *In* "Arid Lands: A Geographical Appraisal" (E. S. Hills, ed.), 461 pp. Methuen, London; UNESCO, Paris.
Welch, J. G., Clark, D. B. and Rutledge, J. J. (1975). *J. Anim. Sci* **41**, 432.
Wijeratne, W. V. S. (1968). *Trop. Agric. Trin.* **45**, 39–45.
Wilson, L. L., Katsigianis, T. S., Dorsett, A. A., Cathopoulis, T. E., Greaves, A. G. and Baylor, J. E. (1980) *Trop. Agric.* **57**, 183–190.
Wilson, P. N. (1957). *E. Afr. Agric. Forest. J.* **23**, 138–147.

17. The Goat in the Humid Tropics

C. Devendra

A. Introduction

Goats are important animals in the humid tropics. In many countries, they outnumber sheep, and their importance is reflected in a variety of functions. This contribution is real in terms of the products of economic significance. Since the distribution of goats and their functional importance vary greatly according to environmental conditions, the humid tropics are considered separately. In this chapter, a comprehensive review is presented on the status of goats, their distribution and importance, recent advances and gaps in our knowledge, and the potential for increasing the contribution from them.

B. The Humid Tropics

It is useful in discussing goats in the humid tropics to keep in mind a definition of the humid tropics. The humid tropics are not well defined, and classifications vary as to the exact definition and extent of climates that might be described as humid and tropical. Humid zones occur in the tropical regions of Africa, South and Central America, Asia, Australia and many island groups throughout the world. The climate varies with latitude, altitude, winds and rainfall. However, the variability of daily and seasonal temperature is small, and day length is almost constant throughout the area. The annual rainfall exceeds the annual loss of water by evaporation. Distribution of tropical areas over the continents and population densities are given in Table I. Blair (1942) has classified the climates within the humid tropics and has distinguished three subtypes: equatorial, monsoon and trade wind. It is estimated that the humid tropical climate extends over approximately 12% of the world's land area (Table II).

Table I. Estimates of continental distribution of a humid tropical climate and population density (FAO, 1976).

Continent	As % of total humid tropics	No. of ha land per person
Africa	20	5·8
Americas	50	7·3
Asia	27	1·2
Oceania	3	14·6

Table II. Estimates of the extent and the population density of the humid tropics (FAO, 1976).

Subtype	As % of world land area	Ha land per person
Equatorial	6·5	4·0
Monsoon	5·0	0·8
Trade wind	0·7	6·5
Humid tropics	12·2	3·8

1. Equatorial

This climate is found between 5 and 7° latitude north and south of the equator. It is characterized by constant heat, rainfall and humidity, an absence of seasonal rhythm and little variation in day length. The mean annual temperature varies around 27°C, and normal daily temperatures vary between 22 and 30°C, with extremes (18 to 35°C) being the exception rather than the rule. Total annual rainfall is within the range 2030 to 3050 mm and is mainly convectional in origin. Humidity is usually high and one or two wet seasons are common.

2. Monsoon

This is a transitional climate, occurring between the equatorial and the wet and dry tropical climate or savannah over an area located 5 to 15° latitude north and south of the equator, except in Asia where it may extend beyond 15° latitude north. The main characteristic of this climate is seasonal difference in rainfall, and usually three seasons are recognizable: cool-dry, hot-dry and hot-wet.

The dry season is usually short. The mean annual temperature varies around 27°C, and total annual rainfall is always above 1500 mm, and often higher than in the equatorial zone. In general, rainfall decreases with increasing distance from the equator. Humidity is high for most of the year.

3. Trade Wind

This is also known as the windward coast or oceanic humid. It occurs on the east coast of continents and on the windward side of mountainous oceanic islands situated 15 to 22° north and south of the equator. Seasonal minimum temperatures are lower than in the equatorial or monsoon zones. Continuous trade winds or local sea breezes alleviate the effect of the humid heat. The main annual temperature is usually below 27°C, and total annual rainfall is always greater than 2030 mm.

Table III. The distribution of goats in the humid tropics (FAO, 1976).

Subtype	Goats (millions)	%
Equatorial	88·4	36·7
Monsoon	106·8	44·3
Trade wind	45·7	19·0
Humid tropics	240·9[a]	100·0

[a] Represents 58·4% of the total world population of 412·6 million goats.

Table IV. Goat equivalents in relation to other livestock in the humid tropics (millions) (FAO, 1976).

Subtype	Goats	Sheep	Cattle	Buffaloes	Total ruminant units[a]
Equatorial	8·8	7·4	182·2	9·1	207·5
Monsoon	10·7	6·6	207·6	85·5	310·4
Trade wind	4·6	5·5	65·1	3·9	79·1
Humid tropics	24·1	19·5	454·9	98·5	597·0
As % of total ruminant units	4·0	3·3	76·2	16·5	100·0

[a] buffalo = 1·0, cattle = 0·8, sheep, goat = 0·1 unit.

C. Population and Production of Goats

Goats are widely distributed in the humid tropics. They account for 73·6% of the total goat population of 327·4 millions in the tropics and subtropics, and 58·4% of the total world population in 1976 of 412·6 million goats (Table III), calculated from FAO (1976). These calculations are for those countries whose climates are classified as humid according to the definition above.

If the population of goats is compared with other livestock in the humid tropics on the basis of ruminant units, it can be seen that they account for about 4% (Table IV).

Most of the goat breeds in the humid tropics, using the classification of Devendra and Burns (1970) on the basis of height at withers, are small (including dwarfs) in size. This is a reflection of their adaptation to this type of environment and poor nutrition. The precise advantage of smaller size is not clear, but it is probably associated with slow growth and a low metabolic rate. In South-East Asia for example, the small Kambing Katjang is widely distributed (Fig. 1).

Fig. 1. The Kambing Katjang, for example, is numerically the most important breed in Malaysia and Indonesia and elsewhere in South-East Asia. The breed is prolifc and twinning is common.

Goats are mainly kept in the humid tropics for subsistence production; there is very little commercial production. Table V presents the estimated productivity of meat, milk and fresh skins by goats in the humid tropics.

D. Ownership

The majority of goats are kept by smallholders or small farmers in rural and rural-fringe areas.

Goats kept on farm households represent 80% of the total goat population in the Philippines (Rumich, 1968).

Table V. Productivity of meat, milk and fresh skins by goats in the humid tropics (10^3 tonnes) (FAO, 1976).

Subtype	Meat	Milk	Fresh skins
Equatorial	256	321	60 482
Monsoon	397	1 854	95 038
Trade wind	146	538	28 237
Humid tropics	799	2 713	183 757
As % of the total	48·6	38·9	58·7

Table VI. Average flock size of goats and sheep owned by small farmers in the South-East Asian region (Devendra, 1979a).

Country	Average size of land area (ha)	Size of flock	
		Goats	Sheep
Indonesia	< 1·0	4–5	3–5
Malaysia	< 1·0	3–4	2–5
Philippines	1–3	4–8	2·0
Thailand	2–4	4–9	3–8

Flock sizes of 2 to 10 goats are common in most countries of South-East Asia (Devendra, 1966b; Devendra and Nozawa, 1976; Table VI).

Raut and Nadkarni (1974) have reported that in Himachal Pradesh, India, the average sizes of migratory and stationary goat flocks were 17 to 74 and 6 to 7 respectively. In the West Indies, ownership of small flocks of goats and sheep is the rule rather than the exception in most of the territories. An example refers to the recent survey in Guyana which revealed that 15 out of 20 goat farms had flock sizes of up to 30 and 5 farms between 31 and 70 goats (Devendra, 1975a). This pattern is similar to that of several countries in the humid regions of South-East Asia and West Africa.

E. Systems of Production

Goats are reared alone or with sheep by small farmers, and this pattern is evident for example, in several countries in South-East Asia and the West Indies. Co-grazing with sheep and or large ruminants has definite advantages, perhaps the most important one of which is effective utilization of the feed resources. Additionally, goats are often used in the humid tropics as leaders of the mixed flock for grazing.

The systems of goat and sheep production can be grouped into four categories:

(a) tethering;

(b) extensive production;
(c) intensive production;
(d) integration into crop agriculture.

The first three systems have been adequately described (Devendra and Burns, 1970). Of these, tethering and extensive production are by far the most common.

1. Tethering

In this system, goats are pegged to a rope about 3 m in length. By shifting the peg once daily, free access to a fresh grazing area is provided. Water is provided when the goats are shifted to a shelter at night. Very little or no concentrates, salt or mineral bricks are provided. Occasionally supplements, household scraps, small quantities of grains or their by-products are given. Tethering is a common practice in South-East Asia, Central America, the Caribbean and elsewhere. The main reasons for tethering are control over the animals to limit damage to crops and *in situ* feeding of crop residues. It is also associated with limiting labour input since farmers are busy with crop cultivation.

2. Extensive Production

In the extensive system, goats graze over large areas or marginal lands which

Fig. 2. Extensive production of goats is very common in the humid tropics. Here, mixed breeds of goats, mainly Kamori type, are shown in Pakistan.

are unsuited for alternative agricultural use. A very low level of unpaid labour represents the main input. Extensive production is probably the most popular system of goat production (Fig. 2). In Guyana, for example, 48% of sheep and goats are managed extensively (Devendra, 1975a). Small flocks of about 5 to 10 animals are common, let loose to browse and graze on waste vegetation, hedges waysides and on marginal land. Often, unpaid

Fig. 3. Extensive grazing is associated with browsing behaviour. Browsing is important in the feeding and nutrition of goats in the humid tropics as the photograph (in India) demonstrates.

family labour including children help in herding goats and sheep together. Very little management is exercised save letting them loose and shutting them in at night. Probably because of cheap family labour and higher returns from this system of management, the flock sizes tend to be relatively large compared to those in the intensive method. For larger herds, usually some form of paid labour is employed. Both tethering and the extensive systems embrace the traditional village system typical in Africa, parts of Central

America and South-East Asia. Goats browse and scavenge on what feeds are immediately available near the farm and the households, for example cassava root peelings and banana skins (Fig. 3).

In the Caribbean, extensive production is perhaps the most traditional method of rearing goats (Devendra, 1971a). Two types of management systems are recognized:

(a) small-scale production of about 5 to 50 goats in urban-fringe areas;

(b) large-scale production of about 50 to 400 goats in rural-fringe areas.

In rice growing countries such as in Thailand, Indonesia, Malaysia and Philippines, goats often graze the stubble in between rice harvests. In Indonesia and Malaysia, it is common to cut leaves of ipil ipil (*Leucaena latislliqua* (L) Gillis), cassava (*Manihot esculenta* Crantz), Jackfruit (*Artocarpus heterophyllys*) or Hibiscus (*Hibiscus Rosasinensis*) to feed goats. Feeding of cut tree leaves is also common in the West Indies. Limited stall feeding of cut grass is provided during shelter. No conservation of forage either of grass or silage is practised.

Nomadism and transhumance are extensive systems of goat production for example in India and West Africa (see Chapter 18).

3. Intensive Production

The intensive system is one in which the goats are fed in confinement with limited access to land. It is by definition a system with a high labour and cash input. In this system cultivated grasses and or by-products are fed *in situ*. Goats may be fed for example, with cut Napier grass (*Pennisetum purpureum*) or Guinea grass (*Panicum maximum*) with or without limited concentrates. Alternatively, both species can be grazed on sown pasture. For Kambing Katjang goats in Malaysia, daily live weight gains of 26 g for Guinea grass *ad libitum* and 76 g for Guinea grass plus concentrates have been recorded (Devendra, 1976b).

In Jamaica, intensive goat production on Pangola grass (*Digitaria decumbens* Stent) with a carrying capacity of 37 to 45 goats per hectare has been demonstrated (Devendra, 1971b). However, intensive production with goats on cultivated fodder is not common.

An avenue that merits exploitation on an intensive basis, is the feeding of agro-industrial by-products like rice straw which is abundantly available in monsoon Asia with molasses-urea supplementation. With meat goats, this system of feeding also has the advantage of control over them. Dairy goats on the other hand are usually stall-fed. A variation of intensive production based on stall feeding is feed lot fattening. This method appears to have had little application with goats, but has been used successfully with sheep. Very much more use can be made of this system especially in environments where agro-industrial by-products are plentiful and easily accessible.

4. Integration into Crop Agriculture

This system can be considered to be either extensive, intensive or semi-intensive, but the fact that it is integrated with crop agriculture justifies a separate identity. Integration of goats into rubber, oil palm and coconut cultivation has been tried, although perhaps not on a large scale, in Sri Lanka, Malaysia, Indonesia and the Philippines.

Some idea of the yield of dry matter and the crude protein content of the herbage undergrowth mainly grass, weeds and legumes in mature rubber and coconut estates in Malaysia is given in Table VII. These results were derived from several sub-samples taken in rubber and coconut estates. Lee *et al.* (1978), for example, have discussed the performance of goats and sheep under rubber in Malaysia. In Fiji, goat raising is entirely in the hands of the Indian settlers, and about 70% of about 67 000 goats are found in areas growing sugarcane. This integration could probably be exploited much more (Devendra, 1978a). The success of this integration depends on a judicious adjustment of stocking rate to available dry matter, and the time of introduction of the animals, avoiding adverse effect on the crop and measuring returns through the crop and animal integration. The benefits of this integration include:

(i) increased soil fertility by return of dung and urine;
(ii) control of waste herbage growth;
(iii) reduced loss of nutrients supplied by fertilizers;
(iv) easier management of the crop;
(v) possibilities of increased crop yield due to (i), (ii) and (iii);
(vi) greater economic returns from both the crop and animal components.

Table VII. Approximate yield and composition of uncultivated undergrowth in mature rubber and coconut plantations in Malaysia.

Plantation	Yield of dry matter (kg/ha)	Crude protein content (%)[a]	Approximate composition
Rubber	480–500	14–16	82% grasses, 14% vines, 3% ferns and 1% shrubs
Coconuts	800–1200	8–12	70–80% grasses, 20–30% shrubs and ferns

[a] On dry matter basis.

F. Level of Production

The present level of productivity of goats in the humid tropics is in general low, and is due to a combination of under-feeding, diseases (especially parasitism), poor husbandry and climatic factors. Of these nutrition is by far the most important. In considering the feeding of goats and effective utilization of feed resources, three objectives are essential:

(a) full exploitation of goats;
(b) a realistic appraisal of the potential level of productivity;
(c) efficient utilization of the available feedstuffs in a system which clearly identifies production and profitability.

Table VIII. Milk yield of various types of goats in tropical and temperate environments (Devendra, 1975b).

Type of goat	Milk yield (kg)	Lactation length (days)	Av. daily yield (kg)
Indigenous (in the tropics)	60–500	126–283	0·5–2·5
Exotic[a] (in the tropics)	119–886	106–344	0·6–3·4
Exotic[a] (in temperate environment)	245–2707	180–730	1·4–4·1

[a] Refers to European breeds.

The evidence for undernutrition and low productivity in goats is seen, for example, in the inferior performance of imported dairy goats in the Caribbean (Chenost and Geoffrey, 1971) where the average fat content was 6 to 7% lower than that recorded for the same breed in France. Devendra (1972a) reported 1·0 to 3·2% lower fat content of Anglo-Nubian goats in Trinidad than in England. The reduced fat content has been attributed to a low intake of dry matter. Low milk yield in West African dwarf goats has also been attributed to poor nutrition (Akinsoyinu *et al.*, 1977).

One other constraint to production of goats is poor genetic quality, as shown in Table VIII. The data, with the exception of that from temperate environments, refer mainly to the humid tropics. Tropical breeds are poorest yielders, while goats in temperate zones can produce twice as much milk or more. Exotic dairy breeds in the tropics appear to be intermediate in milk producing ability. Parallel data for a low level of meat production due to poor feeding and management in rural areas have also been recently reported (Devendra, 1981).

For increased production from goats, improving management must be combined with selective breeding of promising autochthonous breeds and also controlled use of exotic breeds.

G. Breeds

There is a variety of breeds in the humid tropics. The more important of these, especially those which merit breeding and selection, and which can be used further in breeding programmes, have been identified and described (Fig. 4) (Devendra and Burns, 1970). Exploiting goats fully necessitates the recognition of important production characteristics. While the tropics are an important reservoir of well adapted goat breeds, their potential value is not well known. Also, due to uncontrolled breeding, a wide variety of genotypes is found in different environments. The goats of humid South-East Asia for example, including Taiwan, Japan, Korea and the Pacific islands are very diverse, and almost impossible to classify into breed groups for lack of descriptive data. Some distinct breeds and/or types are identifiable, but it is difficult to generalize about their physical characteristics. Nevertheless, certain features are distinct: a definite predominance of black colour especially over brown and white, scimitar-shaped horns, little differentiation between meat and milch breeds. The phylogeny of South-East Asian goats, domestication and productivity have been examined by Devendra and Nozawa (1976).

Fig. 4. Four outstanding breeds in India, from left to right: Jamnapari, Beetal, Barbari and the Black Bengal. India possesses a valuable reservoir of important goat breeds that can be used as improver breeds. Small size is typical of most breeds in the humid tropics.

Fig. 5. The Anglo-Nubian is the most successful European breed in the humid tropics. The photograph illustrates Anglo-Nubian does in Trinidad where daily yields of 1–2 kg have been recorded.

The more important improver breeds that are likely to make a greater contribution within tropical and subtropical regions are listed in Table IX. Anglo-Nubian goats have been imported into several countries in the humid regions (Fig. 5), and Jamnapari goats have been successfully used to improve Kambing Katjang goats in Indonesia and Malaysia.

H. Nutrition and Feeding

1. Dry Matter Intake

The dry matter intake of goats varies in relation to body function. Meat goats have a dry matter intake between 2·5 and 3·0% of live weight. While milch goats may consume up to 8·0% of live weight, in the humid tropics lower intakes are more common. In Trinidad, a dry matter intake of between 5 to 6% of live weight for imported goats was found under experimental conditions (Devendra, 1970).

2. Digestive Efficiency

Grass growth in the humid tropics is very rapid and at maturity fibre content is high (30 to 35% in the dry matter). Many of the roughage residues also have a high fibre content. It has been suggested that one reason for the ability of goats to adapt to tropical conditions is a high digestive efficiency for cellulose (Devendra, 1971c). Data concerning the digestive efficiency of goats compared to other ruminants mainly from the humid tropics but also from the arid and temperate regions have been reviewed (Devendra, 1978b). Of 20 sets of comparisons of goats with one or more species especially sheep or cattle, 12 indicated statistically significant differences in digestibility of dry matter, organic matter, crude fibre and crude protein in favour of goats. Differences of crude fibre digestibility ranged from 3·7 to 29·1% digestibility units. Three of the studies demonstrated that fibre digestibility was higher with poor quality roughages. One implication of higher digestibility of crude fibre is that the metabolizable energy (ME) value in feeds is affected, and it is feasible that goats derive more energy from these feeds than other ruminants. On high fibre diets, therefore, there is a need for separate digestibility coefficients for goats (Devendra, 1978b).

Table IX. Suggested improver breeds (Devendra, 1974a).

Speciality	Breed	Country of origin
High milk yields	Saanen[a]	Temperate, subtropical, wet
	Anglo-Nubian[a]	Temperate, tropical, dry
Medium yields	Jamnapari	Tropical, subtropical, dry
	Barbari	Tropical, dry
	Beetal	Tropical, dry
	Jamnapari	Tropical, subtropical, dry
	Ma T'ou	Subtropical, humid
	Kambing Katjang	Tropical, humid
	Fijian	Tropical, humid
Prolificacy	Black Bengal	Tropical, dry
	Barbari	Tropical, dry
	Malabar[a]	Tropical, humid
	Ma T'ou	Subtropical, humid
Mohair	Angora	Subtropical, humid
Skin	Black Bengal	Tropical, dry

[a] Indicates breed is polled.

3. Nutrient Requirements

(a) *Energy*

The energy requirements for goats in the tropics were first determined in Malaysia (Devendra, 1967a). It was found that daily maintenance requirement was 0·42 ± 0·01 kg digestible organic matter (DOM) which is equivalent to about 6·40 MJ for animals with a live weight of 45·4 kg, similar to the requirement of sheep. This requirement, however, will be considerably higher for goats in outdoor conditions in view of their browsing habits and greater muscular activity. No experimental evidence is available for this, but it is suggested that this value is about 40 to 50% higher than the maintenance requirement for energy of penned goats.

For live weight gain, few data exist of the energy requirements. It has been found in Malaysia (Devendra, 1976b) that for Kambing Katjang goats in the weight range 18 to 26 kg, the cost per kg live weight gain was 2·7 kg DOM. Above 20 kg, these goats were in a fattening phase so that this value is comparable to the figure of 2·5 kg DOM which is applied to sheep. Singh (1969) found that kids of the Barbari and Jamnapari breeds in India grew faster and produced more edible meat when receiving 100 and 125% of the total digestible nutrient (TDN) standard recommended for sheep by Morrison (1956). A similar trend concerning reproductive performance has been reported by Sachdeva *et al.* (1973). Oyenuga and Akinsoyinu (1976) have reported that the cost per kg live weight gain of goats was 21·52 MJ. These findings (Devendra, 1967b; Oyenuga and Akinsoyinu, 1976), are presented in Table X, showing the energy requirements for maintenance and growth.

Table X. The energy and protein requirements for maintenance and growth.

Live weight (kg)	Maintenance (per day)		Requirements for growth (per kg live weight gain)	
	ME (MJ)	D.C.P. (g)	ME (MJ)	D.C.P. (g)
20	6·40	16	21·52	80
30	8·60	24	25·82	85
40	10·02	32	33·51	90
50	12·46	40	36·62	95
60	14·15	48	39·66	100

(b) *Protein*

The protein requirement for maintenance of tropical goats is approximately 0·06 to 0·08% of body weight (Devendra and Burns, 1970). These values are comparable to the recommendation for sheep (A.R.C., 1965). Sachdeva *et al.* (1973) reported that 100 and 125% of the Morrison (1956) recommendation for protein, gave the best reproductive performance. Oyenuga and Akinsoyinu (1976) have reported that the protein requirement of goats weighing 20 kg live weight and gaining 270 g/day was 97·5 g/day. For goats gaining about 150–200 g/day, which is typical of most goats in the humid tropics, the protein requirements have been calculated (Table X). No critical experiments have been reported for the protein requirements during lactation, but Devendra (1970) has established that with milk fat content ranging from 3·5 to 5·5% in goats, the D.C.P. requirements varied from 47 to 73 g per kg of milk.

(c) *Minerals and Vitamins*

Very little is known about mineral and vitamin requirements. Calcium and phosphorus are often in low supply in tropical forages, for example in Caribbean feedstuffs (Devendra, 1977) and deficiency of these elements is apparent in many parts of South-East Asia. Majumdar (1956) has reported that the maintenance requirement for calcium and phosphorus of goats weighing about 45 kg is 6·7 and 3·3 g respectively. It has been estimated that about 13·2 g of sodium chloride is required by each kg of milk product.

I. Diseases

Diseases are one of the main constraints to goat production in the humid tropics. A high incidence of disease is associated with alternating wet and dry seasons as well as the lack of diagnostic and veterinary services. Contagious ecthyma, navel infections, metabolic diseases in the dam and death during parturition cause serious economic losses. In Bangladesh 82·4% of 214 kids born died up to 6 months of age, of which respiratory disorders, gastrointestinal parasitism, and contagious ecthyma were the main causes (Abdur Rahman *et al.*, 1976). Gastrointestinal parasitism is rampant in the humid tropics and the data from Bangladesh also revealed that in adults, 47·8% of 115 deaths were due to this. The most severe effects of diseases and parasites however are inferior performance and infertility. The losses tend to be increased when nutrition is inadequate.

J. Gaps in Current Knowledge

While goats are valuable animals in the humid tropics, knowledge about them is limited. A diversity of breeds exist, but not enough is known about their production potential and the genetic traits that lend themselves to selective breeding. Perhaps more than anywhere else the necessary ingredients for development exist in the tropics: goat breeds, feed resources and cheap labour. Research is needed on target ecological zones, production systems, animal health, breeding and genetics, increasing feed resources and nutritional management, socio-cultural factors and marketing systems (Devendra, 1981). The area that requires much more data is feeding and management. For example, nothing is known presently about the energy requirements for grazing. Associated with this are problems of thermoregulation in goats: effects of humidity and temperature on size, water balance and that of the hormonal system on performance.

Goats are mainly kept at the subsistence level, but little is known about the magnitude of their contribution to subsistence agriculture and human nutrition. Since goats in the tropics are primarily important for meat production, aspects of meat quality are also important.

An area that urgently needs understanding is the place of goats in the total farming enterprise in the humid zones and the advantages or limitations of different production systems. For example, are goats better suited than sheep or cattle or buffaloes to avail themselves of the limited resources? If so, is this consistent with higher biological and economic efficiencies, demand for products, cost–benefit analysis and market outlets? Research on goats in the humid tropics aimed at increasing productivity has the challenging task of alleviating the living standards of the poor people in these areas.

References

Abdur Rahman, Ahmed, M. U. and Mia, A. S. (1976). *Trop. Anim. Hlth Prod.* **8**, 90.

Akinsoyinu, A. O., Mba, A. U. and Olubajo, F. O. (1977). *J. Dairy Res.* **44**, 57–63.

A.R.C. (1965). "The nutrient Requirements of Farm Livestock", No. 2 "Ruminants". Agricultural Research Council, Her Majesty's Stationery Office, London.

Blair, T. A. (1942). "Climatology, General and Regional". Prentice-Hall, New York.

Chenost, M. and Geoffrey, F. (1971). "Proc. 2nd Int. Conf. on Goat Breeding", 17–19th July, 1971, Tours, France, pp. 241–250.

Devendra, C. (1966a). *Z. Tierzücht. Zücht. Biol.* **83**, 72–79.

Devendra, C. (1966b). *Malays. Agric. J.* **45**, 268–274.

Devendra, C. (1967a). *Malays. Agric. J.* **46**, 80–97.

Devendra, C. (1967b). *Malays. Agric. J.* **46**, 98–118.

Devendra, C. (1970). "1st Nat. Sem. Goat Prod.", Barquisemeto, Venezuela, 11–15th Nov., 1970, 11 pp.

Devendra, C. (1971a). "Proc. 2nd Int. Conf. on Goat Breeding", 17–19th July, 1971, Tours, France, pp. 47–54.

Devendra, C. (1971b). *Z. Tierzücht. Zücht. Biol.* **88**, 69–78.

Devendra, C. (1971c). *Trop. Sci.* **13**, 123–132.

Devendra, C. (1972a). *J. Dairy Res.* **39**, 381–385.

Devendra, C. (1974a). *Z. Tierzücht. Zücht. Biol.* **91**, 246–255.

Devendra, C. (1974b). *Span* **17**, 130–132.

Devendra, C. (1975a). *Z. Tierzücht. Zücht. Biol.* **92**, 305–317.

Devendra, C. (1975b). "Proc. Symp. on Bridging the Dairy Gap", Bull. No. 140, Ministry of Agriculture and Rural Development, Kuala Lumpur, pp. 98–111.

Devendra, C. (1976b). Rpt. F. A. O. Expert Group meeting on Livestock Programmes for Small Farmers and Agriculture Labourers in Asia and the Far East, Vol. II, 31st May to 6th June 1976, Bangkok, Thailand, pp. 176–210.

Devendra, C. (1977). *Trop. Agric. Trin.* **54**, 29–38.

Devendra, C. (1978a). "Proc. Integration of Animals with Plantation Crops", 17–19th April, 1978, Penang.

Devendra, C. (1978b). *Wrld Rev. Anim. Prod.* **4** (1), 9–22.

Devendra, C. (1979). *Wld Anim. Rev.* (FAO) **32**, 33–41.
Devendra, C. (1981). *J. Anim. Sci.* **51**, 461–473.
Devendra, C. and Burns, M. (1970). "Goat Production in the Tropics", Tech. Comm. No. 19, Commonw. Bur. Anim. Breed, Genet., Commonwealth Agricultural Bureau, xii + 184 pp.
Devendra, C. and Nozawa, K. (1976). *Z. Tierzücht. Zücht. Biol.* **93**, 101–120.
FAO (1976). "Production Yearbook", Vol. 30, FAO, Rome, 296 pp.
Lee, K. A., Wan Mohd. Wan Embong, Ng, L. H., Ng, F. C. and Phang, A. K. (1978). "Proc. Integration of Animals with Plantation Crops", 17–19th April, 1978, Penang, 13 pp.
Majumdar, B. N. (1956). *In* "Handbook of Animal Husbandry", ICAR, p. 75.
Morrison, F. B. (1956). "Feeds and Feeding" (22nd Edn). Morrison Publishing Co., New York, USA.
Oyenuga, V. A. and Akinsoyinu, A. O. (1976). "1st Int. Symp. on Feed Composition, Animal Nutrient Requirements and Computerization of Diets", Utah, USA, July 11–16th 1976, pp. 505–511.
Raut, R. C. and Nadkarni, U. G. (1974). *Indian J. Anim. Sci.* **44**, 459–665.
Rumich, B. (1968). "The Goats of Indonesia". FAO Regional Office, Bangkok, Thailand (Mimeograph, 22 pp.).
Sachdeva, K. K., Sengar, O. P. S., Singh, S. N. and Lindahl, I. L. (1973). *J. Agric. Sci. Camb.* **80**, 375–379.
Singh, S. N. (1969). Research Progress Rpt., PL. 480 Project A-7-AH-18, ARS, USDA.

18. Socio-Economic Importance of Goat Production

C. Devendra

A. Introduction

Of all the domestic animals of value to man, goats have a particular socio-economic importance. This value goes back several centuries, and is consistent with the fact that next to the dog, goats were the first animals to have been domesticated by man (Zeuner, 1963). There are two aspects to their importance: firstly, to meet man's social, cultural and also recreational needs, and secondly, to provide a supply of meat, milk, fibre and skins. The latter is particularly important in subsistence agriculture in the developing countries. Farmers earn income by the sale of animals for meat, milk and milk products and skins. In addition, the miscellaneous functions of goats also have economic implications. In this chapter we discuss the socio-economic importance of goats in temperate and more particularly in tropical and subtropical regions.

B. The Place of Goats in Farming Systems

1. Temperate Regions

Traditionally, the main purpose of goat-keeping was producing milk to supplement family requirements but also to generate some income. More recently, the trend in some countries is towards commercial milk and cheese production, for example in France, Israel, Italy (Sardinia), Spain, Bulgaria and Australia where goats are kept secondary to the main enterprise or maintained exclusively.

2. Tropical Regions

In many parts of the tropics, goats and other livestock are associated with cropping systems. Livestock are reared for meat, milk, eggs or draft purposes. The socio-economic status of livestock can be demonstrated with data from Malaysia (Table I). Because of the small farm sizes, however, small animals (sheep, goats and poultry) fit better into these systems than cattle who are more demanding of resources. Agriculture in the humid tropics, for example Asia, is characterized by small crop farmers owning a few head of one or more kinds of farm animal. It has been estimated that in Asia, out of a total human population of 2061 million in 1973, 1154 million or 80% of the agricultural population depended for their livelihood on 2 ha or less with an income of US $50, barely sustaining life (FAO, 1975). The characteristics of these small farmers are (Devendra, 1979):

(a) they are crop-oriented subsistence peasants;
(b) they may own one or two cattle or buffaloes but most of them keep sheep or goats or both;
(c) for them, rearing stock is a means of converting crop residues into useful animal products and represents, therefore, an important component of efficiency in the use of existing resources;
(d) landless peasants may grow crops or may be employed for shepherding goats.

While the socio-economic benefits of goat-keeping especially in subsistence agriculture are widely recognized, increasing them is hampered by a variety of constraints such as small size of farms, land alienation policy, poverty of peasants, poor husbandry management and low productivity.

C. Contribution of Goats to Employment, National Income and Gross National Product

Table II demonstrates the contribution of goats to employment, national

Table I. Farmers' response to propositions pertaining to the socio-economic status of livestock rearing in Malaysia[a] (Lai *et al.*, 1975).

Propositions suggested	% response	
	Agree	Disagree
1. Farmers keeping livestock are wealthy people	62·4	37·6
2. Livestock is an important source of farm income	95·8	4·2
3. Livestock cannot be a losing enterprise	98·1	1·9
4. Livestock is necessary to supply meat during festive occasions	83·1	16·9

[a] Based on a total of 425 respondents.

Table II. The contribution of sheep and goats to employment, national income and gross domestic product

Region	Full-time equivalent economically active population dependent on sheep and goats		% contribution to agriculture	Contribution of sheep and goats to Gross Domestic Product	Per capita national income	National income accounted for by sheep and goats	
	No. (10^3)	(%)	(%)	(%)	($ US)	Total ($ US, 10^6)	Per capita ($)
Africa	3133	2·48	3·43	1·20	127	412	1·52
Latin America	41	0·41	1·10	0·17	550	250	0·94
Near East	4006	6·48	11·16	1·12	342	670	3·83
South East Asia	1944	0·47	0·72	0·27	100	307	0·27

Source: 1976 Statistical Yearbook, New York, 1977.

income and gross national product (GNP) in the developing countries. Since under extensive systems of production, goats and sheep are often managed together, both are included in Table II. The full-time equivalent of economically active population dependent on goats and sheep was derived as a product of number and percentage of the population economically active in agriculture (Devendra, 1981). The result indicated that as many as 10 million full-time economically active people were dependent on goats and sheep, notably in the Near East, Africa and South-East Asia. These are equivalent to about 3·5, 1·1 and 0·7%, respectively, of the total labour force in the region. The real number of people actually engaged in rearing these animals is much higher than these figures suggest, especially including part-time engagement of peasant farmers and landless agricultural labourers. In India alone, the latter comprise some 100 million people. The landless agricultural labourers include both nomadic and transhumant societies, who practise shifting agriculture. Rearing goats also creates some employment for landless agricultural labourers.

Some people who keep goats depend for their entire livelihood on them. Examples of this situation are seen in the West Indies, Venezuela, Mexico, the Sahel, Iran, Afghanistan, Pakistan, Nepal and India. In the semi-arid zones of Mexico for example, goat production is the only source of income of about a million people. Likewise in the Sahelian zone, where much of the land is not arable, many millions of people depend on grazing livestock, of which goats are an important part.

The GDP data indicate that Africa is first, followed by the Near East, suggesting that these regions are most dependent on goats and sheep. The contribution of goats and sheep to total and per capita national income was derived by multiplying the total and per capita income by the percentage contribution to agriculture in GDP. This calculation indicates that the Near East was highest, followed by Africa, Latin America and South-East Asia. The differences in per capita incomes are quite marked, and that of the Near East is 14 times higher than that of South-East Asia suggesting that the latter constitutes a very significant target group of the poorest of the poor (Devendra, 1981).

D. Attributes of Goats

To understand the socio-economic contribution of goats, it is essential to bear in mind some of their important attributes. Goats have a higher reproductive rate and higher digestive efficiency for cellulose than sheep (Matthewman, 1977). High digestive efficiency for cellulose and inquisitive feeding behaviour (Devendra, 1978) are particularly significant in areas where the vegetation is coarse such as in many semi-arid and arid regions.

Efficiency of milk and meat production are calculated in Table III. These values are approximate since efficiency varies with the rate of production. From calculations of published data it has been clearly demonstrated that dairy goats with high genetic merit such as the Saanen, British or French Alpine breeds are quite superior to cows, buffaloes and sheep (Devendra, 1975b). This point confirms previous calculations of yield per unit live weight in goats (Devendra and Burns, 1970). The differences were found to be of the order of 8 to 28% higher compared to cows or buffaloes. Spedding (1968) has also reported a higher efficiency of milk production in the goat (185 kg/100 kg DOM) compared to the cow (162 kg) and the ewe (35 kg) in a temperate environment. Gross efficiency of milk production of West African dwarf goats of 18·2% has been reported (Akinsoyinu, 1975) compared to 16·8% for West African tropical sheep (Adu *et al.*, 1974). These considerations have been discussed for the tropics as a whole (Devendra and Burns, 1970) and more recently for the Asian region (Devendra, 1976a).

Goats adapt to various environments and to dehydration and disease in Africa; certain breeds of goats, for example the West African dwarf goat, have a high tolerance to Trypanosomiasis.

Table III. Approximate efficiencies of energy and protein conversion in goats (Devendra, 1976a).

Goats	% Efficiency		Energy cost of protein (g/Mcal ME)[c]
	Energy[a]	Protein[b]	
Milk production (lactation)	24·0	23·7	14·5
Meat production (fattening)			
on grass	4·7	9·1	5·1
on grass + concentrates	6·7	10·2	7·5

[a] Energy expressed as n.c. kcal/100 kcal of metabolizable feed energy consumed.
[b] Proteins are expressed as edible protein per 100 g of feed protein consumed.
[c] Edible protein per Mcal ME.

1. Body Size

Small size is probably the most important goat characteristic for small farmers and landless agricultural labourers. While it is similar to sheep, there are other features in which goats excel. Small size has five important implications:

(a) *Biological:* due to their size, individual goats may be kept on small plots of land and on marginal land. A small number of goats also implies that there are less problems when feed shortages occur. Because of their smaller size, goats are nutritionally more economical than cows (Devendra, 1975c). They require much less and much cheaper food than cows.

Fig. 1. Farmer with his flock of goats on his way home in the evening after grazing in northern India.

(b) *Managerial:* goats occupy little, if any, housing space and are easily managed, they can be cared for by women and children, as is evident in Africa and the countries in South-East Asia. Unpaid family labour therefore forms an important component of goat management (Fig. 1) and skilled labour is unimportant.

(c) *Economic:* with goats there is low investment and small risk of loss by individual death compared to, say, cattle or buffaloes, fast turnover of capital due to early maturity and short generation interval, low labour requirement and efficient use of labour. In the village economy meat and milk are supplied in quantities suitable for immediate family consumption, obviating problems of storage in the tropics. The same applies to sales at local markets where in many countries in South-East Asia, central America and the West Indies relatively high prices for goat meat are an incentive for the sale of surplus males.

In the desert areas, for example in the Sahel, nomadic families consider large numbers of livestock including goats as protection against total loss during droughts as some goats can be sold off quickly when the rains fail.

(d) *Social:* while some people are prejudiced against goat products there are no major religious taboos on goat meat and milk. In some instances goats' milk or cheeses are even preferred to other dairy products. In some countries, such as Ghana (Jollans, 1959), there is a preference for goat. Because of the small size and widespread acceptance of the meat, goats are used extensively in Hindu and Muslim slaughter. Goats are also valuable animals in sacrifice and in some instances the meat and milk is considered to have medicinal properties, for example anti-allergy properties.

(e) *Recreational:* Goats are also small companions for man. In Western countries, goats and kids are sometimes kept as pets, as prize animals to participate in animal shows. In some countries in South-East Asia, bucks are also used for fighting.

E. Socio-economic Aspects of Goat Management

The socio-economic importance of goats is closely associated with their management systems. These systems in turn are dictated largely by the

Fig. 2. Long-legged desert goats in Nigeria. These goats are well adapted to sparse grazing environments, are involved in nomadic and transhumant systems, and make an important socio-economic contribution to subsistence agriculture.

climatic conditions, the topography of the land, plant growth, water systems, presence of parasites such as flies, and social factors.

People keeping goats in the tropics can be classified into five groups:

(1) nomads;
(2) stock owners practising transhumance;
(3) mixed farmers with a relative large number of animals;
(4) crop farmers who own a few animals;
(5) landless agricultural labourers.

Most goats are owned by small farmers, peasants or landless labourers. The farmers are usually crop-oriented subsistence peasants with low incomes, living close to the bread line. They may or may not own cattle or buffaloes (usually not more than one or two head), but most small farmers invariably possess goats or sheep or both. Landless peasants may or may not grow crops or rear animals; they may be hired to shepherd goats for extensive grazing. Rearing goats is a source of investment and income with low risk and minimum management attention (Devendra, 1979). Nomadism and transhumance involve migration of camels, cattle, sheep and goats which is a dominant feature of livestock management especially in the Near East region. In the forest zones where there are very few animals, there is no herding, but in the desert and semi-desert areas, the migrations are spectacular and experienced every year.

1. Nomadism

Nomadism and transhumance, mainly in arid, semi-arid and steppe conditions, are characterized by extensive husbandry.

Many of these are threatened by desertification and Table IV gives an idea

Table IV. Estimates of human population and livelihoods resident in areas undergoing severe desertification (10^3) (United Nations, 1977).

Region	Total population	Urban based	Cropping based	Animal based	Area (km^2)
Mediterranean Basin	9820	2995 (31%)	5900 (60%)	925 (9%)	1 320 000
Sub-Saharan Africa	16 165	3072 (19%)	6014 (37%)	7079 (44%)	6 850 000
Asia and the Pacific	28 482	7740 (27%)	14 311 (54%)	6431 (19%)	4 361 000
Americas	24 079	7683 (32%)	13 417 (50%)	2979 (12%)	17 545 000
Total	78 542	21 490 (27%)	39 642 (51%)	17 414 (22%)	30 076 000

of the number of people in these areas depending on livestock. Management practices are the outcome of centuries of adaptation to peculiar and difficult physical environments; a limited number of crops can be grown, but goat and sheep-rearing is a principal livelihood. In the Near East poor water supply determines the traditional systems of grazing areas with sparse vegetation of mainly weeds.

During the long dry season, the goats are grazed close to water points and are watered either daily or every other day. The hamlets are separate and independent of the camel camps, and each family has its own separate pen for the goats and the sheep. In the Rift Valley, lack of grazing rather than lack of water is the motivation for nomadism. The Turkana nomads of this region divide their stock into animals that browse (camels and goats), and animals that graze (cattle) (Gulliver, 1955). In this system camels and goats are grazed in the plains during the dry season and shifted into the mountains with the onset of rains. Cattle are confined to the mountains during the dry season and moved to the plains to take advantage of better grazing with the first rains.

Animal husbandry in nomadism and transhumance is intimately linked to the social life of the people.

Nomadic livestock management has three features:

(i) *Herd diversification:* different species with their different grazing habits (for example, browsing by goats), reduce the probability of total loss of all animals.
(ii) *Loaning animals and sharing herds:* this enables sharing livestock with others in another area if there is a drought. Also social contacts are strengthened.
(iii) *Movement of herds:* this is an obvious strategy for survival and includes various types of migration: seasonal, short-distance or long-distance disaster migrations.

2. Transhumance

Transhumance or semi-nomadism, while it is also basically a migratory system, differs from nomadism in that usually it also involves some shifting arable cultivation in rain-fed areas or even sedentary systems in villages, rural-fringe areas or at oases during certain seasons of the year (Wilson and Clarke, 1975).

Several patterns of alternating sedentary crop cultivation with migration to grazing areas exist in different parts of the world, mainly in semi-arid areas or areas with extended dry periods:

—alternation between winter quarters in the plains and valleys and summer

mountain grazing areas; there may be an overlap between transumant and nomad people, the latter entering grazing grounds which are left by transhumants for summer grazing;
—winter grazing in desert areas, summer grazing in oases and irrigated cropping areas;
—grazing in plains and valleys which are left during the rainy season when they are flooded and sometimes cannot be used because of diseases risks (tse-tse, parasites).

These movements sometimes follow fixed annual routes. Sometimes the extent of movements from the cropping area is varied according to available feed.

Usually the grazing areas are public land and not individual property but between tribes and families there are traditional rights for grazing and use of watering points. However, because of increasing pressure on ranges from extending crop forest protection areas, these rights are less and less observed and range management is deteriorating. The resulting damage is often attributed to goats (see Chapters 7 and 16).

During the cropping season goats and other stock are herded by children or left to fend for themselves. Sometimes they are entrusted to hired shepherds.

3. Mixed Farmers with a Relatively Large Number of Animals

The mixed farmers are mainly transhumant people who, together with the migrations of goats, practise some form of crop cultivation. The pattern is not very different to shifting agriculture except that the livestock including goats are given as much priority as crop production and occasionally even primary attention.

The Sonjo in Tanzania (Gray, 1963) give particular attention to goats, and their system of goat husbandry is interesting. Goats are divided into two herds: one group in which young animals and does in milk are kept in the village and milked twice a day by the women, and another group comprising adult males and females kept in camps several kilometres away. From these camps the goats are taken by young boys to graze by day and brought back at dusk to the shelter to protect them from predators. Kids are kept in a separate pen and zero grazed by young men gathering branches of trees and green leaves. All males are castrated when 6 months old. Some crops like maize or cassava are grown and these supplement the income earned from goats, and also provide quite a substantial proportion of the food requirement of the people. Additionally, some of the crop residues from crop cultivation are fed to the goats.

In India, the farmers of Andhra Pradesh pay premiums for penning goats in their fields to increase soil fertility. Flock masters are paid 5 Rs per night for tethering 100 goats in the farmer's fields. During this time, a flock master is given three free meals a day by the farmer who benefits from the return of dung and urine. Following the crop harvests, goats are let loose to utilize the crop residues and fend for themselves. This practice continues till the rains set in, following which the crop farming operations commence once again. This system is evident with short-term crops in the West Indies, sorghum and wheat cultivation in Mexico, and the cultivation of rice throughout countries in South-East Africa. Very seldom, however, are the goats fattened or stall-fed with one or more of the crop residues with supplementary feed ingredients in a feeding system. The system of goats grazing cereal stubble land or other crop residues requires minimum labour since the goats are managed by children or unpaid family labour. The grazing of these stubbles is a common feature of livestock management, and complements crop-oriented agriculture.

4. Crop Farmers who own a few Animals

Farmers of this category are the exact opposite of the previous system in that crop production takes precedence over livestock husbandry. Accordingly, this system of livestock management with a few goats in the savannah grassland zones is apparent in the areas nearest to the forest where there are permanent sources of water for the crops. This system is widely practised in Africa, the West Indies, Central America and several countries in South-East Asia. In Cameroon for example, goats are maintained in herds of 20 to 30 as scavenger animals associated with agriculture. One distinct feature of this system is that the farmers do not manage the goats themselves. This is mainly because they attend to crop cultivation. The task is usually entrusted to children when the herd is small (up to about 30 goats), or to hired shepherds when the herds are larger (above 30 goats). During dry weather, the goats wander 4 to 6 km daily to graze, but during the rainy season they are grazed nearer the villages. The goats graze for about 8 hours and are returned to the villages at night.

In many parts of the West Indies, for example Jamaica and Guyana, the goats leave the homestead in the morning, fend for themselves and return on their own in the evening (Devendra, 1975a). When the herd sizes are larger, the goats may be returned to a common kraal for the night. Quite often the hired shepherds care for several herds at the same time so that the goats return to a common kraal. In this system goats also graze in coconut, oil palm or rubber plantations.

F. Economic Aspects of Rearing Goats

Under extensive conditions in the tropics goats make the most efficient use of marginal waste or land. They are comparable to buffaloes in that their value increases with decreasing quality of grazing (Devendra, 1972). Returns from rearing goats may be higher than those from other animals but the magnitude of this advantage will vary in relation to the land use pattern, availability of resources, products demand and market outlets. Quartermain (1973) has discussed the biological efficiency of goats in Zambia in subsistence agriculture and concluded that rearing of goats is a costless enterprise. Since goats are potentially an economic proposition, economic analyses of alternative systems of goat management in selected countries will be presented.

Table V. Model calculations of gross margins in 4 management systems in the humid tropics.

	Type of management system	Approximate gross margins ($ Malaysian)[a]
(A)	Grazing cultivated grass per flock/year	446·90
(B)	Grazing uncultivated grass per flock/year	510·00
(C)	Zero grazing cultivated grass per goat per ha/year	11·00
(D)	Growth and fattening on agricultural by-products	
	(i) value of live weight gain/goat/annum	76·65
	(ii) value of daily live weight gain/goat/annum	36·50

[a] 1 US $ = $2·5 (Malaysian) approximately.

(A) and (B): The calculations are specific to a herd of 10 breeding does, kidding percentage of 100%, 25% culling/annum and cost of cultivated grass 1·8 cents/kg fresh grass. No cost is attached to uncultivated grass and unpaid family labour.

(C): Details in Table VI.

(D): (i) fed on cultivated grass and rice straw/molasses-urea diet in the ratio of 3·1 total dry matter intake; and

(ii) fed exclusively on a rice straw/molasses-urea diet.

Table VI. Comparative economic advantage of goats and cattle on Guinea grass (*Panicum maximum*) in Malaysia (Devendra, 1976b).

Characteristic	Goats	Cattle
Carrying capacity (animals/ha/year)	45·6	5·6
Total daily live weight gain/ha (kg)	3·65	2·66
Value of the daily live weight gain/ha ($)[a]	10·95	6·12
Cost of feeding on grass + labour ($)[b]	1·05	0·73
Gross margin of profits/ha ($)	9·90	5·39
Economic advantage of goats ($):	4·51	

[a] Based on 1 kg live weight costing $3·00 (Malaysian) for goats and or $2·30 (Malaysian) for cattle.

[b] Cost at 1·8 cents/kg fresh weight, and an efficiency of feed conversion of 16 for both species.

1. Malaysia

A model calculation for four alternative systems has been done so as to provide some understanding of the influence of management on productivity and gross margins of return. Restricting the calculations to typical humid tropical conditions, only meat goats are considered in extensive and intensive systems as well as rearing and fattening on by-products integrated with rice cultivation. Calculations and results are summarized in Table V.

A comparative analysis of land utilization by goats and beef cattle has been attempted for conditions specific to Malaysia with zero grazing Guinea grass (*Panicum maximum*) (Table VI). The yield of Guinea grass is assumed as 14 tonnes of dry matter (D.M.) per hectare per year. The adult weights and daily live weight gains of local (crossbred) goats and Kedah-Kelantan cattle are 28 and 227 kg and 80 and 454 g respectively. The daily dry matter requirement for both species is 3% of live weight.

One important aspect of the efficiency of goats for land utilization is the proportion of total costs attributable to feed. Table VII demonstrates that with goats this component is relatively small in comparison to other ruminants.

Table VII. The proportion of total costs attributable to feed ruminants in Malaysia (Devendra, 1975b).

Species and production	Diet[a]	Feeding system	Feed cost as % of total cost
Goats			
Goat meat	(i) Grass	Grazing	2·8
Goat meat	(ii) Grass	Stall feeding	23·0
Milk	(iii) Grass + concentrate	Stall feeding	48·2
Sheep			
Mutton	(i) Grass	Grazing	2·8
	(ii) Grass	Stall feeding	21·0
	(iii) Grass + concentrate	Stall feeding	44·7
Cattle			
Beef	(i) Grass	Grazing	40·7
	(ii) Grass	Stall feeding	50·9
	(iii) Grass + concentrate	Stall feeding	66·0
Milk	(i) Grass + concentrate	Stall feeding limited grazing	59·6
	(ii) Grass + concentrate	Stall feeding	69·5
	(iii) Grass + concentrate	Grazing mainly	15·0
Buffaloes			
Milk	(i) Grass + concentrate	Grazing mainly + some concentrate	11·4

[a] Grass (i) for goats and sheep uncultivated. All others refer to cultivated grass.

2. India

In India, considerable attention has been given to the cost of rearing goats under varying management conditions. In Andhra Pradesh, for example, calculations on the returns from keeping goats suggest that this is very profitable (Sriramaimurthy, 1977). In Rajasthan, Jodha (1966) has reported that based on a 5-year analysis, the net annual income from keeping goats and sheep of a semi-nomadic family was 1600 Rs. The main component of this income was sale of wool and animals. Results of a survey on the economics of feeding and rearing practices of goats and sheep in hilly regions of Himachal Pradesh are presented in Table VIII. Some important conclusions can be drawn from this analysis. Firstly, labour was the main cost component in all systems and in the stationary system this was much higher than in the migratory system, but it was almost exclusively family labour.

Table VIII. Comparative income in terms of components of gross cost from rearing goats and sheep under two systems of management in India (Raut and Nadkarni, 1974).

Components of cost[a]	Migratory system				Stationary system			
	Mandi[b]		Mahsu[b]		Mandi		Mahasu	
	1[c]	2[c]	1	2	1	2	1	2
				GOATS				
Paid labour (%)	35·2	24·0	15·9	11·7	2·9	—	1·2	1·1
Unpaid labour (%)	37·7	36·3	58·2	59·0	83·5	79·7	60·6	66·5
Feed (%)	5·4	2·0	8·3	8·4	1·4	0·8	17·5	8·7
Depreciation (%)	1·4	0·8	1·8	1·4	1·4	7·9	0·7	1·3
Interest on capital (%)	20·3	36·9	15·8	19·5	10·8	11·6	20·0	22·4
Gross cost (Rs/year)[d]	8·2	7·7	18·8	18·3	13·5	18·9	15·4	10·4
Income (%)	72·7	28·7	18·1	11·8	32·4	23·5	23·5	29·4
				SHEEP				
Paid labour (%)	28·4	24·3	16·0	14·3	1·9	1·1	1·6	3·4
Unpaid labour (%)	44·4	39·1	63·6	61·8	86·8	92·2	68·2	75·8
Feed (%)	4·2	2·1	7·9	8·9	2·8	0·8	13·5	7·3
Depreciation (%)	0·8	0·5	1·3	0·9	1·4	1·4	1·8	2·4
Interest on capital (%)	22·2	34·0	11·2	14·1	7·1	4·6	14·9	11·1
Gross cost (Rs/year)	10·0	7·3	22·7	20·9	25·9	35·9	39·8	21·1
Income (%)	21·9	25·6	9·4	11·3	16·9	12·3	16·4	10·0

[a] As % of total cost of production.
[b] Two districts in Himachal Pradesh.
[c] First and second year respectively.
[d] 1 US \$ = 8 Rs (approximately).

Secondly, the feed cost accounted for a relatively low proportion, and was very similar for sheep and goats, migratory and stationary systems. This result is comparable to the figure of 2·8% for goats grazing uncultivated grass that has been reported in Malaysia (Devendra, 1975b). The third and most important conclusion is that the income derived from goats from both systems (11·8 to 72·7%) is substantially higher than that from sheep (10·0 to 25·6%). A higher income from goats was also evident from studies in Malaysia and the US Virgin islands (Park and Parn, 1974) where the average return per family was US $2·01/hour for goats compared to $1·55/hour for sheep.

Table IX. Annual family income from transhumant goat/sheep herding[a] (McDowell, 1976).

	PRs[b]	%
(i) *Annual income*		
Sale of goats	562	37·7
Sale of sheep	676	45·4
Sale of milk	92	6·2
Sale of wool	90	6·0
Sale of skins	42	2·8
Sale of hair	28	1·9
Total cash income	1490	100·0
(ii) *Value of family consumption*		
Meat	510	38·9
Milk	730	55·8
Wool	70	5·3
Total value of family consumption	1310	100·0
(iii) *Gross family income*	2800	
(iv) *Annual production costs*		
Grazing or feed purchase (4 PRs/head)	140	77·7
Veterinary supplies	40	22·3
Total production cost	180	100·0
(v) *Net family income*	2620	

[a] Refers to a transhumant shepherd family owning 10 does and 15 ewes.
[b] Pakistan rupees. 1 US $ = 9·6 PRs.

3. Pakistan

The third study concerns the income to be derived from the transhumant system of rearing goats and sheep by a landless family utilizing crop residues. A breakdown of the family income is given in Table IX. The data were derived from a livestock sector survey conducted in Pakistan in 1974. Sale of wool plus sheep accounted for 51·4%, whereas the sale of goats plus milk accounted for 43·9% of the annual income. About 56% of the value of family consumption was in the form of milk. The net family income was 2620 PRs, equivalent to about US $291, about half of which was cash income. This study confirms the results of an earlier study (Wahid, 1965), which reported that goats contributed about 20 to 40% to the total cash income, and in the more remote parts the contribution was as much as 50%.

4. Cyprus

In Cyprus, where goat production represents 10% of the total agricultural income of an important part of the agricultural population of over 10 000 people, income of three types of sheep and goat herds has been analysed (Table X). Farmers used nearly all their land for crop production. Goats represented 20% of the average flock size and birth rates were one to two kids per doe compared to one lamb per ewe. Sheep and goat husbandry was a family enterprise requiring about 4 to 5 women hours or 8 to 10 man hours per day. In addition cash expenses were incurred for 51 labour hours per animal per year and decreased as the flock size increased. The analysis indicated that the net cash income increased with increasing flock size. The study highlighted the need for selective breeding, improved nutrition and better grazing management, disease control and higher labour efficiency.

Table X. Analysis of incomes of three sizes of goat and sheep herds in Cyprus (£)[a] (Cyprus Agricultural Research Institute, 1964).

Item	No. of sheep and goats		
	56	70	123
Income from goats and sheep	236·0	281·1	521·1
Income from cheesemaking	16·4	75·8	16·2
Sub-total	252·4	356·9	537·3
Income from cash crops, etc.	31·4	47·5	109·0
Income from other occupations	22·2	0·6	0·0
Minus investments	−25·7	−37·6	−131·7
Minus expenditure on fodder crops	−36·4	−45·8	−41·8
Minus rent for land	−24·0	−52·0	−49·1
Minus interest + credit	−26·2	−37·7	−35·1
Total cash income	193·7	231·9	397·6
Less domestic consumption	60·0	59·5	78·2
Net cash income (£)[a]	133·7	172·4	319·4

[a] £1 = US $2 (approximately).

5. Trinidad and Tobago

A survey in Trinidad and Tobago in the West Indies involved milk production from dairy goats who were kept to supply milk for the household, or in some cases, for sale in the immediate neighbourhood (Devendra and Rankine, 1970; Table XI). About 50% of the farms kept one to five goats. Breeds of goats included Saanen, Toggenburg, British Alpine and Anglo-Nubian. The goats were stall-fed with Guinea grass (*Panicum maximum*), Napier grass

Table XI. Annual average income from rearing goats for milk production in Trinidad and Tobago.[a]

	\$ TT[c]	%
(i) *Cost item*		
Costs of animals	1090·00	65·6
Cost of feed	130·00	7·9
Mineral licks	2·43	0·1
Breeding costs	12·70	0·8
Labour	244·00	14·8
Cost of builcings	143·20	8·6
	1622·33	—
Interst charges	36·96	2·2
Total cost	1659·29	100·0
(ii) *Gross income*	2233·00[b]	
Profit (excess ii over i)[a]	573·71	

[a] For a farm with five milking does.
[b] Based on an average annual production of 4808 kg; with a selling price of \$0·455 TT/bottle (0·34 kg); average annual income from the sales of milk of \$811·18 TT.
[c] US \$ 1 = \$2 TT.

(*Pennisetum purpureum*) and Para grass (*Brachiaria mutica*), leaves from the bread fruit tree (*Artocarpus communis*), plantains (*Musa* spp.), hibiscus (*Hibiscus rosa-sinensis*) and sweet potato vines (*Ipomoea batatas*). Usually some concentrates were fed which are purchased. The cost of goats represented the highest cost item (65·6%), followed by labour (14·8%). The profit margin from rearing goats for milk production in Trinidad was \$573·71 TT for five goats, equivalent to US \$286·84.

6. Venezuela

Galéo and Leal (1974), in an arid zone of Venezuela, analysed 83 goat holdings. The average herd composition was 86 head, of which 59·9% females, 1·8% males and 38·3% young stock. Estimated income is given in Table XII.

Table XII. Economic data of 83 goat herds in an arid zone of Venezuela (Galeo and Leal, 1974).

	US \$
Capital invested	1331·22
Gross income from goats	402·91
Gross income from other sources (sales of other agricultural commodities)	559·53
Gross income	962·44
Expenses	41·19
Net income	921·16

7. Peru

Vélez-Nauer and Callacná (1979) analysed economic data of a goat herd of about 2700 head in northern Peru managed under traditional extensive conditions. Expenditures included interest on capital invested in animals, buildings and 44 weekly man hours for management, while income included that from slaughter animals but increase in herd numbers as well. Income per work day in 1976 and 1977 was Soles 478 and 143 against the minimum official wage of Soles 121 and 179 respectively.

G. Contribution to Rural Nutrition

Goats provide a small but significant supply of animal protein and goats, due to their small size, are available to the poor people in rural areas (de Haas and Horst, 1976; Devendra, 1981). The small contribution is especially important for the poor farmers and the landless, the undernourished, the pregnant and nursing mothers and for children. The magnitude of this contribution is unknown as there are unfortunately no data or studies available (Fig. 3). Where there is a tradition for meat and milk consumption, development programmes for goats will have a great impact especially with smallholders in the tropics and the subtropics (Devendra, 1976a). But this task will have to be coupled with education since malnutrition is due to both ignorance and poverty.

H. Gaps in our Knowledge

Much more information is needed on the socio-economic importance of goat production. Details of the magnitude of goats' contribution in any particular ecosystem are required. Any study that involves rearing goats will necessitate a thorough investigation of the social and economic dimensions. Of the two, in the past the social aspects have been least studied. While some cost–benefit analyses of goat-keeping have been done, the information is mainly confined to goats. But since goats and sheep are often run together, the comparative economic benefit of keeping one particular species or perhaps both together, measured also in terms of effective use of resources, needs to be considered. Associated with this, more critical information is required on the magnitude of the benefit to rural nutrition of small regular supplies of meat and milk from goats.

In the arid and semi-arid environments, new knowledge is necessary that will provide the basis for efficient strategies to cope with the stresses and strains of uncertainties imposed by the difficult climate. The primary objective should be that of improving the long-term welfare of the people in these

Fig. 3. Milk production from goats is an important function. For small farmers in rural areas, small daily supplies of milk contribute significantly to the nutrition of the rural community. The photograph illustrates a doe being milked in Mexico.

areas. Goats and livestock should be valued for their quality rather than abundance, in association with control over intensity of land use and flexibility, spreading the resource base more widely. Developments in this connection are bound to be remunerative and also productive for poor people in these regions.

References

Adu, I. F., Olaluku, E. A. and Oyenuga, V. A. (1974). *Niger J. Anim. Prod.* **1**, 151–162.
Akinsoyinu, A. O. (1975). Ph.D. Thesis, Univ. of Ibadan, Nigeria, pp. 206.
Cyprus Agricultural Research Institute (1964). *Farm Mgmt Rpt* No. 3, 47 pp.
deHaas, H. J. and Horst, P. (1979). *Anim. Res. Dev.* **9**, 41–76.
Devendra, C. (1972). *J. Dairy Res.* **39**, 381–385.
Devendra, C. (1975a). *Z. Tierzücht. Zücht. Biol.* **92**, 305–317.
Devendra, C. (1975b). "Proc. Seminar on Livestock Production and the Food Crisis", Bull. No. 141, Ministry of Agriculture and Rural Development, Kuala Lumpur, pp. 18–32.
Devendra, C. (1975c). *Wrld Rev. Anim. Prod.* **11** (1), 46–53.

Devendra, C. (1976a). Rpt FAO Group meeting on Livestock Programmes for Small Farmers and Agricultural Labourers in Asia and the Far East, Vol. II, 31st May to 6th June 1976, Bangkok, Thailand, pp. 176–210.
Devendra, C. (1976b). "Proc. Symp. on Smallholder Livestock Production and Development", Bull. No. 144, Ministry of Agriculture, pp. 29–47.
Devendra, C. (1978). *Wrld Rev. Anim. Prod.* **14** (1), 9–22.
Devendra, C. (1979). "Int. Seminar on the Development of Goats in Asia" 5–11 March 1979, Karnal, India (Mimeograph), 44 pp.
Devendra, C. (1981). *J. Anim. Sci.* **51**, 461–473.
Devendra, C. and Burns, M. M. (1970). "Goat Production in the Tropics". Tech. Comm. No. 19 Commonw. Bur. Anim. Breed. Genet., Commonwealth Agricultural Bureau, xii + 184 pp.
Devendra, C. and Rankine, L. B. (1970). *J. Agric. Sci. Trin. Tobago* **71**, 455–483.
FAO (1975). Statistical Directory of Asia and the Far East. FAO, Rome, 296 pp.
Galéo, F. E. and Leal, D. E. (1974). "Diagnóstico Socio-económico de 83 Criadores de Caprinos del Districto Torres, Edo.Lara". Univ. Centro-Occidental-Escuela de Ciencias Veterinarias-Barquisimeto, Venezuela, 103 pp.
Gray, R. F. (1963). "The Sanjo of Tanganyika". Oxford University Press, London, 181 pp.
Gulliver, P. H. (1955). "The Family Herds". Routledge and Kegan Paul Ltd, London, 271 pp.
Jodha, N. S. (1966). *Agric. Dev. Council paper*, 15 pp.
Jollans, J. L. (1959). *J. W. Afr. Sci. Ass.* **5**, 64–78.
Lai, K. C., Devendra, C., Yusof Hashim, M., Ishak Djafar, M. and Jegathesan, M. (1975). *MARDI Rpt* No. 27, 28 pp.
Matthewman, R. W. (1977). M.Sc. Thesis, University of Reading, England, 174 pp.
McDowell, R. E. (1976). *MUCIA Seminar*, No. 19, 1976, USA (Mimeograph, 24 pp.).
Park, W. L. and Parn, R. L. (1974). *St. Croix U.S. Virgin Is. Agric. Exp. Stn* No. 7, 13 pp.
Quartermain, A. R. (1973). "III Wrld Conf. Anim. Prod.", Vol. 2,1(b) pp. 21–31.
Raut, R. C. and Nadkarni, U. G. (1974). *Indian J. Anim. Sci.* **44**, 459–465.
Spedding, C. R. W. (1968). *Agric. Progress* **44**, 7–23.
Statistical Yearbook (1976). United Nations, New York, 490 pp.
Sriramamurthy, V. (1977). *Annadata Magazine*, Dec. 1977, pp. 6–14.
United Nations (1977). "Desertification: an Overview", United Nations Conf. on Desertification, 29 Aug.–9 Sept. 1977, Nairobi, Kenya, 84 pp.
Vélez Nauer, M. and Callacná, M. A. (1979). *Entwicklung u. Ländlicher Raum* **13** (1), 14–16.
Wahid, A. (1965). *Agric. Pakistan* **16**, 509–534.
Wilson, R. T. and Clarke, S. E. (1975). *Trop. Anim. Hlth Prod.* **7**, 165–187.
Zeuner, F. E. (1963). "A History of Domesticated Animals". Hutchinson, London, 129 pp.

19. Needs and Possibilities for Future Research

P. Charlet

A. The Goat as Experimental Animal

Improving agricultural and especially animal production depends to a great extent on the research devoted to those fields. While many empirical data on goat production have been obtained with efficient techniques, it is necessary to reveal the physiological mechanisms involved, be they genetic, nutritional, pathological or otherwise. This would lead to more efficient and reliable production methods.

In addition to this type of applied research there is the possibility of using goats for basic research on specific aspects of metabolism, cytogenetics, etc., which may contribute to improving our knowledge of basic mechanisms involved in animal production. Results of such studies may be applied to other species. Thus, the goat is being used as a laboratory animal, like the rat, rabbit, cobaya, dog, miniature pig and sheep, which are all used for research where results are applied to other species.

We thus conceive of two different forms of goat research:

(1) applied research for improving goat production techniques;
(2) basic research into more general problems with inter-species applicability.

In both forms, the latter often precedes the former. These investigations are more and more justified both by growing economic interest in the goat in many countries and by its value as a small sized experimental animal, mainly for research on lactation and behaviour.

The content of this book clearly shows that efficient production depends on research in the following fields: pathology, nutrition, reproduction, physiology of adaptation and milk production, genetics, economics, and last but not least animal management. We shall examine the different research fields in this order, referring in each case to both aspects, those of the goat proper and those where the goat is being used as experimental animal.

B. Pathology

1. Brucellosis

Goat milk and fresh-milk products in many countries are considered as the main source of brucellosis in humans. Considering the decrease of registered cases of brucellosis in some areas, it may be assumed that efficient methods for its control are now available. However, in some areas brucellosis still seems to be a major reason for a reluctance to consume goats' milk. Research on brucellosis should therefore receive priority.

2. Chlamydiosis

The situation with this disease, the essential cause of contagious abortion, is very different. Its field control is not yet reliable. Epidemiological research ought to be intensified because this disease is the cause of large losses in many herds.

3. Large Herd Management Diseases

As yet goats have not often been kept in big intensive units like poultry, pigs and cattle, but this will soon occur. However, as in other species, following the development of large goat units, specific diseases will emerge requiring special preventive measures. Large goat production units should therefore be subjected to thorough sanitary control.

4. Parasites

Goat breeding may help to solve some of the problems of meat and milk production in wet tropical or equatorial environments which favour the development of parasites and parasitoses. Research should be undertaken with goats similar to that done in Africa and Australia with *Bos taurus* and *Bos indicus* cattle. In the past, many attempts at increasing the production of dairy goats in tropical areas have been made using temperate breeds for crossbreeding. These efforts often have been unsuccessful because of parasitic diseases. We do not yet know the reaction in tropical areas of the local goat populations whose productivity has been improved by pure breeding or by crossing with improved breeds imported from temperate areas. Owing to its small size, the costs of maintenance of fairly large experimental flocks are lower than that of cattle. Therefore it should be possible to make more accurate studies of the parasite/host interactions, including investigations of the mechanisms of resistance and immunity, by means of group comparisons allowing more precise statistical evaluations.

5. Mycoplasmoses

These are increasing and difficult to control in most domestic species. The goat suffers from a severe disease of this type, contagious agalaxy. It seems that environment and management play an important part in the etiology of this disease. Therefore, it is necessary to work with sufficiently large groups of animals and the goat, like poultry, can be a valuable tool for basic research.

C. Nutrition and Feeding

1. Feeding Standards

While in the past valuable feeding methods were established empirically, we owe the most important recent progress to an ever-improving knowledge of the physiological basis of nutrition. Thus, empirical methods are progressively being replaced by scientific standards derived from research results. The case of the goat is typical and the chapters herein dealing with its feeding clearly show advances made in this field of research and in the application of results. It is often said that rational feeding practices are a summary of our knowledge derived from nutritional studies. However, feeding is more than that because empirical knowledge plays its part when basic mechanisms are not yet known. Therefore, both routes of research, analysis of the mechanism of nutrition as well as the practical aspects of feeding, should be followed.

2. Metabolism

The size of the goat facilitates fundamental research of some physiological mechanisms which until now have not been investigated in spite of their obvious interest. This is particularly the case with problems concerning the metabolism and hormonal interactions. The dynamics of the nutritional balance during successive periods of fetal development and lactation, their influence on the metabolism of carbohydrate, fat and protein and their integrated effect on energy turnover, are examples.

Another area for basic research is the synergies at the cell level of nutrients, vitamins, etc. and their summation at the level of the whole organism. Availability of high yielding animals should allow better analysis of those mechanisms depending on the level of production by determining the different stages of equilibrium; nutritional disorders and diseases like "pregnancy toxaemia" could be cases in point.

3. Feeding Behaviour

The goat exhibits very idiosyncratic feeding behaviour. From practical experience we know of its ability to choose rapidly between various forages of different feed value. The importance of this ability for the goats' adaptation to difficult environments needs to be studied systematically.

4. Feeding under Extensive Conditions

At present, feeding of high yielding goats seems to be particularly important. However, at the same time we must study with well adapted modern methods the feeding of animals on poor pastures, avoiding any assimilation with feeding on intensive pasture. Particular attention should be paid to the fact that the goat is not only a grazer but more a browser.

Among all these topics, it seems that research on the metabolism during gestation and lactation of carbohydrate, fat and protein and its relation to changes in hormone secretion and feeding behaviour, would be most profitable.

Priority will continue to be given to investigations on the assessment of the nutritive value of different feeds and the establishment of new feeding standards based on nutritional research cited above.

D. Reproduction

As Terroine wrote, it is often a mistake to assume physiological similarities where there are anatomical ones. This is especially true for reproduction in mammals. There are, for example, marked differences in the physiology of ovulation between species and breeds. This has stimulated numerous investigators to make an inventory in sheep. Similar research is now in progress on goats.

This is also true of spermatology. It is obvious that further research is essential on artificial insemination which is such a powerful tool for genetic improvement. There is need for investigations on the causes of variation of fertility.

E. Heat Regulation

Studies in various parts of the world have shown the remarkable ability of the goat to adapt to varying climatic conditions and the large range of breeds available in this species. The fact that these abilities are being empirically utilized should not attenuate interest in studying their mechanisms, as in bovine research, i.e. water metabolism, maintenance requirements, respiratory and cardial frequency, characteristics of the skin, hair cover, sweating, the effect of dermic irrigation etc. Variations of these characteristics and their relationship to growth, exercise, milk production and with feeding need to be investigated. The study of the hair cover is important not only because of its role in heat regulation but also for the economic significance of Mohair and Cashmere.

F. The Composition of Milk

In our opinion, the variation of composition of milk has not received the attention it deserves, except by a few workers. Particularly in countries like France or Norway, where about 80% of goat milk is processed into cheese, the relationship between yield and milk composition and different protein proportions, in particular by type of cheese, should be investigated rapidly. Results of such investigations would have immediate consequences for selection and feeding methods.

G. Genetics

Here again the goat seems to be a very good experimental tool allowing us to go further in genetic improvement than is possible from studying such factors as quantitative milk yield. The type of research on physiology and metabolism of lactation indicated above could aim at a better understanding of the various factors involved in lactation. It should be possible to identify the influence of the level of hormonal secretion, the importance of secretory tissues and the individual metabolic disposition. Research on the genetic determination of each of these factors might lead to advanced selection methods. Eventually, it would perhaps be possible to bypass present methods where selection is based on very comprehensive traits whose variation is influenced by a multitude of different factors, and instead consider these factors directly. This may be a futurist view, but any attempt in that direction will surely improve the efficiency of our selection methods and will prevent or explain the slackening of genetic improvement observed in other species like poultry. But in the meantime these views will not deter efforts to improve present selection methods. While in the recent past possibilities of selection for milk yield in both quantity and quality have improved, possibilities should be studied of selecting for other characteristics. For example, body size and udder conformation are important traits. The tendency of some breeds to yield most of their milk at one daily milking may be economically very important. The same is true of the ability to ingest a concentrate ration and at the same time derive an important ration from poor pasture, thus making a specific system of milk production possible.

Selection for meat production either by dual purpose breeds in mixed operation or by specialized breeds should be carefully studied, because a large range of breeds is available, from the very fertile Créole of the Caribbean Islands to the Boer goat of South Africa.

There is need for further research on the possibilities of genetic improvement of fibre (Mohair) and wool (Cashmere).

Finally, the conservation of genetic resources merits attention since some breeds disappear at a very rapid rate. In genetic research, top priority should be given to improving milk and meat production potential and possibly to the analysis of factors involved in lactation.

H. Conclusion

We have tried to indicate some areas of research in animal production where the goat may be used advantageously for three reasons:

(1) the goat's considerable potential for increasing its productivity;
(2) its ability to adapt to a wide range of production conditions;
(3) its potential as an experimental animal, allowing controlled experiments with much larger numbers of animals than would be possible with cattle at the same cost.

Therefore, the goat will be the experimental animal of choice for studies on lactation. Basic research on the main mechanisms of milk production would lead not only to an important improvement in goat production, but also in other species. However, this requires increased communication and maybe even coordination between existing research teams. A maximum of interdisciplinary and international contacts will help achieve the goal. In this way, the goat would no longer be "the cow of the poor" as they say in French, but a high yielding dairy animal, and the preferred tool for advanced research.

Subject Index

C

D